KH93

							0
							2 **He** Helium 4.003
	IIIA	IVA	VA	VIA	VIIA		
	5 **B** Boron 10.81	6 **C** Carbon 12.011	7 **N** Nitrogen 14.007	8 **O** Oxygen 15.999	9 **F** Fluorine 18.998	10 **Ne** Neon 20.17	

IB	IIB	13 **Al** Aluminum 26.98	14 **Si** Silicon 28.09	15 **P** Phosphorus 30.974	16 **S** Sulfur 32.06	17 **Cl** Chlorine 35.453	18 **Ar** Argon 39.948
28 **Ni** Nickel 58.71	29 **Cu** Copper 63.546	30 **Zn** Zinc 65.38	31 **Ga** Gallium 69.735	32 **Ge** Germanium 72.59	33 **As** Arsenic 74.922	34 **Se** Selenium 78.96	35 **Br** Bromine 79.904	36 **Kr** Krypton 83.80
46 **Pd** Palladium 106.4	47 **Ag** Silver 107.868	48 **Cd** Cadmium 112.41	49 **In** Indium 114.82	50 **Sn** Tin 118.69	51 **Sb** Antimony 121.75	52 **Te** Tellurium 127.60	53 **I** Iodine 126.904	54 **Xe** Xenon 131.30
78 **Pt** Platinum 195.09	79 **Au** Gold 196.967	80 **Hg** Mercury 200.59	81 **Tl** Thallium 204.37	82 **Pb** Lead 207.2	83 **Bi** Bismuth 208.98	84 **Po** Polonium (209)	85 **At** Astatine (210)	86 **Rn** Radon (222)

63 **Eu** Europium 151.96	64 **Gd** Gadolinium 157.25	65 **Tb** Terbium 158.93	66 **Dy** Dysprosium 162.50	67 **Ho** Holmium 164.93	68 **Er** Erbium 167.26	69 **Tm** Thulium 168.93	70 **Yb** Ytterbium 173.04	71 **Lu** Lutetium 174.967
95 **Am** Americium (243)	96 **Cm** Curium (247)	97 **Bk** Berkelium (247)	98 **Cf** Californium (251)	99 **Es** Einsteinium (254)	100 **Fm** Fermium (257)	101 **Md** Mendelevium (258)	102 **No** Nobelium (259)	103 **Lr** Lawrencium (260)

CONCEPTUAL Physics

CONCEPTUAL
Physics
FIFTH EDITION

Written & Illustrated by

Paul G. Hewitt

City College of San Francisco

Little, Brown and Company

Boston   Toronto

Library of Congress Cataloging in Publication Data
Hewitt, Paul G.
    Conceptual physics.

    Rev. ed. of: Conceptual physics—a new introduction
to your environment.   4th ed.   1981.
    Includes bibliographies and index.
    1. Physics.   I. Hewitt, Paul G.   Conceptual physics—
a new introduction to your environment. II. Title.
QC23.H56   1985   530   84-21803
ISBN 0-316-35974-2

Library of Congress Catalog Card No. 84-21803

ISBN 0-316-35974-2

9 8 7 6 5 4 3 2

HAL

Published simultaneously in Canada
by Little, Brown & Company (Canada) Limited

Printed in the United States of America

Adaptation of material in Chapters 23 and 25 from R. P. Feynman, R. B. Leighton, and M. Sands, *The Feynman Lectures on Physics,* Volumes I and II. Used by permission of Addison-Wesley Publishing Company, Inc., Reading, MA.
Figure 6.13: Adapted from K. F. Kuhn and J. S. Faughn, *Physics in Your World,* 1980. Reprinted by permission of W. B. Saunders, Publishers.
Figure 32.19: Redrawn from "The Prospects of Fusion Power" by William C. Gough and Bernard J. Eastlund from *Scientific American,* February 1971. Copyright © 1971 by Scientific American, Inc. All rights reserved. Reprinted by permission.
Page 173: Portions of the text in "Scaling," Chapter 10, are based on the following sources: "On Being the Right Size," from *Possible Worlds* by J. B. S. Haldane. Copyright 1928 by Harper & Row, Publishers, Inc. Renewed 1956 by J. B. S. Haldane. By permission of Harper & Row, Publishers, Inc. and Chatto and Windus Ltd. "On Magnitude," from *On Growth and Form* by Sir D'Arcy Wentworth Thompson (New York: Cambridge University Press). Used with permission.

Produced by The Compage Company
Design: The Compage Company
Editorial production: Pearl C. Vapnek
Copyediting: Judy Chaffin
Indexing: James K. Hewitt
Composition: Holmes Typography

Photo Credits

To Frank Oppenheimer
and those like him,
who explore the beauty
that lies in all things

Contents in Brief

Contents in Detail

To the Student

PHYSICS CAN BE AN ENJOYABLE EXPERIENCE --- ESPECIALLY SERIOUS PHYSICS WHEN IT'S PRESENTED IN A NONMATHEMATICAL LANGUAGE AND IN A DOWN-TO-EARTH MANNER. I ENJOY PHYSICS AND HOPE TO CONVEY MY ENTHUSIASM FOR IT AS I'LL BE TALKING WITH YOU ABOUT IT IN THE FOLLOWING 35 CHAPTERS. IF YOU BEGIN BY DISCOVERING THE PHYSICS IN THIS BOOK, YOU'LL SOON FIND, I HOPE, THAT YOU'RE DISCOVERING THE PHYSICS THAT'S IN EVERYTHING YOU DO AND SEE.

SINCE THE PHYSICS IN THIS BOOK IS NOT TREATED AS APPLIED MATHEMATICS, YOU WON'T NEED A CALCULATOR. MOST END-OF-CHAPTER PROBLEMS ARE EXERCISES IN THINKING ABOUT THE MEANING AND IMPLICATIONS OF THE IDEAS OF PHYSICS—IN ENGLISH—NOT VIA ALGEBRAIC MANIPULATIONS AND COMPUTATIONS. BUT IF YOU HAVE A CALCULATOR ANYWAY, SAVE IT. AS YOU FIND THAT PHYSICS IS FASCINATING, YOU MAY TAKE A FOLLOW-UP COURSE. THEN YOU CAN USE YOUR CALCULATOR -- WITH UNDERSTANDING.

ENJOY!

PAUL G. HEWITT

To the Instructor

Physics is the basic science; it is the foundation of chemistry, biology, and all disciplines of science. As such, the study of physics should be part of the educational mainstream, for both science students and nonscience students. Unfortunately, the mathematical language of physics often deters the average nonscience student. But when the ideas of physics are presented conceptually and when formulas are seen to be guides to thinking rather than recipes for algebraic manipulation, our discipline is accessible to all students. And for students who will continue in physics instruction, I am convinced that the ideas of physics should be first understood conceptually before they are used as a base for applied mathematics. This book seeks to build that conceptual base. It is a base for nonscience students and science students alike. For the nonscience student, it is a base from which to view nature more perceptively. For the science student, it is this and a springboard to a greater involvement in physics. A first-semester conceptual overview of Newtonian and modern physics for science majors will help to correct a missing essential in physics education: the practice of conceptualizing before calculating. For nonscience and science students alike, a conceptual way of looking at physics shapes analytical thinking.

Although the overall organization of this edition is essentially that of the previous editions insofar as the sequence from mechanics to astrophysics is concerned, considerable change and new material occur throughout. Part 1, Mechanics, has been substantially revised. It begins with linear motion in Chapter 2 and then goes immediately to projectile motion in Chapter 3. The study of projectiles leads to satellites in the same chapter. This reordering has two advantages. The first is logical progression: the kinematics of vertically falling bodies logically extends to projectiles, which in turn extends to a simple treatment of satellite motion. The second advantage is the early introduction of a topic of much current interest: the space-shuttle missions and other spacefaring activities that have captured the interest and imagination of most of your students. Newton's laws follow in Chapter 4, and logically following Newton's second and third laws is Chapter 5, on momentum. Chapter 6 on energy follows, with a spiral-approach return to satellite motion and escape velocity. Rotational motion, Chapter 7, and the law of universal gravitation, Chapter 8, complete the mechanics sequence. The fourth edition's chapter on power production and the chapter on electronics have been omitted to allow the introduction of new material without inflating the size of the book. The popular section on exponential growth, à la Al Bartlett, now appears on its own as Appendix V. A new chapter on thermodynamics appears in Part 3, Heat. Part 6, Light, has been revamped. Fermat's principle of least time, à la Richard Feynman, now provides your students with an alternative, interesting view of reflection and refraction. Quantum physics has been considerably revised and expanded. Important to the overall flavor of this edition are the many new insights and, I hope, pleasant surprises that are sprinkled in almost every chapter.

There are a wealth of new exercises in this edition. As in the fourth edition, there are more than enough for fresh assignments each semester or quarter without overlap for several terms. Some exercises are moderately simple and are designed to prompt the application of physics to everyday situations. Others are more sophisticated and call for considerable critical thinking. Some are quantitative and involve simple, straightforward calculations that will help your students capture the idea being treated, without requiring algebraic skills. The challenge to your students will be in the conceptual reasoning and critical thinking that are called for in the exercises. Please don't overload your students and blanketly assign *all* exercises to the chapters you cover. If you favor lengthy written assignments, it is more in order to have your students write up all or most of the review questions. These are relatively straightforward and they summarize, in question form, the essentials of the chapter.

As in previous editions, units of measurement are not emphasized. When used, they are almost exclusively expressed in SI (exceptions include such units as calories, grams/cm³, and light-years). Mathematical derivations are avoided in the main body of the text and appear in footnotes or in the appendixes.

More than enough material is included for a one-semester course, which allows for a variety of course designs to fit your taste. These are suggested in the Instructor's Manual, which you'll find to be the most different of instructor's manuals. It contains many lecture ideas and topics not treated in the textbook, as well as teaching tips, suggested step-by-step lectures and demonstrations, information on overhead transparency masters and on new videotaped lectures by the author, and much more to assist you in making Conceptual Physics the most interesting, informative, and worthwhile science course available to your students.

Acknowledgments

For the many suggestions that contributed to this fifth edition, I am indebted to my friend Paul Robinson at Computech, Fresno, California; to my colleagues at City College of San Francisco (CCSF): Jim Conley, Jim Court, Frank Creese, Jerry Hosken, Frank Koehler, Dack Lee, Will Maynez, Dave Wall, and Norman Whitlatch; to Al Bartlett, University of Colorado; Clifford M. Braun, a student at CCSF; Jeffrey J. Braun, Lincoln Land Community College, Indiana; Bill Cary, Madison Memorial High School, Wisconsin; Marshall Ellenstein, Ridgewood High School, Illinois; Gabe Espinda, the Exploratorium, San Francisco; N. J. Farrier, Sinclair Community College, Ohio; Ron Hipschman, the Exploratorium; Lester Hirsch, California State University, Los Angeles; Lillian Lee (Figure 18.3); Tenny Lim (Figure 6.3), California Polytechnic State University, San Luis Obispo; Iain MacInnes, Jordanhill College of Education, Scotland; Mel Mayfield, Austin Peay State University, Tennessee; Frank Oppenheimer, the Exploratorium; Ken Ozawa, California Polytechnic State University, San Luis Obispo; Bob Plumb, California State University, Chico; Ray Sachs, University of California, Berkeley; Richard Stepp, Humboldt State University, California; John Taube and Roger Werner, San Francisco Chapter of Technocracy; Jearl Walker, Cleveland State University; Brian Watson, St. Lawrence University, New York; and Yoshihisa Yoshida, Sagami Women's University, Japan. I am most grateful to the many students, both at CCSF and at the world's most wonderful place to teach physics, the Exploratorium; their feedback was paramount in developing this book, which is in large part a reflection of their participation.

I thank the reviewers of the manuscript: Art Cary, California Polytechnic State University, San Luis Obispo; Horace Coburn, New Mexico State University; Frank Crawford, University of California, Berkeley; Paul Doherty, Oakland University; Neil Fleishon, Southern Oregon State College; Barry Gilbert, Rhode Island College; Roger Hanson, University of Northern Iowa; Sherwood Harrington, CCSF and Astronomical Society of the Pacific; Joseph Klarmann, Washington University; Dack Lee, CCSF; Robert Luke, Boise State University; Joseph Klarmann, Washington University; and Thomas L. Rokoske, Appalachian State University.

I am grateful to those whose own books initially served as principal influences and references: Theodore Ashford, *From Atoms to Stars;* Albert Baez, *The New College Physics—A Spiral Approach;* John N. Cooper and Alpheus W. Smith, *Elements of Physics;* Richard P. Feynman, *The Feynman Lectures on Physics,* Volumes I and II; Kenneth Ford, *Basic Physics;* Eric Rogers, *Physics for the Inquiring Mind;* Alexander Taffel,

Physics: Its Methods and Meanings; UNESCO, *700 Science Experiments for Everyone;* and Harvey E. White, *Descriptive College Physics.*

I am especially grateful to John Hubisz, College of the Mainland, Texas, for reviewing the end-of-chapter exercises and answers, as well as making contributions to the test bank. Special thanks also to my friend and CCSF colleague Annette Rappleyea for improving the test-bank questions and for writing the computer program for the test bank. I thank my photographer-type friends Craig Dawson, Lila Lee, and Dave Vasquez (Figure 5.13) for their many photos that add a nice touch to this edition. A note of appreciation is due my friend Gary Zukav for many discussions on both our similar and our different points of view, the outcome of which is an expanded treatment of quantum physics. Thanks go to my lifelong friend Ernie Brown for designing the new physics logo. For zip-a-tone shading the new drawings and for compiling the index, I thank my son James and, for helping with the index, Lisa Rodriguez. For helping me through all the stages from manuscript through production, I thank most of all Helen Yan (Figures 5.13 and 14.10).

A special note of appreciation is due editor Ron Pullins of Little, Brown, for his very professional concern and assistance. Thanks also to his assistant Pat Bellanca for excellent editorial advice. I am especially indebted to Ken Burke and Pearl C. Vapnek, who produced the book.

San Francisco   *Paul G. Hewitt*

CONCEPTUAL
Physics

By trying to understand the natural world around us, we gain confidence in our ability to determine whom to trust and what to believe about other matters as well. Without this confidence, our decisions about social, political, and economic matters are inevitably based entirely on the most appealing lie that someone else dishes out to us. Our appreciation of the noticings and discoveries of both scientists and artists therefore serves, not only to delight us, but also to help us make more satisfactory and valid decisions and to find better solutions for our individual and societal problems.

Frank Oppenheimer

1 About Science

Science is the body of knowledge about nature that represents the collective efforts, insights, findings, and wisdom of the human race. Science is more than another name for knowledge; it's a human activity, with the function of discovering the orderliness of nature and finding the causes that govern this order. Science had its beginnings before recorded history when people first discovered recurring relationships around them. Through careful observations of these relationships, they began to know nature and, because of nature's dependability, found they could make predictions that gave them some control over their surroundings.

Science made great headway in the sixteenth century when people began asking answerable questions about nature—when they began replacing superstition by a systematic search for order—when experimentation in addition to logic was used to test ideas. Where people once tried to influence natural events with magic and supernatural forces, they now had science to guide them. Advance was slow, however, because of powerful opposition to scientific methods and ideas.

In about 1510 Copernicus suggested that the sun was stationary and that the earth revolved around the sun. He refuted the idea that the earth was the center of the universe. After years of hesitation, he published his findings but died before his book was circulated. His book was considered heretical and dangerous and was banned by the Church for 200 years. A century after Copernicus, the mathematician Bruno was burned at the stake—largely for supporting Copernicus, suggesting the sun to be a star, and suggesting that space was infinite. Galileo was imprisoned for popularizing the Copernican theory and for his other contributions to scientific thought. Yet a couple of centuries later Copernican advocates seemed harmless.

This kind of cycle happens age after age. In the early 1800s geologists met with violent condemnation because they differed with the Genesis account of creation. Later in the same century, geology was safe, but theories of evolution were condemned and the teaching of them forbidden. Every age has one or more groups of intellectual rebels who are persecuted, condemned, or suppressed at the time; but to a later age, they seem harmless and often essential to the elevation of human conditions. "At every crossway on the road that leads to the future, each progressive spirit is opposed by a thousand men appointed to guard the past."[1]

---

[1] From Count Maurice Maeterlinck's "Our Social Duty."

The Scientific Attitude

Human conditions advanced dramatically with the discovery that nature could be analyzed and described mathematically. When the findings of scientific inquiry are expressed in mathematical terms, they are unambiguous and can better be verified or disproved by experiment. The methods of mathematics and experimentation led to the enormous success of science. Galileo, famous scientist of the 1600s, is usually credited with being the Father of the Scientific Method—a method that is extremely effective in gaining, organizing, and applying new knowledge. His method is essentially as follows:

1. Recognize a problem.

2. Guess an answer.

3. Predict the consequences of the guess.

4. Perform experiments to test predictions.

5. Formulate the simplest theory that organizes the three main ingredients: guess, prediction, experimental outcome.

Although this cookbook method has a certain appeal, it has not been the key to most of the breakthroughs and discoveries in science. Trial and error, experimentation without guessing, accidental discovery, and other methods account for much of the progress in science. The success of science has more to do with an attitude common to scientists than with a particular method. This attitude is essentially one of inquiry, experimentation, and humility before the facts. If a scientist holds an idea to be true and finds any contradictory evidence whatever, the idea is either modified or abandoned. In the scientific spirit, the idea must be modified or abandoned in spite of the reputation of the person advocating it. As an example, the greatly respected Greek philosopher Aristotle said that falling bodies fall at a speed proportional to their weight. This false idea was held to be true for more than 2000 years because of Aristotle's compelling authority. In the scientific spirit, however, a single verifiable experiment to the contrary outweighs any authority, regardless of reputation or the number of followers and advocates. In the scientific spirit, argument by appeal to authority is of no value whatever.

Scientists must accept facts even when they would like them to be different. They must strive to distinguish what they see from what they wish to see—since humanity's capacity for self-deception is vast. People have traditionally tended to adopt general rules, beliefs, creeds, theories, and ideas without thoroughly questioning their validity and to retain them long after they have been shown to be meaningless, false, or at least questionable. The most widespread assumptions are the least questioned. Most often, when an idea is adopted, particular attention is given to cases that seem to support it, while cases that seem to refute it are distorted, belittled, or ignored.

The concepts of science are not carved for all time in stone but undergo change. Scientific concepts evolve as they go through stages of redefinition and refinement. This is a strength of science, not a weakness, as some people believe. Similarly, many people feel deeply that it is a sign of weakness to "change our minds." Competent scientists, however, must be expert at changing their minds. They do not easily change their minds about tried and tested principles, however, until confronted with experimental evidence to the contrary or until a conceptually simpler theory forces them to a new point of view. Science seeks not to defend beliefs but to improve them. Better theories are made by those who are honest in the face of fact.

Away from their profession, scientists are inherently no more honest or ethical than other people. But in their profession they work in an arena that puts a high premium on honesty. The cardinal rule in science is that all claims must be testable—they must be susceptible, at least in principle, to being proved wrong. For example, if someone claims that a certain procedure has a certain result, it must in principle be possible to perform a procedure that will either confirm or contradict the claim. To distinguish the claims of pseudoscientists and scientists, ask the following: "If your claim is not true, how would we know?" The scientific claim is the one that can be proved wrong with an experiment. Darwin, for example, claimed that life forms evolve from simpler to more complex forms. This could be proved wrong if paleontologists found that more complex forms of life appeared before their simpler counterparts. Einstein claimed that light is bent by gravity. This claim could be proved wrong if starlight that grazed the sun, and that could be seen during a solar eclipse, were undeflected from its normal path. As it turns out, less complex life forms are found to precede their more complex counterparts, and starlight is found to bend as it passes close to the sun, which supports the claims. If and when a claim is confirmed, it is regarded as useful and a stepping-stone to additional knowledge. None of us has the time or energy or resources to test every claim, so most of the time we must take somebody's word. However, we must have some criterion for deciding whether one person's word is as good as another's and whether one claim is as good as another. The criterion, again, is that the claim must be testable. To reduce the likelihood of error, scientists accept the word only of those whose ideas, theories, and findings are testable—if not in practice, at least in principle. Speculations that cannot be tested are regarded as "unscientific." This has the long-run effect of *compelling* honesty—findings widely publicized among fellow scientists are generally subjected to further testing. Sooner or later, mistakes (and lies) are bound to be found out; wishful thinking is bound to be exposed. The honesty so important to the progress of science thus becomes a matter of self-interest to scientists. There is relatively little bluffing in a game where all bets are called. In fields of study where right and wrong are not so easily established, the pressure to be honest is considerably less.

Question Which of the following is a scientific claim?*

a. The moon is made of green cheese.

b. Intelligent life likely exists elsewhere in the universe.

c. Albert Einstein is the greatest physicist so far in the twentieth century.

The ideas and concepts most important to our everyday lives are largely unscientific; their correctness or incorrectness cannot be determined in the laboratory. Interestingly enough, it seems that people honestly believe their own ideas about things to be correct, and almost everybody is acquainted with people who hold completely opposite views—so the ideas of some (or all) must be incorrect. How do you know whether or not *you* are one of those holding erroneous beliefs? There is a test: before you can be reasonably convinced that you are right about a particular idea, you should be sure that you understand the objections and the positions of your most articulate antagonists. You must find out whether your views are supported by a sound knowledge of opposing ideas or by your *misconceptions* of opposing ideas. You make this distinction by seeing whether or not you can state the objections and positions of your opposition to *their* satisfaction. Even if you can successfully do this, you cannot be absolutely certain of being right about your own ideas, but the probability of being right is considerably higher than if you can't pass this test.

---

*Answer   Only the first statement is scientific. The claim that the moon is made of green cheese not only can be proved wrong, it has in fact been proved wrong. No trace of green cheese was found in the moon rocks and in other tests of the moon's composition. So even though the statement is incorrect, it is nonetheless scientific. The second statement, reasonable or not, is speculation. Although it can be proved correct by the verification of a single instance of intelligent life existing elsewhere in the universe, there is no way to prove the statement wrong if no life is ever found. If we searched the far reaches of the universe for eons and found no life, we would not prove that it doesn't exist around the "next corner." A claim that is capable of being proved right but not capable of being proved wrong is not a scientific claim. Likewise for any principle or concept for which there is no means, procedure, or test whereby it can be proved wrong (if it is wrong). Some pseudo-scientists will not even consider a test for the possible wrongness of their claims. The third statement is an assertion, which has no test for proving its wrongness. If Einstein were not the greatest physicist, how would we know? It is important to note that because the name Einstein is generally held in high esteem, it is a favorite of pseudoscientists and other pretenders of knowledge. So we find it is not surprising that the name of Einstein, like that of Jesus, is cited often by charlatans who wish to bring respect to themselves and their points of view.

Question Suppose in a disagreement between two people you note that person A only states and restates her own point of view, whereas person B clearly states both her own position and that of person A. Who is correct?*

Although the notion of being familiar with counter points of view seems reasonable to most thinking people, just the opposite—shielding ourselves and others from opposing ideas—has been more widely practiced. We have been taught to discredit unpopular ideas without understanding them in proper context. With the 20/20 vision of hindsight, we can see that many of the "deep truths" that were the cornerstones of whole civilizations were shallow reflections of the prevailing ignorance of the time. Many of the problems that have plagued societies stemmed from this ignorance and the resulting misconceptions; much of what they held to be true simply wasn't true. Are we different today?

Science and the Arts The principal values of science and of the arts are quite comparable. In literature we find what is possible in human experience. We can learn about emotions ranging from anguish to love, even if we haven't yet experienced them. The arts do not necessarily give us that experience but describe it to us and suggest what may be in store for us. A knowledge of science similarly tells us what is possible in nature. Scientific knowledge helps us predict possibilities in nature even before these possibilities have been experienced. It provides us with a way of connecting things, of seeing relationships between and among them, and making sense of the myriad natural events we find around us. Science broadens our perspective of the natural environment of which we are a part. A knowledge of both the arts and the sciences makes for a wholeness that affects both the way we view the world and the decisions we make about it and ourselves. A truly educated person is knowledgeable in both arts and science.

Science and Technology Science and technology are different from each other. Science is a method of answering theoretical questions; technology is a method of solving practical problems (which sometimes creates new problems out of the solutions). Science has to do with discovering the facts and relationships among observable phenomena in nature and with establishing theories

---

*__Answer__ Who knows? Person B may have the cleverness of the lawyer who can state various points of view and still be incorrect in her own. We can't be sure about the "other guy." The test for correctness or incorrectness suggested here is not a test of others, but of and for *you*. It can aid your personal development. As you attempt to articulate the ideas of your antagonists, be prepared, like scientists who are prepared to change their minds, to discover evidence counter to your own ideas that will alter your views. Intellectual growth often comes this way.

that serve to organize these facts and relationships; technology has to do with tools, techniques, and procedures for implementing the findings of science.

Another distinction between science and technology has to do with progress. Progress in science excludes the human factor. And this is justly so. Scientists, who seek to comprehend the universe and know the truth with the highest degree of accuracy and certainty, cannot pay heed to their own or other people's likes or dislikes or to popular ideas about the fitness of things. What scientists discover may shock or anger people—as did Darwin's theory of evolution. But even an unpleasant truth is more than likely to be useful; besides, we have the option of refusing to believe it! But hardly so with technology once it is developed: we do not have the option of refusing to hear the sonic boom produced by a supersonic aircraft flying overhead; we do not have the option of refusing to breathe polluted air; and we do not have the option of living in a nonnuclear age. Unlike science, progress in technology *must* be measured in terms of the human factor. Technology must be our slave and not the reverse. The legitimate purpose of technology is to serve people—people in general, not just some people; and future generations, not just those who presently wish to gain advantage for themselves. Technology must be humanistic if it is to lead to a better world.

We are all familiar with the abuses of technology. Many people blame technology itself for the widespread pollution and resource depletion and even social decay in general—so much so that the promise of technology is obscured. That promise is a cleaner and healthier world. Isn't it wiser to combat the dangers of technology with knowledge rather than ignorance? If wise applications of science and technology will not lead to a better world, what will?

Physics—The Basic Science

Science first branches into the study of living things and nonliving things: the life sciences and the physical sciences. Life science branches into such areas as biology, zoology, and botany. Physical science diverges into such areas as astronomy, chemistry, and physics. But physics is more than a part of the physical sciences. Physics is the most fundamental and all-inclusive of the sciences, both life and physical. Physics, essentially the study of matter and energy, is at the root of every field of science and underlies all phenomena. Physics is the present-day equivalent of what used to be called *natural philosophy,* from which most of present-day science arose.

The following chapters represent the findings of those who answered the compelling call to adventure—the expedition in search of the hows and whys of the physical world. Their findings are our legacy. In these chapters we will attempt to develop a conceptual understanding of this legacy as it relates to the phenomena of motion, force, energy, matter, sound, electricity, magnetism, light, and the atom and its nucleus. The analysis of these topics makes up what we call physics, the knowledge of which opens new doors of perception. Our environment is far richer when we are aware of the beauty, harmony, and interplay of the laws of physics around us.

Question Which of the following activities involves the utmost human expression of passion, talent, and intelligence?*

a. art   **b.** literature   **c.** music   **d.** science

In Perspective Only a few centuries ago the most talented and most skilled artists, architects, and artisans of Europe directed their genius and effort to the construction of the great cathedrals. Some cathedrals took more than a century to build, which means the architects and early builders who lived to ripe old ages never saw the finished results of their labors. Whole lifetimes were spent in the shadows of construction that seemed without beginning or end. This enormous focus of human energy was inspired by the desire to transport people's souls to God. The cathedrals were their spaceships of faith, firmly anchored to the earth but pointing to the cosmos. This was during the era of faith, before the time of Galileo and the discovery that the description of motion was consistent with the order and regularity of mathematics. Then followed the time of Newton and the advance of science through the time of Einstein to the present era of physical law. Now the efforts of our most skilled scientists, engineers, artists, and artisans are directed to building the spaceships that will first orbit the earth and then voyage beyond. The time required to build today's spaceships is extremely brief compared to the time spent building the cathedrals of the past. Many people working on today's spaceships were

---

*__Answer__ All of them! The human value of science, however, is the least understood by most individuals in our society. The reasons are varied, ranging from the common notion that science is incomprehensible to people of average ability to the extreme view that science is a dehumanizing force in our society. Most of the misconceptions about science probably stem from the confusion between the *abuses* of science and science itself.

Some people who view science as cold and impersonal seek psychic comfort in mysticism and other counterscientific concepts. Still more embrace a combination of science and mysticism, such as astronomy and astrology, the normal and the paranormal. In straddling science and superstition, they stand with one foot in the twentieth century and the other in the fourteenth. It is unfortunate that they do not see basic science as an enchanting human activity shared by a wide variety of people who, with present-day tools and know-how, are reaching further and finding out more about themselves and their environment than people in the past were ever able to do.

The more you know about science, the more passionate you feel toward your surroundings. There is physics in everything you see, hear, smell, taste, and touch!

alive before Lindbergh made the first solo transatlantic flight. Where will the lives of young people today lead in a comparable time?

We seem to be at the dawn of a major change in our evolution, for as little Jenny suggests in the photo at the beginning of this chapter, we may be like the hatching chicken who has exhausted the resources of its inner-egg environment and is about to break through to a whole new range of possibilities. The earth is our cradle and has served us well. But cradles, however comfortable, are one day outgrown. So with inspiration that in many ways is similar to the inspiration of those who built the early cathedrals, we aim for the cosmos.

Our travels need not be in space alone. We can speculate far beyond. The fundamentals of present-day science have to do not with *why,* but with *how.* How does gravity hold the moon in orbit? How do electrical particles interact? How fast does light travel in a vacuum? The answers to these "how" questions are framed in physical law. We live in the era of physical law. John L. Wheeler, a pioneer and leading authority on quantum physics, speculates that we will soon advance from this era to the next—to the era of *meaning:* to the whys beneath the hows of physical law. In so doing, we should find ourselves closer to the soul of the universe.

This is an exciting time!

Suggested Reading

Cole, K. C. *Sympathetic Vibrations: Reflections of Physics as a Way of Life.* New York: Morrow, 1984.

Florman, S. C. *The Existential Pleasures of Engineering.* New York: St. Martin's, 1976.

Pagels, H. R. *The Cosmic Code: Quantum Physics as the Language of Nature.* New York: Simon & Schuster, 1982.

PART I Mechanics

2 The Study of Motion

More than 2000 years ago, the ancient Greek scientists were familiar with some of the ideas in physics that we study today. They had a very good understanding of some of the properties of light. But they were confused about motion. Probably the first to seriously study motion was Aristotle, the most outstanding philosopher-scientist in ancient Greece. Aristotle attempted to clarify motion by classification.

Aristotle on Motion

Aristotle divided motion into two main classes: *natural motion* and *violent motion*. We shall briefly consider each.

Natural motion was thought to proceed from the "nature" of bodies. In Aristotle's view, every body in the universe had a proper place, determined by this "nature"; and any body not in its proper place would "strive" to get there. Being of the earth, an unsupported lump of clay properly fell to the ground; being of the air, an unimpeded puff of smoke properly rose; being a mixture of earth and air but predominantly earth, a feather properly fell to the ground but not as rapidly as a pure lump of clay. Larger bodies were expected to strive harder; hence, bodies were thought to fall at speeds proportional to their weights: the heavier the body, the faster it was thought to fall.

Natural motion could be either straight up or straight down, as in the case of all things on earth; or it could be circular, as in the case of celestial objects. Unlike up-and-down motion, circular motion could be perceived as being without beginning or end, repeating itself without deviation. Aristotle asserted that celestial bodies were perfect spheres made of a perfect and unchanging substance, which he called *ether*. (The only celestial body with any detectable change or imperfection was the moon, which, being nearest the earth, was thought to be somewhat contaminated by the corrupted earth.)

Violent motion, Aristotle's other class of motion, resulted from pushing or pulling forces. Violent motion was imposed motion. A person pushing a cart or lifting a heavy weight imposed motion, as did someone hurling a stone or winning a tug-of-war. The wind imposed motion on ships. Flood waters imposed it on boulders and tree trunks. The essential thing about violent motion was that it was externally caused and was imparted to objects; they moved not of themselves, but were pushed or pulled.

The concept of violent motion had its difficulties, for the pushes and pulls responsible for it were not always evident. For example, a bowstring moved an arrow until the arrow left the bow; after that, further explanation of the arrow's motion seemed to require some other pushing agent. It was imagined, therefore, that a parting of the air by the moving arrow resulted in a squeezing effect on the rear of the arrow as the air rushed back to prevent a vacuum from forming. The arrow was propelled through the air as a bar of soap is propelled in the bathtub when you squeeze one end of it.

To sum up, Aristotle taught that all motions resulted either from the nature of the moving object or from a sustained push or pull. Provided that a body was in its proper place, it would not move unless subjected to a force. Except for the celestial bodies, the normal state was one of rest.

Aristotle's statements about motion were a beginning in scientific thought, and although he did not consider them to be the final words on the subject, his followers for nearly 2000 years regarded his views to be beyond question. Implicit in the thinking of ancient, medieval, and early Renaissance times was the notion that the normal state of bodies was one of rest. Since it was evident to most thinkers until the sixteenth century that the earth must be in its proper place, and since a force capable of moving the earth was inconceivable, it seemed quite clear that the earth did not move.

Aristotle (384–322 B.C.)

Greek philosopher, scientist, and educator, Aristotle was the son of a physician who personally served the king of Macedonia. At 17 he entered the Academy of Plato, where he worked and studied for twenty years until Plato's death. He then became the tutor of young Alexander the Great. Some eight years later he formed his own school. Aristotle's aim was to systematize existing knowledge, just as Euclid had systematized geometry. Aristotle made critical observations, collected specimens, and gathered together, summarized, and classified almost all existing knowledge of the physical world. His systematic approach became the method from which Western science later arose. After his death, his voluminous notebooks were preserved in caves near his home and were later sold to the library at Alexandria. Scholarly activity ceased through the Dark Ages, and the works of Aristotle were forgotten and lost. Various texts were rediscovered during the eleventh and twelfth centuries and translated into Latin. The Church was the dominant political and cultural force in Western Europe, and after first prohibiting the works of Aristotle, it accepted and incorporated them into Christian doctrine. Any attack on Aristotle was an attack on the Church itself.

Copernicus and the Moving Earth

It was in this climate that the astronomer Copernicus formulated his theory of the moving earth. Copernicus reasoned from his astronomical observations that the earth traveled around the sun. For years he worked without making his thoughts public—for two reasons. The first was that he feared persecution; a theory so completely different from common opinion would surely be taken as an attack on established order. The second reason was that he had grave doubts about it himself; he could not reconcile the idea of a moving earth with the prevailing ideas of motion. Finally, in the last days of his life, at the urging of close friends he sent his *De Revolutionibus* to the printer. The first copy of this famous exposition reached him on the day he died, May 24, 1543.

Most of us know about the reaction of the medieval Church to the idea that the earth traveled around the sun. Because Aristotle's views had become so formidably a part of Church doctrine, to contradict them was to question the Church itself. For many Church leaders, the idea of a moving earth threatened not only their authority but the very foundations of faith and civilization as well. Their fears were well founded. For better or for worse, this new idea was to overturn their conception of the cosmos.

Galileo and the Leaning Tower

Fig. 2.1 Galileo's famous demonstration.

It was Galileo, the foremost scientist of the sixteenth century, who gave credence to the Copernican view of a moving earth. He accomplished this by discrediting the Aristotelian ideas about motion. Although not the first to point out difficulties in Aristotle's views, Galileo was the first to provide conclusive refutation through observation and experiment.

Aristotle's falling-body hypothesis was easily demolished by Galileo. He is said to have dropped objects of various weights from the top of the Leaning Tower of Pisa and compared their falls. Contrary to Aristotle's assertion, he found that a stone twice as heavy as another did not fall twice as fast. Except for the small effect of air resistance, Galileo found that objects of various weights, when released at the same time, fell together and hit the ground at the same time. On one occasion, Galileo allegedly attracted a large crowd to witness the dropping of a light object and a heavy object from the top of the tower. It is said that many observers of this demonstration who saw the objects hit the ground together scoffed at the young Galileo and continued to hold fast to their Aristotelian teachings.

Galileo's Inclined Planes

Aristotle was an astute observer of nature, and he dealt with problems around him rather than with abstract cases that did not occur in his environment. Motion always involved a resistive medium such as air or water. He believed a vacuum to be impossible and therefore did not give serious consideration to motion in the absence of an interacting medium. That's why it was basic to Aristotle that a body requires a push or pull to keep it moving. And it was this basic principle that Galileo denied when he stated that if there is no interference with a moving body, it will keep moving in a straight line forever; no push, pull, or force of any kind is necessary.

Galileo tested this theory of motion by experimenting with the motion of various objects on inclined planes. He noted that balls rolling on downward-sloping planes picked up speed, while balls rolling on upward-sloping planes lost speed (Figure 2.2). From this he reasoned that balls rolling along a horizontal plane would neither speed up nor slow down. In practice, of course, such a rolling ball would slow down. But the ball would finally come to rest not because of its "nature" but because of friction. This idea was supported by Galileo's observation of motion along smoother surfaces: when there was less friction, the motion of bodies persisted for a longer time; the less the friction, the more the motion approached constant speed. He reasoned that in the absence of friction or other opposing forces, a horizontally moving body would continue moving forever.

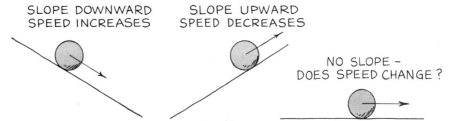

SLOPE DOWNWARD
SPEED INCREASES

SLOPE UPWARD
SPEED DECREASES

NO SLOPE –
DOES SPEED CHANGE ?

Fig. 2.2 Motion of a ball on various planes.

This assertion was supported by a different experiment and another line of reasoning. Galileo placed two of his inclined planes facing each other (Figure 2.3). He observed that a ball released from a position of rest at the top of a downward-sloping plane rolled down and then up the slope of the upward-sloping plane until it almost reached its initial height. He reasoned that only friction prevented it from rising to exactly the same height, for the smoother the planes, the more nearly the ball rose to the same height. Then he reduced the angle of the upward-sloping plane. Again the ball rose to the same height, but it had to go farther. Additional reductions of the angle yielded similar results: to reach the same height, the ball had to go farther each time. He then asked the question, "If I have a long horizontal plane, how far must the ball go to reach the same height?" The obvious answer is "Forever—it will never reach its initial height."[1]

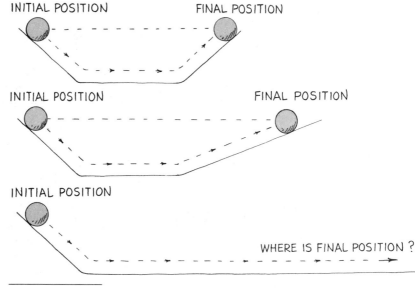

INITIAL POSITION

FINAL POSITION

INITIAL POSITION

FINAL POSITION

INITIAL POSITION

WHERE IS FINAL POSITION ?

Fig. 2.3 A ball rolling down an incline on the left tends to roll up to its initial height on the right. The ball must roll a greater distance as the angle of the incline on the right is reduced.

[1]From Galileo's *Dialogues Concerning the Two New Sciences.*

Galileo analyzed this in still another way. Because the downward motion of the ball from the first plane is the same for all cases, the speed of the ball when it begins moving up the second plane is the same for all cases. If it moves up a steep slope, it loses its speed rapidly. On a lesser slope, it loses its speed more slowly and rolls for a longer time. The less the upward slope, the more slowly it loses its speed. In the extreme case where there is no slope at all—that is, when the plank is horizontal—the ball should not lose any speed. In the absence of retarding forces, the tendency of the ball is to move forever without slowing down. This property of a material object to maintain its state of motion he called **inertia**.

Galileo's concept of inertia discredited the Aristotelian theory of motion. Aristotle did not recognize the idea of inertia because he failed to imagine what motion would be like without friction. In his experience, all motion was subject to resistance, and he made this fact central to his theory of motion. Aristotle's failure to recognize friction for what it is— namely, a force like any other—impeded the progress of physics for 2000 years, until the time of Galileo. An application of Galileo's concept of inertia would show that no force was required to keep the earth in motion. The way was open for Isaac Newton to synthesize a new vision of the universe. Before we go on to Newton, however, we should acquaint ourselves with some of the terms that Galileo introduced to describe motion.

Question	Would it be correct to say that inertia *causes* a body to persist in its state of motion?*

Description of Motion

In Aristotle's description of motion, the *distance* of an object from its proper place was fundamentally important. Galileo broke with the traditional concept and realized that *time* was the important missing ingredient in describing motion. Both distance and time, however, underlie the three ideas needed to describe motion: namely, speed, velocity, and acceleration.

Speed

The most basic property of a moving body is **speed**. By virtue of its motion, a body travels a certain distance in a given time. An automobile,

---

*__Answer__ In a strict sense, no. We don't know *why* bodies exhibit this property. Nevertheless, we call the property to behave in this predictable way *inertia*. We understand many things and have labels and names for these things. There are many things we do not understand, and we have labels and names for these things also. Education consists not so much in acquiring new names and labels but in learning which we understand and which we don't.

Fig. 2.4 A speedometer gives readings in both miles per hour and kilometers per hour.

for example, travels so many kilometers per hour. Speed is simply the ratio of distance traveled per time taken for this travel; that is,

$$\text{Speed} = \frac{\text{distance}}{\text{time}}$$

For example, if we drive a distance of 80 kilometers in a time of 1 hour, we say our speed is 80 kilometers per hour.[2] More precisely, we call this our *average* speed, for the speed of our car during the trip usually varies somewhat. If there is no variation, we can refer to our motion as *constant* speed, which means that *equal distances* are covered in *equal intervals* of time.

The speed that a body has at any one instant is called *instantaneous* speed. It is the speed registered by the speedometer of a car. When we say that the speed of a car at some particular instant is 60 kilometers per hour, we are specifying its instantaneous speed, and we mean that if the car continued moving as fast for an hour, it would travel 60 kilometers. So the instantaneous speed, or speed at a particular instant, is often quite different from the average speed.

We can look at speed in another way. Speed is a *rate* that specifies how rapidly distance is covered. For example, to say that we are traveling at 60 kilometers per hour is to say that it would take 1 hour to cover 60 kilometers, and so on. In these cases, the rate at which we cover distance is 60 kilometers per hour whether we travel for 60 kilometers in 1 hour or the length of a city block in a few seconds. Speed is the rate at which distance is covered.

We have expressed speed in kilometers per hour, but, of course, speed can be expressed in any units of distance divided by any units of time. For motor vehicles (or long distances), the units kilometers per hour (km/hr) or miles per hour (mi/hr or mph) are commonly used. For shorter distances, meters per second (m/s) are often useful units. Table 2.1 shows some comparative speeds in different units.

Table 2.1 Approximate speeds in different units

20 km/hr	= 12.4 mi/hr	= 5.6 m/s
40 km/hr	= 25 mi/hr	= 11 m/s
60 km/hr	= 37 mi/hr	= 17 m/s
80 km/hr	= 50 mi/hr	= 22 m/s
88 km/hr	= 55 mi/hr	= 24 m/s
100 km/hr	= 62 mi/hr	= 28 m/s
120 km/hr	= 75 mi/hr	= 33 m/s

[2]Note that *per* means "divided by." So 80 kilometers per hour means the distance 80 kilometers divided by the time of 1 hour, which can be written 80 km/hr.

1. What is the average speed of a cheetah that sprints 100 m in 4 s? How about if it sprints 50 m in 2 s?*

2. If a car moves with an average speed of 60 km/hr for an hour, it will travel a distance of 60 km. How far would it travel at this rate for 4 hr? For 10 hr?†

Velocity

Loosely speaking, we can use the words *speed* and *velocity* interchangeably. Strictly speaking, however, there is a distinction between the two. When we say that a body travels at a rate of 60 kilometers per hour, we are specifying its speed. But if we say that a body moves at 60 kilometers per hour to the north, we are specifying its velocity. When we describe speed and the *direction* of motion, we are specifying **velocity**.³ Constant velocity implies constant speed and constant direction; that is, motion is unvarying and along a straight-line path. Constant speed and constant velocity are not the same. A body may move at constant speed along a curved path, for example, but it does not move with constant velocity because its direction of motion is changing every instant.

Fig. 2.5 The car on the circular track may have a constant speed, but its velocity is changing every instant. Why?

---

*(Are you reading this before you have formulated a reasoned answer in your thinking? If so, do you also exercise your body by watching others do push-ups? Exercise your thinking: when you encounter the many questions as above throughout this book, *think* before you read the footnoted answer.)

Answer 25 m/s in both cases, which follows from the definition of

$$\text{Average speed} = \frac{\text{distance}}{\text{time}} = \frac{100 \text{ meters}}{4 \text{ seconds}} = \frac{50 \text{ meters}}{2 \text{ seconds}} = 25 \text{ m/s}$$

†**Answer** 240 km and 600 km, respectively. We can see that if we know the average speed and the time of travel, the distance traveled is

$$\text{Distance} = \text{average speed} \times \text{time}$$
$$\text{Distance} = 60 \text{ km/hr} \times 4 \text{ hr} = 240 \text{ km}$$
$$\text{Distance} = 60 \text{ km/hr} \times 10 \text{ hr} = 600 \text{ km}$$

Can you see that this relationship is simply a rearrangement of

$$\text{Average speed} = \frac{\text{distance}}{\text{time}}$$

³A quantity described by both magnitude (how much) and direction (which way) is called a **vector quantity**. Velocity is a vector quantity (see Appendix III). A quantity described only by magnitude, such as speed, is a **scalar quantity**.

Fig. 2.6 We say that a body undergoes acceleration when there is a *change* in its state of motion.

Acceleration

We can change the state of motion of a body by changing its speed, by changing its direction of motion, or by changing both its speed *and* its direction. Any of these changes is a change in velocity. We define the rate of change in velocity as **acceleration**:

$$\text{Acceleration} = \frac{\text{change of velocity}}{\text{time}}$$

We are all familiar with acceleration in an automobile. In driving, we call it "pickup" or "getaway"; we experience it when we tend to lurch toward the rear of the car. Suppose we are driving and in 1 second we steadily increase our velocity from 30 kilometers per hour to 35 kilometers per hour, and then to 40 kilometers per hour in the next second, to 45 in the next second, and so on. We increase our velocity by 5 kilometers per hour each second. This change in velocity is what we mean by acceleration. In this case we would be undergoing an acceleration of 5 kilometers per hour per second. Note that acceleration is not just the total change in velocity; it is the *time rate of change*, or *change per second* in velocity (Figure 2.6).

The term *acceleration* applies to decreases as well as increases in velocity. The brakes of a car, for example, are said to produce large retarding accelerations; that is, there is a large decrease per second in the velocity of the car. We often call this *deceleration*, or *negative acceleration*. We experience deceleration when we tend to lurch toward the front of the car.

Although we cannot accelerate while traveling at constant velocity, it is possible to accelerate while traveling at constant speed. For example, suppose we are driving around a curve at a constant speed of 30 kilometers per hour. Although in this case there will be no change in speed, there will be a change in direction and therefore a change in velocity. This change in velocity is acceleration, which we will feel in response to our inertia as we tend to lurch toward the outer part of the curve. We distinguish speed and velocity for this reason and define *acceleration* as the rate at which velocity changes, thereby encompassing changes both in speed and in direction.

Anyone who has stood in a crowded bus has experienced the difference between velocity and acceleration. Except for the effects of a bumpy road, you can stand with no extra effort inside a bus that moves at constant velocity, no matter how fast it is going. You can flip a coin and catch it exactly as if the bus were at rest. It is only when the bus accelerates—speeds up, slows down, or turns—that you experience difficulty.

Acceleration occurs only when there is a change in a body's state of motion. Just as velocity is the rate at which the position of a body changes, acceleration is the rate at which velocity changes.

Question   In 2.5 s a car increases its speed from 60 km/hr to 65 km/hr while a bicycle goes from rest to 5 km/hr. Which undergoes the greater acceleration? What is the acceleration of each vehicle?*

Falling Bodies   *Velocity and Acceleration*

It was Galileo who first introduced the idea of acceleration. He developed it in describing the motion of falling bodies. Although he had invented the pendulum clock, he could not measure time intervals short enough to measure accurately the velocity of rapidly falling objects. He overcame this difficulty by doing most of his experimental work with inclined planes. This technique enabled him to "slow down" the accelerated motion of rolling balls so that he could more accurately measure the relationships between distance and time.

Galileo verified his guess that the velocity of balls rolling down the inclines increased uniformly with time. He found that the balls picked up the same amount of speed in each successive time interval; that is, the balls rolled with uniform, or constant, acceleration. For example, if we express velocity and time in modern units, a ball rolling down a plane inclined at a certain angle might be found to pick up a velocity of 2 meters per second for each second it rolls. It gains a speed of 2 meters per second each second. Its instantaneous velocity at 1-second intervals, increasing by this amount, is then 0, 2, 4, 6, 8, 10, and so forth, meters per second. We can see that the velocity of the ball at any given time after being released from rest is simply equal to its acceleration multiplied by the time:[4]

$$\text{Velocity acquired} = \text{acceleration} \times \text{time}$$

meters per sec

If we substitute the acceleration of the ball in this relationship, we can see that at the end of 1 second, the ball is traveling at 2 meters per second; at the end of 2 seconds, it is traveling at 4 meters per second; at the end of 10 seconds, it is traveling at 20 meters per second; and so on. The velocity at any time is simply equal to the acceleration multiplied by the number of seconds it has been accelerating.

---

*__Answer__   The accelerations of both the car and the bicycle are the same: 2 km/hr/s.

$$\text{Acceleration}_{\text{car}} = \frac{\text{change in velocity}}{\text{time}} = \frac{65 \text{ km/hr} - 60 \text{ km/hr}}{2.5 \text{ s}} = \frac{5 \text{ km/hr}}{2.5 \text{ s}} = 2 \frac{\text{km}}{\text{hr/s}}$$

$$\text{Acceleration}_{\text{bike}} = \frac{\text{change in velocity}}{\text{time}} = \frac{5 \text{ km/hr} - 0 \text{ km/hr}}{2.5 \text{ s}} = \frac{5 \text{ km/hr}}{2.5 \text{ s}} = 2 \frac{\text{km}}{\text{hr/s}}$$

Here we see that although the velocities involved are quite different, the rates of change of velocities are the same. Each has the same acceleration.

[4]Note that this relationship follows from the definition of acceleration. From $a = v/t$, multiplying both sides of the equation by t gives $v = at$.

Fig. 2.7 The greater the slope of the incline, the greater the acceleration of the ball. What is its acceleration if the incline is vertical?

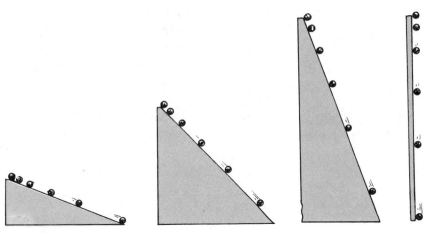

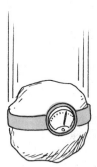

Fig. 2.8 Pretend that a falling rock is equipped with a speedometer. In each succeeding second of fall, we would find that the rock's speed increases by the same amount: 10 m/s. (Table 2.2 shows the speeds we would read at various seconds of fall.)

Table 2.2 Velocity acquired in free fall

Time of fall (seconds)	Velocity acquired (meters/ second)
0	0
1	10
2	20
3	30
4	40
5	50
.	.
.	.
.	.
t	$10t$

Galileo found greater accelerations for steeper inclines. The ball attains its maximum acceleration when the incline is tipped vertically—that is, the acceleration of free fall (Figure 2.7). Regardless of the weight or size, Galileo discovered that in the absence of air resistance, all bodies fall with the same constant acceleration.

Table 2.2 shows what may be observed when air resistance has little or no effect on a falling object. The most important thing to note in these numbers is the way the velocity changes. *During each second of fall, the body gains a velocity of 10 meters per second.* This gain per second is the acceleration. The acceleration of falling bodies under conditions in which air resistance is negligible is approximately equal to 10 meters per second per second or, more accurately, 9.8 meters per second per second.[5] In the case of falling bodies, it is customary to use the letter g to represent the acceleration (because the acceleration is due to *gravity*). Although the value of g varies slightly in different parts of the world, its average value is equal to 9.8 meters per second per second or, in shorter notation, 9.8 meters/second². We round this off to 10 meters/second² in our present discussion and in Tables 2.2 and 2.3 to establish the ideas involved more clearly; multiples of 10 are more obvious than multiples of 9.8. Where accuracy is important, the value of 9.8 meters/second² should be used.

Note in Table 2.2 that the velocity acquired by a falling body is equal to the acceleration multiplied by the time of fall. In shorthand notation,

$$v = gt$$

---

[5] Note the repetition of "per second." From the meaning of the word *acceleration*, the unit of time enters twice—once for the unit of velocity and again for the interval of time in which the velocity is changing. Instead of stating the acceleration of freely falling objects as "9.8 meters per second per second," we can shorten it to "9.8 meters per second²" or, in even briefer notation, "9.8 m/s²," read "9.8 meters per second squared."

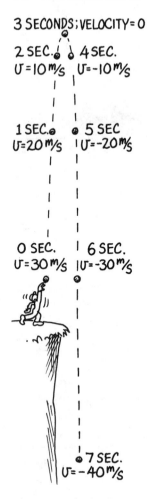

3 SECONDS; VELOCITY = 0

2 SEC.   4 SEC.
$U = 10\,^m/s$ $U = -10\,^m/s$

1 SEC.   5 SEC
$U = 20\,^m/s$ $U = -20\,^m/s$

0 SEC.   6 SEC.
$U = 30\,^m/s$ $U = -30\,^m/s$

7 SEC.
$U = -40\,^m/s$

Fig. 2.9 The rate at which the velocity changes each second is the same.

Fig. 2.10 Pretend that a falling rock is equipped with an odometer. The readings of distance fallen increase with time by $\frac{1}{2}gt^2$ and are shown in Table 2.3.

Here v is the velocity of fall expressed in meters per second when the number of seconds of fall is substituted for t. A more detailed analysis of this, with numerical examples, can be found in Appendix II.

Galileo verified a second reasoned guess with his inclined planes. He reasoned that the distance a uniformly accelerating object travels is proportional to the square of the time. His deduction and an algebraic derivation of this relationship can also be found in Appendix II. We will state here only the results. The distance traveled by a uniformly accelerating body starting from rest is

$$\text{Distance traveled} = \tfrac{1}{2}(\text{acceleration} \times \text{time} \times \text{time})$$

We can express this relationship for the case of a freely falling body in shorthand notation as[6]

$$d = \tfrac{1}{2}gt^2$$

where d is the distance when the time of fall in seconds is substituted for t and squared. If we use 10 meters/second^2 for the value of g, the distance fallen for various times will be as shown in Table 2.3.

Note that an object falls a distance of only 5 meters during the first second of fall, although its velocity at 1 second is 10 meters per second. This may be confusing, for we may think that the object should fall a distance of 10 meters. But for it to fall 10 meters in its first second of fall, it would have to fall at an *average* velocity of 10 meters per second for the *entire* second. It starts its fall at 0 meters per second, and its velocity is 10 meters per second only in the last instant of the 1-second interval. Its average velocity during this interval is the average of its beginning and final velocities, 0 and 10 meters per second. To find the average value of two numbers, we simply add the two numbers and divide by 2. This equals 5 meters per second, which over a time interval of 1 second gives a distance of 5 meters. As the object continues to fall in succeeding seconds, it will fall through ever-increasing distances because its velocity is continuously increasing.

---

[6] d = average velocity × time

$d = \dfrac{\text{beginning velocity + final velocity}}{2} \times \text{time}$

$d = \dfrac{0 + gt}{2} \times t$

$d = \tfrac{1}{2}gt^2$ (See Appendix II for further explanation.)

Question   A cat steps off a ledge and drops to the ground in $\frac{1}{2}$ second. What is its speed on striking the ground? What is its average speed during the $\frac{1}{2}$ second? How high is the ledge from the ground?*

Table 2.3 Distance fallen in free fall

Time of fall (seconds)	Distance fallen (meters)
0	0
1	5
2	20
3	45
4	80
5	125
.	.
.	.
.	.
t	$\frac{1}{2}10t^2$

Fig. 2.11 A feather and a coin fall at equal accelerations in a vacuum.

It is a common observation that all bodies do not fall with equal accelerations. A leaf, a feather, or a sheet of paper, for example, may flutter to the ground slowly. That the air is the factor responsible for those different accelerations can be shown very nicely with a closed glass tube containing light and heavy objects—a feather and a coin, for example. In the presence of air, the feather and coin fall with quite unequal accelerations. But if the air in the tube is evacuated by means of a vacuum pump and the tube is quickly inverted, the feather and coin fall with the same acceleration (Figure 2.11). Although air resistance appreciably alters the motion of falling feathers and the like, the motion of heavier objects like stones and baseballs at ordinary low speeds is not appreciably affected by the air. The relationships $v = gt$ and $d = \frac{1}{2}gt^2$ can be used to a very good approximation for most objects falling in air.

Much of the confusion that arises in analyzing the motion of falling objects comes about because it is easy to get "how fast" mixed up with "how far." When we wish to specify how fast something is falling, we are talking about velocity, which is expressed as $v = gt$. When we wish to specify how far something falls, we are talking about distance, which is expressed as $d = \frac{1}{2}gt^2$. Velocity (how fast) and distance (how far) are entirely different from each other.

A most confusing concept, and probably the most difficult encountered in this book, is "how quickly does how fast change"—acceleration. What makes acceleration so complex is that it is *a rate of a rate*. It is often confused with velocity, which is itself a rate (the rate of change of position). Acceleration is not velocity, nor is it even a change in velocity. Acceleration is the rate at which velocity itself changes.

Please remember that it took people nearly 2000 years from the time of Aristotle to reach a clear understanding of motion, so be patient with yourself if you find that you require a few hours to achieve as much!

---

*Answer   If we round g off to 10 m/s², we find

$$\text{Speed: } v = gt = 10 \text{ m/s}^2 \times \tfrac{1}{2}\text{s} = 5 \text{ m/s}$$

$$\text{Average speed: } \bar{v} = \frac{\text{beginning } v + \text{final } v}{2} = \frac{0 \text{ m/s} + 5 \text{ m/s}}{2} = 2.5 \text{ m/s}$$

We put a bar over the symbol to denote *average* speed—$\bar{v}$:

$$\text{Distance: } d = \bar{v}t = 2.5 \text{ m/s} \times \tfrac{1}{2}\text{s} = 1.25 \text{ m}$$

Or, equivalently,

$$d = \tfrac{1}{2}gt^2 = \tfrac{1}{2} \times 10 \text{ m/s}^2 \times (\tfrac{1}{2}\text{s})^2 = \tfrac{1}{2} \times 10 \text{ m/s}^2 \times \tfrac{1}{4}\text{s}^2 = 1.25 \text{ m}$$

Notice that we can find the distance by either of these equivalent relationships.

a

$$\text{SPEED} = \frac{\text{DISTANCE}}{\text{TIME}}$$

$$\text{SPEED} = \frac{80 \text{ KM}}{1 \text{ HR}} = 80 \text{ KM/HR}$$

SAN FRANCISCO ×

× LIVERMORE

TIME = 1 HOUR

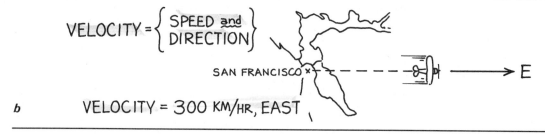

b

$$\text{VELOCITY} = \left\{ \begin{array}{l} \text{SPEED and} \\ \text{DIRECTION} \end{array} \right\}$$

SAN FRANCISCO ×

E

VELOCITY = 300 KM/HR, EAST

ACCELERATION

c

$$\left\{ \begin{array}{l} \text{RATE OF} \\ \text{CHANGE IN} \\ \text{VELOCITY} \end{array} \right\} \text{DUE TO} \left\{ \begin{array}{l} \text{CHANGE IN SPEED} \\ \text{AND/OR DIRECTION} \end{array} \right\}$$

40 KM/HR   80 KM/HR   0 KM/HR

CHANGE IN SPEED
BUT *NOT* DIRECTION

40 KM/HR   40 KM/HR

CHANGE IN DIRECTION
BUT *NOT* SPEED

CHANGE IN SPEED
AND DIRECTION

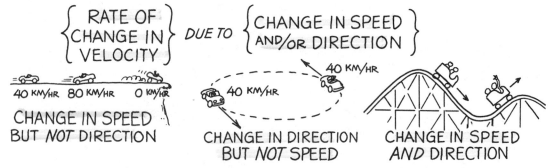

d

$$\text{ACCELERATION} = \frac{\text{CHANGE IN VELOCITY}}{\text{TIME}}$$

TIME = 0, VELOCITY = 0

TIME = 1 s,
   VELOCITY = 9.8 m/s

TIME = 2 s, VELOCITY = 19.6 m/s

$$\text{ACCELERATION} = \frac{19.6 \text{ m/s}}{2 \text{ s}}$$

$$a = 9.8 \frac{\text{m/s}}{\text{s}}$$

$$a = 9.8 \frac{\text{m}}{\text{s}} \text{s}$$

$$a = 9.8 \text{ m/s}^2$$

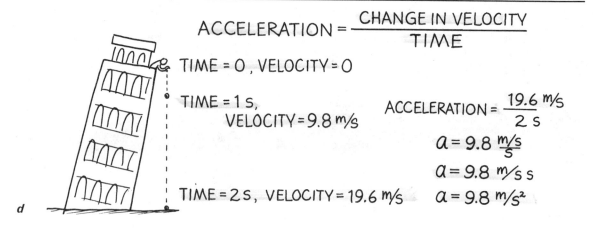

Fig. 2.12 Motion analysis.

Galileo Galilei (1564–1642)

Galileo was born in Pisa in the same year as Shakespeare and the same year that Michelangelo died. He studied medicine at the University of Pisa and then changed to mathematics. He developed an early interest in the mechanics of motion and was soon at odds with his contemporaries who held to Aristotelian ideas on falling bodies. He left Pisa to teach at the University of Padua and became an advocate of the new Copernican theory of the solar system. He was the first to build a telescope and discover mountains on the moon and the moons of Jupiter. Because he published his findings in Italian instead of in the Latin expected of so reputable a scholar, and because of the recent invention of the printing press, his ideas reached a wide readership. He soon ran afoul of the Church and under threat of torture was warned not to teach and hold to Copernican views. He restrained himself publicly for nearly fifteen years and then defiantly published his observations and conclusions, which were counter to Church doctrine. The outcome was a trial during which he was found guilty, and he was forced to renounce his discoveries. By then an old man broken in health and spirit, he was sentenced to perpetual house arrest. Nevertheless, he completed his studies on motion and his writings were smuggled from Italy and published in Holland. He became blind at the age of 74 and died four years later.

Summary of Terms

Inertia The sluggishness or apparent resistance a body offers to changes in its state of motion.

Speed The distance traveled per time.

Velocity The speed of a body and specification of its direction of motion.

Acceleration The rate at which velocity changes with time; the change in velocity may be in magnitude or direction or both.

Review Questions

*Each chapter in this book concludes with a set of review questions and exercises. The **Review Questions** are designed to help you fix ideas and catch the essentials of the chapter material. You'll notice that answers to the questions can be found within the chapters. The **Exercises** stress thinking rather than merely recall of information and call for an* understanding *of the definitions, principles, and relationships of the chapter material. In many cases the intention of particular exercises is to help you to apply the ideas of physics to familiar situations. Unless you cover only a few chapters in your course, you will likely be expected to tackle only a few exercises for each chapter. The large number of exercises is to allow your instructor a wide choice of assignments.*

1. What were the principal reasons five centuries ago for believing that the earth was stationary?

2. What does it mean to say that a material body has inertia?

3. Distinguish between *speed* and *velocity.*

4. If the speedometer of a car indicates a constant speed of 30 kilometers per hour, can you say that the car is *not* accelerating? Why or why not?

5. What is the name of the quantity that describes how fast you change how fast you're going, or how fast you change which direction you're going?

6. How much speed does a body pick up each second when it is undergoing a constant acceleration of 2 meters per second^2? 4 meters per second^2? 9.8 meters per second^2?

7. How much speed does a body pick up in 2 seconds when it is undergoing a constant acceleration of 2 meters per second^2? 4 meters per second^2? 9.8 meters per second^2?

8. What is the maximum acceleration possible for a freely rolling ball on an inclined plane?

9. What is the instantaneous velocity of an object 1 second after it is released from a rest position? What is its average velocity during this 1-second interval? How far will it travel during this time?

10. If an object freely falls from a rest position, what is its instantaneous *speed* at the end of the fifth second? The sixth second?

11. If an object freely falls from a rest position, how *far* will it have fallen at the end of the fifth second? The sixth second?

12. If an object freely falls from a rest position, what is its *acceleration* at the end of the fifth second? The sixth second? For any number of seconds?

Home Projects

1. By any method you choose, determine your average speed of walking.

2. Try this with your friends. Hold a dollar bill so that the midpoint hangs between a friend's fingers and challenge him to catch it by snapping his fingers shut when you release it. He won't be able to catch it! Explanation: From $d = \frac{1}{2}gt^2$, the bill will fall a distance of 8 centimeters (half the length of the bill) in a time of $\frac{1}{8}$ second, but the time required for the necessary impulses to travel from his eye to his brain to his fingers is at least $\frac{1}{7}$ second.

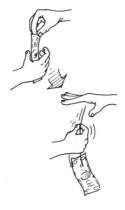

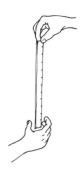

3. You can compare your reaction time with that of your friend by catching a ruler that is dropped between your fingers. Let a friend hold the ruler as shown and you snap your fingers shut as soon as you see the ruler released. The number of centimeters that pass through your fingers depends on your reaction time. You can express this in fractions of a second by rearranging $d = \frac{1}{2}gt^2$. Expressed for time, it is

$$t = \sqrt{2d/g} = 0.045\sqrt{d}$$

where d is expressed in centimeters.

Exercises

1. Contrast the role of a *force* to explain the motion of the earth from an Aristotelian view and a Galilean view.

2. A ball is rolled across the top of a billiard table and slowly rolls to a stop. How would Aristotle interpret this observation? How would Galileo interpret it?

3. One airplane travels due north at 300 kilometers per hour while another travels due south at 300 kilometers per hour. Are their speeds the same? Are their velocities the same? Explain.

4. Cite an example of a body that undergoes acceleration while traveling at constant speed. Is it possible to cite an example of a body undergoing acceleration while traveling at constant velocity? Explain.

5. Can you cite an example wherein the acceleration of a body is opposite in direction to its velocity?

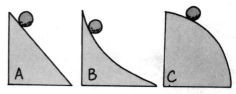

6. On which of the above hills does the ball roll down with increasing speed and *decreasing* acceleration? (Use this example if you wish to explain to someone the difference between speed and acceleration.)

7. What is the *acceleration* of a car that travels at a steady velocity of 100 kilometers per hour for 100 seconds? Explain your answer.

8. One car travels from 0 to 50 kilometers per hour, and later another car travels from 0 to 60 kilometers per hour. From this information can you say which car underwent the greater acceleration? Why or why not?

9. Compare the accelerations of a car and a bicycle for the case where the car increases its speed from 60 to 70 kilometers per hour in the same time that the bicycle goes from rest to 10 kilometers per hour.

10. If you were standing in an enclosed car moving at constant velocity, would you have to lean in some special way to compensate for the car's motion? What if the car were moving with constant acceleration? Explain.

11. Suppose that a freely falling object were somehow equipped with a speedometer. By how much would its speed reading increase with each second of fall?

12. Suppose the freely falling object in the preceding exercise were also equipped with an odometer. Would the readings of distance fallen each second be equal or different for successive seconds? Explain.

13. When a ballplayer throws a ball straight upward, by how much does the speed of the ball decrease each second while ascending? In the absence of air resistance, by how much does it increase each second while descending? How much time is required for rising compared to falling?

14. What is the instantaneous velocity of an object 10 seconds after it is released from a position of rest? What is its average velocity during this 10-second interval? How far will it travel during this time?

15. Someone standing at the edge of a cliff (as in Figure 2.9) throws a ball straight upward with a certain speed and another ball straight downward with the same initial speed. If air resistance is negligible, which ball will have the greater speed when it strikes the ground below?

16. If you drop an object, its acceleration toward the ground is 9.8 meters per second^2. If you instead throw it downward, would its acceleration after throwing be greater than 9.8 meters per second^2? Why or why not?

17. In the preceding exercise can you think of a reason why the acceleration of the object thrown downward *through the air* would actually be *less* than 9.8 meters per second^2?

18. If it were not for air resistance, why would it be dangerous to go outdoors on rainy days?

19. A racing car traveling at 10 meters per second north increases its velocity to 16 meters per second north in a time interval of 3 seconds. What is its average acceleration during this time interval? What is the direction of this acceleration?

20. A fly traveling north at 4 meters per second finds 3 seconds later that it is traveling south at 2 meters per second. What are the magnitude and the direction of its average acceleration during this time interval?

21. Perform one of the few numerical exercises asked for in this book and extend Tables 2.2 and 2.3 from 0 to 5 seconds to 0 to 10 seconds, for the case of no air resistance.

22. In this chapter we studied idealized cases of bodies rolling down frictionless planes and objects falling with no air resistance. Suppose a classmate complains that all this attention focused on idealized cases is valueless because idealized cases simply don't occur in the everyday world. How would you respond to this complaint? How do you suppose the author of this book would respond?

PLEASE LET YOUR PRIMARY GOAL IN LEARNING PHYSICS BE ACQUAINTING YOURSELF WITH THE INFORMATION AND INSIGHTS WITHIN THE CHAPTERS, AND *NOT* PRIMARILY IN DOING THE EXERCISES, WHICH ARE INTENDED AS SOME MENTAL PUSHUPS TO TRY *AFTER* YOU HAVE STUDIED THE CHAPTER MATERIAL!

3 Projectile and Satellite Motion

Projectile Motion

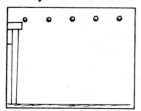

a HORIZONTAL MOTION WITH NO GRAVITY

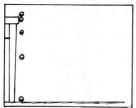

b VERTICAL MOTION ONLY WITH GRAVITY

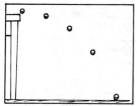

c COMBINED HORIZONTAL AND VERTICAL MOTION

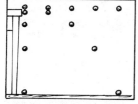

d A SUPERPOSITION OF THE ABOVE CASES

Fig. 3.1 Simulated photographs of a moving ball illuminated with a strobe light.

When a bowling ball rolls across a flat surface, the force of gravity holds it to the surface but does not alter its motion. The ball moves neither with nor against earth gravity; it moves without accelerating. Yet if the ball is held above your head and released, it accelerates as it drops to the ground, for its motion is in the direction of earth gravity. Thus far we have concentrated our attention on motions such as these—objects moving either horizontally or vertically in straight-line paths. But what about objects thrown at various angles to the horizon; what about projectiles?

A **projectile** is any body that is thrown or projected by some force and continues in motion by its own inertia. In the absence of gravity, the motion of a projectile would be simple enough; it would move with constant velocity. But gravity complicates this otherwise simple motion. Because of the force of gravity, a projectile undergoes acceleration, and its motion is different from instant to instant. Its motion is complicated. In order to understand this complicated motion of a projectile, we will separate the motion into two parts—the horizontal part and the vertical part.[1] We will study each of these component parts individually and then combine our findings.

Let's consider a simple projectile: a ball rolled off the edge of a horizontal table. To consider only the horizontal motion, we will do a simple thought experiment. We will "turn off" gravity and watch the motion of the ball as it rolls off the table. A neat way to record its motion is with a strobe light and camera. We will darken the room, set the camera lens open for a time exposure, and set a strobe light so that it blinks at short, equally spaced time intervals. Then we will photograph the position of the ball at different times on the same photographic plate. Our picture of the ball will look something like Figure 3.1*a*. The ball has covered equal distances in equal times in a straight-line path, because no forces act on the ball.

To investigate the vertical component of motion, we will turn gravity back on and take a picture of the ball as it drops from a position of rest at the edge of the tabletop. This time there is no horizontal velocity and it drops straight down. Our "photograph," Figure 3.1*b*, shows that the ball covers successively greater distances in equal intervals of time, which is to

---

[1]A rational way to understand any complicated problem is to separate the problem into its parts and investigate each individually. If the parts can be understood, then the task of understanding the whole is easier—even though the whole may be greater than the sum of its parts—life, for example.

be expected because the ball is accelerating downward in the direction of earth gravity.

Suppose we repeat the first part of our experiment and roll the ball off the edge of the table, but this time in the presence of gravity. A "photograph" of this, Figure 3.1c, reveals that the resulting curved motion of the ball is a combination of the horizontal and vertical motions. During the five flashes of the strobe light, the ball has traveled just as far horizontally as it did with no gravity. While the ball is moving sideways, it is falling, and the combination of these two motions produces a curved path. This is made clearer if we superimpose the three pictures so we can view all three experiments at once (Figure 3.1d). An actual strobe-light photograph of two balls released simultaneously is shown in Figure 3.2. A projectile accelerating vertically while moving at a constant horizontal speed traces out a path called a **parabola**.

Fig. 3.2 A strobe-light photograph of two golf balls released simultaneously from a mechanism that allows one ball to drop freely while the other is projected horizontally.

Question At the instant a horizontally held rifle is fired over a level range, a bullet held at the side of the rifle is released and drops to the ground. Which bullet strikes the ground first—the one fired downrange or the one dropped from rest?*

*****Answer** Both bullets fall the same vertical distance with the same gravitational acceleration g and therefore strike the ground at the same time. Gravity does not take a holiday on moving objects.

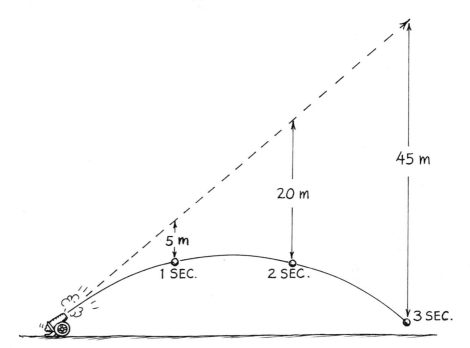

Fig. 3.3 With no gravity, the projectile would follow a straight-line path (dashed line). But because of gravity, it falls beneath this line the same vertical distance it would fall if released from rest. Compare the distances fallen with those given in Table 2.3 in Chapter 2. (With $g = 9.8$ m/s², these distances are more accurately 4.9 m, 19.6 m, and 44.1 m.)

Consider a cannonball shot at an upward angle (Figure 3.3). Pretend for a moment that there is no gravity; then by virtue of inertia, the cannonball will follow the straight-line path shown by the dashed line. But there *is* gravity, so this doesn't happen. What really happens is that the cannonball continually *falls beneath this imaginary line* until it finally strikes the ground. Get this: the vertical distance it falls beneath any point on the dashed line is the same vertical distance it would fall if it were dropped from rest in the same time. This distance is given by $d = \frac{1}{2}gt^2$, which was introduced in the last chapter.

We can put this another way: shoot a projectile skyward and pretend there is no gravity. After so many seconds t, it should be at a certain point along a straight-line path. But because of gravity, it isn't. Where is it? The answer is: it's directly below this point. How far below? The answer in meters is $5t^2$ (or, more accurately, $4.9t^2$). How about that?

Note something else in Figure 3.3. The cannonball moves equal horizontal distances in equal time intervals: that is, its horizontal component of motion is uniform; acceleration does not take place horizontally, only vertically—in the direction of the pull of gravity. The vertical distance it falls below the imaginary straight-line path, however, continually increases with time.

If the effects of air resistance are negligible, every object projected into the air will follow a parabolic path. In practical cases, however, air resistance may be considered negligible only for slowly moving objects with high densities, like a rock or solid ball. For high-speed projectiles such as bullets or cannonballs, air resistance continually slows the projectiles, and the path departs from a parabola (Figure 3.4).

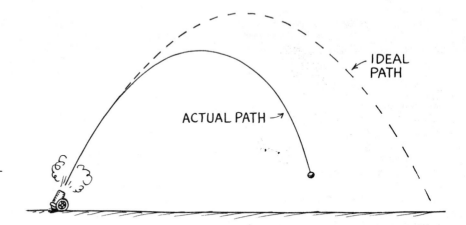

Fig. 3.4 In the presence of air resistance, the trajectory of a high-speed projectile falls short of a parabolic path.

The vertical height and horizontal range of a projectile depend on both the initial speed and the angle of projection of the projectile. The paths of several projectiles all having the same initial velocity but different projection angles are shown in Figure 3.5. The figure neglects the effects of air resistance, so the paths are all parabolas. The maximum height is obtained when the projection is straight up; and the maximum horizontal distance,

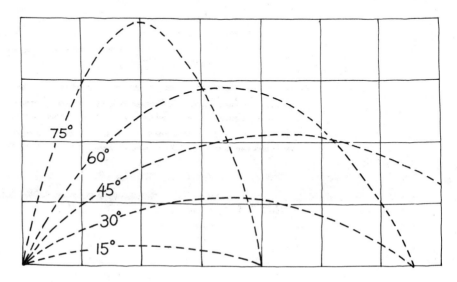

Fig. 3.5 Ranges of a projectile shot at the same speed at different projection angles.

when the angle of projection is 45 degrees (Figure 3.6). Note from Figure 3.5 that the same horizontal distance, or range, can be obtained from two different projection angles. This is true of all pairs of angles that add up to 90 degrees. An object thrown into the air at an angle of 30 degrees, for example, will land just as far downrange as if it were thrown at the same speed at an angle of 60 degrees. For the steeper angle, of course, the object remains in the air for a longer time.

Fig. 3.6 Maximum range is attained when a ball is batted at an angle of 45°. (For common cases in sports in which the weight of the projectile is significant in comparison to the force of the projection, the applied force does not produce the same velocity for different projection angles, and maximum range occurs for angles less than 45°.)

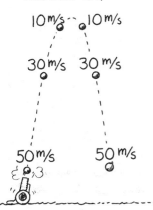

If air resistance is negligible, a projectile will rise to its maximum height in the same time it takes to fall from that height to the ground (Figure 3.7). This is because its deceleration by gravitation while going up is the same as its acceleration by gravitation while coming down. The speed it loses while going up is therefore the same as the speed it gains while coming down. So the projectile arrives at the ground with the same speed it had when it was projected from the ground.

We'll now see that if an object is projected fast enough, it will fall all the way around the earth and be an earth **satellite**.

Fig. 3.7 Without air resistance, speed lost while going up equals speed gained while coming down; time going up equals time coming down.

Question   A projectile is shot at an angle into the air. If air resistance is negligible, what is its downward acceleration? Its horizontal acceleration?*

---

*__Answer__   Its downward acceleration is g because the force of gravity is downward; its horizontal acceleration is zero because no horizontal forces act on it. We will see in the next chapter that the direction of acceleration is always the same as the direction of the net force.

Earth Satellites

Suppose we drop a stone from rest and observe its motion. It will fall to the ground in a straight-line path. If we move our hand horizontally and drop the stone, we observe that it follows a curved path as it falls to the ground (Figure 3.8). By repeating this experiment with faster horizontal motions, we conclude that the faster the initial horizontal motion of the stone when dropped (or thrown), the wider the arc of the curved path.

Fig. 3.8 The greater the stone's horizontal motion when released, the wider the arc of its curved path.

What would happen if the horizontal motion were so fast that the stone's curved path matched the curvature of the earth? The answer is that the stone would fall *around* the earth rather than *into* it. If there were no air resistance or other obstruction, the stone would orbit the earth and be an earth satellite. This is easier to see if we consider the same procedure on a smaller earth (Figure 3.9). It is also easy to see that the stone doesn't have to travel as fast to orbit a smaller and less massive planet.

Fig. 3.9 If the speed of the stone and the curvature of its trajectory are great enough, the stone may become a satellite.

How fast would the stone have to be thrown in order to orbit the earth? The answer to this depends on the rate at which the stone falls and the degree to which the earth curves. Recall from the last chapter that a stone will accelerate 9.8 meters/second² and fall a vertical distance of 4.9 meters during the first second of fall. It so happens that the earth curves away from a line tangent to its surface a vertical distance of 4.9 meters for every 8000 meters of surface (Figure 3.10). This means that if you were swimming in a calm ocean, you would be able to see only the top of a 4.9-meter mast on a ship 8 kilometers away. Now if the stone can be thrown fast enough to travel a horizontal distance of 8 kilometers during the time it falls a vertical distance of 4.9 meters, it will follow the curvature of the earth. It therefore must travel at 8 kilometers per second. At this speed it will coast in earth orbit. Convert this to kilometers per hour, and you'll get about 29,000 kilometers per hour (or 18,000 miles per hour)—the speed you see quoted in newspaper articles when satellites are launched (actually somewhat less at higher altitudes).

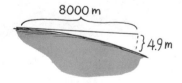

Fig. 3.10 Earth's curvature—not to scale!

This idea is sufficiently important to merit another example. Suppose we borrow a laser from a physics lab, set it up at the seashore, and direct a beam of light out across the surface of the ocean. Because of the curvature of the earth, we would find that 8 kilometers from the seashore the beam would be 4.9 meters above the water. (Actually, the water would be 4.9 meters below the beam.) Suppose further that we had a supercannon with a fantastically high muzzle velocity. If we fired a projectile along the beam of light, 1 second later it would be 4.9 meters below the beam. (Remember, gravity does not take a holiday on moving objects; they are pulled toward the earth whether moving or not.) If the projectile is to miss hitting the water, it must get 8 kilometers downrange during the time it falls 4.9 meters. It takes any object 1 second to fall 4.9 meters. So the muzzle velocity must be at least 8 kilometers per second, the minimum speed necessary for close earth orbit if we neglect air resistance. In practice, a satellite must stay at least 150 or so kilometers above the earth's surface to keep from burning up like a "falling star" against the friction of the atmosphere.

Fig. 3.11 (*a*) The force of gravity on the bowling ball is at 90° to its direction of motion, so it has no components of force to pull it forward or backward, and the ball rolls at constant speed. (*b*) The same is true even if the bowling alley is larger and remains "level" with the curvature of the earth. (*c*) If the ball moves at 8 km/s and encounters no air resistance, it needs no alley for support and will continue on its "level" course; it will be in orbit.

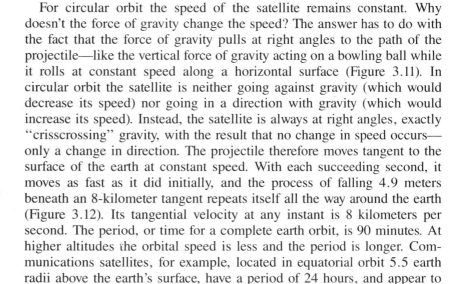

a   *b*   *c*

For circular orbit the speed of the satellite remains constant. Why doesn't the force of gravity change the speed? The answer has to do with the fact that the force of gravity pulls at right angles to the path of the projectile—like the vertical force of gravity acting on a bowling ball while it rolls at constant speed along a horizontal surface (Figure 3.11). In circular orbit the satellite is neither going against gravity (which would decrease its speed) nor going in a direction with gravity (which would increase its speed). Instead, the satellite is always at right angles, exactly "crisscrossing" gravity, with the result that no change in speed occurs—only a change in direction. The projectile therefore moves tangent to the surface of the earth at constant speed. With each succeeding second, it moves as fast as it did initially, and the process of falling 4.9 meters beneath an 8-kilometer tangent repeats itself all the way around the earth (Figure 3.12). Its tangential velocity at any instant is 8 kilometers per second. The period, or time for a complete earth orbit, is 90 minutes. At higher altitudes the orbital speed is less and the period is longer. Communications satellites, for example, located in equatorial orbit 5.5 earth radii above the earth's surface, have a period of 24 hours, and appear to

Fig. 3.12 A projectile moving horizontally at 8 km/s falls 4.9 m beneath successive 8-km tangents every second.

hover motionless above the same point on earth. The more distant moon has a period of 27.3 days. A satellite in higher orbit takes longer and moves slower.[2]

All this was deduced by Isaac Newton, who stated that a satellite was simply a projectile with the proper speed. We will see in Chapter 8 how Newton developed his law of gravitation by considering the moon to be moving as a simple projectile. Closer to home, he considered a cannon fired atop a high mountain (Figure 3.13). A cannonball fired horizontally with a small velocity falls to the ground along a parabolic path. But if fired fast enough, its path would be a circle and the cannonball would coast around the earth again and again (provided the cannoneer and the cannon got out of the way). Newton calculated the required speed to be equivalent to 8 kilometers per second, and since such a cannon-muzzle velocity was clearly impossible, he did not foresee people's launching artificial satellites (he did not consider multistage rockets).

Fig. 3.13 ". . . the greater the velocity . . . with which [a stone] is projected, the farther it goes before it falls to the earth. We may therefore suppose the velocity to be so increased, that it would describe an arc of 1, 2, 5, 10, 100, 1000 miles before it arrived at the earth, till at last, exceeding the limits of the earth, it should pass into space without touching."—Isaac Newton, *System of the World.*

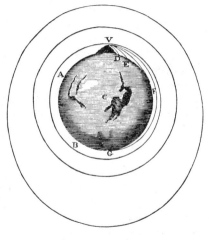

Elliptical Orbits

If a projectile just above the drag of the atmosphere is given a horizontal speed somewhat greater than 8 kilometers per second, it will "overshoot" a circular path and trace an oval-shaped path, called an **ellipse**. An ellipse is a very specific curve. It is the path of a point that moves so that the sum of its distances from two fixed points (called *foci*) is constant. In the case of a satellite orbiting a planet, the center of the planet is at one focus; the other focus is not occupied by anything in particular.

---

[2]The speed of a satellite in circular orbit is given by $v = \sqrt{\dfrac{GM}{d}}$, and the period of satellite motion is given by $T = 2\pi \sqrt{\dfrac{d^3}{GM}}$, where G is the universal gravitational constant (see Appendix IV), M is the mass of the earth (or whatever body the satellite orbits), and d is the distance of the satellite measured from the center of the earth or parent body.

An ellipse is easily constructed by using a pair of tacks (one at each focus), a loop of string, and a pencil, as shown in Figure 3.14. The closer the foci are to each other, the more the ellipse approximates a circle. A circle is actually a special case of an ellipse, in which both foci are at the same point, the center of the circle.

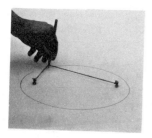

Fig. 3.14 A simple method for constructing an ellipse.

Fig. 3.15 The shadows cast by the ball are all ellipses, one for each lamp in the room. The point at which the ball makes contact with the table is the common focus of all three ellipses.

Whereas the speed of a satellite is constant in a circular orbit, speed varies in an elliptical orbit. This is because a projectile given a speed greater than 8 kilometers per second overshoots a circular path and moves against the force of gravity. It therefore loses speed. Like a rock thrown into the air, it slows to a point where it no longer recedes, and it begins to fall back toward the earth. The speed it lost in receding is regained as it falls back toward the earth, and it finally crosses its original path with the speed it had initially (Figure 3.16). So it repeats the procedure over and over, tracing out an elliptical orbit each cycle.

Fig. 3.16 Elliptical orbit. An earth satellite that has a speed somewhat greater than 8 km/s overshoots a circular orbit (*a*) and travels away from the earth. Gravitation slows it to a point where it no longer leaves the earth (*b*). It falls toward the earth gaining the speed it lost in receding (*c*) and overshoots as before in a repetitious cycle.

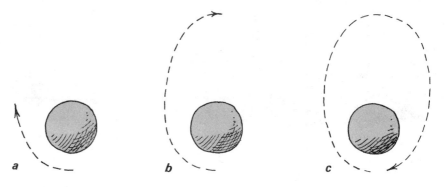

a   *b*   *c*

Fig. 3.17 (*a*) The parabolic path of the cannonball is part of an ellipse that extends within the earth. The earth's center is the far focus. (*b*) All paths of the cannonball are ellipses. For less than orbital speeds, the center of the earth is the far focus; for circular orbit, both foci are at the earth's center; for greater speeds, the near focus is the earth's center.

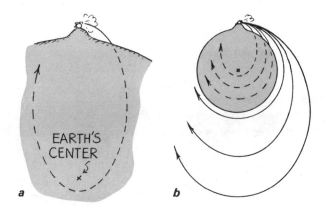

Interestingly enough, the parabolic path of a projectile such as a tossed baseball or a cannonball is actually a tiny segment of a thin ellipse that extends within and just beyond the center of the earth (Figure 3.17*a*). In Figure 3.17*b*, we see several paths of cannonballs fired from "Newton's mountain." All ellipses have the center of the earth as one focus. As muzzle velocity is increased, the ellipses are less eccentric (wider); and when muzzle velocity reaches 8 kilometers per second, the ellipse rounds into a circle and does not intercept the earth's surface. The cannonball coasts in circular orbit. At greater muzzle velocities, the orbiting cannonball traces the familiar external ellipse.

Putting a payload into earth orbit requires control over the speed and direction of the rocket that carries it above the atmosphere. A rocket initially fired vertically is intentionally tipped from the vertical course; then, once above the drag of the atmosphere, it is aimed *horizontally*, whereupon the payload is given a final thrust to 8 kilometers per second or more. This is shown in Figure 3.18, where for the sake of simplicity the payload is the entire single-stage rocket. We see that with the proper tangential velocity it falls around the earth, rather than into it, and becomes an earth satellite.

Fig. 3.18 The initial thrust of the rocket pushes it up above the atmosphere. Another thrust to a horizontal speed of at least 8 km/s is required if the rocket is to fall around rather than into the earth.

Escape Velocity   What would happen if a rocket remained on a vertical course instead of tipping horizontally? Then the old saying "What goes up must come down" would hold true, for just as surely as a stone thrown skyward will be returned by gravity, a vertically fired rocket will be returned by gravity (unless, as we shall see, its speed is great enough). We know that a cannonball fired horizontally at 8 kilometers per second from "Newton's mountain" would find itself in orbit, but if the cannonball were instead fired at the same speed *vertically*, it would rise to some maximum height, then momentarily come to a stop, and then fall back to earth.

In today's spacefaring age, it is more accurate to say "What goes up *may* come down," for there is a critical speed in excess of which an object is able to outrun gravity and escape the earth. This critical speed is called the **escape velocity**. From the earth's surface the escape velocity is 11.2 kilometers per second (25,000 miles per hour). Fire a cannonball or any projectile skyward at less than 11.2 kilometers per second and it will fall back to earth. If air resistance is negligible, it will return with the same speed given to it on projection. Get it going above the drag of the earth's atmosphere at 11.2 kilometers per second and it will just barely escape the earth and not return.[3] At speeds greater than 11.2 kilometers per second, it will escape the earth with speed to spare.

If a rocket leaves the earth at 11.2 kilometers per second, it can escape the earth and go, say, to the moon. Interestingly enough, with an ample period of sustained power, it could go to the moon with lesser speeds. Escape velocity of 11.2 kilometers per second is required for escape from the earth only in the case of unpowered flight. If a spaceship is a simple projectile, or even a rocket with a relatively short burning time, it must be propelled to 11.2 kilometers per second near the earth's surface. Gravity is "diluted" by distance from the earth, so away from the earth's surface escape velocity is less.[4] At 1 earth radius above the surface, for example, escape velocity is only 7.9 kilometers per second. A rocket propelled to this altitude with any speed need be moving only 7.9 kilometers per second to escape the earth. Of course, a spaceship under continuous power could escape the earth at any velocity above zero, providing it has enough fuel and time.

---

[3] Interestingly enough, this might well be called the *maximum falling velocity,* because any object, however far away, released from rest and allowed to fall to earth by virtue of only the earth's gravity would never exceed 11.2 km/s.

[4] The value of escape velocity from any body is given by $v = \sqrt{\dfrac{2GM}{d}}$, where G is the gravitational constant (Appendix IV), M is the mass of the attracting body, and d is the distance from its center. (At the surface of the body, d would simply be the radius of the body.)

We will return to the topic of escape velocity when we learn about energy in Chapter 6, and to the topic of how gravity is diluted with distance when we discuss the inverse-square law in Chapter 8.

The escape velocities of various bodies in the solar system are shown in Table 3.1. We see that escape velocity from the surface of the sun is an impressive 620 kilometers per second; and from the sun at a distance equaling that of the earth's orbit, 42.5 kilometers per second. So an object projected from the earth at 11 kilometers per second may escape the earth; but if the speed is less than 42.5 kilometers per second, it will not be free of the sun and, rather than recede forever, will take up an orbit about the sun.[5]

Table 3.1 Escape velocities at the surface of bodies in the solar system

Astronomical body	Mass (earth masses)	Radius (earth radii)	Escape velocity (kilometers/second)
Sun	330,000	109	620
Sun (at a distance of the earth's orbit)		23,000	42.5
Jupiter (pole)	318	10.5	62.0
Jupiter (equator)		11.2	60.0
Saturn (pole)	95.2	8.5	37.7
Saturn (equator)		9.5	35.6
Neptune	17.3	3.4	25.4
Uranus	14.5	3.7	22.4
Earth	1.0	1.0	11.2
Venus	0.82	0.96	10.4
Mars	0.11	0.525	5.2
Mercury	0.054	0.380	4.3
Ganymede (Jupiter's satellite)	0.026	0.395	2.9
Moon	0.0123	0.273	2.4

In summary, a satellite near the surface of the earth must have a tangential velocity between 8 and 11.2 kilometers per second. Below 8 kilometers per second, it will fall into the earth; above 11.2 kilometers per

---

[5]The probe *Pioneer 10*, however, has escaped the solar system even though its launching speed on March 2, 1972, was only 15 km/s. This is because it was directed into the path of oncoming Jupiter and whipped about its great gravitational field. It picked up speed in the process just like the speed of a ball encountering an oncoming bat is increased when it departs from the bat. Its speed of departure from Jupiter was increased enough to exceed the sun's escape velocity at the distance of Jupiter. It passed through the orbit of Pluto in 1984, and, barring an accidental collision with an asteroid or whatever, it will wander indefinitely through interstellar space. Like a note in a bottle cast into the sea, *Pioneer 10* contains information about the earth that should be of interest to extraterrestrials, if it is ever washed up and found on some distant "seashore."

Questions

1. Since the earth is gravitationally attracted to the sun, why doesn't it simply fall into the sun?*

2. Why doesn't the ever-present gravitational force that acts on a satellite moving in a circular orbit change its speed? Why does gravitational force change the speed of a satellite moving in an elliptical orbit?†

3. Is the following statement true or false: If a spaceship from earth is to reach the moon, it must be launched with a speed of at least 11.2 km/s.‡

second, it will escape the earth and orbit the sun or some other body; above 42.5 kilometers per second, it will escape the solar system altogether.

To go to a destination such as the moon, a rocket must reach a minimum speed of 11.2 kilometers per second if its rocket engines burn out when still close to the earth. It is interesting to note that the accuracy with which a rocket reaches its destination does not depend on its staying on a planned trajectory or on its getting back on that trajectory if it strays off course. The rocket makes no effort to return to an original path. Instead, it in effect asks, "Where am I now with respect to where I want to go, and what's the best way to get there from here given my present situation?" With the aid of high-speed computers the answers to these questions are used to find a *new* path. Corrective thrusters orient the rocket to this new path, and this process is repeated over and over again all the way to the goal.[6]

*Answer   If the earth were somehow stopped in its tracks and at momentary rest with respect to the sun, it would indeed fall into it. Like any satellite, because of its tangential speed it falls around the sun, continually "missing" it rather than falling into it.

†Answer   In circular orbit the distance of a satellite from the earth is constant; it moves neither against nor with the gravity of the earth, much like a bowling ball that rolls along a horizontal surface. A satellite in elliptical orbit, however, is moving either away from or toward the earth and, except at the apogee (farthest distance) and perigee (closest distance), moves with some component of gravitational force along its direction of motion. Its speed increases when moving with gravity and decreases when moving against gravity.

‡Answer   Escape velocity is required for escape from a planet only where unpowered flight is concerned. If the spaceship is a simple projectile, or even a rocket with a relatively short burning time, the statement is true. The statement is false, however, if the spaceship is under constant power, for it can move any finite distance from earth at any velocity above zero, as long as it has enough fuel and time.

[6]Is there a lesson to be learned here? If you find that you are "off course," you may, like the rocket, find it more fruitful to take a course that leads to your goal *as best plotted from your present position and circumstances* rather than to try to get back on the course you plotted from a previous position and under, perhaps, different circumstances.

Summary of Terms

Projectile   Any body that is projected by some force and continues in motion by virtue of its own inertia.

Parabola   The curved path followed by a projectile under the influence of gravitational attraction only.

Satellite   A projectile or small celestial body that orbits a larger celestial body.

Ellipse   The closed oval-shaped curve wherein the sum of the distances from any point on the curve to both foci is a constant. When the foci are together at one point, the ellipse is a circle. The farther apart the foci, the more eccentric the ellipse.

Escape velocity   The velocity that a projectile, space probe, etc., must reach to escape the gravitational influence of the earth or celestial body to which it is attracted.

Review Questions

1. Can a projectile be considered a freely falling body? What is its downward acceleration? Its horizontal acceleration?

2. How far below an initial straight-line path will a projectile fall in 1 second?

3. Does your answer to the preceding question depend on the projection angle or the initial speed of the projectile? Explain.

4. All things being equal, what angle will yield the maximum horizontal range for a projectile?

5. A projectile hurled into the air at an angle of 50 degrees lands downrange. For the same initial speed, what other projection angle would yield the same range?

6. Why is it incorrect to say that a satellite does not fall?

7. Can a satellite be considered a freely falling body?

8. What is the minimum speed for an earth satellite near the earth's surface? The maximum speed?

9. What exactly is an ellipse? What shape does it assume when both foci are together?

10. Is the speed of a satellite constant or variable when in circular orbit? In elliptical orbit?

11. If a projectile is launched from the earth's surface at a speed of 10 kilometers per second and maintains a *vertical* path, will it become an earth satellite? Explain.

12. What minimum speed must be imparted to a projectile at the surface of the earth in order that it escape the earth forever?

Exercises

1. A baseball is first dropped straight downward from an initial rest position, and then it is thrown upward at an angle. What is its downward acceleration in both cases? Its horizontal acceleration? Can it be considered a projectile in both cases? Explain.

2. If you throw a rock off a cliff, how does the horizontal component of its motion compare for all points along its trajectory?

3. If you are standing in a uniformly moving bus and drop a ball from your outstretched hand, you'll see its path as a vertical straight line. How will the path appear to a friend standing at rest at the side of the road? (This question involves an important idea we have not treated yet—frames of reference.)

4. Why does a high jumper approach the jump at a relatively slow running speed whereas a long jumper runs as fast as possible for the jump?

5. Suppose you are on a ledge in the dark and wish to estimate the height of the ledge above the ground. So you drop a stone over the edge and hear it strike the ground in 1 second. What is the height of the ledge? How could you use the stone to estimate the height if you weren't close enough to the edge to simply drop the stone? Explain.

6. A friend claims that bullets fired by some high-powered rifles travel for many meters in a straight-line path without dropping. Another friend disputes this claim and states that all bullets from any rifle drop beneath a straight-line path a vertical distance given by $\frac{1}{2}gt^2$ and that the curved path is apparent at low velocities and less apparent at high velocities. Now it's your turn: Will all bullets drop the same vertical distance in equal times? Explain.

7. The boy on the tower throws a ball 20 meters downrange as shown. What is his pitching speed?

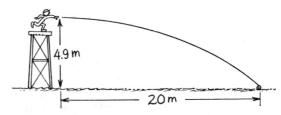

8. At what angle should you hold a garden hose so that the stream of water will go farthest?

9. If you have ever watched the launching of an earth satellite, you may have noticed that the rocket departs from a vertical course and continues its climb at an angle. Why?

10. Why are satellites normally sent into orbit by firing them in an easterly direction (the direction in which the earth spins)?

11. What is the shape of the orbit when the velocity of the satellite is everywhere perpendicular to the force of gravity?

12. Since the moon is gravitationally attracted to the earth, why doesn't it simply crash into the earth?

13. Would the speed of a satellite in close circular orbit about the moon be greater than, equal to, or less than 8 kilometers per second?

14. The orbiting space shuttle travels at 8 kilometers per second with respect to the earth. Suppose it projects a capsule rearward at 8 kilometers per second with respect to the shuttle. Describe the path of the capsule with respect to the earth.

15. The orbital velocity of the earth about the sun is 30 kilometers per second. If the earth were suddenly stopped in its tracks, it would simply fall radially into the sun. Devise a plan whereby a rocket loaded with radioactive wastes could be fired into the sun for permanent disposal. How fast and in what direction with respect to the earth's orbit should the rocket be fired?

16. What is the maximum possible speed of impact upon the surface of the earth for a faraway body initially at rest that falls to earth by virtue of the earth's gravity only?

17. If Pluto were somehow stopped short in its orbit, it would fall into rather than around the sun. How fast would it be moving when it hits the sun?

18. If a pair of satellites at different altitudes but moving in the same direction are seen from the earth to pass overhead, which of the two overtakes the other?

19. Many schools and hospitals could have been constructed with the funds used in the last decade to probe and explore the solar system. How many of these schools and hospitals do you speculate would in fact have been constructed if the various space missions were not funded?

20. Do a bit of research and find out why the locations called *L4* and *L5* are the likely locations for the first projected large-scale inhabited space facilities to orbit the earth.

REMEMBER, REVIEW QUESTIONS PROVIDE YOU WITH A SELF CHECK OF WHETHER OR NOT YOU GRASP THE CENTRAL IDEAS OF THE CHAPTER. THE EXERCISES ARE EXTRA "PUSHUPS" FOR YOU TO TRY AFTER YOU HAVE AT LEAST A FAIR UNDERSTANDING OF THE CHAPTER AND CAN HANDLE THE REVIEW QUESTIONS.

Isaac Newton (1642–1727)

Isaac Newton was born prematurely and barely survived on Christmas Day, 1642, the same year that Galileo died. Newton's birthplace was his mother's farmhouse in Woolsthorpe, England. His father died several months before his birth, and he grew up under the care of his mother and grandmother. As a child he showed no particular signs of brightness, and at the age of 14½ he was taken out of school to work on his mother's farm. As a farmer he was a failure, preferring to read books he borrowed from a neighboring druggist. An uncle sensed the scholarly potential in young Isaac and prompted him to study at the University of Cambridge, which he did for five years, graduating without particular distinction.

A plague swept through London, and Newton retreated to his mother's farm—this time to continue his studies. At the farm, at the age of 23, he laid the foundations for the work that was to make him immortal. Seeing an apple fall to the ground led him to consider the force of gravity extending to the moon and beyond, and he formulated the law of universal gravitation (which he later proved); he invented the calculus, an indispensable mathematical tool in science; he extended Galileo's work and formulated the three fundamental laws of motion; and he formulated a theory of the nature of light and showed with prisms that white light is composed of all the colors of the rainbow. It was his experiments with prisms that first made him famous.

When the plague subsided, Newton returned to Cambridge and soon established a reputation for himself as a first-rate mathematician. His mathematics teacher resigned in his favor and Newton was appointed the Lucasian professor of mathematics. He held this post for twenty-eight years. In 1672 he was elected to the Royal Society, where he exhibited the world's first reflector telescope, which can still be seen preserved at the library of the Royal Society in London with the inscription: "The first reflecting telescope, invented by Sir Isaac Newton, and made with his own hands."

It wasn't until Newton was 42 that he began to write what is generally acknowledged as the greatest scientific book ever written, the *Principia Mathematica Philosophiae Naturalis*. He wrote the work in Latin and completed it in eighteen months. It appeared in print in 1687 and wasn't printed in English until 1729, two years after his death. When asked how he was able to make so many discoveries, Newton replied that his solutions to problems were not by sudden insight but by continually thinking very long and hard about them until he worked them out.

At the age of 46, his energies turned somewhat from science when he was elected a member of Parliament. He attended the sessions in Parliament for two years and never gave a speech. One day he rose and the House fell silent to hear the great man. Newton's "speech" was very brief; he simply requested that a window be closed because of a draft.

A further turn from his work in science was his appointment as warden and then as master of the mint. Newton resigned his professorship and directed his efforts toward greatly bettering the workings of the mint, to the dismay of counterfeiters who flourished at that time. He maintained his membership in the Royal Society and was elected president, and re-elected each year for the rest of his life. At the age of 62, he wrote *Opticks*, which summarized his work on light. Nine years later he wrote a second edition to his *Principia*.

Although Newton's hair turned gray at 30, it remained full, long, and wavy all his life, and unlike others in his time he did not wear a wig. He was a modest man, was very sensitive to criticism, and never married. He remained healthy in body and mind into old age. At 80, he still had all his teeth, his eyesight and hearing were sharp, and his mind was alert. In his lifetime he was regarded by his countrymen as the greatest scientist who ever lived. In 1705 he was knighted by Queen Anne. Newton died at the age of 85 and was buried in Westminster Abbey along with England's kings and heroes.

Newton showed that the universe ran according to natural laws that were neither capricious nor malevolent—a knowledge that provided hope and inspiration to scientists, writers, artists, philosophers, and people of all walks of life, and that ushered in the Age of Reason. The ideas and insights of Isaac Newton truly changed the world and elevated the human condition.

4 Newton's Laws of Motion

**Newton's
First Law
of Motion**

In 1642, the year that Galileo died, Isaac Newton was born. Twenty-three years later, Newton developed his famous laws of motion, which completed the overthrow of the Aristotelian ideas that had dominated the thinking of the best minds for 2000 years. We will consider these laws in order: the first, sometimes called the law of inertia, comes from Newton's *Principia*.

Law 1: *Every body continues in its state of rest, or of uniform motion in a straight line, unless it is compelled to change that state by forces impressed upon it.*

The key word in this law is *continues:* a body *continues* to do whatever it happens to be doing unless a force is exerted upon it. If it is at rest, it *continues* in a state of rest. If it is moving, it *continues* to move without turning or changing its speed. In short, the law says that a body does not accelerate of itself; acceleration must be imposed against the tendency of a body to retain its state of motion. This neatly summarizes our discussion of Chapter 2: recall that the tendency of a body to resist a change in motion was what Galileo called **inertia**.

Every body possesses inertia; how much depends on the amount of matter in the substance of a body— the more matter, the more inertia. In speaking of how much matter a body has, we use the term *mass*. The greater the mass of a body, the greater its inertia. **Mass** is a measure of the inertia of a body.

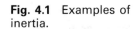

Fig. 4.1 Examples of inertia.

WHY WILL THE COIN DROP INTO THE GLASS WHEN A FORCE ACCELERATES THE CARD ?

WHY DOES THE DOWNWARD MOTION AND SUDDEN STOP OF THE HAMMER TIGHTEN THE HAMMERHEAD ?

WHY IS IT THAT A SLOW CONTINUOUS INCREASE IN THE DOWNWARD FORCE BREAKS THE STRING ABOVE THE MASSIVE BALL, BUT A SUDDEN INCREASE BREAKS THE LOWER STRING ?

Mass corresponds to our intuitive notion of **weight**. We say something has a lot of matter if it is heavy. That's because we are accustomed to measuring the quantity of matter in a body by its gravitational attraction to the earth. But mass is more fundamental than weight; it is a fundamental quantity that completely escapes the notice of most people. There are times, however, when weight corresponds to our unconscious notion of inertia. For example, if you are trying to determine which of two small objects is the heavier, you might shake them back and forth in your hands or move them in some way instead of lifting them. In doing so, you are judging which of the two is more difficult to move, seeing which is the more resistant to a change in motion. You are really making a comparison of the inertias of the objects. It is easy to confuse the ideas of mass and weight, mainly because they are directly proportional to each other. If the mass of an object is doubled, its weight is also doubled; if the mass is halved, its weight is halved. But there is a distinction between the two. We can define each as follows:

Mass: *The quantity of matter in a body. More specifically, it is the measurement of the inertia or sluggishness that a body exhibits in response to any effort made to start it, stop it, or change in any way its state of motion.*

Weight: *The force upon a body due to gravity.*

In the United States, the quantity of matter in an object has commonly been described by its gravitational pull to the earth, or its weight. This has usually been expressed in *pounds*. In most of the world, however, the measure of matter is commonly expressed in a mass unit, the **kilogram**. At the surface of the earth, the mass of a 1-kilogram brick weighs 2.2 pounds. In the metric system of units, the unit of force is the **newton**, which is equal to a little less than a quarter pound (like the weight of a quarter-pounder hamburger *after* it is cooked). A 1-kilogram brick weighs 9.8 newtons (9.8 N).[1] Away from the earth's surface, where the influence of gravity is less, a 1-kilogram brick weighs less. It would also weigh less on the surface of planets with less gravity than the earth. On the moon, for example, where the gravitational force on a body is only $\frac{1}{6}$ as strong as on earth, 1 kilogram weighs about 1.6 newtons (or 0.36 pound). On more

---

[1]So 2.2 lb equals 9.8 N, or 1 N is approximately equal to 0.2 lb—about the weight of an apple. In the metric system it is customary to specify quantities of matter in units of mass (in grams or kilograms) and rarely in units of weight (in newtons). In the United States and places that have been using the British system of units, however, quantities of matter have customarily been specified in units of weight (in pounds); and the unit of mass, called the *slug*, is not well known. See Appendix I for more about systems of measurement.

massive planets it would weigh more. But the mass of the brick is the same everywhere. The brick offers the same resistance to speeding up regardless of whether the earth, moon, or anything at all is attracting it. In a space capsule located between the earth and the moon, where gravitational forces cancel one another, a brick still has mass. If placed on a scale, it wouldn't weigh anything, but its resistance to a change in motion is the same as on earth. Just as much force would have to be exerted by an astronaut in the space capsule to shake the brick back and forth as would be required to shake it back and forth while standing on earth. Except for friction effects, it would be just as hard to push a Cadillac limousine across a level surface on the moon as on earth. The difficulty of lifting it against gravity (weight) is something else. Mass and weight are different from one another (Figures 4.2 and 4.3).

It is also easy to confuse mass and size—or, more specifically, **volume**. Perhaps this is because when we think of a massive object, we think of a big object—that is, a voluminous object that occupies much space. But an object can be massive—a lead storage battery, for example—without occupying much space at all. This confusion might also occur because mass and volume are often proportional to each other, at least for the same material. For example, 2 kilograms of sugar will fill a bag of twice the volume as 1 kilogram of sugar. But this does not mean that mass *is* volume. Because 2 kilograms of sugar have twice the sweetening power of 1 kilogram, we don't say that mass *is* sweetening power. Two loaves of bread may have the same mass but quite unequal volumes. You can always compress a loaf of bread and change its volume, but the mass doesn't change; it contains the same amount of matter. So you see that mass is neither weight nor volume.

Although Galileo introduced the idea of inertia, Newton grasped its significance. The law of inertia defines natural motion and tells us what kinds of motion are the result of impressed forces. Whereas Aristotle maintained that the forward motion of an arrow through the air required an impressed force, Newton's law of inertia instead tells us that the behavior of the arrow is natural; constant speed along a straight line (or, simply, constant velocity) requires no force. And whereas Aristotle and his followers held that the circular motions of heavenly bodies were natural and moving without impressed forces, the law of inertia clearly states that in the absence of forces of some kind the planets would not move in the divine circles of ancient and medieval astronomy but would move instead in straight-line paths off into space. Newton maintained that the curved motion of the planets was evidence of some kind of force. We shall see in the next two chapters that his search for this force led to the law of gravitation.

Fig. 4.2 An anvil in outer space, between the earth and moon, for example, may be weightless; but it is not massless.

Fig. 4.3 The astronaut in space finds it is just as difficult to shake the "weightless" anvil as it would be on earth. If the anvil is more massive than the astronaut, which shakes more—the anvil or the astronaut?

Question   Would it be easier to lift a Cadillac limousine on the earth or to lift it on the moon?*

Newton's Second Law of Motion

Fig. 4.4 The greater the mass, the greater the force must be for a given acceleration.

Every day we see bodies that do not continue in a constant state of motion: things initially at rest later may move; moving objects may follow paths that are not straight lines; objects in motion may stop. Most of the motion we observe is accelerated motion and is the result of one or more impressed forces. The overall net force, whether it be from a single source or a combination of sources, produces acceleration. The relationship of acceleration to force and inertia is given in Newton's second law.

Law 2: *The acceleration of a body is directly proportional to the net force acting on the body and inversely proportional to the mass of the body and is in the direction of the net force.*

In summarized form, this is

$$\text{Acceleration} \sim \frac{\text{net force}}{\text{mass}}$$

In symbol notation, this is simply

$$a \sim \frac{F}{m}$$

We shall use the wiggly line $\sim$ as a symbol meaning "is proportional to." We say that acceleration a is directly proportional to the overall net force F and inversely proportional to the mass m. By this we mean that if F increases, a increases; but if m increases, a decreases. With appropriate units of F, m, and a, the proportionality may be expressed as an exact equation:

$$a = \frac{F}{m}$$

FORCE OF HAND ACCELERATES THE BRICK

TWICE AS MUCH FORCE PRODUCES TWICE AS MUCH ACCELERATION

TWICE THE FORCE ON TWICE THE MASS GIVES THE SAME ACCELERATION

Fig. 4.5 Acceleration is directly proportional to force.

A body is accelerated in the direction of the force acting on it. Applied in the direction of the body's motion, a force will increase the body's speed. Applied in the opposite direction, it will decrease the speed of the

---

*__Answer__   A Cadillac limousine would be easier to lift on the moon because the force of gravity is less on the moon. When you *lift* an object, you are contending with the force of gravity (its weight) in addition to its inertia (its mass). Although its mass is the same on either the earth or the moon, its weight is only $\frac{1}{6}$ as much on the moon, and only $\frac{1}{6}$ as much effort is required to lift it there. To move it horizontally, however, you are not pushing against gravity. When mass is the only factor, equal forces will produce equal accelerations whether the object is on the earth or the moon.

body. Applied at right angles, it will deflect the body. Any other direction of application will result in a combination of speed change and deflection. *The acceleration of a body is always in the direction of the net force.*

A **force**, in the simplest sense, is a push or a pull. Its source may be gravitational, electrical, magnetic, or simply muscular effort. In the second law, Newton gives a more precise idea of force by relating it to the acceleration it produces. He says in effect that *force is anything that can accelerate a body.* Furthermore, he says that the larger the force, the more acceleration it produces. For a given body, twice the force results in twice the acceleration; three times the force, three times the acceleration; and so forth. Acceleration is directly proportional to force (Figure 4.5).

The body's mass has the opposite effect (Figure 4.6). The more massive the body, the less its acceleration. For the same force, twice the mass results in half the acceleration; three times the mass, one-third the acceleration. Increasing the mass decreases the acceleration. For example, if we put identical Ford engines in a Cadillac and a Volkswagen, we would expect quite different accelerations even though the driving force in each car is the same. The Cadillac with its greater mass has a greater resistance to a change in velocity than does the Volkswagen. Consequently, the Cadillac requires a more powerful engine to achieve comparable acceleration. To attain the same acceleration, a larger mass requires a correspondingly larger force. We say that acceleration is inversely proportional to mass.[2]

The acceleration of a body, then, depends *both* on the magnitude of the net force and on the mass of the body.

FORCE OF HAND ACCELERATES THE BRICK

THE SAME FORCE ACCELERATES 2 BRICKS ½ AS MUCH

3 BRICKS, ⅓ AS MUCH ACCELERATION

Fig. 4.6 Acceleration is inversely proportional to force.

Question   In Chapter 2 acceleration was defined to be the time rate of change of velocity; that is, $a = $ (change in v)/time. Are we in this chapter saying that acceleration is instead the ratio of force to mass; that is, $a = F/m$? Which is it?*

---

*__Answer__   Acceleration *is* the time rate of change of velocity and is *produced by* a force. How much $\dfrac{\text{force}}{\text{mass}}$ (the cause) results in the rate $\dfrac{\text{change in } v}{\text{time}}$ (the effect).

[2]Mass is operationally defined as the proportionality constant between force and acceleration in Newton's second law, rearranged to read $m = \dfrac{F}{a}$. A 1-unit mass is that which requires 1 unit of force to produce 1 unit of acceleration. So 1 kg is the amount of matter that 1 N of force will accelerate 1 m/s². In British units, 1 slug is the amount of matter that 1 pound of force will accelerate 1 ft/s². We shall see later that mass is simply a form of concentrated energy.

Net Force

Newton's second law relates the acceleration of a body to the net force exerted upon it. Let's examine more carefully what we mean by *net* and consider what happens when more than one force is exerted on a body. If you and a friend pull in the same direction with equal forces on an object, the object will have twice the acceleration than if you pulled alone. If, however, you each pull with equal forces but in opposite directions, the object will not accelerate. Because they are oppositely directed, the forces on the object cancel one another. One of the forces can be considered to be the negative of the other, and they add algebraically to zero.

When forces are applied to an object in the same or opposite directions, the acceleration of the object is found to be proportional to the algebraic sum of the forces. If the forces are in the same direction, they are simply added; if in opposite directions, they are subtracted (Figure 4.7). For example, if you pull on an object with a force of 20 newtons and your friend pulls in the opposite direction with a force of 15 newtons, the resulting acceleration is the same as if you pulled alone with a force of 5 newtons. The resulting 5 newtons is the net force. It is the net force that accelerates things.

If two or more forces are pulling at some angle to one another so they are neither in the same nor opposite directions, they add geometrically. This more complicated case is treated in Appendix III.

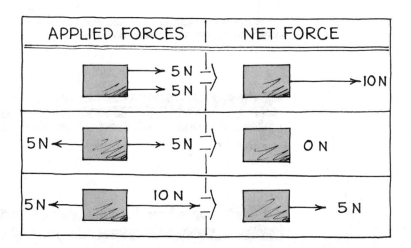

Fig. 4.7 Net force.

Friction

Even when a single force is applied to an object, the net force is usually less than the applied force. This is because of friction. **Friction** is the result of the mutual contact of irregularities in the surfaces of sliding objects. The irregularities act as obstructions to motion. Even surfaces that appear to be very smooth have irregular surfaces when viewed microscopically. The atoms cling together at many points of contact. As sliding commences, the atoms snap apart or are torn from one surface by the other. There is also friction in fluids, because a body moving through a fluid must push aside some of the fluid.

The direction of the frictional force is always in a direction opposing motion. Thus, if an object is to move at constant velocity, a force equal to the opposing force of friction must be applied. The two forces will exactly cancel one another. We say the net force is zero; hence, the acceleration is zero. But zero acceleration does not mean zero velocity. Zero acceleration means that the object will maintain the velocity it happens to have, neither speeding up nor slowing down nor changing direction.

Interestingly enough, the friction force is appreciably greater for an object that is on the verge of sliding than when the same object is sliding. Physicists and engineers distinguish between **static friction** and **sliding friction**. Let's look at this more closely. Consider a 50-kilogram crate that rests on a rough floor. Friction will come into play only when the crate begins to slide. At rest with no applied force, the friction force is zero. But suppose you push horizontally on the crate, say, with a force of 50 newtons. Still the crate does not begin sliding, but there now exists a friction force of 50 newtons between the floor and the crate. This force acts in a direction opposite to your push; hence, the net force on the crate is zero, and it remains at rest. (If the crate were instead resting on a friction-free air table, our 50-newton push would accelerate the crate.) Suppose we increase our push, and when we get up to 100 newtons, the crate is just on the verge of sliding. Then the friction force has built up to 100 newtons also. If we push with more force, the crate breaks loose and slides. Once the crate is sliding, the irregularities of the surfaces exert less force on each other than when it is at rest, and the friction force reduces to, say, 75 newtons. We can reduce our applied force to 75 newtons, and the crate will slide at uniform speed across the floor. If we persist with our 100-newton push, the crate will accelerate and gain speed.

Interestingly enough, it so happens that the force of friction depends not on the speed but only on the surfaces involved and how hard they are pressed together. The force of friction will be 75 newtons regardless of the speed of the sliding crate.

The difference between static friction and sliding friction is important when braking your car in an emergency stop. It is very important that you not jam on the brakes so as to make the tires lock in place. When tires lock, they slide, providing less friction than if they are made to roll to a stop. While the tire is rolling, its surface does not slide along the road surface and friction is static friction, and therefore greater than sliding friction. The difference between static and sliding friction is also apparent when your car takes a corner too fast. Once the tires start to slide, the frictional force is reduced and off you go! A skilled driver keeps her tires below the threshold of breaking loose into a slide.

75-N FRICTION 75-N
FORCE   APPLIED FORCE

Fig. 4.8 The crate slides to the right because of an applied force of 75 N. A friction force of 75 N opposes motion and results in a zero net force on the crate, so the crate slides at constant velocity (zero acceleration).

Questions

1. The force of sliding friction between the 50-kg crate and the floor is 75 N, so an applied force of 75 N will keep it sliding. How fast will it slide?*

2. What will be the acceleration of the sliding crate if we apply a force of 100 N? How fast will it slide?†

Newton's Laws and Falling Bodies

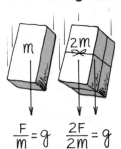

$$\frac{F}{m} = g \qquad \frac{2F}{2m} = g$$

Fig. 4.9 The ratio of weight (*F*) to mass (*m*) is the same for all bodies in the same locality; hence, their accelerations are the same in the absence of air resistance.

Case 1: Free Fall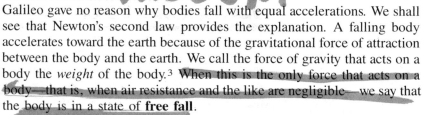

Galileo gave no reason why bodies fall with equal accelerations. We shall see that Newton's second law provides the explanation. A falling body accelerates toward the earth because of the gravitational force of attraction between the body and the earth. We call the force of gravity that acts on a body the *weight* of the body.[3] When this is the only force that acts on a body—that is, when air resistance and the like are negligible—we say that the body is in a state of **free fall**.

A heavier body is attracted to the earth with more force than a light body. The double brick in Figure 4.9, for example, is attracted with twice as much gravitational force as the single brick. Why then, as Aristotle supposed, doesn't the double brick fall twice as fast? The answer is that the acceleration of a body depends not only on the force—in this case, the weight—but on the mass as well. Whereas force tends to accelerate things, mass tends to resist the acceleration. So twice the force acting on twice the inertia produces the same acceleration as half the force acting on half the inertia. Both accelerate equally. The acceleration due to gravity is *g*.

*****Answer**   The speed of the sliding crate will depend on how long it was accelerating after it broke loose from its rest position. If the slightly more than 100 N applied force that got it sliding to begin with were quickly reduced to 75 N, the speed would be less than if the reduction of applied force were more gradual. Suppose, for example, that the speed of the crate just as the applied force is reduced to 75 N is 1 m/s. The crate would continue moving at a constant velocity of 1 m/s as long as the applied force equaled the friction force. If the crate were pushed to 2 m/s when the forces became equal and opposite, it would have a constant velocity of 2 m/s.

†**Answer**   From Newton's second law, we find the acceleration:

$$a = \frac{F_{net}}{m} = \frac{100 \text{ N} - 75 \text{ N}}{50 \text{ kg}} = \frac{25 \text{ N}}{50 \text{ kg}} = \tfrac{1}{2} \text{m/s}^2$$

The speed of the crate increases by $\tfrac{1}{2}$ m/s for each second we continue to push ($v = at$). How fast it will slide depends on the amount of time it is pushed. At the end of 2 seconds of acceleration, its speed will be 1 m/s; at 5 seconds, its speed will be $2\tfrac{1}{2}$ m/s; and so on.

Note that we have used the 75 N value of friction in our formula, which is the value for sliding. The 75 N will remain constant as long as motion continues. If the applied force is taken away and the crate again comes to rest, then the value of the friction force will have to again build up to 100 N before sliding starts. Sliding friction is less than static friction.

---

[3] Weight and mass are directly proportional to each other, and the constant of proportionality is *g*. We see that Weight = *mg*; so 9.8 N = (1 km)(9.8 m/s²).

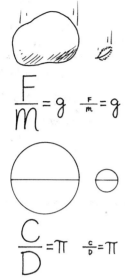

$$\frac{F}{m} = g \quad \frac{F}{m} = g$$

$$\frac{C}{D} = \pi \quad \frac{c}{D} = \pi$$

Fig. 4.10 The ratio of weight (F) to mass (m) is the same for the large rock and the small feather; similarly, the ratio of circumference (C) to diameter (D) is the same for the large and the small circle.

The constant ratio of weight to mass for freely falling objects is similar to the constant ratio of circumference to diameter for circles, which is π. The ratio of weight to mass is the same for both heavy and light objects, just as the ratio of circumference to diameter is the same for both large and small circles (Figure 4.10). The acceleration of free fall is g. We use the symbol g, rather than a, to denote that acceleration is due to gravity alone.

The acceleration of free fall is independent of weight. A boulder 100 times more massive than a pebble falls at the same acceleration as the pebble because although the force on the boulder (its weight) is 100 times greater than the force (or weight) on the pebble, its mass (resistance to a change in motion) is 100 times that of the pebble. The greater force is needed to offset the equally greater mass.

Case 2: Nonfree Fall—Air Resistance

Objects falling in a vacuum are one thing, but what of the practical cases of objects falling in air? Although a feather and a stone will fall equally fast in a vacuum, they fall quite differently in the presence of air. How do Newton's laws apply to objects falling in air? The answer is that Newton's laws apply for *all* falling bodies, freely falling or falling in the presence of resistive forces. The accelerations, however, are quite different for the two cases. The important thing to keep in mind is the idea of *net force*. In a vacuum or in cases where air resistance can be neglected, the net force is the weight because it is the only force acting on a falling object. In the presence of air resistance, however, the net force is the difference between the weight and the force of air resistance.[4]

The force of air resistance acting on a falling body depends on two things. First, it depends on the size of the falling object—that is, on the amount of air the object must plow through in falling. Second, it depends on the speed of the falling object; the greater the speed, the greater the number of air molecules an object encounters per second and the greater the force of molecular impact. Air resistance depends on the size and the speed of a falling object.

A feather dropped in the air accelerates very briefly and then "floats" to the ground at constant velocity. This happens because the air resistance acting against the feather quickly builds up to counteract the weight. The

---

[4]In mathematical notation,

$$a = \frac{F_{\text{net}}}{m} = \frac{mg - R}{m}$$

where mg is the weight, and R is the air resistance. Note that when $R = mg$, $a = 0$; then, with no acceleration, the object falls at *constant* velocity. With elementary algebra we can go another step and get

$$a = \frac{F_{\text{net}}}{m} = \frac{mg - R}{m} = g - \frac{R}{m}$$

We see that the acceleration a will always be less than g if air resistance R impedes falling. Only when $R = 0$ does $a = g$.

net force on the feather is then zero, and no further acceleration takes place. The feather is very light and has a relatively large surface area, so it doesn't have to fall very fast before the air resistance equals its weight. The same idea applies to all objects falling in air. As a falling object gains speed, the force of air resistance builds up until it equals the weight of the falling object. When this happens, the *net* force becomes zero and the object no longer accelerates; it falls at a constant speed. Acceleration terminates, and we say the object has reached its **terminal velocity**. For a feather, terminal velocity is a few centimeters per second, whereas for a skydiver it is about 200 kilometers per hour. A skydiver varies terminal velocity by varying the position of fall. Head or feet first is a way of encountering less air resistance and attaining maximum terminal velocity. Minimum terminal velocity is attained by spreading oneself out like a flying squirrel.

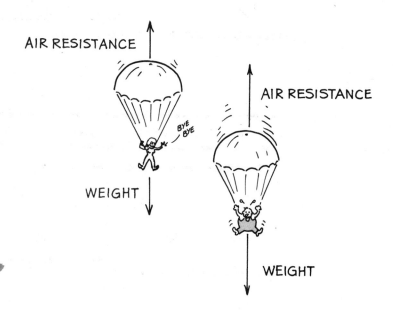

Fig. 4.11 The heavier parachutist must fall faster than the lighter parachutist for air resistance to cancel his greater weight.

Consider a man and woman parachuting together from the same altitude (Figure 4.11). Suppose that the man is twice as heavy as the woman and that their chutes are initially opened and are the same size. The same size chute means that at equal speeds the air resistance is the same on each. Who gets to the ground first—the heavy man or the lighter woman? The answer is that the person with the greater terminal velocity gets to the ground first. At first we might think that because the chutes are the same, the terminal velocities for each would be the same, and therefore both would reach the ground together. This doesn't happen because air resistance also depends on speed. The man will not reach his terminal velocity until air resistance against his chute builds up to a value that equals his weight. And before this happens, the air resistance against the chutes of each will first build up to equal the weight of the woman. Cancelations of forces occur first for her, while he continues to accelerate to enough speed to produce an amount of air resistance equaling his greater weight.[5] Terminal velocity is greater for the heavier person, with the result that the heavier person reaches the ground first.

A baseball will fall faster in air than a tennis ball for the same reasons as stated for the parachutists. The difference in the rates of fall is not as pronounced, but the difference is there. The principle is obvious for an exaggerated case, say, the falling of a bird's feather and a cast-iron feather of the same size. Acceleration terminates when air resistance builds up to the weight of the falling object.

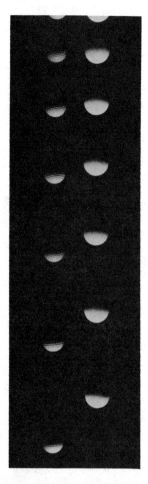

Fig. 4.12 A stroboscopic study of a golf ball and a styrofoam ball falling in air. The air resistance is negligible for the heavier golf ball, and it accelerates essentially at g. Air resistance is not negligible for the lighter styrofoam ball, which sooner reaches its terminal velocity.

Newton's Third Law of Motion

A force in the simplest sense is a push or a pull. More specifically, a force is not a thing in itself but is an *interaction* between one thing and another. We pull on a cart and it accelerates. A hammer hits a stake and drives it into the ground. One body interacts with another body. Which exerts the force and which receives the force? Newton's answer to this is that neither force has to be identified as "exerter" or "receiver"; he believed nature is symmetrical and concluded that both bodies must be treated equally. For example, when we pull the cart, the cart in turn pulls on us, as evidenced perhaps by the tightening of the rope wrapped around our hand. And the hammer exerts a force on the stake but is itself brought to rest in the process. The same force that drives the stake also decelerates the hammer. Such observations led Newton to his third law—the law of action and reaction.

Law 3: *Whenever one body exerts a force on a second body, the second body exerts an equal and opposite force on the first*.

Newton's third law is often stated: "To every action there is always opposed an equal reaction." It is important to emphasize that the action and reaction forces act on *different* bodies. They never act on the same body.

---

[5] Terminal velocity for the man will be somewhat less than double that of the half-as-heavy woman because the retarding force of the air is not exactly proportional to the velocity.

↑ PUSH OF ATHLETE

WEIGHT OF BARBELL

Fig. 4.13 Action-reaction: the action force is the upward push of the athlete on the barbell; the reaction force is the weight of the barbell pushing down on the athlete.

The action and reaction forces make up a *pair* of forces. Forces always occur in pairs. There is never a single force in any situation. For example, in walking across the floor, we push against the floor, and the floor in turn pushes against us. Likewise, the tires of a car push against the pavement, and the pavement pushes back on the tires. When we swim, we push the water backward, and the water pushes us forward. The reaction forces, those acting in the direction of our resulting accelerations, are what account for our motion in these cases. These forces depend on friction; a person or car on ice, for example, may not be able to exert the action force to produce the needed reaction force by the ice.

It doesn't matter which force we call action and which we call reaction as long as we remember that neither exists without the other. In our first example, we may call our push against the floor the action force and the floor pushing back the reaction force. We could just as well say the push of the floor against us is the action force, in which case our push against the floor is the reaction force.

Be careful, however, to distinguish forces exerted *on* a body and *by* a body. If you call the force that acts *on* a body the action force, the reaction force will be a force exerted *by* the body on whatever other body is involved in the interaction.

Questions

1. Can a body possess force?*

2. Since action and reaction are exactly equal and oppositely directed, how can there ever be a net force on a body?†

3. Can you identify the action and reaction forces in the case of a body falling in a vacuum?‡

---

*__Answer__ No, a force is not something a body *has*, like mass, but is an interaction between one body and another. Depending on the circumstances, a body may possess the capability of exerting a force on another body, but it cannot possess force as a thing in itself.

†__Answer__ The net force on any body is a combination of forces that act *on* the body, say, the action forces. The reaction forces *by* the body act on something else. For example, when you kick a can, the can kicks back on you. But your foot provides the only force exerted on the can. The can flies through the air because it experienced only the action force; the reaction force was exerted *by* the can, not *on* the can. The reaction force acts on your foot, so the motion of your foot is altered—which is of little interest from the can's point of view. The can recognizes only what happens to it, not what it does to your foot. You can't cancel a force that acts on a body with a force that the body in turn exerts on something else.

‡__Answer__ To identify a pair of action-reaction forces in any situation, first identify the pair of interacting bodies involved. Something is interacting with something. In this case the whole is interacting (gravitationally) with the falling body. So the world pulls downward on the body (call it action), and the body in turn pulls upward on the world (reaction).

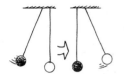

Fig. 4.14 The force imparted by the dark ball moves the white ball. The reaction force by the white ball stops the dark ball.

The idea of a falling body pulling on the earth may seem puzzling. The pull of the earth on the body is less puzzling because the acceleration of 9.8 meters/second² is quite noticeable. The same force acting on the huge mass of the earth, however, produces an acceleration so small it cannot be measured. But it is there. A less extreme example might make this clearer. When a rifle is fired, the force exerted on the bullet is exactly equal to the reaction force exerted on the rifle; hence, the rifle kicks. Since the forces are equal, one might expect the kick to be considerably more than it is.

ACTION: TIRE PUSHES ON ROAD   REACTION: ROAD PUSHES ON TIRE

ACTION: ROCKET PUSHES ON GAS   REACTION: GAS PUSHES ON ROCKET

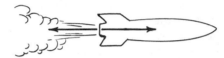

ACTION: MAN PULLS ON SPRING   REACTION: SPRING PULLS ON MAN

Fig. 4.15 Action and reaction forces. Note that when action is *A* exerts force on *B*, the reaction is then simply *B* exerts force on *A*.

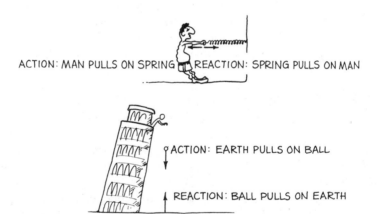

ACTION: EARTH PULLS ON BALL

REACTION: BALL PULLS ON EARTH

But in analyzing changes in motion, Newton's second law reminds us that we must also consider the masses involved. Suppose we let F represent both the action and reaction force, m the mass of the bullet, and m the mass of the more massive rifle. The accelerations of the bullet and the rifle are then found by taking the ratio of force to mass. The acceleration of the bullet is given by

$$\frac{F}{m} = a$$

while the acceleration of the recoiling rifle is

We see why the change in motion of the bullet is so huge compared to the change of motion of the rifle. A given force divided by a small mass produces a large acceleration, while the same force divided by a large mass produces a small acceleration. We have used different-sized symbols to indicate the differences in relative masses and resulting accelerations. If we used similarly exaggerated symbols to represent the acceleration of the earth when reacting to a falling body, the symbol m for the earth's mass would be larger than the page. This huge quantity divided into the force F, the weight of the falling body, would result in a microscopic a to represent the acceleration of the earth toward the falling body.

We can see that the earth accelerates slightly in response to a falling body in still another way by considering the exaggerated examples a through e in Figure 4.16. The forces between bodies A and B are equal and opposite in each case. If acceleration of body A is unnoticeable in a, then it is more noticeable in b where the difference between the masses is less extreme. In c, where both bodies have equal mass, acceleration is as evident on body A as it is on B. Continuing, we see the acceleration of A becomes even more evident in d and even more so in e. When you step off the curb, the street comes up ever so slightly to meet you.

Using Newton's third law, we can understand how a helicopter gets its lifting force. The whirling blades are shaped to force air particles downward (action), and the air in turn forces the blades upward (reaction). This upward reaction force is called *lift*. When lift equals the weight of the craft, it is able to hover in midair. When lift is greater, the helicopter climbs upward.

This is true for birds and airplanes. Birds fly by pushing air downward. The air in turn pushes the bird upward. When soaring, the wing must be shaped so that moving air particles are deflected downward. Slightly tilted wings that deflect oncoming air downward produce the lift on an airplane. Air must be pushed downward continuously to maintain lift. This supply of air is obtained by the forward motion of the aircraft which results from propellers that push air backward. When the propellers push air backward, the air in turn pushes the propellers forward. For jet aircraft, the jet pushes exhaust gases backward and the exhaust gases in turn push the jet forward. We will learn later that the curved surface of a wing is an airfoil, which enhances the lifting force.

When you push against a wall, the wall in turn pushes against you. You may be reluctant to think of the wall as *really* pushing against you, for the notion that inanimate things push and pull has probably escaped you all your life. If so, you should begin to see your interactions with things in a different light. Let's look at this wall-push idea more closely. Suppose we introduce something like a sheet of paper between your hand and the wall, and you push against the paper with a force of 20 newtons. The paper doesn't accelerate; it remains stationary, indicating that the net force

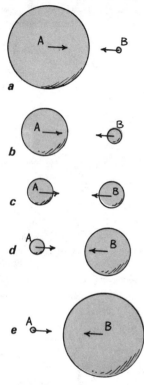

Fig. 4.16 Which body falls toward the other-- A or B? Do they both fall to each other at rates that relate to their relative masses?

acting on it is zero. So *something* has to be pushing the paper toward you with a force of 20 newtons—and that can only be the wall.

Consider the case of pulling. If you fasten a hook to a wall so you can pull on the wall, the wall in turn will pull back on you. This force can be measured with a rope and spring scale (Figure 4.17). Suppose you pull on the rope with a force of 500 newtons. Then a tension of 500 newtons is transmitted through the rope and spring scale to the wall. But this tension does not accelerate the rope! That's because the wall is also pulling on the rope, in the opposite direction. Your pull and the equal but oppositely directed pull of the wall on the rope cancel each other, producing a zero net force on the rope as a whole. The equal pulls of 500 newtons at each end establish a tension of 500 newtons *within* the rope, which is read on the spring scale. Note that there is a distinction between the net force exerted *on* the rope (which would accelerate the rope) and the tension *within* the rope.

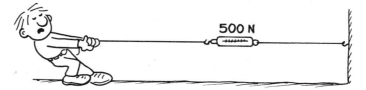

Fig. 4.17 A tug-of-war with a hook on the wall.

To understand better that the wall pulls on the rope just as hard as you do, let's look at this another way. Suppose you untie the rope from the wall and place the end in the hands of a strongman instead, and suppose the strongman holds the rope as rigidly as the wall. When you pull again with a force of 500 newtons, wouldn't the strongman have to pull equally hard in the opposite direction just to "hold the rope still" while you pulled on it (Figure 4.18)? And isn't the tension in the rope as read on the spring scale the same as before? Think about this. In this case it is easy to see that the rope is being pulled at both ends. But the strongman who "holds the rope" is doing just what the wall was doing when it "held the rope." Both the strongman and the wall pull on the rope. So you see, when you pull on an object, the object really does pull back!

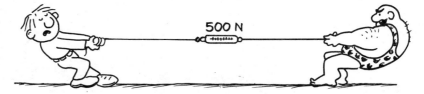

Fig. 4.18 A tug-of-war with a strongman.

The person who wins in a tug-of-war is not the person who pulls harder on the rope. It is the person who *pushes* harder against the ground. For example, in Figure 4.18 each man pushes against the ground with a force of 500 newtons, and the ground in turn pushes against each man with 500 newtons, which just counterbalances the 500-newton pulls that each experiences with the taut rope. Neither man can exert more force on the rope than he can exert against the ground (consider a tug-of-war on frictionless ice, for example). The person to win the tug-of-war is thus the person who pushes harder against the ground to produce a net force on the two-person system. To win a tug-of-war, you can (1) maintain your stance until your opponent tires and lessens his push against the ground; (2) push harder against the ground than your opponent and maintain rope tension; or (3) push harder against the ground than your opponent and increase rope tension. In each case there is a net force and the desired acceleration.

Question   Apply Newton's third law to the tug-of-war. If action is you pulling on the rope, is the reaction force the ground pushing back on you or your opponent pulling back on the rope?*

We see Newton's third law at work everywhere. A fish pushes the water backward with its fins, and the water in turn pushes the fish forward. The wind pushes against the branches of a tree, and the branches in turn push back on the wind and we have whistling sounds. Forces are interactions between different things. Every contact requires at least a twoness; there is no way that a body can exert a force on nothing. Forces, whether large shoves or slight nudges, always occur in pairs, each of which is opposite to the other. Thus, we cannot touch without being touched.

Fig. 4.19 You cannot touch without being touched—Newton's third law.

*__Answer__   Neither. If action is you pulling on the rope, reaction is the rope pulling on you. In situations like this one, there are usually many action-reaction pairs.

Summary of Newton's Three Laws

An object at rest tends to remain at rest; an object in motion tends to remain in motion at constant speed along a straight-line path. This tendency of objects to resist a change in motion is called *inertia*. Mass is a measure of inertia. Objects will change their states of motion only in the presence of a net force.

When a net force is impressed upon an object, it will accelerate. The acceleration is directly proportional to the net force and inversely proportional to the mass. Symbolically,

$$a \sim \frac{F}{m}$$

Acceleration is always in the direction of the net force. When objects fall in a vacuum, the net force is simply the weight, and the acceleration is g. (We use the symbol g rather than a to denote that acceleration results from gravity alone.) When objects fall in air, the net force is equal to the weight minus the force of air resistance, and the acceleration is less than g. When the force of air resistance equals the weight of a falling object, acceleration terminates, and the object falls at constant velocity.

Forces are interactions between bodies. Whenever one body exerts a force—say, action—on a second body, the second body exerts an equal and opposite force—reaction—against the first. We see that forces always occur in pairs. Since they act on different bodies, action and reaction never cancel out.

Summary of Terms

Mass   The quantity of matter in a body. More specifically, it is the measurement of the inertia or sluggishness that a body exhibits in response to any effort made to start it, stop it, or change in any way its state of motion.

Weight   The force due to gravity on a body.

Volume   The quantity of space a body occupies.

Force   Any influence that can cause a body to be accelerated, measured in newtons (in *pounds* in the British system).

Friction   The resistive forces that arise to oppose the motion or attempted motion of a body past another with which it is in contact.

Free fall   Motion under the influence of gravitational pull only.

Review Questions

1. Clearly distinguish among *mass*, *weight*, and *volume*.

2. Does a 2-kilogram brick have twice as much inertia as a 1-kilogram brick? Twice as much mass? Twice as much weight (when weighed in the same vicinity)?

3. If we say that one quantity is *proportional* to another quantity, does this mean they are *equal* to each other? Explain briefly, using mass and weight as an example.

4. What is the weight of a 1-kilogram rock?

5. What is the mass of a rock that weighs 9.8 newtons at the surface of the earth?

6. Suppose you have two identical cans, one filled with lead and the other empty, and you are far in space at a place where the two cans are "weightless." How can you tell which has the greater mass?

7. In what kind of path would the planets move if suddenly no force acted on them?

8. Suppose a body is being propelled along a straight-line path by an impressed force. By how much would its pickup in speed increase if the force were doubled? How about if its mass were also doubled?

9. Why is it more difficult to slide a crate from a position of rest across the floor than it is to keep it in motion once it is sliding?

10. If the force of friction acting against a sliding body is 10 newtons, how much force must be applied to maintain a constant velocity? What is the resulting net force in this case? The acceleration?

11. What is meant by *free fall?* Why doesn't a heavy object accelerate more than a lighter object when both are in a state of free fall?

12. What is the net force acting on a 10-newton freely falling body? What is the net force when it encounters 4 newtons of air resistance? 10 newtons of air resistance?

13. Why does a heavy parachutist fall faster than a lighter parachutist who wears the same size parachute?

14. In a tug-of-war, what is the net force acting *on* the rope when the participants each pull with opposing forces of 500 newtons? What is the tension force *within* the rope?

15. Consider hitting a baseball with a bat. If we call the force on the bat against the ball the *action* force, identify the *reaction* force.

Home Projects

1. Ask a friend to drive a small nail into a piece of wood placed on top of a pile of books on your head. Why doesn't this hurt you?

2. Drop a sheet of paper and a coin at the same time. Which reaches the ground first? Why? Now crumple the paper into a small, tight wad and again drop it with the coin. Explain the difference observed. Will they fall together if dropped from a second-, third-, or fourth-story window? Try it and explain your observations.

3. Drop two balls of different weight from the same height, and for small velocities they practically fall together. Will they roll together down the same inclined plane? If each is suspended from an equal length of string, making a pair of pendulums, and displaced through the same angle, will they swing back and forth in unison? Try it and see.

4. The net force acting on an object and the resulting acceleration are always in the same direction. You can demonstrate this with a spool. If the spool is pulled horizontally to the right, which way will it roll?

5. Hold your hand like a flat wing outside the window of a moving automobile. Then slightly tilt the front edge upward and notice the lifting effect. Can you see Newton's laws at work here?

Exercises

Please do not be intimidated by the large number of exercises in this and other meatier chapters. If your course is to cover many chapters, your instructor will likely assign only a few exercises from each.

1. In terms of inertia, what is the disadvantage of a lightweight camera when snapping the shutter? Why is a massive tripod preferred by most photographers?

2. Your empty hand is not hurt when it bangs lightly against a wall. Why is it hurt if it does so while carrying a heavy load? Which of Newton's laws is most applicable here?

3. In tearing a paper towel or plastic bag from a roll, why is a sharp jerk more effective than a slow pull?

4. Each of the chain of bones forming your spine is separated from its neighbors by disks of elastic tissue. What happens, then, when you jump heavily on your feet from an elevated position? Can you think of a reason why you are a little taller in the morning than at night? (*Hint:* Think about the hammerhead in Figure 4.1.)

5. Before the time of Galileo and Newton it was thought by many learned scholars that a stone dropped from the top of a tall mast on a moving ship would fall vertically and hit the deck behind the mast by a distance equal to how far the ship had moved forward while the stone was falling. In light of your understanding of Newton's laws, what do you think about this?

6. The chimney of a stationary toy train consists of a vertical spring gun that shoots a steel ball a meter or so straight into the air—so straight, in fact, that the ball always falls back into the chimney. Suppose the train moves at *constant speed* along a straight track. Do you think the ball will still return to the chimney if shot from the moving train? How about if the train *accelerates* along a straight track? How about if it moves at constant speed on a *circular* track? Why are your answers different?

7. What is your own mass in kilograms? Your weight in newtons?

8. Gravitational force on the moon is only $\frac{1}{6}$ that of the gravitational force on the earth. What would be the weight of a 10-kilogram body on the moon and on the earth? What would its mass be on the moon and on the earth?

9. If a mass of 1 kilogram is accelerated 1 meter per second^2 by a force of 1 newton, what would be the acceleration of 2 kilograms acted on by a force of 2 newtons?

10. How much acceleration does a 747 jumbo jet of mass 30,000 kilograms experience in takeoff when the thrust for each of four engines is 30,000 newtons?

11. We know that a rocket becomes progressively easier to accelerate as it travels through space. Why is this so? (*Hint:* About 90 percent of the mass of a newly launched rocket is fuel.)

12. Does a stick of dynamite contain force?

13. If we find a body that is not moving even though we know it to be acted on by a force, what inference can we draw?

14. When your car moves along the highway at constant velocity, the net force on it is zero. Why, then, do you continue running your engine?

15. A "shooting star" is usually a grain of sand from outer space that burns up and gives off light as it enters the atmosphere. What exactly causes this burning?

16. In the attempt to ascend a steep road from a standstill position in a vehicle, why is it important that the wheels not spin against the pavement?

17. Why is no air resistance experienced when one is soaring in a balloon?

18. What is the net force on an apple that weighs 1 newton when you hold it at rest above your head? What is the net force on it when you release it?

19. Can a dog wag its tail without the tail in turn "wagging the dog"? (Consider a dog with a relatively massive tail.)

20. When the athlete in Figure 4.13 holds the barbell overhead, the reaction force is the weight of the barbell as stated in the caption. How does this force vary for the case where the barbell is accelerated upward? Downward?

21. Use Newton's third law to explain why when standing on a weighing scale you cannot decrease your weight by pulling upward on your bootstraps.

22. Why can you exert greater force on the pedals of a bicycle if you pull up on the handlebars?

23. If the earth exerts a force of 1000 newtons on an orbiting communications satellite, how much force does the satellite exert on the earth?

24. Your weight is the result of a gravitational force of the earth on your body. What is the corresponding reaction force?

25. As you stand on a floor, does the floor exert an upward force against your feet? How much force does it exert? Why are you not moved upward by this force?

26. Consider the two forces acting on the person who stands still, namely, the downward pull of gravity and the upward support of the floor. Are these forces equal and opposite? Do they comprise an action-reaction pair? Why or why not?

27. If you stand with your weight evenly distributed over two bathroom scales, what will be the reading on each scale compared to your weight? (If you have access to a pair of scales, try this.)

28. The little girl hangs at rest from the ends of the rope as shown. How does the reading on the scale compare to her weight?

29. Two 100-newton weights are attached to a spring scale as shown. Does the scale read 0, 100, or 200 newtons, or give some other reading? (*Hint:* Would it read any differently if one of the ropes were tied to the wall instead of to the hanging 100-newton weight?)

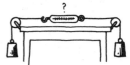

30. The strongman will push the two initially stationary freight cars of equal mass apart before he himself drops to the ground. Is it possible for him to set either of the cars in greater motion than the other? Why or why not?

31. Suppose two carts, one twice as massive as the other, fly apart when the compressed spring that joins them is released. How fast does the heavier cart roll compared to the lighter cart?

32. If a Mack truck and a Volkswagen have a head-on collision, upon which vehicle is the impact force greater? Which vehicle undergoes the greater change in its motion? Explain your answers.

33. Two people of equal mass attempt a tug-of-war with a 12-meter rope while standing on frictionless ice. When they pull on the rope, they each slide toward each other. How do their accelerations compare, and how far does each person slide before they meet?

34. Suppose in the preceding exercise that one person has twice the mass of the other. How far does each person slide before they meet?

35. A popular game in a games tournament is a modified tug-of-war, in which each person squats on a wooden block and attempts to upset the balance of the other by pulling on the rope that each holds. Which is more beneficial in this game—strength or mass? Why?

36. A horse pulls a heavy wagon with a certain force. The wagon, in turn, pulls back with an opposite but equal force on the horse. Doesn't this mean the forces cancel one another, making acceleration impossible? Why or why not?

37. If the effects of friction are negligible, which will roll down a hill faster—a Cadillac or a Volkswagen? Explain.

38. How does the weight of a falling body compare to the air resistance it encounters just before it reaches terminal velocity? After?

39. Why is it that a tennis ball dropped from the top of a 50-story building will hit the ground no faster than if it were dropped from the twentieth story?

40. As an object falls faster and faster through the air, where air resistance is a factor, does its acceleration increase, decrease, or remain constant?

41. When Galileo dropped two balls from the top of the Leaning Tower of Pisa, air resistance was *not* really negligible. Assuming both balls were the same size, yet one much heavier than the other, which ball struck the ground first? Why?

42. What is the acceleration of a rock at the top of its trajectory when thrown straight upward? (Is your answer consistent with Newton's second law?)

43. In the absence of air resistance, if a ball is thrown vertically upward with a certain initial speed, on returning to its original level it will have the same speed. When air resistance *is* a factor affecting the ball, will the ball fall to the ground faster, the same, or slower than if there were no air resistance? Why? (Consider the "principle of exaggeration" in formulating your answer; that is, consider the exaggerated case of a feather, not a ball, because the effect of air resistance on the feather is more pronounced and therefore easier to visualize.)

44. If a ball is thrown vertically into the air in the presence of air resistance, would you expect the time during which it rises to be longer or shorter than the time during which it falls? (Consider the "principle of exaggeration.")

45. In the check question on page 52 we considered the case of applying a force of 100 newtons to a 50-kilogram crate and sliding it across the floor against a constant friction force of 75 newtons. Recall that the resulting net force of 25 newtons accelerates the crate $\frac{1}{2}$ meter per second^2. (*a*) Show by substituting appropriate values into the formula for Newton's second law that by doubling the applied force, the acceleration is increased by five times. (*b*) How much applied force would be necessary to only double the acceleration?

5 Momentum

In Chapter 2 we introduced Galileo's idea of inertia, and in Chapter 4 we showed how this idea was incorporated into Newton's laws of motion. We discussed inertia in terms of objects at rest and objects in motion. In this chapter we will concern ourselves only with the inertia of moving objects. When we combine the ideas of inertia and motion, we are dealing with momentum. *Momentum* refers to moving things.

Impulse and Momentum

We all know that a heavy truck is harder to stop than a small car moving at the same speed. We state this fact by saying that the truck has more momentum than the car. By **momentum** we mean inertia in motion, or, more specifically, the product of the mass of an object and its velocity; that is,

$$\text{Momentum} = \text{mass} \times \text{velocity}$$

Or, in shorthand notation,

$$\text{Momentum} = mv$$

We can see from the definition that a moving body can have a large momentum if either its mass or its velocity is large, or if both its mass and velocity are large. The truck has more momentum than the car moving at the same speed because its mass is larger. We can see that a huge ship moving at a small velocity can have a large momentum, as can a small bullet moving at a high velocity. And, of course, a huge object moving at a high velocity, such as a massive truck rolling down a steep hill with no brakes, has a huge momentum, whereas the same truck at rest has no momentum at all.

Fig. 5.1 Why are the engines of a super-tanker normally cut off 25 km from port?

Fig. 5.2 A large change in momentum in a short time requires a large force.

When a bullet or a truck crashes into a wall, a large force is exerted against the wall. Where does this force come from? It comes from a *change* in velocity. The force of impact is proportional to the rate of change in velocity of the moving object. And the more massive the moving object, the greater the force; so the force of impact is also proportional to the mass of the moving object.

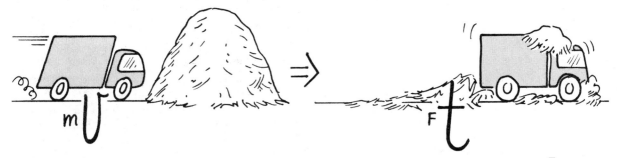

Fig. 5.3 A large change in momentum in a long time requires a small force.

These ideas are consistent with Newton's second law, $a = F/m$, as discussed in Chapter 4. There we spoke about the acceleration of masses produced by forces. A ball accelerates when a force is exerted on it. How fast it finally moves, however, depends on something besides its mass and the impressed force. The final velocity depends on the *time* involved. A long, sustained force pushes the ball to a greater velocity than the same force applied briefly. Newton's second law can be expressed in another way to make the time factor more evident by replacing the term for acceleration with its definition, change in velocity per time.[1] Then we get

---

[1]Newton's second law states that $F = ma$, and since $a = $ change in velocity/time, we can say $F = m$(change in v/t). Simple algebraic rearrangement gives $Ft = $ change in mv, or $Ft = \Delta mv$.

force × time is equal to the change in mass × velocity. We have already defined *mass × velocity* as momentum. We call the quantity *force × time* **impulse**. So we can say

$$\text{Impulse} = \text{change in momentum}$$

Or, simply,

$$Ft = \text{change in } mv$$

We can shorten this expression further by introducing the delta symbol Δ (Greek letter *D*), which stands for "change in" (or "difference in"):

$$Ft = \Delta mv$$

which reads, "force multiplied by the time-during-which-it-acts = change in momentum."

To change the momentum of a body, we should consider the impulse, that is, the amount of force *and* the time of contact. A golfer hits a ball with great force to impart momentum to the ball; but to get maximum momentum, he follows through on his swing, extending the time of contact of the force upon the ball. A large force multiplied by a large time is a greater impulse, which produces a greater change in the momentum of the ball. Any time we wish to impart the greatest momentum possible to an object, we simply apply the greatest force possible *and* extend as much as possible the time of contact.

The forces involved in impulses are not steady forces but usually vary from instant to instant. For example, the golf club initially exerts zero force on the ball until it comes in contact; the force increases rapidly as the ball is distorted (Figure 5.4) and then diminishes as the ball comes up to speed and returns to its original shape. When we speak of such impact forces in this chapter, we mean the *average* force of impact.

A golf ball sits initially on a tee with no momentum; the application of an impulse produces momentum. Consider the opposite case: a body that initially has momentum until stopped by an impulse. Suppose a car traveling at high speed crashes against a massive concrete abutment. The large momentum is expended in a very short time. A little thought will show that if an object having a large momentum is stopped suddenly, the force required to stop it must be very large. Compare the different results for a high-speed car crashing into a concrete wall and into a haystack. In both cases the momentum of the car is the same, so the impulse to stop it in each case is the same. But the times of impact are different. When the car hits a concrete wall, the impact time is short, so the average force of impact is huge. When it hits a haystack, however, the impulse is extended over a longer time, and the impact force is considerably smaller.

Fig. 5.4 Impact force against a golf ball.

Fig. 5.5 Cassy imparts a large impulse to the bricks in a short time and produces a considerable force.

The idea of short time of contact explains why a karate expert can sever a stack of bricks with the blow of her bare hand (Figure 5.5). She brings her arm and hand swiftly against the bricks with considerable momentum. This momentum is drastically reduced when she delivers an impulse to the bricks. The impulse is the force of her hand against the bricks multiplied by the time her hand makes contact with the bricks. By swift execution she makes the time of contact as short as possible and, correspondingly, makes the force of impact huge.

Question   A golfer imparts to a ball a large force for a long time for best results, whereas a karate expert imparts a large force for a short time for best results. Isn't there a contradiction here?*

Often we wish to change the momentum of an object with as small a force as possible. A boxer confronted with a high-momentum punch is interested in minimizing the force of impact. If he cannot avoid being hit, at least he has a choice of relative magnitudes for F and t in providing the impulse that will absorb and change the incoming momentum of his opponent's punch. The force of impact is lessened by extending the time of contact; hence, he "rides" or "rolls with the punch." If he is hit when approaching his opponent, the time of contact is reduced, resulting in an

*__Answer__   There is no contradiction because the best results for each are different. The best result for the golfer is maximum *momentum*, produced by maximum force over a maximum time. The best result for the karate expert is maximum *force*, produced by maximum momentum delivered in minimum time.

increased force (Figure 5.6). This force is further increased because of the additional impulse produced when his momentum of approach is stopped. This increased impulse and short time of impact result in forces that account for many knockouts.

Fig. 5.6 In both cases the impulse by the boxer's jaw reduces the momentum of the punch. (a) The boxer is moving away when the glove hits, thereby extending the time of contact. Much of the impulse therefore involves time. (b) The boxer is moving into the glove, thereby lessening the time of contact. Much of the impulse therefore involves force.

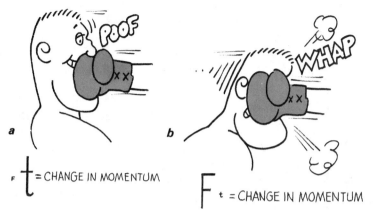

A wrestler thrown to the floor tries to extend his time of arrival on the floor by relaxing his muscles and spreading the crash into a series of impacts, as foot, knee, hip, ribs, and shoulder fold onto the floor in turn.

A person jumping from an elevated position to a floor below bends his knees upon making contact, thereby extending the time during which his momentum is being reduced by 10 to 20 times that of a stiff-legged, abrupt landing. Such knee bending reduces the forces experienced by the bones by 10 to 20 times.

A person is better off falling on a wooden floor than a concrete floor. This is because a wooden floor with "give" allows for a longer time of impact and therefore a lesser force of impact than a concrete floor with little give. A safety net used by acrobats provides an obvious example of small impact force over a long time to provide the required impulse to reduce the momentum of fall.

Questions   **1.** Does a moving body have impulse?*

2. Does a moving body have momentum?†

---

*__Answer__   No, impulse is not something a body *has*, like momentum. Impulse is what a body can *provide* when it interacts with some other body. A body cannot possess impulse just as it cannot possess force.

†__Answer__   Yes, but like velocity, in a relative sense; that is, with respect to a frame of reference, usually taken to be the earth's surface. The momentum possessed by a moving body with respect to a stationary point on earth is different from the momentum it possesses with respect to another moving body.

Conservation of Momentum

Recall from Chapter 4 that Newton's second law tells us that if we want to accelerate an object, we must apply a force on it. We say much the same thing in this chapter when we say that to change the momentum of an object we must apply an impulse on it. In whatever case, the force or impulse must be exerted on the object by something outside the object. Internal forces don't count. For example, the molecular forces within a baseball have no effect on the momentum of the baseball, just as a person sitting inside an automobile pushing against the dashboard has no effect in changing the momentum of the automobile. This is because these forces are internal forces, that is, forces that act and react within the bodies themselves. An outside, or external, force acting on the baseball or automobile is required for a change in momentum. If no external force is present, then no change in momentum is possible.

When a bullet is fired from a rifle, the forces present are internal forces. The total momentum of the system comprising the bullet and rifle therefore undergoes no net change (Figure 5.7). By Newton's third law of action and reaction, the force exerted on the bullet is equal and opposite to the force exerted on the rifle. The forces acting on the bullet and rifle act for the same time, resulting in equal but oppositely directed momenta. The recoiling rifle has just as much momentum as the speeding bullet. Although both the bullet and rifle by themselves have gained considerable momentum, the bullet and rifle together as a system experience no net change in momentum. Before firing, the momentum was zero; after firing, the net momentum is still zero. No momentum is gained and no momentum is lost.

Fig. 5.8 The machine gun recoils from the bullets it fires and climbs upward.

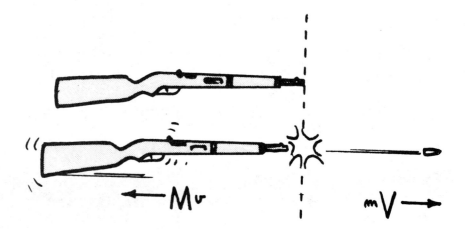

Fig. 5.7 The momentum before firing is zero. After firing, the *total,* or *net,* momentum is still zero, because the momentum of the rifle cancels the momentum of the bullet.

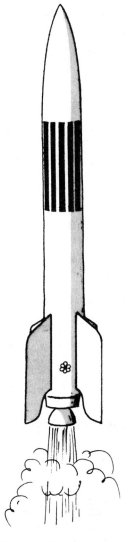

Two important ideas are to be learned from the rifle-and-bullet example. The first is that momentum, like force, is a quantity that is described by both magnitude and direction; we measure both "how much" and "which way." Quantities that are described by both magnitude and direction are called **vector quantities**. Examples of vector quantities are velocity, acceleration, and force. When any of these act in the same direction, they are simply added; when they act in opposite directions, they are subtracted. (Vectors are treated in more detail in Appendix III.) The second important idea to be taken from the rifle-and-bullet example is the idea of **conservation**. When a physical quantity remains unchanged during a process, that quantity is said to be *conserved*. We say momentum is conserved.

If we extend the idea of a rifle recoiling or "kicking" from the bullet it fires, we can understand rocket propulsion. Consider a machine gun recoiling each time a bullet is fired. The momentum of recoil increases by an amount equal to the momentum of each bullet fired. If the machine gun is free to move, mounted on a cart and track, for example, it will accelerate away from its target as the bullets are fired. A rocket accomplishes acceleration by the same means. It is continually "recoiling" from the ejected exhaust gases. Each molecule of exhaust gas can be thought of as a tiny bullet shot out of the rocket. If we consider the total system of rocket and exhaust gases, then, like the rifle and bullet, the total momentum undergoes no net change. But we are not usually interested in the total momentum in such a case, and we concern ourselves with only the momentum of the rocket.

A common misconception is that a rocket is propelled by the exhaust gases pushing against the atmosphere. This is equivalent to saying that a gun kicks because the bullet pushes against the atmosphere—not true for either the recoiling gun or the recoiling rocket. In fact, a rocket works best above the atmosphere, where there is no air resistance to restrict speed. A rocket simply recoils from the gases it ejects. It will recoil best where air resistance is absent.

Question

What happens to the speed of a fighter aircraft chasing another when it opens fire? What happens to the speed of the pursued aircraft when it returns the fire?*

Fig. 5.9 The rocket recoils from the "molecular bullets" it fires and climbs upward.

*__Answer__ The momentum of the chasing aircraft is reduced by an amount equal to the momentum of the forward-directed bullets it fires. Its speed is therefore reduced. The speed of the aircraft being chased, however, is increased because its momentum is increased by an amount equal to the momentum it gives to the bullets it fires from its rear. The overall effect is to increase the separation between the two aircraft.

Collisions

The total momentum of a system of colliding bodies is unchanged before, during, and after the collision. This is because the forces that act during the collision are internal forces—forces acting and reacting within the system itself. There is only a redistribution or sharing of whatever momentum exists before the collison.

When a moving billiard ball makes a head-on collision with another billiard ball at rest, the moving ball comes to rest and the other ball moves with the speed of the colliding ball. We call this an **elastic collision**; the colliding objects rebound without lasting deformation or the generation of heat. Whatever the initial motions of colliding objects, their motions after rebound are such that they have the same total momentum (Figure 5.10).

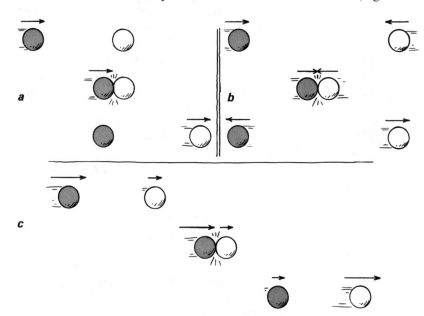

Fig. 5.10 Elastic collisions. (a) A dark ball strikes a ball at rest. (b) A head-on collision. (c) A collision of balls moving in the same direction. In each case, momentum is transferred from one ball to the other.

In any collision, we can say

Total momentum before collision = total momentum after collision

This is true even when the colliding objects become entangled during the collision. This is an **inelastic collision**, characterized by deformation or the generation of heat or both. Consider, for example, the case of a freight car moving along a track and colliding with another freight car at rest (Figure 5.11). If the freight cars are of equal mass and are coupled by the collision, can we predict the velocity of the coupled cars after impact? Suppose the single car is moving at 10 meters per second, and we consider the mass of each car to be m. Then, from the conservation of momentum,

$$(\text{Total } mv)_{\text{before}} = (\text{total } mv)_{\text{after}}$$
$$(m \times 10)_{\text{before}} = (2m \times ?)_{\text{after}}$$

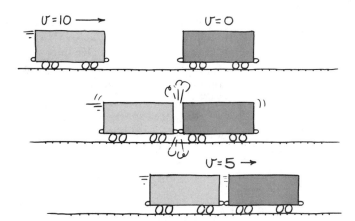

Fig. 5.11 Inelastic colli-sion. The momentum of the freight car on the left is shared with the freight car on the right after collision.

Since twice as much mass is moving after the collision, the velocity must be half as much as the velocity before collision, or 5 meters per second. Both sides of the equation are then equal.

Notice the importance of direction in these cases. We see that momentum, like force, is a vector quantity. Recall from Chapter 4 the importance of direction in combining forces: we add forces acting along the same line algebraically to find the *net* force. When two forces act in the same direction, we simply add them; but when they oppose each other, one of the forces is considered negative, and we combine them by subtraction. Likewise with momentum. Note the inelastic collisions shown in Figure 5.12. If A and B are moving with equal momenta in opposite directions (A and B colliding head-on), then one of these is considered to be negative, and the momenta add algebraically to zero. After collision, the coupled wreck remains at the point of impact, with zero momentum.

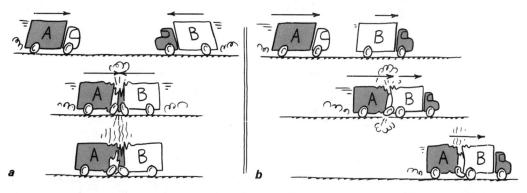

a   *b*

Fig. 5.12 More inelas-tic collisions. The net momentum of the trucks before and after collision is the same.

If A and B are moving in the same direction (A catching up with B), the net momentum is simply the addition of their individual momenta.

If A, however, moves east with, say, 10 more units of momentum than B moving west (not shown in the figure), after collision the coupled wreck moves east with 10 units of momentum. The wreck will finally come to a rest, of course, because of the external force of friction by the ground. The time of impact is short, however, and the impact force of the collision is so much greater than the external friction force that momentum in practice is conserved. The total momentum just before the trucks collide (10 units) is equal to the combined momentum of the crumpled trucks just after impact. When external force is absent—as it is, for example, when space vehicles dock in space—the 10 units of momentum will persist indefinitely.

Bouncing

An interesting consequence of momentum conservation occurs when bouncing takes place. Consider a golf ball that collides with a massive bowling ball at rest (Figure 5.13). If the collision is perfectly elastic so that the golf ball bounces with only a tiny loss of speed, the bowling ball recoils with nearly *twice* the momentum (not twice the speed) of the incident golf ball. This is consistent with the law of momentum conservation, for if the initial momentum of the golf ball is positive, then after bouncing, it is negative. The negative momentum of the golf ball is compensated by the greater momentum of the bowling ball. The net momentum before and after collision is the same. This fact was employed with great success in California during the gold rush days with the invention of the Pelton wheel, a turbine with blades shaped so as to cause incident water or steam to bounce and make a U-turn. By this means the moving fluid imparts nearly twice the momentum to the turbine.

The greater momentum change that occurs in bouncing collisions results from a correspondingly greater impulse. We see this in the impact of a rubber-nosed dart against a wooden block (Figure 5.14). When the dart is equipped with a sharp nail and makes impact against the wooden block, the impulse is not great enough to tip the block over. When the nail is removed to expose the rubber nose of the dart and it swings with the same momentum and *bounces* from the block, the nearly doubled impulse knocks the block over. (If a flower pot ever falls from a shelf and hits you on the head, you may be in trouble. If it bounces from your head, you're certainly in trouble.) We'll treat bouncing further in the next chapter.

Vector Addition for Momentum

The net momentum before and after any collision is unchanged, even when the colliding objects move at an angle to each other. Expressing the net momentum when different directions are involved requires a technique called *vector addition*. The momentum of each object is expressed as a **vector**, an arrow drawn to scale and pointing in the direction of motion; the net momentum is found by combining the vectors geometrically. We will not treat this more complicated case in any detail here but will show a simple example, Figure 5.15, to convey the idea.

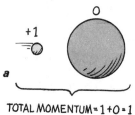

BEFORE COLLISION

0

+1

a

TOTAL MOMENTUM = 1 + 0 = 1

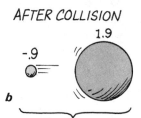

AFTER COLLISION

1.9

-.9

b

TOTAL MOMENTUM = -.9 + 1.9 = 1

Fig. 5.13 Elastic collision involving bouncing. The net momentum of the balls before and after collision is the same. This means the large ball has nearly twice the momentum (not the speed) of the incident small ball, because the small ball bounces back with negative momentum.

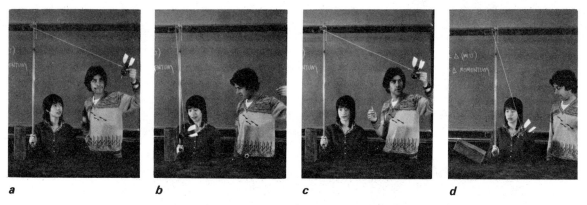

a   *b*   *c*   *d*

Fig. 5.14 (*a*) Dave releases the dart, which is brought to rest when it sticks into the wooden block (*b*). Dave then removes the metal point from the dart to expose its rubber nose and (*c*) again releases it from the same height. (*d*) The dart *bounces* from the block, which this time is knocked over. Dave says, "The impulse is greater for bouncing because the dart is more than simply stopped—it is thrown back out again." In terms of momentum, Helen adds, "If the dart hits with, say, positive momentum, then it bounces with negative momentum. The *change* in momentum from positive to negative in bouncing is greater than from positive to zero in sticking. The greater change in momentum results in greater impulse."

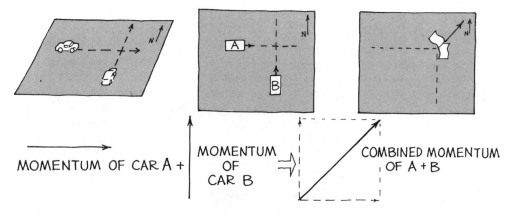

MOMENTUM OF CAR A + MOMENTUM OF CAR B COMBINED MOMENTUM OF A + B

Fig. 5.15 Momentum is a vector quantity.

In the figure, car A has a momentum directed due east, and car B's momentum is directed due north. If their individual momenta are equal in magnitude, then their combined momentum after collision will be in a northeast direction. Just as the diagonal of a square is not equal to the sum of two of the sides, the magnitude of the resulting momentum will *not* simply equal the sum of the two momenta before collision.[2]

[2]The diagonal of a rectangle can be calculated by the Pythagorean theorem: the square of the diagonal is equal to the sum of the squares of each side. Take the square root of that quantity, and you have the diagonal value.

Figure 5.16 shows a falling bomb exploding into two pieces. The momenta of the bomb fragments combine by vector addition to equal the original momentum of the falling bomb.

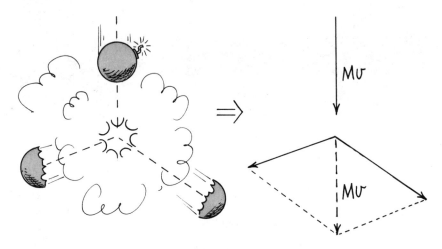

Fig. 5.16 When the bomb bursts, the momenta of its fragments add up (by vector addition) to the original momentum.

We introduce these more complicated cases not as a matter for further study but to acknowledge more general cases and to acknowledge that although the idea of momentum conservation is elegantly simple, its application to more complicated collisions can be difficult (especially if you're not on to vector addition).

Fig. 5.17 Momentum is conserved for colliding billiard balls and for colliding nuclear particles in a liquid hydrogen bubble chamber. In (a), the billiard ball A strikes billiard ball B, which was initially at rest. In (b), proton A collides with protons B, C, and D. The moving protons leave tracks of tiny bubbles.

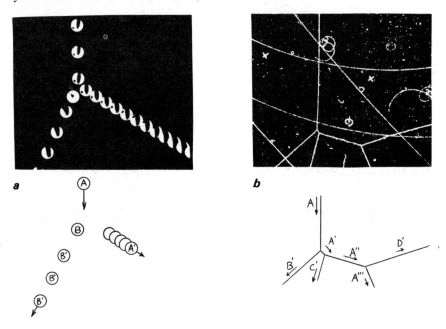

Whatever the nature of a collision or however complicated, the total momentum before, during, and after remains unchanged. This extremely useful concept enables us to learn much from collisions without regard to the form of the forces that interact in collisions. We will see in the next chapter that energy as well as momentum is conserved. By applying momentum and energy conservation to the collisions of atomic nuclei as observed in bubble and streamer chambers, we can compute the masses of elementary particles of matter. We obtain this information by measuring momenta and energy before and after collisions. The forces of the collision processes, however complicated, need not be of concern.

Summary of Terms

Momentum   The product of the mass of a body and its velocity.

Impulse   The product of the force acting on a body and the time during which it acts.

Relationship of impulse and momentum   Impulse is equal to the change in the momentum of the body the impulse acts on. In symbol notation,

$$Ft = \text{change in } mv$$

Conservation of momentum   When no external net force acts on an object or a system of objects, no change of momentum takes place. Hence, the momentum before an event involving only internal forces is equal to the momentum after the event:

$$mv_{(\text{before event})} = mv_{(\text{after event})}$$

Vector quantity   A quantity that has both magnitude and direction. Examples are force, velocity, acceleration, and momentum.

Vector   An arrow drawn to scale used to represent a vector quantity.

Review Questions

1. Which has the greater momentum—a heavy truck at rest or a moving roller skate?

2. Which has the greater momentum when they move at the same speed—a heavy truck or a roller skate? Which requires the greater stopping force?

3. In answering the preceding question, perhaps you stated that the truck required more stopping force. Make an argument that the roller skate could require more stopping force. (Consider relative times.)

4. What is the effect of extending the time of contact (following through) when a ballplayer strikes a ball?

5. In a crash or collision, why is it advantageous for an occupant to extend the time during which the collision is taking place? (How does this differ from the example in the previous question?)

6. The momentum of the rifle and bullet shown in Figure 5.7 is conserved. Why do we not say the *velocities* are conserved? What would happen if the mass of the bullet were equal to the mass of the rifle?

7. Can a rocket be propelled in a region completely devoid of an atmosphere? Explain.

8. What does it mean to say that momentum is "conserved" in a collision?

9. Railroad car A rolls at a certain speed against car B of the same mass, which is at rest. If the collision were perfectly elastic, compare the motions of the cars after collision.

10. In the preceding question, compare the motions of the cars before and after collision if the collision is inelastic.

11. Compare the impulses for the cases of sticking versus bouncing during a collision.

12. What is a vector?

Home Project

When you get a bit ahead in your studies, cut classes some afternoon and visit your local pool or billiards parlor and bone up on momentum conservation. Note that no matter how complicated the collision of balls, the momentum along the line of action of the cue ball before impact is the  same as the combined momentum of all the balls along this direction after impact, and that the components of momenta perpendicular to this line of action cancel to zero after impact, the same value as before impact in this direction. This is more clearly seen when rotational skidding— English—is not imparted to the cue ball. When English is imparted by striking the cue ball off center, rotational momentum, which is also conserved, somewhat complicates analysis. But regardless of how the cue ball is struck, in the absence of external forces, both linear and rotational momentum are always conserved. Pool or billiards offers a first-rate exhibition of momentum conservation in action.

Exercises

1. To bring a supertanker to a stop, its engines are typically cut off about 25 kilometers from port. Why is it so difficult to stop or turn a supertanker?

2. In terms of impulse and momentum, why are padded dashboards safer in automobiles?

3. In terms of impulse and momentum, why are nylon ropes, which stretch considerably under stress, favored by mountain climbers?

4. Why is it important that an airplane wing be designed so that it deflects oncoming air downward?

5. The apple that is said to have dropped on the head of Isaac Newton may have weighed about 1 newton. The force of impact, however, would have been considerably greater than 1 newton. Why?

6. When an apple falls from a tree and strikes the ground without bouncing, what becomes of its momentum?

7. If a ball is projected upward from the ground with ten units of momentum, what is the momentum of recoil of the world? Why do we not feel this?

8. Why does a judo expert slap the mat when she has been thrown to the floor?

9. Why is a punch more forceful with a bare fist than with a boxing glove?

10. A boxer tires more readily when he misses his opponent than when he hits him. Why is this so?

11. Railroad cars are loosely coupled so that there is a noticeable time delay from the time the first and last cars are moved from rest by the locomotive. Discuss the advisability of this loose coupling and slack between cars from the point of view of impulse and momentum.

12. If only an external force can change the state of motion of a body, how can the internal force of the brakes bring a car to rest?

13. A fully dressed person is at rest in the middle of a pond on perfectly frictionless ice and must get to shore. How can this be accomplished?

14. If we throw a ball horizontally while standing on roller skates, we roll backward with a momentum that matches that of the ball. Will we roll backward if we go through the motions of throwing the ball, but instead hold onto it? Explain.

15. Using examples, show that situations explained by Newton's third law can also be explained by the conservation of momentum.

16. Your friend says that the law of momentum conservation is violated when a ball rolls down a hill and gains momentum. What do you say?

17. Why is it difficult for a fire fighter to hold a hose that ejects large amounts of high-speed water?

18. Would you care to fire a gun that has a bullet ten times as massive as the gun? Explain.

19. A sailboat is stalled on a lake on a windless day. The skipper's only piece of auxiliary equipment is a large fan, which he has set up so as to blow air into the sail. Will this technique get the boat to shore?

20. In the preceding exercise, should the skipper spread out the sail so as to catch all the air blown into it by the fan, or should he take the sail down altogether? Defend your answer.

21. When you are traveling in your car at highway speed, the momentum of a bug is suddenly changed as it splatters onto your windshield. Compared to the change in momentum of the bug, by how much does the momentum of your car change?

22. If a Mack truck and a Volkswagen have a head-on collision, which vehicle will experience the greater force of impact? The greater impulse? The greater change in its momentum? The greater acceleration?

23. Would a head-on collision between two cars be more damaging to the occupants if the cars stuck together or if the cars rebounded upon impact?

24. Suppose there are three astronauts outside a spaceship, and two of them decide to play catch with the third man. All the astronauts weigh the same on earth and are equally strong. The first astronaut throws the second one toward the third one and the game begins. Describe the motion of the astronauts as the game proceeds. How long will the game last?

25. Suppose you throw a ball against a wall and catch it on the rebound. How many impulses were applied to the ball? Why is the impulse against the wall greatest?

26. In terms of impulse, why will the bowling ball in Figure 5.13 move with nearly twice the momentum of the incident golf ball?

27. In reference to Figure 5.5, how will the impulse at impact differ if Cassy's hand bounces back upon striking the bricks? In any case, how does the force she exerts on the bricks compare to the force exerted on her hand?

28. Light consists of tiny "corpuscles" called *photons* that possess momentum. This can be demonstrated with a radiometer, shown in the sketch. Metal vanes painted black on one side and white on the other are free to rotate about the point of a needle mounted in a vacuum. When photons are incident on the black surface, they are absorbed; when photons are incident upon the white surface, they are reflected. Upon which surface is the impulse of incident light greater, and which way will the vanes rotate? (They rotate in the opposite direction in the more common radiometers where air is present in the glass chamber; your instructor can tell you why.)

29. A *deuteron* is a nuclear particle of unique mass made up of one proton and one neutron. Suppose it is accelerated up to a certain very high speed in a cyclotron and directed into an observation chamber, where it collides and sticks to a target particle that is initially at rest and then is observed to rebound at exactly half the speed of the incident deuteron. Why do the observers state that the target particle is itself a deuteron?

30. A billiard ball will stop short when it collides head-on with a ball at rest. The ball cannot stop short, however, if the collision is not exactly head-on—that is, if the second ball moves at an angle to the path of the first. Do you know why? (*Hint:* Consider momentum before and after the collision along the initial direction of the first ball and also in a direction perpendicular to this initial direction.)

6 Energy

Perhaps the idea most central to all of science is that of energy. The combination of energy and matter makes up the universe: matter is the substance, and energy is the mover of the substance. The idea of matter is easy to grasp. Matter is stuff that we can see, smell, and feel. It has mass and occupies space. Energy, on the other hand, is more abstract. We cannot see, feel, taste, or smell energy. The only time energy is evident is when it changes. Energy is a concept—the intangible combination of physical properties and interactions evident in various familiar forms, such as heat, light, and electricity, or in the tiny nucleus of the atom. We are at the beginning of an age in which people will have control of inexhaustible amounts of energy. This control of energy can bring promise for a peaceful future or it can be the means of the world's destruction. In this critical time, it is imperative that we understand more about this prime mover. We will begin by introducing a related concept: work.

Work

In the last chapter we saw that changes in a body's motion have to do not only with force but also with how long the force acts. And when we considered "how long," we considered duration, or time. We called the quantity *force × time* impulse. But "how long" needn't mean duration. It can mean distance as well. When we consider the quantity *force × distance*, we are speaking about a wholly different quantity: **work**. Work is a way of changing energy from one form to another.

When we lift a load against the earth's gravity, work is done. The higher we lift the load or the heavier the load, the more work we do. Two things enter into every case where work is done: (1) the exertion of a *force* and (2) the *movement* of something by that force. For the simplest case, where the force is constant and the motion takes place in a straight line in the direction of the force, we say

The work done on an object by an applied force is defined and measured by the product of the force and the distance through which the object moves.

Work = force × distance moved in the direction of the force or, simply,

$$W = Fd$$

If we lift two loads one story high, we do twice as much work as lifting one load, because the weight, or *force*, is twice as great. Similarly, if we lift a load two stories, we do twice as much work because the *distance* is twice as great.

Fig. 6.1 Work is done in lifting the barbell. If the weightlifter were taller, he would have to expend proportionally more energy to press the barbell over his head.

Fig. 6.2 He may expend energy when he pushes on the wall, but if it doesn't move, no work is performed on the wall.

Note that our definition of work involves both a force *and* a distance. We may push against a wall for hours and become really tired, but if the wall does not move—that is, if the distance the force moves is zero—we have done no work on the wall (Figure 6.2). Work may be done on the muscles by stretching and contracting, which is force × distance on a biological scale, but this work is not done on the wall.

The unit of measurement in which work is expressed contains units of force and units of distance, the newton-meter, called the **joule** (rhymes with *pool*). One joule of work is done when a force of 1 newton is exerted over a distance of 1 meter, like lifting an apple over your head. For larger values, we speak of thousands of joules, kilojoules (kJ), or millions of joules, megajoules (MJ). The weightlifter in Figure 6.1 does work on the order of kilojoules, and the energy contents of petroleum fuels are rated in megajoules.

Power

Note that the definition of work says nothing about the time during which work is done. The same amount of work is done in carrying a load up a flight of stairs whether we walk up or run up. So why is it that we are more tired after running up that flight in a few seconds than after walking up in a few minutes? To distinguish such cases, we introduce a measure of how fast work is done, which we call **power**. Power is simply equal to work divided by the time during which the work is done:

$$\text{Power} = \frac{\text{work}}{\text{time}}$$

Table 6.1 Potential energies of various fuels (in megajoules per kilogram)

Hydrogen	142
Methane	55
Natural gas	52
Gasoline	48
Kerosene	46
Charcoal (pure)	34
Coal	20–33
Alcohol	27
Wood	13
Dynamite	5
Gunpowder	3

An engine of much power can do work rapidly. For example, the engines of most cars will provide the work necessary to drive the cars to the top of a steep hill. But different engines will do the same task in different times. An engine that will drive a car up the hill in a short time develops more power than an engine that requires a longer time to do the same amount of work.

Another way to look at power is this: a liter of gasoline can do a specified amount of work, but the power produced when we burn it can be any amount, depending on how *fast* it is burned. The liter may produce 50 units of power for $\frac{1}{2}$ hour in an automobile or 60,000 units of power for $1\frac{1}{2}$ seconds in a Boeing 747.[1]

The unit of power measured in joules per second is the watt (in honor of James Watt, the eighteenth-century developer of the steam engine): 1 watt of power is expended when 1 joule of work is done in 1 second; 1 kilowatt equals 1000 watts. In the United States we have customarily rated engines in units of horsepower and electricity in kilowatts, but actually either may be used. The present conversion to the metric system will see automobiles rated in kilowatts; 1 horsepower is the same as $\frac{3}{4}$ kilowatt, so an engine rated at 134 horsepower is a 100 kW engine.

Mechanical Energy

Fig. 6.3 The potential energy of Tenny's drawn bow equals the work (average force × distance) she did in drawing the arrow into position. When released, the potential energy of the drawn bow will become the kinetic energy of the arrow.

Work is done in lifting the heavy ram of a pile driver, and, as a consequence, the ram acquires the property of being able to do work on a body beneath when it falls. When work is done in winding a spring mechanism, the spring acquires the ability to do work on an assemblage of gears to run a clock, ring a bell, or sound an alarm. And when a battery is charged, it may in turn do the work of a wound spring. In each case, something has been acquired. When work is done on an object, something is given to the object, which, in many cases, enables it to do work. This "something" may be a physical separation of attracting bodies; it may be a compression of atoms in the material of a body; or it may be a rearrangement of electric charges in the molecules of a body. This something that enables a body to do work is called **energy**. Like work, energy is measured in joules. It appears in many forms, which we will discuss in subsequent chapters. We shall give attention here to the two forms of mechanical energy: potential energy and kinetic energy.

Potential Energy

An object may store energy by virtue of its position. Such stored energy is called **potential energy** (PE) because when energy is in the stored state, an object has the potential for doing work. A stretched or compressed spring, for example, has potential energy. When a bow is drawn, energy is stored in the bow (Figure 6.3). A stretched rubber band has potential energy because of its position; if it is part of a slingshot, it is capable of doing work.

[1]From J. Holdren and P. Herrera, *Energy* (San Francisco: Sierra Club, 1971).

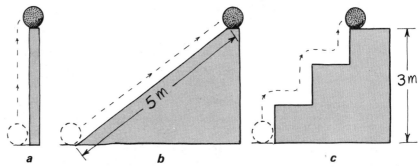

Fig. 6.4 The potential energy of the 10-N ball is the same (30 J) in all three cases because the work done in elevating it 3 m is the same whether it is (a) lifted with 10 N of force, (b) pushed with 6 N of force up the 5-m incline, or (c) lifted with 10 N up each 1-m stair. No work is done in moving it horizontally (neglecting friction).

The chemical energy in fuels is potential energy, for it is actually energy of position when looked at from a microscopic point of view. This energy is available when the positions of electrical charges within and between molecules are altered, that is, when a chemical change takes place. Potential energy is possessed by any substance that can do work through chemical action. The energy of coal, gas, electric batteries, and foods is potential energy.

Potential energy may be due to an elevated position of a body. Water in an elevated reservoir and the heavy ram of a pile driver when lifted have energy because of position. The energy of elevated positions is usually called **gravitational potential energy**.

The measure of the gravitational potential energy of an elevated body is the work done against gravity in lifting it. The upward force required is equal to the weight of the body *mg,* and the work done in lifting it through a height *h* is given by the product *mgh*; so we say

$$\text{Gravitational potential} = \text{weight} \times \text{height}$$

$$= mgh$$

We say a body at the height *h* has potential energy *mgh* relative to the original position and is independent of the path taken to raise it (Figure 6.4).

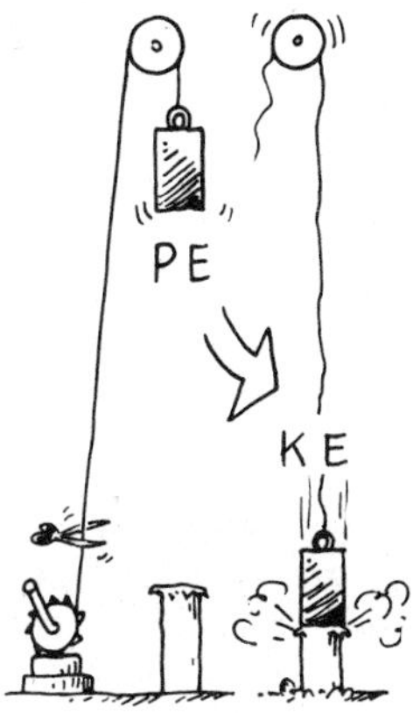

Kinetic Energy

If we push on an object, we can set it in motion. More specifically, if we do work on an object, we can change the energy of motion of that object. If an object is in motion, by virtue of that motion it is capable of doing work. We call energy of motion **kinetic energy** (KE). The kinetic energy of an object is equal to half its mass multiplied by its velocity squared:

$$\text{KE} = \tfrac{1}{2} m v^2$$

It can be shown that the kinetic energy of a moving body is equal to the work it can do in being brought to rest:[2]

$$\text{Net force} \times \text{distance} = \text{kinetic energy}$$

Fig. 6.5 The potential energy of the elevated ram is converted to kinetic energy when released.

PE

KE

---

[2] If we multiply both sides of $F = ma$ (Newton's second law) by d, we get $Fd = mad$. Since $d = \tfrac{1}{2} at^2$, we can say $Fd = ma(\tfrac{1}{2} at^2) = \tfrac{1}{2} m(at)^2$; and substituting $v = at$, we get $Fd = \tfrac{1}{2} mv^2$.

Or, in shorthand notation,

$$Fd = \tfrac{1}{2} mv^2$$

Notice that velocity is squared, which means that if the velocity of an object is doubled, its kinetic energy is quadrupled ($2^2 = 4$), enabling the object to do four times as much work. Accident investigators, for example, are well aware that an automobile traveling at 80 kilometers per hour has four times as much kinetic energy as an automobile traveling at 40 kilometers per hour. This means that a car traveling at 80 kilometers per hour will skid four times as far when its brakes are locked as a car traveling at 40 kilometers per hour. This is because the velocity is squared for kinetic energy.

We can interpret the above formula to mean that the kinetic energy of a moving body is equal to the work required to bring it to rest. We can also interpret it to mean that the work done on a body initially at rest will set it in motion with kinetic energy equal to the work done. Either way, we can say that the work done is equal to the change in kinetic energy:

$$\text{Work} = \Delta KE$$

Questions

1. When the brakes of a car traveling at 90 km/hr are locked, how much farther will the car skid compared to locking the brakes at 30 km/hr?*

2. Can a body have energy?†

3. Can a body have work?‡

Kinetic energy underlies various forms of energy that appear to be more different from one another than they actually are. For example, heat involves the kinetic energy of tiny molecules of which matter is composed. If the kinetic energy of molecules is made to vary in rhythmic patterns, we have sound. Electricity involves the kinetic energies of electrons, and even light energy originates from the vibratory motion of electrons within the atom. Whatever phase of energy we study, we find that an understanding of mechanical energy underlies an understanding of energy in its other forms.

****Answer**   Nine times farther; the car has nine times as much energy when it travels three times as fast: $\tfrac{1}{2} m(3v)^2 = \tfrac{1}{2} m9v^2 = 9(\tfrac{1}{2} mv^2)$.

†**Answer**   Yes, but in a relative sense. For example, an elevated body may possess potential energy relative to the ground below, but none relative to a point at the same elevation. Similarly, the kinetic energy that a body has is with respect to a frame of reference, usually taken to be the earth's surface. We will see that a body has "energy of being"—the congealed energy that comprises its mass. Read on!

‡**Answer**   No, unlike momentum or energy, work is not something that a body *has*. Work is something that a body *does* to some other body. A body can *do* work only if it has energy.

Fig. 6.6 The downhill "fall" of the roller coaster results in its roaring speed in the dip, and this kinetic energy sends it up the steep track to the next summit.

Conservation of Energy

Recall from Chapter 2 Galileo's inclined-plane experiments, in which he rolled balls down inclines. Galileo found that when a ball was released and allowed to roll from a particular point down one incline and up another, it never reached a level higher than its origin (Figure 6.7).

Fig. 6.7 The released ball will roll to its original height.

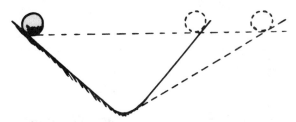

The same effect can be demonstrated with a pendulum. An extended pendulum bob released from a position of rest will rise no higher than its starting level when it swings to the other side. The bob swings back and forth beneath this level. This is true even if the string is obstructed by a peg, as shown in Figure 6.8. When the ball or pendulum bob in Figures 6.7 and 6.8 is at the starting level, it has potential energy. When released, potential energy diminishes as the height diminishes, and the energy is converted to energy of motion. At the lowest part of the inclined plane or swing, the ball or bob has minimum potential energy and maximum kinetic energy. At this point, the maximum amount of potential energy has been converted to kinetic energy. In the ideal case where friction is negligible, the kinetic energy gained is exactly equal to the potential energy

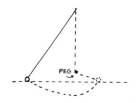

Fig. 6.8 The pendulum bob will swing to its original height whether or not the peg is present.

Fig. 6.9 Energy transitions in a pendulum.

lost. The transfer of energy continues as the kinetic energy is transformed again into potential energy when the ball or bob reaches its original height. The process continues as potential energy changes to kinetic energy and then back to potential energy and so on in a cycle (Figure 6.9).

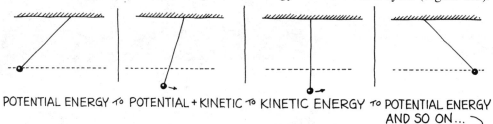

POTENTIAL ENERGY ↗ POTENTIAL + KINETIC ↗ KINETIC ENERGY ↗ POTENTIAL ENERGY
AND SO ON... ⟩

We find in practice that the swinging pendulum bob or the rolling ball doesn't quite reach its original level. Some of the energy is expended in overcoming friction. The less friction there is, however, the closer the bob or ball comes to its starting level. But energy is not lost, even in the presence of friction. Careful examination would show that air molecules and the molecules in the swinging string or the inclined plane have gained kinetic energy—exactly the amount "lost" by the swinging bob or rolling ball. We call molecular kinetic energy *heat*. In any case, no energy is lost; it is simply transformed from one form to another.

The study of the various forms of energy and of the transformation of one form of energy into another has led to one of the great generalizations in physics: the law of **conservation of energy**.

Energy cannot be created or destroyed; it may be transformed from one form into another, but the total amount of energy never changes.

Energy conservation is more evident where friction is practically absent, like above the atmosphere—in the motion of satellites.

Energy Conservation in Satellite Motion

Recall from Chapter 3 that satellites are in a continual state of free fall as they fall around (orbit) a parent body. Everywhere in its orbit a satellite has kinetic and gravitational potential energy with respect to the body it orbits. The sum of the kinetic and potential energies will be a constant for the orbital system. The simplest case occurs with circular orbits.

Consider a satellite in circular orbit about the earth. The radial distance between the earth's center and the satellite doesn't change. Because it moves at a constant altitude, its potential energy is a constant. By the conservation of energy, its kinetic energy must also be constant (Figure 6.10). In circular orbit the speed of a satellite does not change, and it moves with the same kinetic energy everywhere along its path. The reason its kinetic energy doesn't change is that the force of gravity does

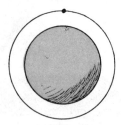

Fig. 6.10 Both the KE and the PE are constant for a satellite in circular orbit.

no work on the satellite. We know that work is done only when a force acts on a body along its direction of motion. In circular orbit the satellite criss-crosses the force of gravity at right angles, moving in a direction neither with nor against it, so there is no component of force along the direction in which the satellite moves.[3] No work is done, so there is no change in kinetic energy (Figure 6.11).

Fig. 6.11 (a) No work is done by the force of gravity that acts on a bowling ball that rolls along a horizontal surface, because the force acts in a direction perpendicular to the direction of the ball's motion. Likewise in (b): no work is done by the force of gravity on a satellite in circular orbit.

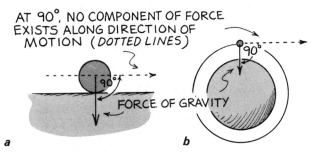

AT 90°, NO COMPONENT OF FORCE EXISTS ALONG DIRECTION OF MOTION (DOTTED LINES)

FORCE OF GRAVITY

a   *b*

Consider a satellite in elliptical orbit (Figure 6.12). Both speed and radial distance vary. The gravitational potential energy is greatest when the satellite is farthest away (at the apogee) and least when the satellite is closest (at the perigee). Correspondingly, the kinetic energy of the satellite is least when the potential energy is most and most when the potential energy is least. At every point in orbit, the sum of the kinetic and potential energies is the same.

Fig. 6.12 Both KE and PE vary for a satellite in elliptical orbit. Similar to the situation with a pendulum, when PE is maximum, KE is minimum, and vice versa. Note that for the position shown the direction of motion is not perpendicular to the force of gravity, and a component of gravitational force does work to change the KE of the satellite.

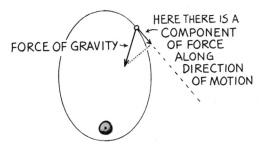

FORCE OF GRAVITY →

HERE THERE IS A COMPONENT OF FORCE ALONG DIRECTION OF MOTION

At all points along the orbit, except the apogee and perigee, there is a component of gravitational force that does work to change the kinetic energy of the satellite. When the satellite gains altitude and goes against gravity, negative work decreases its kinetic energy. This decrease continues to the apogee. Once over the apogee, the satellite moves in a direction with gravity, and positive work is done to increase kinetic energy. This increase continues until the satellite shoots past the perigee to repeat the cycle as its kinetic energy decreases and potential energy increases. The greater the potential energy, the less the kinetic energy.

---

[3] Vector components are treated in Appendix III.

Questions

1. The orbital path of a satellite is shown in the sketch. In which of the marked positions A through D does the satellite have the greatest KE? Greatest PE? Greatest total energy?*

2. If the satellite is moving clockwise, in which position is its KE decreasing? In which position is it gaining KE?†

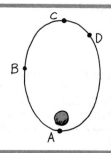

Escape Velocity

Consider the extreme case: the potential energy of a body infinitely far from the earth's surface. This amount of energy would be equal to the work required to move the body against the force of gravity from the earth's surface to a distance infinitely far away. We might think at first that the potential energy would be infinite because the distance is infinite. But gravity diminishes with distance. (We will learn exactly how in Chapter 8.) Most of the work done in launching a rocket, for example, occurs near the earth's surface, where gravity is strongest. It turns out that the value of potential energy for a 1-kilogram mass infinitely far away is 60 million joules; so to put a payload infinitely far from the earth's surface requires at least 60 million joules of energy per kilogram of load. Now a kinetic energy of 60 million joules per kilogram corresponds to a velocity of 11.2 kilometers per second whatever the mass involved. This is the value of escape velocity from the surface of the earth. If we give a payload any more than this energy at the surface of the earth or, equivalently, any more speed than 11.2 kilometers per second, then, neglecting air resistance, the payload will escape from the earth, never to return. As it continues outward, its potential energy increases and its kinetic energy decreases; its speed becomes less and less, though it is never reduced to zero. The payload outruns the gravity of the earth: it escapes.

---

*Answer   KE is maximum at the perigee, A; PE is maximum at the apogee, C; total energy (KE + PE) is the same along all points in the orbital path.

†Answer   Its KE decreases at B because it is moving away from the earth against gravity. At D it is moving toward the earth with gravity, so its KE increases. In these positions it isn't moving *directly* against or with gravity but, more properly, at a non-90° angle to the gravity vector that everywhere points toward the center of the earth. The satellite slows down at B and speeds up at D, just as a baseball thrown at an angle skyward from the surface of the earth slows down while ascending and speeds up while descending. While ascending, a component of earth gravity along and against its direction of motion does negative work, which decreases its KE; and while descending, a component of gravity along and in the direction of its motion does positive work, which increases its KE. (Vector components are treated in Appendix III.)

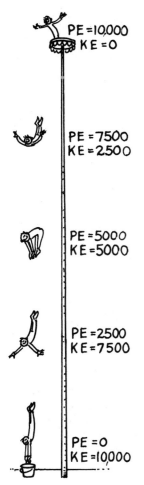

PE =10,000
KE =0

PE =7500
KE =2500

PE =5000
KE =5000

PE =2500
KE =7500

PE =0
KE =10,000

In any closed system, whether it be as simple as a satellite in orbit or as complex as a galaxy in explosion, there is one quantity that doesn't change—energy. It may change form, but the total energy score stays the same.

This energy score takes into account the fact that the atoms that make up matter are themselves congealed bundles of energy. The sun shines because part of its mass is being transformed to radiant energy. In nuclear reactors, part of the masses of atoms is transformed to energy. Early in the twentieth century, Albert Einstein stated that whenever mass is transformed into energy (or vice versa), the transformation is in accord with his celebrated equation $E = mc^2$, where E is the amount of energy appearing or disappearing, m is the mass disappearing or appearing, and c is the speed of light. This equation is most often associated with nuclear reactions because a measurable mass change of about 1 part in 1000 takes place, but it applies as well to the tiny mass change of about 1 part in 10^9 when a match is struck and lit. $E = mc^2$ applies to all forms of energy.

Thermonuclear reactions in the sun convert mass to radiant energy, a tiny part of which reaches the earth and falls on plants. A tinier part of this part ultimately becomes coal. Another part supports life in the food chain that begins with plant-eating creatures, and part of *this* part ultimately becomes oil. Part of the energy from the sun goes into evaporating water from the ocean, to form clouds and eventually to return to the earth as rain that may be trapped behind a dam. By virtue of its position the water has energy that may drive a turbine, where it will be transformed to electrical energy and then distributed to houses, where it is used to light, heat, cook, and operate machines and gadgetry. So we see that energy is transformed from one form to another.

Fig. 6.13 A circus diver at the top of a pole has a potential energy of 10,000 J. As he dives, his potential energy converts to kinetic energy. Note that at successive positions one-fourth, one-half, three-fourths, and all the way down, the total energy is constant. (Adapted from K. F. Kuhn and J. S. Faughn, *Physics in Your World*, Philadelphia: Saunders, 1980.)

Machines—An Application of Energy Conservation

At the same time that work is done on a machine, work is done by a machine. If the heat energy resulting from frictional forces is small enough to neglect, the work output will, in good approximation, be equal to the work input. We can say

$$\text{Work input} = \text{work output}$$

Or, since Work = force × distance,

$$(Fd)_{\text{input}} = (Fd)_{\text{output}}$$

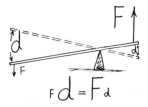

Fig. 6.14 The lever.

The simplest machine is the lever. By applying a small force through a large distance, we can exert a large force through a correspondingly smaller distance (Figure 6.14). Although the output force can be many times greater than the input force, the energy or work output can be no greater than the input energy. Otherwise, the law of energy conservation would be violated.

A child uses the principle of the lever in jacking up the front end of an automobile. By exerting a small force through a large distance, she is able to provide a large force acting through a small distance. Consider the ideal example illustrated in Figure 6.15. Every time she pushes the jack handle down 25 centimeters, the car rises only a hundredth as far but with 100 times as much force.

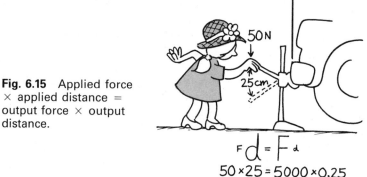

Fig. 6.15 Applied force × applied distance = output force × output distance.

Fig. 6.16 Applied force × applied distance = output force × output distance.

A block and tackle, or system of pulleys, is a simple machine that multiplies force at the expense of distance. One can exert a relatively small force through a large distance and lift a heavy weight through a relatively short distance. With the ideal pulley system such as that shown in Figure 6.16, the man pulls 10 meters of rope with a force of 50 newtons and lifts 500 newtons a vertical distance of 1 meter. The energy the man expends in pulling the rope is numerically equal to the increased potential energy of the 500-newton block.

Any machine that multiplies force does so at the expense of distance. Likewise, any device that multiplies distance, like the construction of your forearm and elbow, does so at the expense of force. No machine or device can put out more energy than is put into it. No machine can create energy; it can only transform it from one form to another.

Efficiency

The three previous examples were of ideal machines; 100 percent of the work input was transformed to work output. An ideal machine would operate at 100 percent efficiency. In practice, this doesn't happen, and we can never expect it to happen. In any transformation some energy is dissipated to molecular kinetic energy—heat.

If the girl in Figure 6.15 puts in 100 units of work and increases the potential energy of the car by 60 units, the jack is 60 percent efficient; 40 units of work have been used to do work in overcoming the friction force, and these appear as heat. If we put in 100 units of work on the lever of Figure 6.14 and get out 98 units of work, the lever is 98 percent efficient, and we waste only 2 units of work input in producing heat. The higher the efficiency of a machine, the lower the proportion of energy wasted as heat. Inefficiency exists whenever energy is transformed from one form to another. **Efficiency** can be expressed by the ratio

$$\text{Efficiency} = \frac{\text{work done}}{\text{energy used}}$$

Table 6.2   Approximate efficiencies for some energy transformations

System	Energy in	Energy out	Efficiency (percent)
Incandescent lamp	Electrical	Light	5
Steam locomotive	Chemical	Mechanical	8
Fluorescent lamp	Electrical	Light	20
Solar cell	Light	Electrical	25
Automobile engine	Chemical	Mechanical	25
Coal-fired power plant	Chemical	Electrical	30
Nuclear power plant	Nuclear	Electrical	30
Diesel engine	Chemical	Mechanical	38
Steam turbine	Heat	Mechanical	47
Fuel cell	Chemical	Electrical	60
Small electric motor	Electrical	Mechanical	63
Home oil furnace	Chemical	Heat	85
Large steam boiler	Chemical	Heat	88
Dry cell battery	Chemical	Electrical	90
Large electric motor	Electrical	Mechanical	92
Electric generator	Mechanical	Electrical	99

An automobile engine is a machine that transforms chemical energy stored in fuel into mechanical energy. The bonds between the molecules in the petroleum fuel break up when the fuel burns by reacting with the oxygen in the air. Carbon atoms then bond with oxygen to form carbon monoxide, where the bonds have less energy stored in them than the original bonds. Some of the remaining energy goes into running the engine. We'd like all this energy converted into mechanical energy; that is, we'd like an engine that is 100 percent efficient. Unfortunately, even the best-designed engines are unlikely to be more than 30 percent efficient. Some of the energy converted to heat goes into the cooling system and is wasted through the radiator to the air. Some goes out with the hot exhaust gases, and nearly half is wasted in the friction of the moving engine parts. In addition to these inefficiencies, the fuel doesn't even burn completely, and a certain amount of fuel energy goes unused.

Question   Suppose a miracle car has a 100 percent efficient engine and burns fuel that has an energy content of 40 megajoules per liter. If the air drag and overall friction forces on this car traveling at highway speed is 2000 newtons, what is the upper limit in distance per liter it could be driven at this speed?*

The inefficiency that accompanies transformations of energy can be looked at in this way: in any transformation there is a dilution of *useful energy* to a state of wasted energy. The amount of usable energy decreases with each transformation and ultimately becomes heat. When we study thermodynamics, we'll see that the energy in heat is useless for doing work unless it can be converted to a lower temperature. Heat is the graveyard of useful energy.

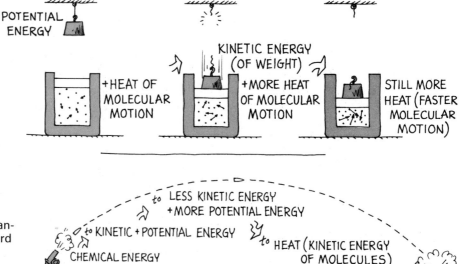

Fig. 6.17 Energy transitions. The graveyard of kinetic energy is heat.

Comparison of Kinetic Energy and Momentum†   Kinetic energy and momentum are both quantities having to do with the motion of an object. But they are different. For example, the momentum of one body in collision with another may add or subtract. Like velocity, momentum is a vector quantity and is directional and capable of being canceled entirely. But kinetic energy is a **scalar quantity**, like mass, and can never be canceled. The momenta of two cars just before a head-on

*****Answer**   20,000 m, or 20 km (force × distance = energy content of fuel; 2000 N × 20,000 m = 40 × 10⁶ J = 40 MJ) in the ideal case where the entire available energy content of the fuel is converted to mechanical energy. The point is that even with a perfect engine, there is an upper limit of fuel economy dictated by the conservation of energy.

†This section can be skipped in light treatments of mechanics.

collision may cancel to zero, and the combined wreck after collision will have the same zero value for momentum; but the kinetic energies add, as evidenced by the deformation and heat after collision. Or the momenta of two firecrackers approaching each other may cancel, but when they explode, there is no way their energies can cancel. Energies transform to other forms; momenta do not. The vector quantity momentum is different from the scalar quantity kinetic energy.

Another difference is the velocity dependence of the two. Whereas momentum is proportional to velocity (mv), kinetic energy is proportional to the square of velocity ($\frac{1}{2}mv^2$). An object that moves with twice the velocity of another object of the same mass has twice the momentum but four times the kinetic energy. It can provide twice the impulse to a body it encounters but do four times as much work.

Consider a lead bullet fired into a very large block of wood (Figure 6.18). When the bullet strikes, the impulse tends to knock the block over as the momentum of the bullet is transferred to the block. If the same bullet strikes with twice the velocity, it has twice the capability of knocking the block over; that is, if the first bullet is just barely able to knock the block over, the twice-as-fast bullet will deliver twice the impulse and be able to knock over a geometrically similar yet twice-as-massive block. But the twice-as-fast bullet has four times the kinetic energy and will penetrate four times as far. It delivers twice the wallop but four times the damage.

If our bullet is made of rubber instead of lead so that it bounces from the block instead of penetrating, it will be even more effective than a lead bullet in knocking the block over. Let's see why. If the momenta are the same for the lead and rubber bullets, the change in momenta and the accompanying impulses in being brought to rest are the same; but the story does not end there for the rubber bullet. It bounces, which means its change in momentum and accompanying impulse is even greater, for it is not only brought to rest but also "thrown back out again." If it bounces elastically with no loss in speed, then the change in momentum and impulse is doubled. It hits with twice as much wallop as a penetrating bullet of the same momentum. But it does not penetrate and, in this sense, it does no damage. This is an idealized case. In practice, such collisions are less than perfectly elastic. But now you have a better idea of why rubber bullets are used by police in riots when knockdowns with minimum injury are desired.

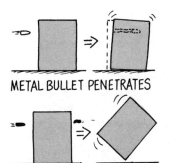

METAL BULLET PENETRATES

RUBBER BULLET BOUNCES

Fig. 6.18 Compared to the metal bullet with the same momentum, the rubber bullet is more effective in knocking the block over because it bounces on impact. The rubber bullet therefore undergoes the greater change in momentum and thereby imparts the greater impulse or wallop to the block. Which bullet does more damage?

Question   Suppose a hunter is confronted with a charging bear. Which will be more effective in knocking the bear down—a rubber or lead bullet of the same momentum?*

---

*__Answer__   A hunter will more effectively knock a bear down with a rubber bullet, but he'd better be good at running when the bear gets up.

Suppose you're carrying a football and are about to be slammed into and tackled by an approaching player whose oncoming momentum is equal but opposite to your own. The combined momenta of you and the approaching player before impact cancel to zero, and, sure enough, on impact you both stop short in your tracks. The wallop or jarring exerted on each of you is the same. This is the case whether you are tackled by a slow-moving, heavy player or a fast-moving, light player. If the product of his mass and his velocity matches yours, you're stopped short. Stopping power is one thing, but what about damage? Every football player knows it

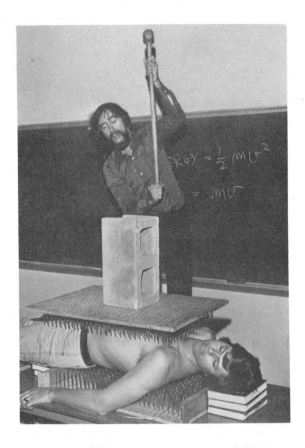

Fig. 6.19 The author puts kinetic energy and momentum into the hammer, which strikes the block that rests on fellow physics instructor Paul Robinson, who is bravely sandwiched between beds of nails. Paul is not harmed. Why? Except for the flying cement fragments, every bit of the momentum of the hammer at impact is imparted to Paul, and subsequently to the table and the earth that supports him. But the momentum only provides the wallop; the energy does the damage. Most of the kinetic energy never gets to him, for it goes into smashing the block apart and into heat. What energy remains is distributed over the more than 200 nails that make contact with his body. The driving force per nail is not enough to puncture the skin.

hurts more to be stopped by a fast-moving, light player than by a slow-moving, heavy player. Why? Because a light player moving with the same momentum has more kinetic energy. If he has the same momentum as a heavy player but is only half as massive, he has twice the velocity. Twice the velocity doesn't give him four times the energy because his mass is half, but it gives him twice the kinetic energy of the heavier player.[5] He does twice the work on you, tends to penetrate twice as far into you, and generally does twice the damage to you. Watch out for the fast-moving, little guys!

Energy for Life   Every plant, animal, and living cell is a living machine and, like any other machine, needs an energy supply. The fuels for living organisms on this planet are various hydrocarbon compounds that react with oxygen to release energy. Like the petroleum fuel previously discussed, the chemical bonds between hydrogen, carbon, and oxygen in foods store more energy than ends up in the bonds of resulting carbon dioxide and water products after digestion. The energy difference sustains life.

The principal difference between the combustion of food in digestion and the combustion of fossil fuels in mechanical engines is the rate at which the reactions take place. In digestion the reaction is much slower, and energy is released as it is required. Like the burning of fossil fuels, the reaction is self-sustaining once started as carbon combines with oxygen to form carbon dioxide. The reverse process is more difficult. Only green plants can make carbon dioxide combine with water to produce hydrocarbon compounds such as sugar. This process is *photosynthesis* and is energized by sunlight. Sugar is the simplest food, and all others—carbohydrates, proteins, and fats—are also synthesized compounds of carbon, hydrogen, and oxygen. How very fortunate that more energy is locked in the bonds of these hydrocarbons than in their products after combustion!

We see efficiency at work in the food chain. Larger creatures feed on smaller creatures, who in turn eat smaller creatures, and so on down the line to land plants and ocean plankton that are nourished by the sun. Moving each step up the food chain involves inefficiency. In the African bush, 10 kilograms of grass may produce 1 kilogram of gazelle. However, it will take 10 kilograms of gazelle to sustain 1 kilogram of lion. We see that each energy transformation along the food chain contributes to overall inefficiency. Interestingly enough, some of the largest creatures on the planet, the elephant and the blue whale, eat far down on the food chain. And more and more humans are considering substances such as krill or yeast as efficient sources of nourishment.

---

[5]Note that $\frac{1}{2}(m/2)(2v)^2 = mv^2$, which is twice the value $\frac{1}{2}mv^2$ of the heavier player of mass m and speed v.

Summary of Terms

Work   The product of the force exerted and the distance through which the force moves:

$$W = Fd$$

Power   The time rate of work:

$$\text{Power} = \frac{\text{work}}{\text{time}}$$

Energy   The property of a system that enables it to do work.

Potential energy   The stored energy that a body possesses because of its position with respect to other bodies.

Kinetic energy   Energy of motion, described by the relationship

$$\text{Kinetic energy} = \tfrac{1}{2}mv^2$$

Conservation of energy   Energy cannot be created or destroyed; it may be transformed from one form into another, but the total amount of energy never changes.

Conservation of energy and machines   The work output of any machine cannot exceed the work input. In an ideal machine, where no energy is transformed into heat,

$$\text{Work input} = \text{work output}$$
$$(Fd)_{\text{input}} = (Fd)_{\text{output}}$$

Efficiency   The percent of the work put into a machine that is converted into useful work output.

Review Questions

1. A force sets a body in motion. When the force is multiplied by the time of its application, we call the quantity *impulse*, which changes the *momentum* of that body. What do we call the quantity *force × distance*, and what quantity does this change?

2. Cite an example where a force is exerted on a body without doing any work on the body.

3. Which requires more work—lifing a 50-kilogram sack a vertical distance of 2 meters or lifting a 25-kilogram sack a vertical distance of 4 meters?

4. If both sacks in the preceding question are lifted their respective distances in the same time, how does the power required for each compare? How about for the case where the lighter sack is moved its distance in half the time?

5. What will be the kinetic energy of an arrow shot from a bow having a potential energy of 40 joules?

6. An apple hanging from a limb has gravitational potential energy. If the apple falls, what becomes of this energy just before it hits the ground? After it hits the ground?

7. If the speed of an object is doubled, by how much does its kinetic energy increase?

8. Which has the greater kinetic energy—a car traveling at 30 kilometers per hour or a half-as-heavy car traveling at 60 kilometers per hour?

9. Gravity changes the speed of a projectile all along its trajectory, but if the projectile is in circular orbit, its speed is not changed by gravity. Why is this so?

10. Does the escape velocity for a rocket leaving the planet Earth depend on the mass of the rocket?

11. Compared to work input, how much work is put out by a machine that operates at 30 percent efficiency?

12. The energy we require for existence comes from the chemically stored potential energy in food, which is transformed into other forms by the process of digestion. What happens to a person whose work output is less than the energy he consumes? Whose work output is greater than the energy he consumes? Can an undernourished person perform extra work without extra food? Briefly discuss.

Home Projects

1. Fill two mixing bowls with water from the cold tap and take their temperatures. Then run an electric or hand beater in the first bowl for a few minutes. Compare the temperatures of the water in the two bowls.

2. Pour some dry sand into a tin can with a cover. Compare the temperature of the sand before and after vigorously shaking the can for a couple of minutes.

Exercises

1. Why is it easier to stop a lightly loaded truck than a heavier one that has equal speed?

2. Which requires more work to stop—a light or a heavy truck moving with the same momentum?

3. The design of road dividers that separate opposing traffic lanes used to consist of metal rails mounted as shown on the left. The newer design is of concrete with a cross section very deliberately shaped as shown at the right. Why is this new design more effective in slowing down a car that sideswipes it?

4. To determine the potential energy of Tenny's drawn bow (Figure 6.3), would it be an underestimate or an overestimate to multiply the force with which she holds the arrow in its drawn position by the distance she pulled it? Why do we say the work done is the *average* force × distance?

5. When a rifle with a long barrel is fired, the force of expanding gases acts on the bullet for a longer distance. What effect does this have on the velocity of the emerging bullet? (Do you see why long-range cannons have such long barrels?)

6. You and a flight attendant toss a ball back and forth in an airplane in flight. Does the KE of the ball depend on the speed of the airplane? Carefully explain.

7. Can a body have energy without having momentum? Explain. Can a body have momentum without having energy? Defend your answer.

8. At what point in its motion is the KE of a pendulum bob a maximum? At what point is its PE a maximum? When its KE is half its maximum value, how much PE does it have?

9. This question is typical on some driver's license exams: A car moving at 50 kilometers per hour skids 15 meters with locked brakes. How far will the car skid with locked brakes at 150 kilometers per hour?

10. A physics instructor demonstrates energy conservation by releasing a heavy pendulum bob, as shown in the sketch, and allowing it to swing to and fro. What would happen if in his exuberance he gave the bob a slight shove as it left his nose? Why?

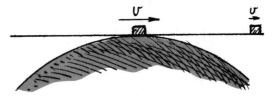

11. Discuss the design of the roller coaster shown in the sketch in terms of the conservation of energy.

12. Suppose that you and two classmates are discussing the design of a roller coaster. One classmate says that each summit must be lower than the previous one. Your other classmate says this is nonsense, for as long as the first one is the highest, it doesn't matter what height the others are. What do you say?

13. A satellite orbits the earth. The point in orbit farthest from the earth is the *apogee*, and the point nearest the earth is the *perigee*. With respect to the earth, at which of these points does the satellite have the greater gravitational PE? The greater KE? The greater total energy? Explain.

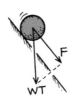

14. In the sketch on the left, a ball gains KE when rolling down a hill because work is done by the component of weight (*F*) that acts in the direction of motion. Sketch in the similar component of gravitational force that does work to change the KE of the satellite on the right.

15. A rocket coasts in an elliptical orbit around the earth. To attain the greatest amount of KE for escape using the least amount of fuel, should it fire its engines at the apogee or at the perigee? (*Hint:* Let the formula $Fd = \Delta KE$ be your guide to thinking. Suppose the thrust F is brief and the same in either case. Then consider the distance d the rocket would travel during this brief burst at the apogee and at the perigee.)

16. Suppose that a body is set sliding with a speed less than escape velocity on an infinite frictionless plane in contact with the surface of the earth as shown. Describe its motion. (Will it slide forever at constant velocity? Will it slide to a stop? What will it do?)

17. Does a car burn more gasoline when its lights are turned on? Does your answer depend on whether or not the engine is running? Defend your answer.

18. Matter can be recycled without loss of substance. Can energy be similarly recycled? Explain.

19. An inefficient machine is said to "waste energy." Does this mean that energy is actually lost? Explain.

20. You tell your friend that no machine can possibly put out more energy than is put into it, and your friend states that a nuclear reactor puts out more energy than is put into it. What do you say?

21. Your monthly electric bill is likely expressed in kilowatt-hours, a unit of energy delivered by the flow of 1 kilowatt of electricity for 1 hour. How many joules of energy do you get when you buy 1 kilowatt-hour?

22. No other concept in physics has the economical significance that energy does. Do you think it a sensible idea that paper money be "energy certificates" rather than certificates of precious metals reserves?

23. This may seem like an easy question for a physics type to answer: With what force does a rock that weighs 10 newtons strike the ground if dropped from a rest position 10 meters high? This question does not have a straightforward numerical answer. Why?

24. Your friend is confused about ideas discussed in Chapter 4 that seem to contradict ideas discussed in this chapter. For example, in Chapter 4 we learned that the net force is zero for a car traveling along a level road at constant velocity, and in this chapter we learned that work is done in such a case. "How can work be done when the net force equals zero?" asks you friend. What is your explanation?

25. In the absence of air resistance, a ball thrown vertically upward with a certain initial KE will return to its original level with the same KE. When air resistance *is* a factor affecting the ball, will it return to its original level with the same, less, or more KE? Does your answer contradict the conservation of energy? Defend your answer.

26. You're on a rooftop and throw one ball downward to the ground below and another upward. The second ball, after rising, falls and also strikes the ground below. If your downward and upward initial speeds are the same, how do the speeds of the balls compare upon striking the ground? (Use the idea of energy conservation to arrive at your answer.)

27. Scissors for cutting paper have long blades and short handles, whereas metal-cutting shears have long handles and short blades. Bolt cutters have very long handles and very short blades. Why is this so?

28. In the hydraulic machine shown, it is observed that when the small piston is pushed down 10 centimeters, the large piston is raised 1 centimeter. If the small piston is pushed down with a force of 100 newtons, how much force is the large piston capable of exerting?

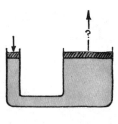

29. Discuss the fate of the physics instructor sandwiched between the beds of nails (Figure 6.19) if the block were less massive, unbreakable, and the beds contained fewer nails.

30. A golf ball is hurled at and bounces from a very massive bowling ball that is initially at rest. After collision, which of the two balls has the greater momentum? The greater kinetic energy?

31. Consider the swinging balls in the photo. If Paul releases two balls, momentum is conserved and two balls pop out the other side with the same momentum as the released balls at impact. But momentum would also be conserved if one ball popped out at twice the speed. Can you explain why this never happens? (And why this exercise is in Chapter 6 rather than Chapter 5?)

32. Consider the inelastic collision between the freight cars in the last chapter (Figure 5.11). The momentum before and after the collision is the same. The KE, however, is less after the collision than before the collision. How much less, and what becomes of this energy?

33. How many kilometers per liter will a car obtain if its engine is 25 percent efficient and it encounters an average that the energy content of gasoline is 40 megajoules per liter.

34. What is the efficiency of a pulley system that will raise a 1000-newton load a vertical distance of 1 meter when 3000 joules of effort are involved?

35. What is the efficiency of the body when a cyclist expends 1000 watts of power to deliver mechanical energy to her bicycle at the rate of 100 watts?

36. Lovemaking consumes from about 500 to 1500 kilojoules of energy, depending on fervor. How much mass would a very fervid three-rounds-per-week person lose in a year if no additional food were eaten to match his or her sexual appetite? (Assume that the energy content of 1 gram of fat is 40 kilojoules.)

7 Rotational Motion

Which moves with the greater speed on a merry-go-round—a horse near the outside rail or a horse near the inside rail? Which part of a phonograph record passes faster under the stylus—a piece at the beginning of the record or one near the end? Ask different people these questions, and you'll get different answers. That's because it's easy to get translational speed (as was discussed in Chapter 2) confused with rotational speed. All parts of the merry-go-round and the phonograph turntable have the same rotational speed, measured in revolutions per minute (RPMs), but all parts do not have the same translational speed, which is measured in meters per second. The farther away something is from the axis of rotation, the greater is its translational speed. When a row of people locked arm in arm at the skating rink makes a turn, the motion of "tail-end Charlie" is evidence of this greater speed. We distinguish between rotational speed and translational speed.

In Chapter 4 we treated Newton's laws of motion and applied the ideas of inertia, force, and acceleration to cases involving translational motion—that is, to motion of bodies as a whole without regard to any rotation they might undergo. In this chapter we extend these ideas to rotational motion, to the rotation of a body about some axis, whether inside or outside the body itself. A body can rotate about more than one axis at a time. When the axis is inside the rotating body, it is common to say the body rotates; and when the axis is outside, it is common to say the body revolves. For example, the earth rotates daily about an axis through its center, while it simultaneously revolves yearly about another axis in the sun, which in turn revolves about the center of the Milky Way.

Rotational Inertia

Just as an object at rest tends to stay at rest and an object in motion tends to remain moving in a straight line, *an object rotating about an axis tends to remain rotating about the same axis unless interfered with by some external influence.* (We shall see shortly that this external influence is properly called a *torque*.) The property of a body to resist changes in its rotational state of motion is called **rotational inertia**.[1] Bodies that are rotating tend to remain rotating, while nonrotating bodies tend to remain nonrotating. In the absence of outside influences, a rotating top keeps rotating, while a top at rest stays at rest.

Like inertia in the translational case, the rotational inertia of a body also depends on the mass of the body. But unlike the case for translational motion, rotational inertia depends on the distribution of the mass with respect to the axis of rotation. The greater the distance between the bulk of

---

[1]Often called *moment of inertia*.

Fig. 7.1 Rotational inertia depends on the distribution of mass with respect to the axis of rotation.

EASY TO ROTATE   DIFFICULT TO ROTATE

an object's mass and its axis of rotation, the greater the rotational inertia. Industrial flywheels, for example, are constructed so that most of their mass is concentrated along the rim. Once rotating, they have a greater tendency to remain rotating than they would if their mass were concentrated nearer the axis of rotation. It follows that the greater the rotational inertia of an object, the harder it is to change the rotational state of that object. If rotating, it is difficult to stop; if at rest, it is difficult to rotate. This fact is employed by a circus tightrope walker when he carries a long pole to aid balancing. Much of the mass of the pole is far from the axis of rotation, its midpoint. The pole therefore has considerable rotational inertia. If the tightrope walker starts to topple over, a tight grip on the pole would require that the pole rotate. But the rotational inertia of the pole resists this, giving the tightrope walker a sufficient margin of time to readjust his balance. The longer the pole, the better. And if our tightrope walker has no pole, he can at least increase the rotational inertia of his body by extending his arms full length.

ROTATIONAL INERTIA - THAT'S THE TICKET !

Fig. 7.2 The tendency of the pole to resist rotation aids the acrobat.

Fig. 7.3 Short legs have less rotational inertia than long legs. An animal with short legs has a quicker stride than one with long legs, just as a short pendulum swings to and fro faster than a long one.

Fig. 7.4 You bend your legs when you run to reduce rotational inertia.

Fig. 7.5 A solid cylinder rolls down an incline faster than a ring, whether or not they have the same mass or outer diameter. A ring has greater rotational inertia compared to its mass than a cylinder does.

A long pendulum has a greater rotational inertia than a short pendulum and therefore swings back and forth more slowly than a short one. When we walk, we allow our legs to swing with the help of gravity, that is, at a pendulum rate. Just as a long pendulum takes a long time to swing to and fro, a person having long legs tends to walk with slower strides than a person having short legs. This is also evident in animals; giraffes, horses, and ostriches run with a slower gait than dachshunds, mice, and bugs.

Ever tried running with your legs straight? It's difficult. When running, you bend your legs to reduce their rotational inertia so you can rotate them back and forth more quickly. Bet you never looked at it that way!

Because of rotational inertia, a solid cylinder starting from rest will roll down an incline faster than a ring or hoop. The ring has its mass concentrated farthest from its axis of rotation and therefore is harder to start rolling. (That is, a ring has a greater tendency to resist a change in its state of rotation.) Any solid cylinder will beat any ring on the same incline. This doesn't seem plausible at first thought, but remember that any two objects, regardless of mass, will fall together when dropped. They will also slide together when released on an inclined plane. When rotation is introduced, the object with the larger rotational inertia *compared to its own weight* has the greater resistance to a change in its motion. Hence, any disk will roll faster than any ring on the same incline. Try it and see!

Torque

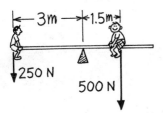

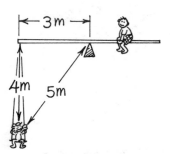

Fig. 7.6 No rotation is produced when the torques balance each other.

Fig. 7.7 The lever arm is still 3 m.

A **torque** (rhymes with *fork*) is the rotational counterpart of force. Force tends to change the motion of things; torque tends to twist or change the state of rotation of things. If you want to make a body move, apply a force. If you want to make a body spin, apply a torque. Torque differs from force just as rotational inertia differs from translational inertia: both involve distance from the axis of rotation. In the case of torque, this distance is called the *lever arm*. We define *torque* as the product of this lever arm and the force that tends to produce rotation about this particular distance:

$$\text{Torque} = \text{lever arm} \times \text{force}$$

Torques are intuitively familiar to youngsters playing on a seesaw. Kids can balance a seesaw even when their weights are unequal. Weight alone doesn't produce rotation. Torque does, and a child soon learns that the distance he sits from the pivot point is every bit as important as weight (Figure 7.6). The torque produced by the boy on the right tends to produce clockwise rotation, while torque produced by the boy on the left tends to produce counterclockwise rotation. If the torques are equal, making the net torque zero, no rotation is produced.

Suppose that the seesaw is arranged so that the half-as-heavy boy is suspended from a 4-meter rope hanging from his end of the seesaw (Figure 7.7). He is now 5 meters from the fulcrum, and the seesaw is still balanced. We see that the lever-arm distance is still 3 meters and not 5 meters. *The lever arm about any axis of rotation is the perpendicular distance from the axis to the line along which the force acts.* This will always be the shortest distance between the axis of rotation and the line along which the force acts.

This is why the stubborn bolt shown in Figure 7.8 is more likely to turn when the applied force is perpendicular to the handle, rather than at an oblique angle as shown in the first figure. In the first figure the lever arm is shown by the dotted line and is less than the length of the wrench handle. In the second figure the lever arm is equal to the length of the wrench handle. In the third figure the lever arm is extended with a pipe to produce a greater torque.

Fig. 7.8 Although the magnitudes of the force are the same in each case, the torques are different.

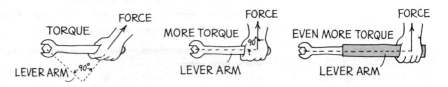

Question

If a pipe effectively extends a wrench handle to three times its length, by how much will the torque increase for the same applied force?*

---

*__Answer__ Three times.

Center of Mass and Center of Gravity

Throw a baseball into the air, and it will follow a smooth parabolic trajectory. Throw a baseball bat spinning into the air, and its path is not smooth; its motion is wobbly, and it seems to wobble all over the place. But, in fact, it wobbles about a very special place, a point called the **center of mass**.

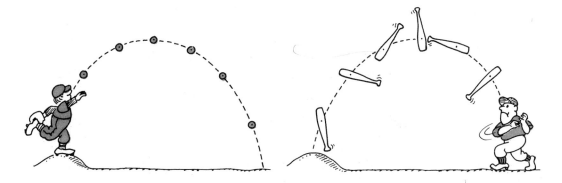

Fig. 7.9 The center of gravity of the baseball and that of the bat follow parabolic trajectories.

For a given body, the center of mass is the average position of all the particles of mass that constitute the body. For example, a symmetrical object such as a ball can be thought of as having all its mass concentrated at its center; by contrast, an irregularly shaped body such as a baseball bat has more of its mass toward one end. The center of mass of a baseball bat therefore is toward the heavier end. A cone has its center of mass exactly one-fourth of the way up from its base.

Center of gravity is a term more popularly used to express center of mass. The center of gravity is simply the average position of weight distribution. Since weight and mass are proportional to each other, center of gravity and center of mass refer to the same point of a body.[2] The physicist prefers to use the more general term *center of mass*, for a body has a center of mass whether or not it is under the influence of gravity. However, we shall use either term to express this concept, and favor the term *center of gravity* when consideration of weight is involved.

---

[2]For very large bodies such as the moon, where the farthest part is in a region of weaker gravitation than the nearest part, the center of gravity is not located at the center of mass.

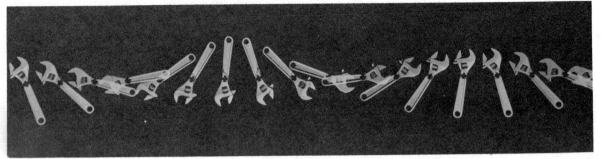

Fig. 7.10 The center of mass of the spinning wrench follows a straight-line path.

The multiple-flash photograph (Figure 7.10) shows a top view of a wrench sliding across a smooth horizontal surface. Note that its center of gravity, indicated by the dark mark, follows a straight-line path, while other parts of the wrench wobble as they move across the surface. Since there is no external force acting on the wrench, its center of gravity moves equal distances in equal time intervals. The motion of the spinning wrench is the combination of the translational motion of its center of gravity and the rotational motion about its center of gravity.

Locating the Center of Gravity

The center of gravity of a uniform object such as a meter stick is at its midpoint, for the stick acts as though its entire weight were concentrated there. Supporting that single point supports the whole stick. Balancing an object provides a simple method of locating its center of gravity. In Figure 7.11 the many small arrows represent the pull of gravity all along the meter stick. All these can be combined into a resultant force acting through the center of gravity. The entire weight of the stick may be thought of as acting as this single point. Hence, we can balance the stick by applying a single upward force in a direction that passes through this point.

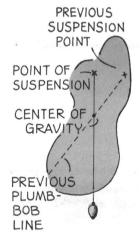

Fig. 7.12 Finding the center of gravity for an irregularly shaped object.

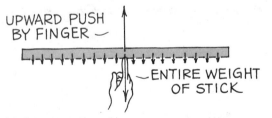

Fig. 7.11 The weight of the entire stick behaves as if it were concentrated at its center.

The center of gravity of any freely suspended object lies directly beneath (or at) the point of suspension (Figure 7.12). If a vertical line is drawn through the point of suspension, the center of gravity lies somewhere along that line. To determine exactly where it lies along the line, we have only to suspend the object from some other point and draw a second vertical line through that point of suspension. The center of gravity lies where the two lines intersect.

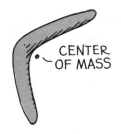

Fig. 7.13 The center of mass can be outside the mass of a body.

The center of mass of an object may be a point where no mass exists. For example, the center of mass in a ring or a hollow sphere is at the geometrical center where no matter exists. Similarly, the center of mass of a boomerang is outside the physical structure, not within the material making up the boomerang (Figure 7.13).

Fig. 7.14 Dwight Stones executes a "Fosbury flop" to clear the bar while his center of gravity passes beneath the bar.

Stability

The location of the center of gravity is important for stability (Figure 7.15). If we drop a line straight down from the center of gravity of a body of any shape and it falls inside the base of the body, it is in stable *equilibrium;* it will balance. If it falls outside the base, it is unstable. Why doesn't the

Fig. 7.15 The center of mass of the L-shaped object is located where no mass exists. In (a), the center of mass is above a point of support, so the object is stable. In (b), it is not above a point of support, so the object is unstable and will topple over.

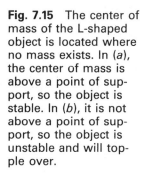
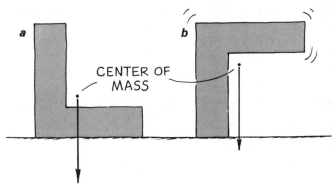

famous Leaning Tower of Pisa topple over? As we can see in Figure 7.16, a line from the center of gravity of the tower falls inside its base, so the Leaning Tower has stood for several centuries.

CENTER OF
GRAVITY

Fig. 7.16 The center of gravity of the Leaning Tower of Pisa lies above a point of support, and therefore the tower is in stable equilibrium.

To reduce the likelihood of tipping, it is usually advisable to design objects with a wide base and low center of gravity. The wider the base, the higher the center of gravity must be raised before the object will tip over.

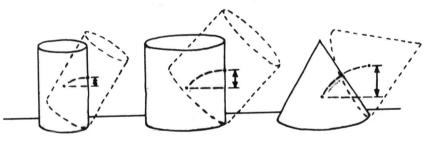

Fig. 7.17 The vertical distance that the center of gravity is raised in tipping determines stability. An object with a wide base and low center of gravity is most stable.

Fig. 7.18 When you stand, your center of gravity is somewhere above the area bounded by your feet. Why do you keep your legs far apart when you have to stand in the aisle of a bumpy-riding bus?

When you stand erect (or lie flat), your center of gravity is within your body. Why is the center of gravity lower in an average woman than in an average man of the same height? Is your center of gravity always at the same point in your body? Is it always inside your body? What happens to it when you bend over?

If you are in reasonable shape, you can bend over and touch your toes without bending your knees—providing you are not standing with your back against a wall. Ordinarily, when you bend over and touch your toes, you extend your lower extremity as shown in Figure 7.19, such that your center of gravity is above a point of support, your feet. If you attempt to do this when standing against a wall, however, you cannot counterbalance yourself, and your center of gravity soon protrudes beyond your feet as at the right in Figure 7.19. You are off-balance and you rotate.

Fig. 7.20 To maintain balance, Sylvia must keep her center of gravity directly above the small area bounded by her left foot.

Fig. 7.19 You can lean over and touch your toes without falling over only if your center of gravity is above the area bounded by your feet.

You rotate because of an unbalanced torque. This is evident in the two L-shaped objects shown in Figure 7.21. Both are unstable and will topple unless otherwise fastened to the level surface. It is easy to see that even if both shapes have the same weight, the one on the right is less stable. This is because of the greater lever arm and, hence, greater torque.

Fig. 7.22 A pigeon jerks its head and neck back and forth with each step and thereby keeps its center of gravity always above the supporting foot.

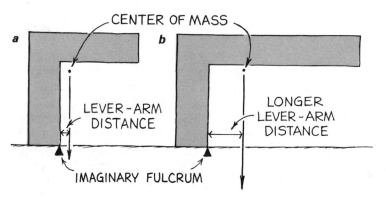

Fig. 7.21 The greater torque acts on the figure in (b) for two reasons. What are they?

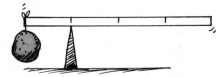

Question

A uniform meter stick supported at the 25-cm mark balances when a 1-kg rock is suspended at the 0-cm end. What is the mass of the meter stick?*

Centripetal and Centrifugal Force

Fig. 7.23 The force exerted on the whirling can is toward the center.

Any force that causes an object to follow a circular path is called a **centripetal force**. *Centripetal* means "center-seeking," or "toward the center." The force of a rotating carnival centrifuge on an occupant inside is a center-directed force; if it ceased to act, the occupant would no longer be held in a circular path.

If we whirl a tin can on the end of a string, we find that we must keep pulling on the string (Figure 7.23). The string transmits the centripetal force, which pulls the can into a circular path. Gravitational and electrical forces can be transmitted across empty space to produce centripetal forces. The moon, for example, is held in an almost circular orbit by gravitational force directed toward the center of the earth. The orbiting electrons in atoms experience an electrical force toward the central nuclei.

Centripetal force is *not* a new kind of force but is simply the name given to *any* force, whether string tension, gravitational, electrical, or whatever, that is directed at right angles to the path of the moving body and produces circular motion.

Fig. 7.24 (*a*) When a car goes around a curve, there must be a force pushing the car toward the center of the curve. (*b*) A car skids on a curve when the centripetal force (friction of road on tires) is not great enough.

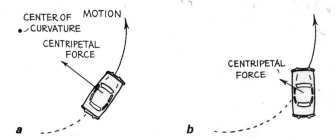

When an automobile rounds a corner, the friction between the tires and the road is the centripetal force that holds the car in a curved path (Figure 7.24). If this friction is not great enough, the car fails to make the curve and the tires slide sideways; we say the car skids.

Centripetal force plays the main role in the operation of a centrifuge. A familiar example is the spinning tub in an automatic washer (Figure 7.25). In its spin cycle, the tub rotates at high speed and produces a centripetal

Fig. 7.25 The clothes are forced into a circular path, but the water is not.

****Answer**   1 kg. Why? The system is in equilibrium, so any torques must be balanced: the torque that results from the weight of the rock is balanced by the equal but oppositely directed torque that results from the weight of the stick at its center of gravity, the 50-cm mark. The support force at the 25-cm mark is applied midway between the rock and the center of gravity of the stick, so the lever arms about the support point are equal (25 cm). This means that the weights and hence the masses of the rock and stick must also be equal.

force on the wet clothes, which are forced into a circular path against the inner wall of the tub. The tub exerts great force on the clothes, but the holes in the tub prevent the exertion of the same force on the water in the clothes—and the water escapes. Strictly speaking, the clothes are forced away from the water; the water is not forced away from the clothes. Think about that.

Fig. 7.26 The centripetal force (adhesion of mud on the spinning tire) is not great enough to hold the mud on the tire, so it spins off in straight-line directions.

In these examples, we have described circular motion as a center-directed force. Sometimes an outward force is attributed to circular motion. This outward force is called a **centrifugal force**.[3] *Centrifugal* means "center-fleeing," or "away from the center." In our first example, we considered the whirling tin can; it is a common misconception to state that a centrifugal force pulls outward on the can. If the string holding the

Fig. 7.27 Large centripetal forces on the wings of an aircraft enable it to fly in circular loops. The acceleration away from the straight-line path that the aircraft would follow in the absence of centripetal force is often several times greater than the acceleration due to gravity, *g*. For example, if the centripetal acceleration is 98 m/s² (ten times as great as 9.8 m/s²), we say the aircraft is undergoing 10 *g*'s. The seat presses against the pilot with a force ten times greater than his weight. Typical fighter aircraft are designed to withstand accelerations on the order of 8 or 9 *g*'s. The pilot as well as the aircraft must withstand these accelerations. When an aircraft is performing an 8-*g* turn, for example, the force required to hold the blood in the pilot's brain must be eight times the ordinary force provided by the heart. Without it, blood may rush out of the pilot's brain and cause a "blackout."

---

[3]Both centrifugal and centripetal forces depend on the mass (*m*), tangential speed (*v*), and radius (*r*) of curvature of the circularly moving body. The exact relationship is $F = mv^2/r$. Angular speed is usually symbolized by ω, and is measured in radians per second (where 0.1 rad/s = 1 RPM). Using this unit, the relationship is $F = mr\omega^2$.

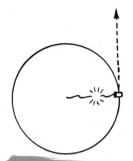

Fig. 7.28 When the string breaks, the whirling can moves in a straight line, tangent to its circular path.

whirling can breaks (Figure 7.28), it is commonly stated that a centrifugal force pulls the can from its circular path. But the fact is that when the string breaks, the can goes off in a straight-line, tangential path because *no* force acts on it. The idea may be confusing and merits more explanation.

If you are riding in a car that suddenly stops short, you pitch forward against the dashboard. You wouldn't say that a force pushed you forward. You hit the dashboard because of the *absence* of a force, which seat belts would have provided. By the same token, if you are in a car that rounds a sharp corner, you pitch outward against the inside of the car door, not because of some outward or centrifugal force but because there is *no* centripetal force holding you in circular motion (which seat belts would have provided). In the absence of force, you tend to go in a straight line while the car door curves, crosses your straight-line path, and intercepts you. There is no centrifugal or centripetal force banging you against the car door, even though it is common to say this is so.

When we swing a tin can into a circular path by a string, no force pulls the can outward; that is, there is no centrifugal force acting on the can. There is only an inward force, the centripetal force—the inward pull of the string—acting on the can. By Newton's third law, we can say that the can pulls back on the string and that there is therefore an outward force acting

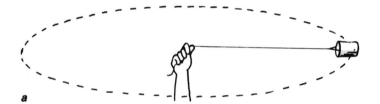

a

Fig. 7.29 (a) The only force that is exerted *on* the whirling can (neglecting gravity) is directed *toward* the center of circular motion. This is called *centripetal force*. No outward force is exerted on the can.

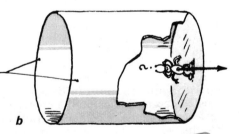

b

(b) If a bug is inside the whirling can, it is held in circular motion by the centripetal force provided by the bottom of the can against its feet. But the bug regards this as a response to a force directed *away from* the center of circular motion. It calls this outward force a *centrifugal force*, which is as real to it as gravity. We call it a *fictitious force* because it has no agent.

on the string but not on the can. No centrifugal force acts on the can. Now suppose there is a bug inside the whirling can (Figure 7.29). The pressing of the bottom of the can against its feet provides the centripetal force that holds it in a circular path and, neglecting gravity, is the only force acting on the bug. From our outside stationary frame of reference, there is no centrifugal force exerted on the bug, just as there was no centrifugal force exerted on the person who lurched against the car door. The "centrifugal-force effect" is attributed not to any real force but to inertia—the tendency of a moving body to follow a straight-line path. But try telling that to the bug!

Centrifugal Force in a Rotating Frame

Our **frame of reference**[4] very much influences our view of nature. From one frame of reference, a body may have a particular velocity, while from another frame, it does not. From one frame of reference, a body may have potential energy, while from another frame, it has none. And we have just seen from a nonrotating frame of reference that the force that holds a body in circular motion is not a centrifugal force but a centripetal force. In the case of our little bug, this was the force of the bottom of the can exerted on its feet. No other force was pressing on the bug—certainly not the reaction to the centripetal force—for this was exerted on the can (action and reaction never act on the same body).

But nature seen from the frame of reference of the rotating body is different. In the rotating frame, centrifugal force is not a reaction force. It is not the force exerted by the bug on the can but is a force *on the bug itself.* In the rotating frame, centrifugal force is not a reaction to centripetal force but is a force in its own right, like the pull of gravity. But there is a fundamental difference: gravitational force is an interaction between one mass and another. The gravity we experience is our interaction with the earth. But centrifugal force has no such agent—it has no interaction counterpart. For this reason, we rank it as a *fictitious force* and not a real force like gravity, electromagnetism, and the nuclear forces. Nevertheless, to observers in a rotating system, centrifugal force seems to be, and is interpreted to be, a very real force. Just as from the surface of the earth we find gravity ever present, so within a rotating system centrifugal force is ever present.

Fig. 7.30 The spin of the earth sets up a centrifugal force that slightly decreases our weight. Like the outside horse on the merry-go-round, we have the greatest tangential speed farthest from the earth's axis, at the equator. Centrifugal force is therefore maximum for us when we are at the equator and zero at the poles, where we have no tangential speed. So, strictly speaking, if you want to lose weight, walk toward the equator!

[4]A nonaccelerating frame of reference called an *inertial* frame of reference. Newton's laws are seen to hold exactly in an inertial frame.

Question   Why are oppositely directed centripetal and centrifugal forces exactly equal in magnitude?*

Humans today live on the outer surface of this spherical planet and are held there by gravity. The planet has been the cradle of humankind. But we do not stay in the cradle forever. We are becoming a spacefaring people. Many humans tomorrow will likely live not on the outside surfaces of planets, but on the inside surfaces of huge, lazily rotating colonies in space. They will be held to the inner surfaces by centrifugal force.

Simulated Gravity   People living in space colonies orbiting at moonlike distances would be in a continuous state of free fall toward the earth. But they needn't feel weightless. Consider some ants inside the inner tube of a bicycle wheel (Figure 7.31). If the wheel is tossed into the air, there will be no support force for the ants, and they will experience weightlessness. But what if the wheel is spinning? The ants will then feel themselves thrown to the far surface of the inner tube like the bug pressed to the rotating can in Figure 7.29. If the wheel spins at the proper rate, the ants will feel themselves

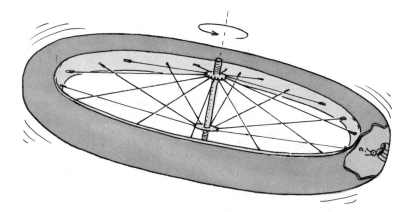

Fig. 7.31 Even if the wheel is in free fall, the ants inside will feel a force like gravity if the wheel spins. To them, the direction "up" is toward the spin axis.

*Answer   In a nonrotating frame of reference, centripetal and centrifugal forces comprise an action-reaction pair and are therefore equal in magnitude. In the case of a rock whirling at the end of a string, the string provides a centripetal force by pulling inward on the rock (action) while the rock pulls outward on the string (reaction). In this sense we could call the pull of the rock on the string a centrifugal force; however, the term *centrifugal force* is appropriate only in rotating frames. From the rotating frame, the centrifugal force is exerted on the rock itself, not by the string but by some fictitious agent. In this case there is no reaction force to the fictitious centrifugal force.

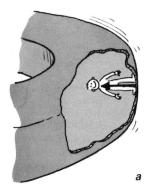

a

Fig. 7.32 (*a*) As seen from rest outside the rotating system, the floor presses against the man (action) and the man presses back on the floor (reaction). The force of the floor is the only force exerted on the man and is directed toward the axis— a centripetal force.

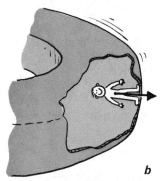

b

(*b*) As seen from inside the rotating system, in addition to the man-floor pair of forces, there is a centrifugal force that is exerted on the man at his center of gravity. It seems as real as gravity but, unlike gravitational force, it has neither an agent nor a reaction counterpart. It is therefore a fictitious force.

pressing against the inner tube with a force just equaling their weight. Even though the wheel is in a state of free fall, the spin effect neatly simulates gravity for the ants, who perceive the direction up as toward the spin axis and down as radially outward from the axis. It is the same for any life forms in a rotating habitat orbiting in space.

Any body undergoing uniform circular motion, whether a bug in a whirling can, ants in a spinning wheel, or inhabitants inside the rim of a rotating cylinder or rotating torus, undergoes acceleration at every moment. Although speed may be constant, the direction of motion changes continually, and the body undergoes centripetal acceleration directed toward the axis of rotation. Viewed from the rotating frame, the acceleration is centrifugal and is directed outward from the axis. In either case the magnitude of this acceleration is directly proportional to the product of the radial distance and the square of the angular speed.[5] So for a given speed in revolutions per minute, the acceleration varies only with the distance from the rotational axis. In a system that rotates at 5 RPM, for example, at a radial distance of 18 meters, the acceleration is 5 meters/second^2; and at twice this distance, 36 meters, the acceleration is 10 meters/second^2— approximately 1 *g*. So inhabitants in a cylinder or torus of radius 36 meters (72 meters in diameter) that rotates at 5 RPM would experience earth-normal gravity. People would stand erect, walk, run, or do push-ups on the inner surface of their environment just as would their counterparts below on the outer surface of the earth.

The designs for orbiting human habitats in space that simulate earth-normal gravity call for the appropriate relations of size and rotational rate. A space structure of small diameter would rotate at a relatively greater rate than would a structure of a large diameter in order to produce the same acceleration. The 72-meter–5-RPM structure we have just considered would produce the proper gravity, but it would make a poor design, for two reasons. First, the rate of rotation in itself is uncomfortable. Sensitive and delicate organs in our middle ears sense rotation, and although there appears to be no difficulty at a single revolution per minute or so, many people find difficulty in adjusting to rotational rates above 2 or 3 RPMs (although some easily adapt to 10 or so RPMs). Second, the high rotational rate and the small radial distances involved would cause unequal forces across the body. Standing erect, a person's head would be nearer the rotational axis than his feet. Consequently, the person's head would accelerate slightly less than his midsection, which would accelerate a bit less

---

[5]In shorthand notation, $a = r\omega^2$; that is, the centrifugal acceleration *a* in m/s^2 is equal to the radius *r* in meters multiplied by the square of the angular speed ω in units of radians per second, where 0.1 rad/s is approximately 1 RPM. So if we express ω in the more familiar RPMs, $a = r(0.1\omega)^2 = 0.01\, r\omega^2$.

than his feet. The person would literally be light-headed. The greater a person's height is, compared to the radial distance, the greater the internal stresses. If the radius of the habitat is very much greater than the height of its occupants, such stresses are negligible.

For the earliest habitats that simulate earth-normal gravity, overall economy appears to dictate a rotation of 1 RPM. This corresponds to a structure about 2 kilometers in diameter, an immense structure compared to today's space-shuttle vehicles. Present plans are for much smaller initial space stations that will not rotate at all. The inhabitants will adjust to living in a micro-*g* environment. Larger rotating habitats, such as the one depicted in Figure 7.33, may follow later.

The variety of fractions of *g* possible from the rim of a rotating habitat to the micro-*g* region at the rotational axis holds promise for a most different and (at this writing) yet unexperienced environment. We would

Fig. 7.33 Artist's rendering of the interior of a space colony that would be occupied by a few thousand people.

enjoy ballet at $\frac{1}{4}$ g; swimming, diving, and acrobatics at $\frac{1}{10}$ g; sports as we know them, three-dimensional soccer, and new sports not yet conceived in low-g states; and, quite likely, right there along the axis in micro-g—Honeymoon Hotel.

Angular Momentum

Just as a mass moving in a straight line has momentum (usually called *translational*, or *linear*, momentum), a mass moving in a circular path has angular momentum. **Angular momentum** is a measure of the *rotational property of motion*. The planets orbiting the sun, a rock whirling at the end of a string, and the tiny electrons whirling about their central nuclei in atoms, all have angular momentum. In addition to mass m and speed v, angular momentum depends on the distance r between the mass and the axis about which it rotates. If this radial distance is large compared to the size of the rotating body, angular momentum is simply equal to the product of mass, speed, and radial distance.[6] In shorthand notation,

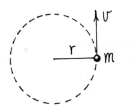

Fig. 7.34 A body of mass m whirling in a circular path of radius r with a speed v has angular momentum mvr.

$$\text{Angular momentum} = mvr$$

Just as changing the linear momentum of a body requires an external net force, changing the angular momentum of a body requires an external net torque. We will restate Newton's first law of inertia for rotating systems in terms of angular momentum.

A body or system of bodies will maintain its state of angular momentum unless acted upon by an unbalanced external torque.

The greater the angular momentum, the greater the torque required to change the angular momentum, either in magnitude or direction. We all know that it is difficult to balance on a bicycle that is at rest. The wheels have no angular momentum. If our center of gravity is not above a point of support, a slight torque is produced and we tumble over. When the bicycle is moving, however, the angular momentum of the wheels requires greater torque to produce any change in the magnitude or direction of spin. This makes it easier to balance on a moving bicycle.

Conservation of Angular Momentum

Just as momentum is conserved for bodies moving in a straight line, angular momentum is conserved for bodies in rotation. Providing that no unbalanced external torque acts on a rotating system, the angular momentum of that system will not change. In other words, the numerical quantity mvr at one time will equal the numerical quantity mvr at any time, with no change in the direction of the rotational axis.

---

[6]When the mass of a body is nonlocalized and nearer to the axis of rotation, as with a cylindrical grindwheel, for example, angular momentum is found from the more general definition: the product of a body's rotational inertia and its angular speed.

Figure 7.35 interestingly illustrates conservation of angular momentum. Suppose that a man with negligible rotational inertia stands on a turntable with weights in each hand. With arms fully extended horizontally, he is first set rotating slowly. When he draws the weights in toward his chest, thereby decreasing r, the speed v increases. If he decreases r by half, for example, the speed doubles. If he again extends his arms, the speed decreases to its original value. In practice, though, the rotational inertia of the man's body is far from negligible. He can obtain results almost as dramatic without the use of the heavy weights, for he decreases his own rotational inertia when he draws his arms in, which increases his angular speed. In either case, angular momentum is conserved. This experiment is best appreciated by the turning person who feels changes in speed by what seems to be a mysterious force. A figure skater feels this principle in action when she starts to whirl with her arms and perhaps one leg extended and then draws her arms and leg in to obtain a greater rotational speed. Whenever a rotating body contracts, its angular speed increases.

$$m v r = m \overline{v} r$$

Fig. 7.35 Conservation of angular momentum: when the radius of the whirling weights is decreased, the angular speed is correspondingly increased.

An astronaut taking a space walk from his orbiting capsule must take the conservation of angular momentum into account. If he keeps physical contact with his space capsule by means of a tether line, and if any rotational motion exists and he is drawn back to the ship by having the tether line pulled in, the decreasing r would be accompanied by an increasing v, and he would find himself whirling around the capsule at a

Fig. 7.36 Experiment with a falling cat.

dangerous rate. To correct this, he uses a gas-spurting jet gun, which he carries in his hand. Space walks with a jet pack that has no physical connection to the capsule avoid this problem, of course.

If a cat is held upside down and dropped (Figure 7.36), the cat has a constant angular momentum about its center of gravity. While falling, the cat rearranges its limbs and tail, thereby changing its rotational inertia. Repeated reorientations of the body configuration result in the head and tail rotating one way and the feet the other so that the feet are downward when the cat strikes the ground. Similarly, the body of an acrobat is given a certain angular momentum in the initial spring of a somersault. This angular momentum is conserved throughout the process as the acrobat controls angular velocity by variations in the body's rotational inertia (Figure 7.37).

Fig. 7.37 Rotational rate is controlled by variations in the body's rotational inertia as angular momentum is conserved during a forward somersault.

Ocean tides moving around the earth experience friction at the ocean bottom. As a result, the earth slows in its daily rotation just as an automobile's wheels do when brakes are applied. According to the law of conservation of angular momentum, the slowing down of the earth and the lessening of angular momentum are accompanied by an equal increase of the angular momentum of the moon in its orbital motion around the earth. This increase in the moon's angular momentum results in the moon's increasing distance from the earth and a decrease in its speed. This increase of distance amounts to one-quarter of a centimeter per rotation. Have you noticed that the moon is getting farther away from us lately? Well, it is; each time we see another full moon, it is one-quarter of a centimeter farther away!

Summary of Terms

Rotational inertia   That property of a body to resist any change in its state of rotation: if at rest, the body tends to remain at rest; if rotating, it tends to remain rotating and will continue to do so unless interrupted.

Torque   The product of force and lever-arm distance, which tends to produce rotation.

Center of mass   The average position of mass or the single point associated with a body where all its mass can be considered to be concentrated.

Center of gravity   The average position of weight or the single point associated with a body where the force of gravity can be considered to act.

Equilibrium   The state of a body when not acted upon by a net force or net torque. A body in equilibrium may be at rest or moving at uniform velocity; that is, it is not accelerating.

Centripetal force   A center-seeking force that causes an object to follow a circular path.

Centrifugal force   An outward force due to rotation. In an inertial frame of reference, it is fictitious in the sense that it doesn't act on the rotating body but on whatever supplied the centripetal force; it is the reaction to centripetal force. In a rotating frame of reference, it *does* act on the rotating body and is fictitious in the sense that it is not an interaction with an agent or entity such as mass or charge but is a force in itself that is solely a product of rotation; it has no reaction-force counterpart.

Frame of reference   A rigid framework relative to which positions and movements may be measured.

Inertial frame of reference   An unaccelerated frame in which Newton's laws hold exactly.

Angular momentum   A measure of an object's rotation about a particular axis. For an object that is small compared to the radial distance, it is the product of mass, speed, and radial distance of rotation.

Conservation of angular momentum   When no external torque acts on an object or a system of objects, no change of angular momentum takes place. Hence, the angular momentum before an event involving only internal torques is equal to the angular momentum after the event.

$$mvr_{\text{(before event)}} = mvr_{\text{(after event)}}$$

Suggested Reading

Brancazio, P. J. *Sport Science.* New York: Simon & Schuster, 1984.

Clarke, A. C. *Rendezvous with Rama.* New York: Harcourt Brace Jovanovich, 1973. This is the first science fiction novel to seriously consider habitation *inside* a spinning space facility.

NASA, *Space Settlements: A Design Study.* Washington, D.C.: U.S. Government Printing Office, 1977.

O'Neill, G. K. *The High Frontier.* New York: Morrow, 1976.

Review Questions

1. How does a long pole help a tightrope walker to balance?

2. Why do people with long legs generally walk with a slower stride than people with short legs?

3. Distinguish between a *force* and a *torque*.

4. Is the net torque changed when one of the persons on a seesaw stands or hangs instead of sits? (Does the weight of the person or the lever-arm distance change?)

5. Is the center of mass of a body necessarily located at a place where mass physically exists? Explain.

6. How far can a body be tipped before it topples over?

7. Why doesn't the Leaning Tower of Pisa topple?

8. Distinguish between *centripetal force* and *centrifugal force*. Which of Newton's laws tells us that these oppositely directed forces are equal in magnitude?

9. What kind of force is exerted on your clothes when in a spin cycle in a washer—centripetal or centrifugal force?

10. How can gravity be simulated in an orbiting space colony?

11. Why is it easier to balance on a bicycle when it is moving?

12. When an ice skater spins with arms outstretched and pulls them inward, does the speed at which she spins increase? Does her angular momentum increase?

Home Projects

1. Fasten a fork, spoon, and wooden match together as shown. The combination will balance nicely—on the edge of a glass, for example. This happens because the center of gravity actually "hangs" below the point of support.

2. Stand with your heels and back against a wall and try to bend over and touch your toes. You'll find you have to stand away from the wall to do so without toppling over. Compare the minimum distance of your heels from the wall with that of a friend of the opposite sex. Who can touch their toes with their heels nearer to the wall—women or men? On the average and in proportion to height, which sex has the lower center of gravity?

3. Impress your friends with this one. Tell them that their wishes are your commands if they can do the following: stand facing a wall with toes against the wall and simply stand unaided on tiptoes for a couple of seconds. When they find it can't be done, their wishes will not be your commands, but you should at least explain why they failed. Can you?

4. Rest a meter stick on two fingers as shown. Slowly bring your fingers together. At what part of the stick do your fingers meet? Can you explain why this always happens, no matter where you start your fingers?

5. Swing a pail of water around rapidly in a circle at arm's length and the water will not spill. Why?

6. Place the hook of a wire coat hanger over your finger. Carefully balance a coin on the straight wire on the bottom directly under the hook. You may have to flatten the wire with a hammer or fashion a tiny platform with tape. With practice you can swing the hanger and balanced coin back and forth, and then in a circle. Centripetal force holds the coin in place.

7. If you ride a motorcycle, note that the spinning wheels act as gyroscopes wherein a turn to the right is initiated by twisting the handlebars to the left! The heavier the wheel, the more pronounced the effect. It can also be noted on a lightweight bicycle if it moves at a high speed. The explanation is somewhat complex and may merit a visit to your instructor during his or her office hours.

Exercises

1. Which moves faster on a merry-go-round—a horse on the inner rail or one on the outer rail?

2. Does a phonograph needle ride faster or slower over the groove at the beginning or the end of the record? If fidelity increases with speed, what part of the record produces the highest fidelity?

3. Suppose the first and last selections on a phonograph record are 3-minute cuts. Which, if either, of these cuts is wider on the record? (That is, which contains more grooves along a radial direction?)

4. If you use large-diameter tires on your car, how will your speedometer reading differ?

5. Why are the front wheels located so far out in front on the racing vehicle?

6. Which will roll down a hill faster—a cylinder or a sphere?

7. Why do buses and heavy trucks have large steering wheels?

8. Which is easier for turning stubborn screws—a screwdriver with a thick handle or one with a long handle? Provide an explanation.

9. When you pedal a bicycle, maximum torque is produced when the pedal sprocket arms are in the horizontal position, and no torque is produced when they are in the vertical position. Explain.

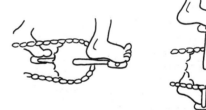

10. Why is the middle seating most comfortable in a bus traveling on a bumpy road?

11. Explain why a long pole is more beneficial to a tightrope walker if it droops.

12. Why is the center of mass of the solar system not at the geometrical center of the sun?

13. Why is the wobbly motion of a single star an indication that the star has a planet or system of planets?

14. Why must you bend forward when carrying a heavy load on your back?

15. Why is it easier to carry the same amount of water in two buckets, one in each hand, than in a single bucket?

16. Is it necessary that weights be placed in the middle of the pans of an equal-arm balance for accurate measurements? Why or why not?

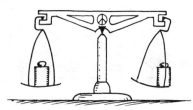

17. Nobody at the playground wants to play with the obnoxious boy, so he fashions a seesaw as shown so he can play with himself. Explain how this is done.

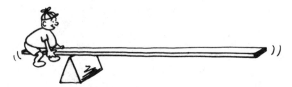

18. Using the ideas of torque and center of gravity, explain why a ball rolls down a hill.

19. How can the three bricks be stacked so that the top brick has maximum horizontal displacement from the bottom brick? For example, stacking them like the dotted lines suggest would be unstable and the bricks would topple. (*Hint:* Start with the top brick and work down. At every interface the center of gravity of the bricks above must not extend beyond the end of the supporting brick.)

20. Why is it dangerous to roll open the top drawers of a fully loaded file cabinet that is not secured to the floor?

21. Describe the comparative stabilities of the three objects shown in Figure 6.17 in terms of work and potential energy.

22. The centers of gravity of three trucks parked on a hill are shown by the X's. Which truck(s) will tip over?

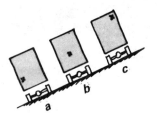

23. Why is less effort required in doing sit-ups when your arms are extended in front of you? Why is it more difficult when your arms are placed in back of your head?

24. The rock has a mass of 1 kilogram. What is the mass of the measuring stick if it is balanced by a support force at the one-quarter mark?

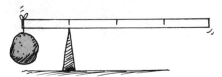

25. A long track balanced like a seesaw supports a golf ball and a more massive billiard ball with a compressed spring between the two. When the spring is released, the balls move away from each other. Does the track tip clockwise, counterclockwise, or remain in balance as the balls roll outward? What principles do you use for your explanation?

26. The value of *g* at the earth's surface is about 9.8 meters per second^2. How would this value change if the earth rotated faster about its polar axis?

27. When a rocket is launched toward the equator from a northern (or southern) latitude, it lands west of its "intended" longitude. Why? (*Hint:* Consider a flea jumping from the middle to the outer edge of a moving phonograph record.)

28. If you should buy a quantity of gold in Mexico and weigh it carefully on a spring balance, would the same quantity of gold weigh more, less, or the same if weighed on the same spring balance in Alaska? Defend your answer.

29. A motorcyclist is able to ride on the vertical wall of a bowl-shaped track as shown. His weight is counteracted by the friction of the wall on the tires (vertical arrow). Is it centripetal or centrifugal force that acts on the motorcycle? On the wall?

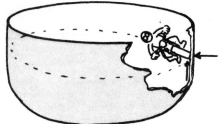

30. Your friend says that a satellite in circular orbit does not accelerate, as evidenced by its constant velocity. Your other friend says it does accelerate, as evidenced by the centripetal force supplied by gravity. What do you say?

31. Does the centripetal force that acts on an object that rotates in a circular path do work on the object? What is your explanation, and what evidence can you cite to support your answer?

32. In an inertial frame of reference. whatever centripetal force is exerted on a body is part of an action-reaction pair of forces. Is this also true of centrifugal force in a rotating frame of reference? Defend your answer.

33. Explain why it is that the faster the earth spins, the less one weighs, whereas the faster a space colony spins, the more one weighs.

34. A space habitat rotates at an appropriate rate to provide a comfortable acceleration of 1 g, earth-normal gravity, for its inhabitants. If it is rotated at twice the angular speed, by how much will the "weight" of the inhabitants increase?

35. Consider a too-small space habitat that consists of a rotating cylinder of radius 4 meters. If a man standing inside is 2 meters tall and his feet are at 1 g, what is the g force at the elevation of his head? (Do you see why projections call for large habitats?)

36. If the variation in g between one's head and feet is to be less than $\frac{1}{100}$ g, then compared to one's height, what should be the minimum radius of the space habitat?

37. A basketball player wishes to balance a ball on his fingertip. Will he be more successful with a spinning ball or a stationary ball? What physical principle supports your answer?

38. You sit in the middle of a large, freely rotating turntable at an amusement park. If you crawl toward the outer rim, does the rate of rotation increase, decrease, or remain unchanged? What physical principle supports your answer?

39. As more and more skyscrapers are built on the surface of the earth, does the day tend to become longer or shorter? What physical principle supports your answer?

40. A sizable quantity of earth is washed down the Mississippi River and deposited in the Gulf of Mexico. What effect does this tend to have on the length of a day?

41. If the polar ice caps melted, how would this affect the length of the day?

42. A toy train is initially at rest on a track fastened to a bicycle wheel, which is free to rotate. How does the wheel respond when the train moves clockwise? When the train comes to a stop? When the train backs up? Does the angular momentum of the wheel-train system change during these maneuvers? How would the resulting motions depend on the relative masses of the wheel and train?

43. Why does a typical small helicopter have a small propeller on its tail?

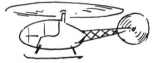

44. How would the length of a day be affected if all the world's inhabitants walked in an easterly direction? When they stopped walking?

45. We believe our galaxy was formed from a huge cloud of gas and particles. The original cloud was far larger than the present size of the galaxy, was more or less spherical, and was rotating very much more slowly than the galaxy is now. In the sketch we see the original cloud and the galaxy as it is now (seen "edgewise"). Explain how the law of gravitation and the conservation of angular momentum contribute to the galaxy's present shape, and why it rotates faster now than when it was a larger, spherical cloud.

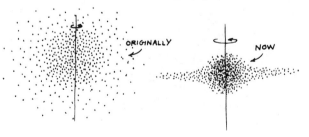

Castle in the Pyrenees, painting by René Magritte © ADAGP, Paris, 1980, photo © Gabriel D. Hackett

8 Gravitation

For thousands of years people have looked into the nighttime sky and wondered about the stars. Having only their naked eyes, they neither saw nor dreamed that the stars are greater in number than all the grains of sand on all the deserts and beaches of the world. Our ancestors envisioned the earth at the center of the universe and considered the stars to be fixed on a great revolving crystal sphere. They distinguished the planets from the stars and placed them on inner spheres with more complicated motions. They believed that the complicated motions and the positions of the planetary spheres influenced earthly events.[1] In the fifteenth century Copernicus discovered that the motions of the planets were much simpler if they were considered to be circling about the sun rather than the earth. The Copernican view was revolutionary, and there were great debates about whether the planets revolved around the sun or not.

The approach to these debates by the Danish astronomer Tycho Brahe was quite different from that of his contemporaries. Rather than carrying on with philosophical arguments, Brahe believed it would be better to measure accurately the positions of the planets in the sky and to chart their motions. This was a magnificent idea—that in formulating theories, one's efforts are best spent gathering facts and making careful measurements rather than arguing over philosophical speculations. Following this idea, Brahe built the first great observatory. Although telescopes had not yet been invented, he built huge brass protractorlike quadrants and cataloged the motions of the planets so accurately that his measurements are still used today. He recorded the planets' positions for 20 years to $\frac{1}{60}$ of a degree. Just before his death he entrusted one of his contemporaries, Johannes Kepler, with editing and publishing his planetary tables. From the data, Kepler discovered some very remarkable and simple laws regarding planetary motion.

Kepler's Laws

Today we know what Brahe and Kepler did not know, that the planets of the solar system are simply falling continuously around the sun (discussed in Chapter 3). Kepler, to learn how the planets moved, performed the enormous task of transforming Brahe's earthbound observations into a path in space such as might be seen by a stationary observer outside the solar system. His conviction that the planets would circle perfectly on perfect circles around the sun was shattered after years of effort. He found the paths to be ellipses.

Kepler also found that the planets do not go around the sun at a uniform speed but move faster when they are nearer the sun and more slowly when they are farther from the sun. They do this in such a way that an imaginary

---

[1]Terminology from earlier times has carried over to the present in many areas; for example, in political science we speak of "spheres of influence."

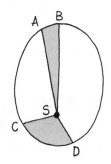

Fig. 8.1 Equal areas are swept out in equal intervals of time.

line or spoke joining the sun and the planet sweeps out equal areas of space in equal times. The triangular-shaped area swept out during a month when a planet is orbiting far from the sun ("triangle" ASB in Figure 8.1) is equal to the triangular area swept out during a month when the planet is orbiting closer to the sun ("triangle" CSD in Figure 8.1).

Ten years later Kepler discovered a third law. He had spent these years searching for a connection between the sizes of the planets' orbits and their periods (the time required for a planet to make a complete revolution around the sun). From Brahe's data Kepler found that the squares of the periods are proportional to the cubes of their average distances from the sun.[2] He discovered this by noting that the fraction R^3/T^2 is the same for all the planets, where R is the planet's average orbit radius and T is the planet's period measured in earth days. Thus, **Kepler's laws of planetary motion** are:

Law 1: *Each planet moves in an elliptical orbit with the sun at one focus.*

Law 2: *The line from the sun to any planet sweeps out equal areas of space in equal time intervals.*

Law 3: *The squares of the times of revolutions (periods) of the planets are proportional to the cubes of their average distances from the sun.* *($T^2 \sim R^3$ for all planets.)*

Kepler's laws apply not only to planets but also to moons or any satellite in orbit around any body. Except for Pluto (which Kepler had no knowledge of), the elliptical orbits of the planets are very nearly circular. Only the precise measurements of Brahe showed the slight differences. Kepler had no idea *why* the planets traced elliptical paths about the sun and no general explanation for the mathematical relationships that he had discovered. He was familiar with Galileo's ideas about inertia and accelerated motion, but he failed to apply them to his own work. As a consequence, Kepler's thinking about forces was concerned with forces he supposed were directed in the same direction as the planets' paths to keep them moving. He never appreciated the concept of inertia. Galileo, on the other hand, never appreciated Kepler's work and held to his conviction that the planets moved in circles.[3] Further understanding of planetary motion required someone who could integrate the findings of these two great scientists. This task fell to Isaac Newton.

---

[2]The period of a satellite is given by $T = 2\pi \sqrt{\frac{R^3}{GM}}$ (footnote from Chapter 3, page 34). Can you see that squaring this equation gives $T^2 \sim R^3$, Kepler's third law?

[3]It is not easy to look at familiar things through the new insights of others. We tend to see only what we have learned to see or wish to see. Galileo reported that many of his colleagues were unable or refused to see the moons of Jupiter when they peered skeptically through his telescopes. Galileo's telescopes were a boon to astronomy, but more important than a new instrument to see things better was a new way of seeing what was already there. Is this true today?

**Newton's
Development
of the Law
of Gravitation**

From the time of Aristotle, the circular motions of heavenly bodies were regarded as natural. The ancients believed that the stars, planets, and moon moved in divine circles, free from any impressed forces. As far as the ancients were concerned, this circular motion required no explanation.

Newton recognized that a force of some kind must be acting on the planets; otherwise, their paths would be straight lines. From the law of inertia, he knew this force was not directed along the path of the planets as Kepler had speculated but was, instead, in the direction in which the planets are curved—toward the sun. His analysis of Kepler's second law showed that the origin of this force was the sun. From Kepler's third law, he deduced that this force varies with distance in the special way the force decreases with the square of the distance. A planet twice as far from the sun would be pulled toward the sun with $\frac{1}{4}$ the force; if it were three times as far away, the force would decrease to $\frac{1}{9}$ as much, and so forth. This relationship is called the *inverse-square law*, which we will treat in detail shortly.

What sort of force could act from one place to another without contact? Newton was aware of such a force. When an apple falls to the ground, it is accelerated from its position on the tree. He wondered if the same force extended to the moon, holding it in an elliptical path around the earth, a path similar to a planet's path around the sun. This stroke of intuition— conceiving the same force to hold both the moon and apple—was a revolutionary break with the prevailing notion that there were two sets of natural laws, one for earthly events and another altogether for motions in the heavens.

To test this theory, Newton compared the fall of an apple with the "fall" of the moon. He realized that the moon falls in the sense that *it falls away from the straight line it would follow if there were no forces acting on it.* Because of its tangential speed, it "falls around" the round earth. Now the moon is sixty times farther from the center of the earth than an apple at the earth's surface (Figure 8.2). The apple will fall nearly 5 meters in its first second of fall or, more accurately, 4.9 meters. If the inverse-square theory is correct, the moon should fall $(\frac{1}{60})^2$ of 4.9 meters, which is approximately 1.4 millimeters.

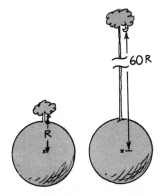

Fig. 8.2 An apple falls 4.9 m during the first second of fall from a tree at the earth's surface. Newton asked how far a "moon apple" would fall in the same time if it were sixty times farther from the center of the earth. If the inverse-square law applies, the distance should be 1.4 mm.

The distance the moon travels in a complete orbit ($2\pi R$) divided by the time it takes (approximately 27 days) gives us the tangential speed of the moon. If we were in a mathematical mood, we could neglect gravity for a moment and calculate how far along a tangent the moon would travel in 1 second, then bring gravitation back and figure with a little geometry how

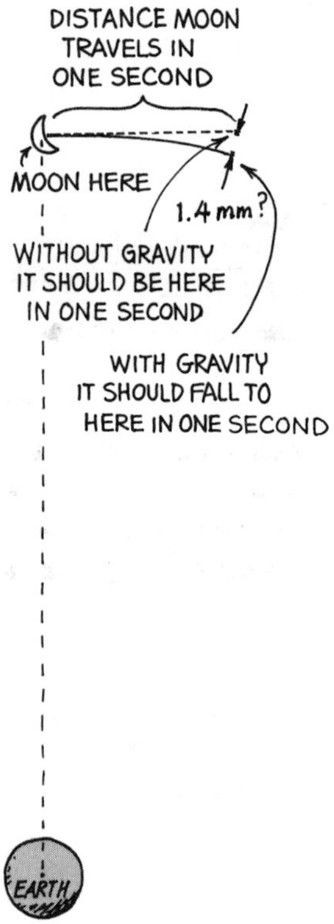

DISTANCE MOON
TRAVELS IN
ONE SECOND

MOON HERE

1.4 mm?

WITHOUT GRAVITY
IT SHOULD BE HERE
IN ONE SECOND

WITH GRAVITY
IT SHOULD FALL TO
HERE IN ONE SECOND

EARTH

Fig. 8.3 If the force that pulls apples off trees also pulls the moon into orbit, the circle of the moon's orbit should fall 1.4 mm below a point along a straight-line tangent where the moon would otherwise be after 1 s.

far the circle of the moon's orbit falls below this point on the tangent line (Figure 8.3). We'd find this distance to be about 1.4 millimeters.

Newton carefully made similar calculations and was disappointed to end up with a large discrepancy. Recognizing that brute fact must always win over a beautiful hypothesis, he placed his papers in a drawer where they remained for 15 years. During this period he founded and developed the field of geometric optics for which he first became famous. When he finally returned to the moon problem at the prodding of his astronomer friend Edmund Halley and used a corrected figure for the distance between earth and moon, he obtained excellent results. Only then did he publish what is one of the most far-reaching generalizations of the human mind: the **law of universal gravitation**.[4] Newton didn't discover gravity per se; he discovered that gravity was universal, that every object in the universe is attracted to every other object in the universe.

Newton's Law of Universal Gravitation

Everything pulls on everything else in a beautifully simple way that involves only mass and distance. Newton stated that every body attracts every other body with a force that for any two bodies is directly proportional to the mass of each body and that is inversely proportional to the square of the distance between them.

This statement can be expressed symbolically as

$$F \sim \frac{m_1 m_2}{d^2}$$

where m_1 is the mass of one body, m_2 is the mass of the other, and d is the distance separating them.[5] The greater the masses m_1 and m_2, the greater the force of attraction between them.[6] The greater the distance d between the bodies, the weaker the force of attraction, but weaker as the inverse square of the distance of separation.

We can better understand how gravity is diluted with distance by considering how paint from a paint gun spreads with increasing distance (Figure 8.4). Suppose we position a paint gun at the center of a sphere with a radius of 1 meter, and a burst of paint spray travels 1 meter to

---

[4]This is a dramatic example of the painstaking effort and cross-checking that go into the formulation of a scientific theory. Contrast Newton's approach with the failure to "do one's homework," the hasty judgments, and the absence of cross-checking that so often characterize the pronouncement of less-than-scientific theories.

[5]When the proportionality constant G is included, the relationship can be expressed as the equation $F = Gm_1m_2/d^2$. G is called the universal gravitational constant and stands for the actual measured force between two particles, each with a mass of 1 kg, when their centers are separated by 1 m. There is more about this in Appendix IV.

[6]Note the different role of mass here. Thus far we have treated mass as a measure of inertia, which is called *inertial mass*. Now we see mass as a measure of gravitational force, which in this context is called *gravitational mass*. It is experimentally established that the two are equal, and, as a matter of principle, the equivalence of inertial and gravitational mass is the foundation of Einstein's general theory of relativity.

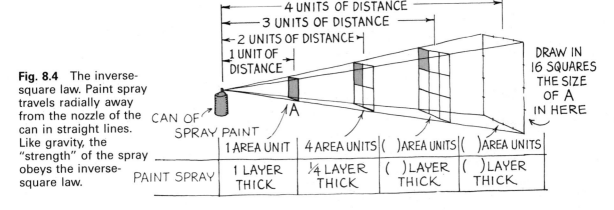

Fig. 8.4 The inverse-square law. Paint spray travels radially away from the nozzle of the can in straight lines. Like gravity, the "strength" of the spray obeys the inverse-square law.

produce a square patch of paint that is 1 millimeter thick. How thick would the patch be if the experiment were done in a sphere with twice the radius? If the same amount of paint travels in straight lines for 2 meters, it will spread to a patch twice as tall and twice as wide. The paint would be spread over an area four times as big, and its thickness would be only $\frac{1}{4}$ millimeter. Can you see from the figure that for a sphere of radius 3 meters the thickness of the paint patch would be only $\frac{1}{9}$ millimeter? Can you see that the thickness of the paint decreases as the square of the distance increases? This is known as the **inverse-square law**. The inverse-square law holds for gravity and for all phenomena wherein the effect from a localized source spreads uniformly throughout the surrounding space: the electric field about an isolated electron, light from a match, radiation from a piece of uranium, and sound from a cricket.

In using Newton's equation for gravity, we measure distances from the centers of bodies. Note in Figure 8.5 that the apple that normally weighs 1 newton at the earth's surface weighs only $\frac{1}{4}$ as much when it is twice the

Fig. 8.5 If an apple weighs 1 N at the earth's surface, it weighs only $\frac{1}{4}$ N when it is twice as far from the center of the earth because the gravitational pull is only $\frac{1}{4}$ as strong. At three times the distance, it weighs only $\frac{1}{9}$ N. What would it weigh at four times the distance? Five times?

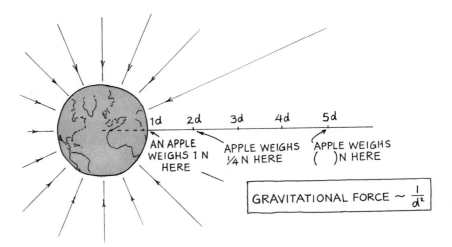

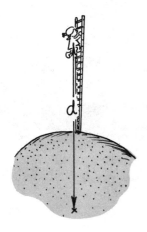

distance *from the earth's center.* The greater the distance from the earth's center, the less the weight of a body. A child that weighs 300 newtons at sea level will weigh only 299 newtons atop Mt. Everest (Figure 8.6). But no matter how great the distance, the earth's gravitational force does not drop to zero. Even if you were transported to the far reaches of the universe, the gravitational influence of home would still be with you. It may be overwhelmed by the gravitational influences of nearer and/or more massive bodies, but it is there. The gravitational influence of every body, however small or however far, is exerted through all of space.[7] Isn't that nice?

Fig. 8.6   According to Newton's equation, her weight (not mass) decreases as she increases her distance from the earth's center (not surface).

Question   By how much does the gravitational force between two bodies decrease when the distance between them is doubled? Tripled? Increased tenfold?*

Gravitation is the dominant force in the universe; it crushes stars to make them shine and determines the motions of celestial bodies. This is because of the enormous masses involved. Yet gravitation is actually the weakest of the four fundamental forces in nature. The other three are *electromagnetic* forces and *strong* and *weak nuclear* forces.[8] Electromagnetic forces obey the inverse-square law, but unlike gravitation they can be repulsive as well as attractive. At large distances, the attracting and repelling interactions usually balance themselves, but at atomic-size distances, these contrary forces don't quite balance. At the atomic level, the attractive electromagnetic force is dominant and is responsible for such things as the binding together of atoms to form molecules. Strong and weak nuclear forces dominate at distances smaller than the size of the atomic nucleus and are negligible outside the nucleus. Although gravitation is the weakest of these forces, between large aggregates of atoms, such as you and the earth, the resulting force is quite noticeable—your weight.

---

*Answer**   One-fourth, one-ninth, one-hundredth.

[7]Paul A. M. Dirac, the 1933 Nobel Prize winner in physics, put it this way: "Pick a flower on earth and you move the farthest star!"

[8]At incredibly tiny distances, less than 10^{-29} cm, strong and weak nuclear forces and electromagnetic forces appear to be merely components of one unified force. Physicists today are getting closer to completing Einstein's lifelong and unfinished task—showing that all four forces in nature are aspects of one unified force.

Weight and Weightlessness

The force of gravity, like any force, causes acceleration. Bodies under the influence of gravity accelerate toward each other—unless they're already in contact with each other. We are almost always in contact with the earth, and for this reason we think of gravity as primarily concerned with weight.

When we stand on a spring balance such as a bathroom scale, we measure the apparent gravitational attraction between the earth and ourselves—our weight. (Weight is modified slightly by the earth's rotation.) If we stood on a bathroom scale in a moving elevator, we would find that our weight would vary. If the elevator accelerated upward, our weight would increase; the bathroom scale would push harder against our feet. If the elevator accelerated downward, our weight would decrease; push or support of the scale would decline. If the elevator cable broke and the elevator fell freely, the reading on the scale would go to zero. According to the reading, we would be weightless. Would we really be weightless? We can answer this question only if we first say what we mean by *weight*.

NORMAL WEIGHT

Fig. 8.7 The sensation of weight equals the force with which you press against the supporting floor. If the floor accelerates up or down, your weight varies.

GREATER THAN NORMAL WEIGHT

LESS THAN NORMAL WEIGHT

ZERO WEIGHT

We define the weight of a body as the force it exerts against the supporting floor or the weighing scales. According to this definition, you are as heavy as you feel; so in an elevator that accelerates downward, the supporting force of the floor is less and you weigh less. If the elevator is in free fall, your apparent weight is zero (Figure 8.7). Even in this weightless condition, however, there is still a gravitational force acting on you, which is evidenced by your downward acceleration. But gravitational attraction in this case is not sensed as weight because of the absence of a support force.

Fig. 8.8 Both are "weightless."

Consider an astronaut in orbit. The astronaut is in a state of apparent **weightlessness**. He feels weightless because he is not supported by anything (Figure 8.8). There would be no compression in the springs of a bathroom scale placed beneath his feet because the bathroom scale is falling as fast as he is. Any objects that are released fall together with him and remain in his vicinity, unlike what happens on the ground. All the local effects of gravity are eliminated for him. The organs of his body respond as though the gravitational forces were absent, and this gives the sensation of weightlessness. The astronaut experiences the same sensation in orbit that he would experience in a falling elevator; in both cases he would be in a state of free fall.

It is interesting to note that the astronaut is still under the influence of a gravitational force. That's why we say he is in a state of *apparent* weightlessness. To be truly weightless, he would have to be far out in space, well away from the earth, sun, and other attracting bodies, where gravitational forces are negligible. In this truly weightless condition, any motion would be in a straight-line path rather than in the curved path of closed orbit.

The space station depicted by NASA artists in Figure 8.9 is in a state of apparent weightlessness. The shuttle, station facility, and astronauts all accelerate equally toward earth, somewhat less than 1 *g* because of their altitude. This acceleration is not sensed at all; with respect to the earth, the astronauts experience zero *g*. Besides earth gravity, the spacecraft is subjected to slight disturbances from venting, attitude control, and drag that varies with altitude and solar activity. These sometimes constitute micronewtons of force, so we say the astronauts are in a micro-*g* environment (Figure 8.9).

Fig. 8.9 The half-dozen or so inhabitants in this proposed laboratory and docking facility will continually experience weightlessness in their micro-*g* environment.

Question   In what sense is being infinitely far away from all celestial bodies like stepping off a table?*

Tides

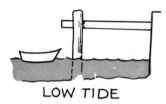

LOW TIDE

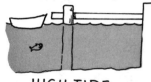

HIGH TIDE

Fig. 8.10   Ocean tides.

Fig. 8.11   A ball of taffy remains spherical when all parts of it are pulled equally in the same direction. When one side is pulled more than the other, however, the base shape is distorted.

People have always thought there was a connection between the ocean tides and the moon, but no one could offer a satisfactory theory explaining the two high tides per day. Newton showed that the ocean tides are caused by the *differences* in the gravitational pulls by the moon on opposite sides of the earth. The moon pulls harder on the side of the earth nearest it and not so hard on the side of the earth farthest from it. This is simply because the force of gravity gets weaker with increased distance. To understand why the difference in gravitational pulls by the moon on opposite sides of the earth produces tides, pretend you have a large, several-kilogram spherical glob of taffy. If you pull in one direction on every kilogram of the taffy with exactly the same amount of force, it will accelerate but still remain in the same spherical shape. What happens if you pull one side of the taffy a little bit harder than you pull the middle, and the middle a little bit harder than you pull the taffy on the opposite side? Can you see that you would stretch the taffy? It would no longer be perfectly spherical, but elongated; it would be an ellipsoid (Figure 8.11). And this is what happens to this big ball we are living on. The side nearest the moon is attracted with more force than the earth's center, so water there is pulled away from the earth into a tidal bulge. The earth's center in turn is attracted with more force than the side farthest away, so the earth in effect is pulled away from the water on this distant side, which forms the second tidal bulge. These bulges of water are held by the difference in the moon's gravitational pulls on the near and far sides of the earth while the solid earth rotates daily beneath them, giving us two high tides for each 24-hour rotation.[9] The depressions between the bulges are the two low tides.

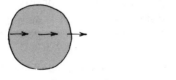

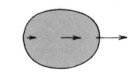

*****Answer**   In both cases you'd experience weightlessness. Far from all celestial bodies you'd be truly weightless, for you'd be away from all gravitational influences; stepping from a table, you'd momentarily experience apparent weightlessness because there would be a momentary lapse of support force.

[9]While the earth spins, the moon moves in its orbit and appears at its same position overhead every 25 hr, so the two-high-tide cycle is at about 25-hr intervals. Tidal friction of the ocean against the ocean floor causes a lag, often up to $\frac{1}{4}$ day, so high tides do not correspond to a time when the moon is directly overhead.

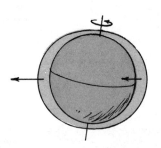

Fig. 8.12 Two tidal bulges remain relatively fixed with respect to the moon while the earth spins daily beneath them.

Fig. 8.13 When the attractions of the sun and moon are in conjunction, spring tides occur. When the attractions of the sun and moon are at right angles to one another, neap tides occur.

The sun also contributes to tides, although it is less than half as effective as the moon in raising tides. This often seems puzzling when it is realized that the sun pulls the earth with a force about 180 times stronger than the moon's pull. Why, then, doesn't the sun cause tides 180 times greater than, or at least as great as, the lunar tides? The answer is that the sun's pull on the near side of the earth is not appreciably stronger than its pull on the far side of the earth. Compared to the 150-million-kilometer (usually remembered as 93-million-mile) distance to the sun, the 12,740-kilometer difference between near and far sides of the earth (its diameter) is tiny. So although the gravitational pull of the sun is very great, its pull is not very different on each side of the earth and only slightly elongates the earth's shape, producing only small tidal bulges. Although the moon's gravitational pull is much less, the *difference* between the lunar pull on the near and far sides of the earth is more appreciable. The extra 12,740-kilometer distance to the part of the earth farthest from the moon is $\frac{1}{30}$ of the 384,000-kilometer distance from the earth to the moon. This 12,740 kilometers is, however, only about $\frac{1}{12,000}$ of the 150-million-kilometer distance from the earth to the sun. For this reason, the difference in gravitational force from the moon easily ''outtides'' the sun— even though the sun pulls harder than the moon.

Ocean tides are complicated because of the presence of interfering land masses and friction with the ocean bottom. Because of these complications, the tides break up into smaller "basins of circulation," where a tidal bulge travels around the basin like a circulating wave that travels around a small basin of water when it is tilted properly. There is always a high tide someplace in the basin, although at a particular locality it may be hours away from an overhead moon. In midocean the variation in the water level—the range of the tide—is usually a meter or two. This range varies in different parts of the world; it is greatest in some Alaskan fjords and is most notable in the basin of the Bay of Fundy, between New Brunswick and Nova Scotia in southeast Canada, where tidal differences sometimes exceed 15 meters. This is largely due to the ocean floor, which funnels shoreward in a V-shape. The tide often comes in faster than a person can run. Don't dig clams near the water's edge at low tide in the Bay of Fundy!

When the sun's and the moon's tides coincide, as at times of the new and full moon, we have high tides higher than usual followed by low tides lower than usual. These are called *spring* tides. (Spring tides have nothing to do with the spring season.) At times of the first and third quarters of the moon, the sun's and the moon's tides are out of step, and we have lower high tides and higher low tides than usual. These are called *neap* tides.

The earth is not absolutely rigid. Consequently, the moon-sun tidal forces produce earth tides as well as water tides. Twice each day the solid surface of the earth rises and falls as much as a quarter of a meter! Earthquakes and volcanic eruptions have a higher probability of occurring when the earth is experiencing an earth spring tide.

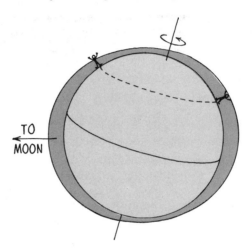

Fig. 8.14 Inequality of the two high tides per day. Because of the earth's tilt, a person in the Northern Hemisphere may find the tide nearest the moon much lower (or higher) than the tide half a day later. Inequalities of tides vary with the positions of the moon and the sun.

We live at the bottom of an ocean of air that also experiences tides. These tides are very small compared to ocean tides because a unit volume of air is only about a thousandth as massive as a unit volume of water and because the tide-generation forces vary with the mass of the body being attracted. So variations in atmospheric pressure due to atmospheric tides are very small. Most often they are obscured by thermal variations. More important are the tidal effects above the atmosphere in the region called the *ionosphere*, where displacements of ionized air in the upper layers produce electric currents that alter the magnetic field surrounding the earth. These are magnetic tides, which in turn regulate the degree to which cosmic rays penetrate into the lower atmosphere. The highs and lows of magnetic tides are greatest when the atmosphere is having its spring tides. This time of greatest fluctuations affects the ionic composition of the lower atmosphere, which in turn is evidenced in subtle changes in the behavior of living things—us, for example. Have you noticed that some of your friends are a bit weirder at the time of the full moon?

Questions

1. Would there be tidal bulges on the moon if it were covered with water? How many? How often would high and low tides occur?*

2. Which pulls with the greater force on the earth's waters—the sun or the moon?†

Gravitational Field

The earth pulls on the moon. We regard this as *action at a distance*, the bodies interacting even though they are not in contact. We can look at this in a different way: we can regard the moon as interacting with the **gravitational field** of the earth. We can think of gravity in terms of a *force field*. The properties of space surrounding any mass can be considered to be so altered that another mass introduced to this region will experience a force. This alteration of space is the gravitational field, the magnitude of which is equal to the force per unit of mass located in that space. It is common to think of distant rockets and the like as interacting with gravitational fields rather than with the masses of the earth and other bodies responsible for these fields. The field plays an intermediate role in our thinking about the forces between bodies.

Fields have both magnitude and direction, and they can be pictorially represented with vector arrows. In Figure 8.15 the vectors represent the gravitational field about the earth. All the vectors point in the direction of the force a body in the region would experience, a force that would be directed toward the center of the earth. In Figure 8.15a the vectors farther away are shorter, showing that the pull on any object decreases with increasing distance from the earth. Ideally we can think of a vector drawn at every point in space. Figure 8.15b shows a more simplified representation of this. Where the lines are farther apart, the field is weaker.

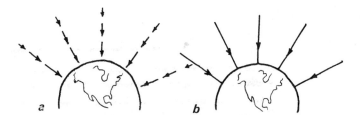

Fig. 8.15 Two ways to represent the earth's gravitational field.

---

*Answer   There would be two tidal bulges for the same reason two tidal bulges occur on the earth. (The earth pulls differentially on the nearest and farthest parts of the moon. As a result, there *are* two bulges on the slightly elliptically shaped solid surface of the moon.) Since the moon takes 28 days to make a single revolution about its own axis (as well as about the earth-moon axis), the same part of its surface faces the earth; therefore, the tidal bulges would be stationary upon the moon's surface, and no high or low tides would occur.

†Answer   The sun. (If you missed this, go back and reread from page 134.)

We may think of the earth's gravitational field as being exterior to the earth, but the field concept extends to the earth's interior as well. Imagine a tunnel bored through the earth, from one side to the other and passing straight through the center.[10] Would a gravitational force be exerted on a body located midway through the tunnel at the very center of the earth? The answer is no, because the body would be pulled equally in every direction and the forces would cancel. The gravitational field at the center of the earth is therefore zero. But away from the center, opposing gravitational forces would not cancel to zero. As we move from the center of the earth, the gravitational field increases and is maximum at the surface. If the composition of the earth were of uniform density, the field inside would increase in direct proportion to the distance from the center; halfway to the surface, for example, the gravitational field would be half that at the surface. We have not stated why this is so, but perhaps your instructor will provide the explanation. In any event, the gravitational field intensity for regions inside and outside a planet is graphed in Figure 8.16.

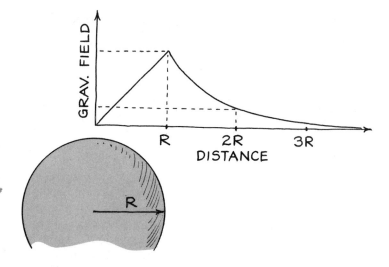

Fig. 8.16 The gravitational field intensity inside a planet of uniform composition and density is directly proportional to the radial distance from its center and is maximum at its surface. Outside, it is inversely proportional to the square of the distance from its center.

Imagine a room at the center of a planet. The room would be gravity-free because of the cancelation of gravity in every direction. Interestingly enough, the size of the room doesn't change this fact—even if the room constitutes most of the volume of the planet! Imagine a hollow planet like

---

[10]If an object were dropped into such a bottomless well, it would accelerate and attain a speed at the center equal to the speed of a circularly moving satellite in close orbit. After passing the center, it would then decelerate the rest of the way to its initial zero speed at the opposite surface. This trip would take 42 min. (the same time it would take for an earth satellite in close orbit to reach the same place). The object would oscillate back and forth in the tunnel; if it started its drop from the same altitude as the satellite, both would keep the same rhythm. How about that!

a huge basketball (Figure 8.17). Complete cancelation of gravity will occur anywhere inside the cavity. Consider a body at point P, for example, which is twice as far from the left side of the planet as it is from the right side. If gravity depended only on distance, a body at P would be attracted only $\frac{1}{4}$ as much to the left side as to the right side (according to the inverse-square law). But gravity also depends on mass. The same solid angle subtends regions A and B, and we see that whatever the angle, A will have four times the area and therefore four times the mass as region B. Since $\frac{1}{4}$ of 4 is equal to 1, a body at P is as attracted to the farther but more massive region A as it is to the closer but less massive region B. Cancelation occurs. More thought will show that cancelation will occur anywhere inside a planetary shell that has uniform thickness and composition. A gravitational field would exist at and beyond its outer surface and would behave as if all the mass of the planet were concentrated at its center—the center of gravity; but everywhere inside the hollow part, there would be no gravitational field. Occupants inside would be completely weightless.

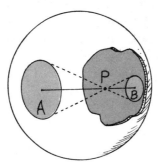

Fig. 8.17 The gravitational field anywhere inside a spherical shell of uniform thickness and composition is zero, because the field components from all the particles of mass in the shell cancel one another. A body at point P, for example, is attracted just as much to the larger but farther region A as it is to the smaller but closer region B.

Although gravity can be canceled inside a body or between bodies, it cannot be shielded in the same way that electricity and magnetism can. Electricity and magnetism have repelling as well as attracting forces that enable shielding, but gravitation only attracts and therefore cannot be shielded. Eclipses provide convincing evidence for this. The moon is in the gravitational field of both the sun and the earth. During a lunar eclipse the earth is directly between the moon and the sun, and any shielding of the sun's field by the earth would result in a deviation of the moon's orbit. Even a very slight shielding effect would accumulate over a period of years and show itself in the timing of subsequent eclipses. But there have been no such discrepancies; past and future eclipses are calculated to a high degree of accuracy using only the simple law of gravitation. No shielding effect in gravitation has ever been found.

The idea of a force field has wider application in the study of electromagnetism. It also plays a central role in gravitation as explained by Albert Einstein's general theory of relativity.

Einstein's Theory of Gravitation

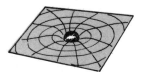

Fig. 8.18 Warped space-time. Space near a star is curved in four dimensions in a way similar to the two-dimensional surface of a waterbed when a heavy ball rests on it.

A model for gravity quite unlike Newton's was presented by Einstein in his general theory of relativity. Einstein perceived a gravitational field as a geometrical warping of four-dimensional space and time; he realized that bodies put dents in space and time just as a massive ball placed in the middle of a large waterbed dents the two-dimensional surface (Figure 8.18). The more massive the ball, the greater the dent or warp. If we roll a marble across the top of the bed but away from the ball, the marble will roll in a straight-line path. But if we roll the marble near the ball, it will curve as it rolls across the indented surface of the waterbed. If the curve closes on itself, the marble will orbit the ball in an elliptical or circular path. Speaking loosely, we can say from a Newtonian view that the marble curves because it is attracted to the ball; whereas from an Einsteinian view, the marble curves not because of any force but because the surface on which it moves curves. In Chapter 34 we will treat Einstein's theory of gravitation in more detail.

Black Holes

Suppose you were indestructible and could somehow stand on the surface of a star. Your weight would depend both on your mass and the star's mass and on the distance between the star's center and your belly button. If the star were to burn out and collapse to half size with no change in its mass, your weight at the surface, determined by the inverse-square law, would be 4 times greater (Figure 8.19). If the star collapsed to a tenth its size, your weight at the surface would be 100 times as much, and so on. The gravitational field at the star's surface would become stronger as the star shrank. It would be more and more difficult for a starship to leave such a shrinking star. Escape velocity would increase with continued shrinking. If a star such as our sun collapsed to a radius of 3 kilometers, the escape velocity from its surface would be equal to the speed of light. A sunship would have to travel as fast as light to escape its pull. If the radius shrank to less

Fig. 8.19 If a star collapses to half its radius and there is no change in its mass, gravitation at its surface would be increased by 4.

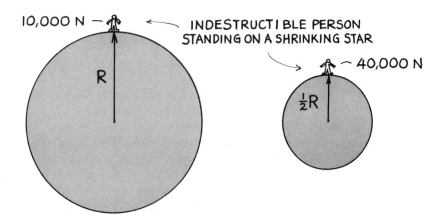

10,000 N — INDESTRUCTIBLE PERSON STANDING ON A SHRINKING STAR

40,000 N

R

½R

than 3 kilometers, escape velocity would be greater than the speed of light. Light itself couldn't escape. Nothing could. The sun would be invisible—it would be a **black hole**.

The sun in fact has too little mass to experience such a collapse, but when some stars with masses greater than four suns reach the end of their nuclear resources, they undergo collapse; and unless rotation is high enough, the collapse continues until the stars reach infinite densities. Gravitation near the surfaces of these shrunken stars is so enormous that light cannot escape from them. They have crushed themselves out of visible existence. Black holes are completely invisible.

A black hole is no more massive than the star from which it collapsed, so the gravitational field in regions at and greater than the original star's radius is no different after the star's collapse than before. But closer distances near the vicinity of a black hole are nothing less than the collapse of space itself, with a surrounding warp into which anything that passes too close—light, dust, or a spaceship—is drawn. Astronauts could enter the fringes of this warp and with a powerful spaceship still escape. After a certain distance, however, they could not—and they would disappear from the observable universe.

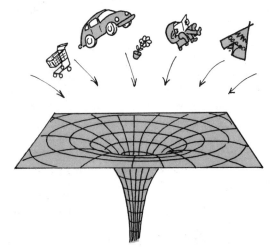

Fig. 8.20 Anything that falls into a black hole is reduced to subatomic particles.

It has been a matter of some conjecture that material drawn into a black hole may reappear elsewhere through a "white hole." What vanishes in one place might be spewed out at another. Even if this were true, however, black holes would be no less frightening to space voyagers. Everything that falls into a black hole is reduced to elementary particles, and everything would emerge as particles. If black holes are a gate from one realm to another, they are a gate through which nothing can pass intact (Figure 8.21).

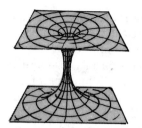

Fig. 8.21 A black hole may be the portal to another universe.

Although a black hole cannot be seen, its presence can be detected by its gravitational influence on neighboring stars. For example, it could be part of a binary star system, in which case a nearby luminous star would appear to be orbiting about an empty point. Several black holes have been tentatively identified. The first was Cygnus X-1,72; material from a blue, supergiant binary companion accelerating into the nonvisible star has been found to be emitting X-ray radiation. Similar evidence exists in the binary Circinus X-1 (3U 1516-56). Other massive black holes are thought to exist at the centers of globular clusters, and present speculation is that they may be the driving engines for the very powerful and distant cosmic objects called *quasars*. We will return to black holes in Chapter 35.

Universal Gravitation

We all know that the earth is round. But why is the earth round? It is round because of gravitation. Everything attracts everything else, and so the earth has attracted itself together as far as it can! Any "corners" of the earth have been pulled in; as a result, every part of the surface is equidistant from the center of gravity. This makes it a sphere. Therefore, we see from the law of gravitation that the sun, the moon, and the earth are spherical because they have to be (rotational effects make them slightly ellipsoidal).

If everything pulls on everything else, then the planets must pull on each other. The force that controls Jupiter, for example, is not just the force from the sun; there are also the pulls from the other planets. Their effect is small in comparison to the pull of the much more massive sun, but it still shows. When Saturn is near Jupiter, its pull disturbs the otherwise smooth ellipse traced by Jupiter. Both planets "wobble" about their correct orbits. This wobbling is called a *perturbation*. Early in the nineteenth century, unexplained perturbations were observed for the planet Uranus. Even when the influences of the other planets were taken into account, Uranus seemed to be behaving strangely. Either the law of gravitation was failing at this great distance from the sun or some unknown influence such as another planet was perturbing Uranus. An Englishman and a Frenchman, J. C. Adams and Urbain Leverrier, assumed Newton's law to be valid and independently calculated where an eighth planet should be located to account for the perturbation. At about the same time, both sent letters to their respective observatories with instructions to search a certain area of the sky. The request by Adams was delayed by misunderstandings at Greenwich, but Leverrier's request to the director of the Berlin observatory was heeded immediately. The planet Neptune was discovered that very night!

Other perturbations of the planet Uranus led to the prediction and discovery of the ninth planet, Pluto. It was discovered in 1930 at the Lowell Observatory in Arizona. Pluto takes 248 years to make a single revolution about the sun, so no one will see it in its discovered position again until the year 2178.

The perturbations of double stars and the shapes of distant galaxies are evidence that the law of gravitation is true at larger distances. Over still larger distances, gravitation underlies the fate of the entire universe.

The universe most likely originated in the explosion of a primordial fireball some 20 billion years ago. Condensed from the clashing gas of this explosion, the galaxies today continue to rush away from one another. This expansion may go on indefinitely, or it may eventually be overcome by the combined gravitation of all the galaxies and come to a stop. All matter would then fall back into a single unity, again presumably to explode, producing a new universe. We will not know whether the expansion is indefinite or cyclic until we have a better estimate of the total mass of the universe. If the mass is not great enough to halt expansion, the galaxies are at or have exceeded escape velocity and the expansion is indefinite. If the mass is greater than most present-day estimates, the escape velocity of the universe is greater than the outward velocity of the galaxies, contraction will be inevitable, and the universe will implode (the Big Crunch) and most probably reexplode (the Big Bounce), repeating the cycle. The period of oscillation for the universe is estimated to be 80 billion years or so. If the universe does oscillate, who can say how many times this process has repeated itself? We know of no way a civilization could leave a trace of ever having existed, for all the matter in the universe is reduced to bare subatomic particles during such an event. Formation of the elements, stars, galaxies, and life again takes place. All the laws of nature, such as the law of gravitation, are then rediscovered by the higher evolving life forms. And then students of these laws read about them, as you are doing now. Think about that!

Few theories have affected science and civilization as much as Newton's theory of gravitation. The successes of Newton's ideas ushered in the so-called Age of Reason or Century of Enlightenment; for Newton had demonstrated that by observation and reason and by employing mechanical models and deducing mathematical laws, people could uncover the very workings of the physical universe. How profound that all the moons and planets and stars and galaxies have such a beautifully simple rule to govern them; namely,

$$F \sim \frac{m_1 m_2}{d^2}$$

The deduction of this simple rule is one of the major reasons for the success in science that followed, for it provided hope that other phenomena of the world might also be described by equally simple laws.

This hope nurtured the thinking of many scientists, artists, writers, and philosophers of the 1700s. One of these was the English philosopher John Locke, who argued that observation and reason, as demonstrated by Newton, should be our best judge and guide in all things and that all of nature and even society should be searched to discover any "natural laws" that might exist. Using Newtonian physics as a model of reason, Locke and his followers modeled a system of government that found adherents in the thirteen British colonies across the Atlantic. These ideas culminated in the Declaration of Independence and the Constitution of the United States of America.

Summary of Terms

Kepler's laws of planetary motion

Law 1: Each planet moves in an elliptical orbit with the sun at one focus.

Law 2: The line from the sun to any planet sweeps out equal areas of space in equal time intervals.

Law 3: The squares of the times of revolution (or years) of the planets are proportional to the cubes of their average distances from the sun ($R^3 \sim T^2$ for all planets).

Inverse-square law A law relating the intensity of an effect to the inverse square of the distance from the cause:

$$\text{Intensity} \sim 1/\text{distance}^2$$

Gravity follows an inverse-square law, as do the laws of electric, magnetic, light, sound, and radiation phenomena.

Law of universal gravitation Every body in the universe attracts every other body with a force that for two bodies is proportional to the masses of the bodies and inversely proportional to the square of the distance separating them:

$$F \sim \frac{m_1 m_2}{d^2}$$

Weightlessness A condition wherein apparent gravitational pull is lacking.

Gravitational field The space surrounding a massive body in which another massive body experiences a force of attraction.

Black hole The configuration of a massive star that has undergone gravitational collapse, in which gravitation is so intense that even its own light cannot escape.

Suggested Reading

Einstein, A., and L. Infeld. *The Evolution of Physics*. New York: Simon & Schuster, 1938.

Gamow, G. *Gravity*. Science Study Series. Garden City, N.Y.: Doubleday (Anchor), 1962.

Valens, E. G. *The Attractive Universe: Gravity and the Shape of Space*. New York: World Publishing, 1969. A delightful book for the lay person, heavily illustrated with photographs by Berenice Abbott.

Review Questions

1. Who gathered the data that showed that the planets travel in elliptical paths around the sun? Who discovered this fact? Who explained this fact?

2. How can we say that the moon is falling when it doesn't get any closer to the earth?

3. The moon "falls" 1.4 millimeters each second. Does this mean that it gets 1.4 millimeters closer to the earth each second? Would it get closer if its tangential speed were reduced? Explain.

4. Newton was not the discoverer of gravity. But he was the discoverer of something about gravity. What?

5. By how much does the gravitational force between two bodies decrease when the distance between them is doubled? Tripled? Increased tenfold?

6. What are the four fundamental forces found in nature?

7. If you were in a car that drove off the edge of a cliff, why would you feel weightless? Would gravity still be acting on you in this state?

8. Which exerts more force on the earth's oceans—the sun or the moon? Which contributes more to ocean tides—the sun or the moon? Why are your answers different?

9. Would ocean tides exist if the gravitational pull of the moon (and the sun) were equal on all parts of the world?

10. What is a gravitational field, and how can its presence be detected?

11. Why is a black hole invisible?

12. Why is the earth "round"? Why isn't it exactly spherical?

Home Projects

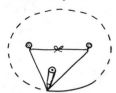

1. Draw an ellipse with a loop of string, two tacks, and a pencil or pen. Try different tack spacings for ellipses of different eccentricities.

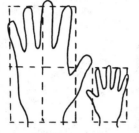

2. Hold your hands outstretched, one twice as far from your eyes as the other, and make a casual judgment as to which hand looks bigger. Most people see them to be about the same size, while many see the nearer hand as slightly bigger. Almost nobody upon casual inspection sees the nearer hand as four times as big. But by the inverse-square law, the nearer hand should appear twice as tall and twice as wide, and therefore occupy four times as much of your visual field, as the farther hand. Your belief that your hands are the same size is so strong that you likely overrule this information. Now if you overlap your hands slightly and view them with one eye closed, you'll see the nearer hand as clearly bigger. This raises an interesting question: What other illusions do you have that are not so easily checked?

Exercises

1. Gravitational force acts on all bodies in proportion to their masses. Why, then, doesn't a heavy body fall faster than a light body?

2. At which of the indicated positions does the satellite in elliptical orbit experience the greatest gravitational force? The greatest orbital speed? The greatest acceleration?

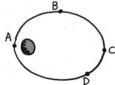

3. The moon is gravitationally attracted to the sun with more than twice the force with which it is gravitationally attracted to the earth. Why, then, does the moon orbit about the earth?

4. Which planets—those closer to the sun than the earth or those farther from the sun than the earth— have a period greater than one earth year?

5. What are the magnitude and the direction of the gravitational force that acts on a man who weighs 700 newtons at the surface of the earth?

6. The weight of an apple near the surface of the earth is 1 newton. What is the weight of the earth in the gravitational field of the apple?

7. The earth and the moon are attracted to each other by gravitational force. Does the more massive earth attract the less massive moon with a force that is greater, smaller, or the same as the force with which the moon attracts the earth?

8. If the mass of the earth somehow increased, with all other factors remaining the same, would your weight also increase? (*Hint:* Let the equation for gravitational force guide your thinking.)

9. A small light source located 1 meter in front of a 1-square-meter opening illuminates a wall behind. If the wall is 1 meter behind the opening (2 meters from the light source), the illuminated area covers 4 square meters. How many square meters will be illuminated if the wall is 3 meters from the light source? 5 meters? 10 meters?

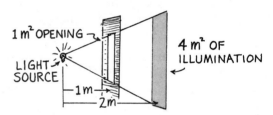

10. If you want to make a profit in buying precious material by weight at one altitude and selling it at another altitude for the same price per body weight, should you buy or sell at the higher altitude location? (Assume weighing is done on a spring scale.)

11. The planet Jupiter is more than 300 times as massive as earth, so it might seem that a body on the surface of Jupiter would weigh 300 times as much as on earth. But it so happens a body would scarcely weigh 3 times as much on the surface of Jupiter as it would on the surface of the earth. Can you think of an explanation for why this is so? (*Hint:* Let the terms in the equation for gravitational force guide your thinking.)

12. Why do you experience a definite queasy sensation in your insides when you are in a state of weightlessness?

13. If you were in a freely falling elevator and you dropped a pencil, you'd see the pencil hovering. Is the pencil falling? Explain.

14. Explain why the following reasoning is wrong. "The sun attracts all bodies on the earth. At midnight, when the sun is directly below, it pulls on an object in the same direction as the pull of the earth on that object; at noon, when the sun is directly overhead, it pulls on an object in a direction opposite to the pull of the earth. Therefore, all objects should be somewhat heavier at midnight than they are at noon." (*Hint:* Relate this exercise to the hovering pencil in the preceding exercise.)

15. If the mass of the earth increased, your weight would correspondingly increase. But if the mass of the sun increased, your weight would not be affected at all. Why?

16. Most people today know that the ocean tides are caused by the gravitational influence of the moon. And most people therefore think that the gravitational pull of the moon on the earth is greater than the gravitational pull of the sun on the earth. What do you think?

17. If somebody tugged on your shirt sleeve, it would likely tear. But if all parts of your shirt were tugged equally, no tearing would occur. How does this relate to tidal forces?

18. Why aren't high ocean tides exactly 12 hours apart?

19. With respect to spring and neap ocean tides, when are the lowest tides? That is, when is it best for digging clams?

20. Whenever the ocean tide is unusually high, will the following low tide be unusually low? Defend your answer.

21. The Mediterranean Sea has very little sediment churned up and suspended in its waters mainly because of the absence of any substantial ocean tides. Why do you suppose the Mediterranean Sea has practically no tides? Similarly, are there tides in the Black Sea? Great Salt Lake? Your county reservoir? A glass of water? Explain.

22. The human body is about 80 percent water. Is it likely that the moon's gravitational pull could cause biological tides, cyclic changes in water flow among the fluid compartments of the body? (Would the moon's pull on the near and far parts of the body differ?)

23. Since a "tidal bulge" sweeps around the earth in approximately 24 hours, does this mean that the incoming tide at the seashore comes in faster than a jet plane (which can almost beat the time zones)?

24. The value of g at the earth's surface is about 9.8 meters per second^2. What is the value of g at a distance of twice the earth's radius?

25. If the earth were of uniform composition, what would the value of g be inside the earth at half its radius?

26. If the earth were of uniform composition, would your weight increase or decrease at the bottom of a deep mine shaft? Defend your answer.

27. It so happens that no actual decrease in weight is found even in the deepest mine shafts. What does this tell us about the density of the earth's composition?

28. If the earth shrank in size, all other factors remaining the same, would escape velocity from its surface be greater, less, or the same as it presently is? Explain.

29. Which requires more fuel—a rocket going from the earth to the moon or a rocket coming from the moon to the earth? Why?

30. If you could somehow tunnel inside a star, would your weight increase or decrease? If, instead, you somehow stood on the surface of a shrinking star, would your weight increase or decrease? Why are your answers different?

31. If our sun shrank in size to become a black hole, show from the gravitational force equation that the earth's orbit would not be affected.

32. If the earth were hollow but still had the same mass and same radius, would your weight in your present location be more, less, or the same as it is now? Explain.

33. We say on page 138 that gravity cannot be shielded and on the same page that gravitational components cancel to zero inside a uniform shell. Is there a contradiction here? Why or why not?

34. If a massive object landed on the surface of the uniform shell of a hollow planet, would occupants inside the planet sense a gravitational attraction to it? Defend your answer.

35. Strictly speaking, you weigh a tiny bit less when you are in the lobby of a massive skyscraper. Why is this so?

36. A person falling into a black hole would be killed by tidal forces before ever encountering the hole itself. Explain why this is so.

PART 2 Properties of Matter

ATOMS ARE SO TINY THAT I INHALE BILLIONS OF TRILLIONS WITH EACH BREATH, NEARLY A TRILLION TIMES MORE ATOMS THAN THE TOTAL POPULATION OF PEOPLE SINCE TIME ZERO!! IN EACH BREATH, I INHALE BILLIONS OF ATOMS EXHALED BY EVERY PERSON WHOSE BREATHS ARE NOW EVENLY MIXED IN THE AIR. SO THERE'S ATOMS IN MY BODY FROM EVERY PERSON WHO EVER LIVED. NOT YET FROM NEWBORN BABIES FAR AWAY --- BUT I'VE BEEN BREATHING, SWEATING, AND VAPORIZING LIKE EVERYONE, SO THOSE BABIES ARE MADE OF ATOMS THAT HAVE BEEN A PART OF ME. IN THIS SENSE, AT LEAST, WE'RE ALL ONE!

9 The Atomic Nature of Matter

Sometime when you have a few quiet moments, take a fantasy trip. The only ticket required is a fertile imagination. Imagine that you fall off your chair in slow motion, and while falling to the floor, you also slowly shrink in size. What would such a trip be like? What would you see? As you topple off the chair and approach the floor, you brace yourself for impact against the smooth, solid surface. And as you get nearer and nearer to it, becoming smaller and smaller all the while, you note that the floor is not as smooth as you supposed it to be, for great cracks appear. These are the microscopic irregularities found in all apparently smooth surfaces. In falling into one of these cracks, which look like canyons as you continue to shrink in size, you again brace yourself for impact against the canyon floor only to find that the bottom of the canyon is itself a myriad of cracks and crevices. Falling into one of these crevices and becoming still smaller, you note that the solid walls have given way to nebulous surfaces that throb and pucker. If you could see greater detail, you would notice that the throbbing surfaces consist of hazy blobs, mostly spherical, some egg-shaped, some larger than others, and all oozing into each other, making up long chains of complicated structures. Falling still farther, you again brace yourself for impact as you approach one of these cloudy spheroids closer and closer, smaller and smaller, and—wow!—you find you have penetrated into a new universe. You fall into a sea of emptiness, occupied by occasional specks that whirl past at unbelievably high speeds. You are in an atom, as empty of matter as the solar system. You have found that the solid floor you have fallen to is, except for specks of matter here and there, empty space. If you continue falling, you might fall many kilometers before making a direct hit with a subatomic speck.

All matter, however solid it appears, is made up of tiny building blocks, which themselves are mostly empty space. These are atoms—the atoms that combine to form molecules that in turn combine to form the compounds and substances of which matter is composed. In this chapter we will study the atomic nature of matter. We will follow this up in succeeding chapters by investigating the properties of matter in the solid, liquid, gaseous, and plasma states.

Atoms All the innumerable substances that occur—shoes, ships, mice, people, stars, everything we can think of—can be analyzed into their constituent **atoms**. Everything is made of atoms. One might think that an incredible number of different kinds of atoms exist to account for the rich variety of

149

substances we find. But the number is surprisingly small. The great variety of substances results, not from any great variety of atoms, but from the great variety of ways in which a few types of atoms can be combined—just as in a color print three colors can be combined to form almost every conceivable color. To date (1985) we know of 109 distinct atoms. These are called the chemical **elements**. Of these, only 90 kinds are found to occur naturally; the others are made in the laboratory with high-energy atomic accelerators and nuclear reactors. These heaviest elements are too unstable (radioactive) to occur naturally in any appreciable amounts.

Hydrogen was apparently the original element and still constitutes over 90 percent of the atoms in the known universe. Elements heavier than hydrogen are manufactured in the deep interiors of stars, where enormous temperatures and pressures cause the fusion of hydrogen atoms into more complex elements. With the exception of some of the hydrogen, all the elements that occur in nature are remnants of stars that exploded long before the solar system came into being.

These star remnants are the building blocks of all matter. And all matter, however complex, living or nonliving, is some combination of these elements. From a pantry having about 100 bins, each containing a different element, we have all the materials needed to make up any substance occurring in the universe. About a dozen elements compose most of the things we see every day; the majority of elements are not found in great abundance, and some are exceedingly rare.[1] Living things, for example, are composed primarily of four elements: carbon (C), hydrogen (H), oxygen (O), and nitrogen (N). The letters in the parentheses represent the chemical symbols for these elements.

Atoms are ageless. Atoms in your body have existed since the beginning of time, cycling and recycling among innumerable forms, both nonliving and living. When you breathe, for example, only part of the atoms that you inhale are exhaled in your next breath. The remaining atoms are taken into your body to become part of you, and they later leave your body by various means. You don't "own" the atoms that make up your body; you borrow them. We all share from the same atom pool as atoms forever migrate around, within, and throughout us. So some of the atoms in the nose you scratch today could have been part of your neighbor's ear yesterday! Not only are we all made of the same *kinds* of atoms, we are also made of the *same* atoms—atoms that cycle from person to person as we breathe, sweat, and vaporize.

---

[1]Most common substances are formed out of combinations of two or more of these most common elements: hydrogen (H), carbon (C), nitrogen (N), oxygen (O), sodium (Na), magnesium (Mg), aluminum (Al), silicon (Si), phosphorus (P), sulfur (S), chlorine (Cl), potassium (K), calcium (Ca), and iron (Fe).

Fig. 9.1 She is made of stardust—in the sense that the carbon, oxygen, nitrogen, and other atoms that make up her body originated in the deep interior of ancient stars that have long since exploded.

Atoms are small, so small that there are about as many atoms of air in your lungs at any moment as there are breaths of air in the atmosphere of the whole world.[2] Since every breath of air you exhale is uniformly mixed in the atmosphere (in about 6 years), every person in the world breathes an average of one of your exhaled atoms in a single breath—for *each* breath you exhale! Considering the many thousands of breaths people exhale, there are many atoms in your lungs at any moment that were once in the lungs of every person who ever lived. We are literally breathing each other.

Yet it is still difficult to imagine how small atoms are. Atoms are so small, in fact, that the question "What do they look like?" has no meaning. Atoms do not look like anything; they have no visible appearance. We could stack microscope on top of microscope and never "see" an atom. This is because light travels in waves, and atoms are smaller than the wavelengths of visible light. The size of a particle visible under the highest magnification must be larger than the wavelength of light. We can better understand this by considering an analogy of water waves. The wavelength of the water waves is simply the distance between the crests of

---

[2]There are about 10^{22} atoms in a liter of air at atmospheric pressure and a total of about 10^{22} liters of air in the atmosphere.

Fig. 9.2 Information about the boat is revealed by passing waves because the distance between wave crests is small compared to the size of the boat.

Fig. 9.3 Information about the grass shoots is not revealed by passing waves because the distance between wave crests is large compared to the size of the blades of grass.

Fig. 9.4 The strings of dots are chains of thorium atoms as revealed by a scanning electron microscope.

successive waves. Consider the ship in Figure 9.2. The ship is much larger than the crest-to-crest distance, or wavelength, of the waves incident on it. Information about the ship is easily revealed by its influence on the passing waves. Consider Figure 9.3. Waves are incident on the blades of grass. The grass shoots are much smaller than the incident waves, which pass by as if the grass were not there. Only if the blades of grass were thicker— that is, wider than the distance between wave crests—would the waves carry information regarding details of the grass. In the same way, waves of visible light are too coarse compared to the size of an atom to reveal details of the size and shape of atoms. Atoms are incredibly small.

The first somewhat direct evidence for the existence of atoms was unknowingly discovered in 1827 by a Scottish botanist, Robert Brown, who was studying the spores of pollen under a microscope. He noticed that the spores were in a constant state of agitation, always moving, always jumping about. At first he thought that the spores were moving by virtue of their animate nature, but later he found the motion to be shared by inanimate dust particles and grains of soot. This perpetual jiggling of particles—called **Brownian motion**—is the result of bombardment by neighboring particles and atoms.

More direct evidence is available today. A photograph of individual atoms is shown in Figure 9.4. The photograph was not made with visible light but with an electron beam. An electron beam, such as that which sprays the picture on your television screen, is a stream of particles. But particles have wave properties,[3] and a high-energy electron beam can have an associated wavelength more than 1000 times smaller than the wavelength of visible light. (In an electron microscope, the beam is focused not by mirrors and lenses but by magnetic fields.) The historic photograph in Figure 9.4 was taken with a very powerful yet extremely thin (50 billionths of a centimeter) electron beam in a scanning electron microscope developed in 1970 by Albert Crewe at the University of Chicago's Enrico Fermi Institute. It is the first high-resolution photograph of individual atoms.

---

[3]More about the wave properties of particles in Chapters 29 and 30.

In 1978 refinements in technique enabled Crewe and M. C. Isaacson to make movies of individual uranium atoms (Figure 9.5). In this technique uranium and other heavy atoms to be photographed are deposited on an ultrathin carbon film. The film is then bombarded with the narrow electron beam, which then scatters off the atoms. The microscope translates the scattering pattern into images of the atoms and assigns them colors according to various characteristics. Color movies of individual atoms, revealing their motions and behaviors, enable advances in a wide range of fields, from metallurgy to medicine.

Question   Are there literally atoms that were once a part of Albert Einstein in the brain matter of *all* your classmates?*

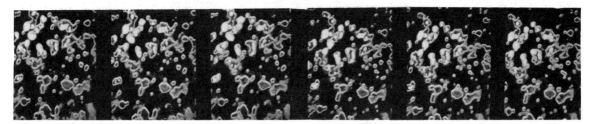

Fig. 9.5   Movie frames of actual atoms, showing firsthand how they move and interact.

Molecules   Atoms combine to form **molecules** (Figure 9.6). Two atoms of hydrogen (H_2) combine with a single atom of oxygen (O) to produce a water molecule (H_2O). A molecule may be as simple as the two-atom combination of oxygen (O_2) or nitrogen (N_2) or as complex as the double helix of deoxyribonucleic acid (DNA), which consists of millions of atoms and is the basic building block of life.

Fig. 9.6   Models of simple molecules. The atoms in a molecule are not just mixed together but are joined in a well-defined way.

*__Answer__   Yes, and from Charlie Chaplin, too, although the configurations of these atoms with respect to others are now quite different! The next time you have one of those days when you feel like you'll never amount to anything, take comfort in the thought that the atoms that now compose you will live forever in the bodies of all the people on earth who are yet to be.

A molecule itself can be subdivided into atoms that have chemical properties of their own. You can do a simple experiment to see this for yourself. Place two wires that are connected to the terminals of an ordinary battery into a glass of salted water (Figure 9.7). Position the wires on opposite sides of the glass so they don't touch, and you will see bubbles of gas forming on the wires. If you collect these gases, you will find that they have entirely different chemical properties from water vapor. One of the gases is hydrogen, and the other is oxygen. If the gases are mixed and ignited with a match, a quick fire or explosion results. The explosion is the violent combination of hydrogen and oxygen to again form water. The amount of energy that is suddenly released is the same amount of energy that was gradually drawn from the battery to separate the gases.

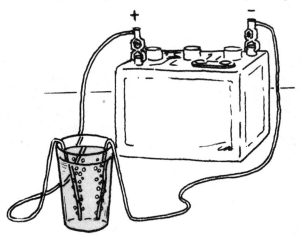

Fig. 9.7 When electric current passes through salted water, bubbles of oxygen form at the left wire and bubbles of hydrogen form at the right wire.

The energy released in common chemical burning comes from the combining of carbon and oxygen atoms to form carbon monoxide (CO) or carbon dioxide (CO_2). Carbon atoms attract oxygen much more than oxygen attracts oxygen or carbon attracts carbon. Oxygen and carbon snap together and produce energy like a slamming screen door produces sound. We get heat from the combination of oxygen and carbon, which is ordinarily in the form of molecular motion of the hot gas, but in certain circumstances the heat is so great that it generates light. This is how we get flames.

If the burning is slow enough, like burning wood in a fireplace, the carbon will attach itself to two oxygen atoms and form carbon dioxide (CO_2). If the burning takes place very rapidly, like in an automobile engine, toxic carbon monoxide (CO) is formed. (Wouldn't the air be healthier if cars burned gasoline more slowly?) These and many other arrangements release large amounts of energy and produce flames and explosions, depending on the reactions. Any process in which atoms rearrange to form different molecules is called a **chemical reaction**.

Molecules are too small to be seen with optical microscopes, but exceedingly small quantities of them are evident to our sense of smell. Our olfactory organs clearly discern noxious gases such as sulfur dioxide, ammonia, and ether. The smell of perfume is the result of molecules that rapidly evaporate from the liquid to the gaseous state and bumble around haphazardly in the air until some of them accidentally work their way into our noses. Certainly they are not drawn to noses. They are just a few of the billions of jostling molecules that, in their aimless wanderings, happen to bumble up into the nose. You can get an idea of the speed of molecules in the air when you are in your bedroom and smell food almost immediately after the oven door in the kitchen has been opened.

More direct evidence of molecules is shown in the electron microscope photograph of virus molecules composed of thousands of atoms (Figure 9.8). These giant molecules are still too small to be seen with visible light. So we see again that an atom or a molecule that is invisible to light may be visible to a shorter-wavelength electron beam.

We often say that seeing is believing. But we don't always have to see something to believe in its existence. We don't have to look inside a closed cardboard box that is too heavy to lift to know that something is in it. And we could measure its mass, for example, without benefit of light. Atoms and molecules are too small to be seen with light, but this doesn't prevent us from measuring their masses with great precision.

Fig. 9.8 Electron microscope photograph of virus molecules.

Molecular and Atomic Masses

The earliest determination of molecular and atomic masses was made with the measurements of gases at the beginning of the nineteenth century. As an example of how this is done, consider the gases of hydrogen and oxygen that are collected when water is decomposed by electricity. The mass of water before decomposition is equal to the sum of the masses of the two gases collected. Figure 9.9 shows that twice as much hydrogen gas as oxygen gas is collected. Water is composed of two parts of hydrogen to one part of oxygen, hence the chemical formula H_2O. When the masses of these two gases are compared to each other, we find that for every 2 grams of hydrogen formed, 16 grams of oxygen are formed. Since there are twice as many hydrogen molecules as oxygen, it follows that the oxygen molecule is 16 times heavier than the hydrogen molecule. The key assumption in this analysis is that twice the number of hydrogen molecules were formed as were oxygen molecules. Our observation was that twice the *volume* of hydrogen gas was formed. Our assumption, then, is that equal volumes of gas (at the same temperature and pressure) contain the same number of molecules.

This was first proposed in 1811 by the Italian physicist Amedeo Avogadro. **Avogadro's principle** states that equal volumes of gas at the same temperature and pressure contain equal numbers of molecules—regardless of the size of the molecules. This statement makes sense only

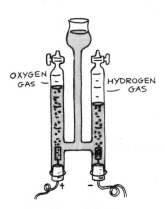

Fig. 9.9 In the electrolysis of water, twice the volume of hydrogen gas as of oxygen gas is collected.

when it is realized that in a gas the molecules are very far apart when compared with the sizes of the molecules themselves. Water vapor (steam), for example, occupies about 1700 times the volume of the same mass of liquid. If the molecules are just touching each other in the liquid, they must be quite far apart in the gas.

The theory of relative masses of molecules follows from Avogadro's hypothesis. One simply weighs equal volumes of gases and compares the mass of one sample with another. The ratio of these masses of gas is identical to the ratio of the masses of the individual molecules making up the gases. For example, if gas A is twice as heavy as an equal volume of gas B, then molecule A has twice the mass of molecule B. If the chemical formula for the molecule is known, the relative masses of individual atoms can be determined.

To compare the masses of the elements, one atom is chosen as a standard and the masses of the other atoms are expressed relative to that standard. Our present standard of atomic mass is the common carbon atom, which is assigned an arbitrary mass of 12.0000. One-twelfth of the mass of carbon is called the **atomic mass unit**, abbreviated amu. The mass of a carbon atom therefore is 12 amu. Hydrogen, the lightest atom, has a mass of 1 amu; oxygen is 16 amu; uranium, one of the heaviest, is 238 amu.

Elements, Compounds, and Mixtures

Certain solids such as gold, liquids such as mercury, and gases such as neon are composed of a single element. The substances themselves are called **elements**, for they cannot be broken down into simpler substances by chemical means. Certain other solids such as crystals of common table salt, liquids such as water, and gases such as methane are made up of elements that are chemically combined. These are called **compounds**. Salt is a compound of the elements sodium (Na) and chlorine (Cl). Although sodium is a yellowish metal that reacts violently with water, and chlorine is a poisonous greenish gas, the compound of these two elements is the harmless white crystals (NaCl) you sprinkle on your potatoes. Water (H_2O) is a compound of the elements hydrogen and oxygen, and methane (CH_4) is a compound of hydrogen and carbon. Compounds are formed only when the elements react chemically and bond with each other.

Substances mixed together without combining chemically are called **mixtures**. Hydrogen and oxygen gases form a mixture until ignited, whereupon they form the compound water. Sand combined with salt is a mixture. The most common mixture consists of nitrogen and oxygen together with a small percentage of carbon dioxide and traces of rare gases such as neon. It is the air we breathe.

Question   A mixture of hydrogen and oxygen gases when ignited will form the compound water, H_2O. How many grams of oxygen should be mixed with 1 g of hydrogen to get pure water?*

Atomic Structure   Nearly all the mass of an atom is concentrated in the **nucleus**, which occupies only a few quadrillionths the volume of an atom. The nucleus therefore is extremely dense. If bare atomic nuclei could be packed against each other into a lump 1 centimeter in diameter (about the size of a large pea), the lump would weigh 133,000,000 tons! Huge electrical forces of repulsion prevent such close packing of atomic nuclei. Each nucleus is electrically charged and repels other nuclei. Only under special circumstances are the nuclei of two or more atoms squashed into contact, and when this happens, a violent reaction takes place. This is what happens in a hydrogen bomb, and this reaction will be discussed in Chapter 32.

The principal building block of the nucleus is the **nucleon**, which in turn is composed of fundamental particles called **quarks**. When the nucleon is in its electrically neutral state, it is a *neutron;* when it is in its electrically charged state, it is a *proton*. All protons are identical; they are copies of one another. Likewise with neutrons; each neutron is like every other neutron. The known atomic nuclei are composed of from 1 to about 260 nucleons. The lighter nuclei have roughly equal numbers of protons and neutrons; more massive nuclei have more neutrons than protons. Protons have a positive electric charge. Positive electric charges repel other positive charges but attract negative charges. So like kinds of electrical charges repel one another, and unlike charges attract one another. It is the positive charge of the protons in the nucleus that attracts a surrounding cloud of negatively charged particles called **electrons** to make an atom.

Electrons make up the flow of electricity in electrical circuits. They are exceedingly light, almost 2000 times lighter than nucleons, and contribute very little mass to the atom. An electron in one atom is identical to any electron in or out of any other atom. Electrons are negatively charged and repel other electrons, but are attracted to the positively charged nucleus.

---

*__Answer__   8 g. Each H_2O molecule has two atoms of H for each atom of O. The relative masses of H and O are 1 and 16. Water has 2 g of hydrogen for every 16 g of oxygen. When two or more elements combine to form a compound, they always do so in the same proportion by mass. This is called the **law of definite proportions**, and for water it's always 8 g oxygen to 1 g hydrogen.

The number of protons in the nucleus is electrically balanced by an equal number of electrons whirling in the surrounding cloud. The atom itself is electrically neutral, so it normally doesn't attract or repel other atoms. But when atoms are close together, the negative electrons of one atom may at times be closer to the positive nucleus of another atom, which results in a net attraction between the atoms. This is why some atoms combine to form molecules.

Like the solar system, the atom is mostly empty space. The nucleus and surrounding electrons occupy only a tiny fraction of the atomic volume. If it weren't for the electrical forces of repulsion between the electrons of neighboring atoms, solid matter would be much more dense than it is. We and the solid floor are mostly empty space, because the atoms making up these and all materials are themselves mostly empty space. These same electrical repulsions prevent us from falling through the solid floor. Electrical forces keep atoms from caving in on each other under pressure. Atoms too close will repel (if they don't combine to form molecules), but when atoms are several atomic diameters apart, the electrical forces on each other are negligible.

The main characteristic that distinguishes atoms from one another is the number of electrons about the nucleus (or, equivalently, the number of protons in the nucleus). Hydrogen is the simplest element and consists of a single electron orbiting a single proton. Helium has two electrons orbiting a pair of protons and neutrons. Lithium has three electrons, and the list continues with each successive element having an additional proton in the nucleus and an additional electron in the cloud of orbiting electrons (Figure 9.10). Equal numbers of protons and electrons make the atom electrically neutral.

Fig. 9.10 The classical model of the atom consists of a tiny nucleus surrounded by electrons that orbit within spherical shells. As the size and charge of nuclei increase, electrons are pulled closer, and the shells become smaller.

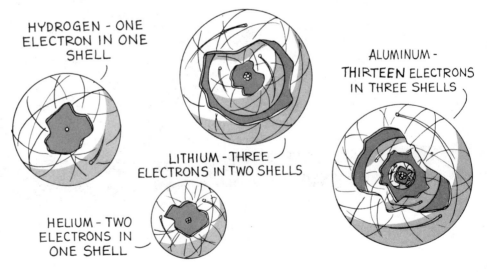

HYDROGEN - ONE ELECTRON IN ONE SHELL

HELIUM - TWO ELECTRONS IN ONE SHELL

LITHIUM - THREE ELECTRONS IN TWO SHELLS

ALUMINUM - THIRTEEN ELECTRONS IN THREE SHELLS

Atoms are classified by their **atomic number**, which is the same as the number of protons in the nucleus. Hydrogen has atomic number 1, helium 2, and so forth, in sequence to naturally occurring uranium with atomic number 92. The numbers continue through the artificially produced transuranic elements, at this writing up to 106. The arrangement of elements by their atomic numbers makes up the **periodic table**.[4] See the front endpapers of this book.

We find that electrons revolve about the nucleus in orbits or perhaps in "waves," sweeping out concentric shells at various distances from the nucleus.[5] In the innermost shell there are at most two electrons; in the second shell, eight electrons; in the third shell, eighteen. In atoms having more than eighty-six electrons, there are seven shells altogether. These shells are indicated in the periodic table by rows. Note that the uppermost row consists of only two elements, hydrogen and helium. The electrons in helium complete the innermost shell. The second row contains eight elements, beginning with lithium and ending with neon, whose electrons complete the second shell, and so on. The order in which successive shells (and subshells) are filled becomes more complicated for the heavier elements.

In the periodic table the elements are arranged vertically on the basis of similarity in electronic configuration, which is then reflected in similarities in the physical and chemical properties of the elements and their compounds. It is the configuration of electrons in the outer shell of the atom that quite literally gives life and color to the world. The configuration of the outer electrons dictates whether and how atoms bond to become molecules, the melting and freezing temperatures, and the electrical conductivity, as well as the taste, texture, appearance, and color of substances.

Antimatter

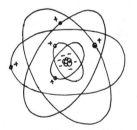

Whereas matter is composed of atoms with positively charged nuclei and negatively charged electrons, **antimatter** is composed of atoms with negative nuclei and positive electrons, or *positrons* (Figure 9.11). Positrons were first discovered in 1932, in cosmic rays bombarding the earth's atmosphere. Today antiparticles of all types are regularly produced in laboratory experiments involving nuclear physics. A positron has the same mass as an electron and the same magnitude of charge but of opposite sign. *Antiprotons* have the same mass as protons but are negatively charged. Antiparticles have the same mass as particles and differ only in their electromagnetic properties. All subatomic particles have been

Fig. 9.11 An atom of antimatter has a negatively charged nucleus surrounded by positrons.

---

[4]The enormous human effort and ingenuity that went into finding regularity and system in the chemical elements make a fascinating atomic detective story that can be found in the books suggested at the end of the chapter.

[5]The wave nature of electrons is discussed in Chapter 30.

found to have corresponding antiparticles. There are even antiquarks. Gravitational force does not distinguish between matter and antimatter. Also, there is no way to tell whether something is made of matter or antimatter by the light it emits. Only by contact could we tell, for when matter and antimatter meet, they mutually annihilate each other, and all the material is converted into radiant energy. This process, more so than any other known, results in the maximum energy output per gram of substance. It's $E = mc^2$ with a 100 percent mass conversion. (Nuclear fission and fusion, in contrast, convert less than 1 percent of the matter involved.)

Obviously there cannot be both matter and antimatter in our immediate environment, for something made of antimatter would be completely transformed to energy as soon as it touched matter. And it would transform an equal amount of normal matter into energy in the process. If the moon were made of antimatter, for example, a flash of energy would result as soon as one of our spaceships touched it. Both the spaceship and an equal amount of the antimatter moon would disappear in a burst of energy. So we know the moon is not antimatter. But what about other galaxies? It may be that the elemental building blocks in other parts of the universe consist of antimatter just as our part consists of matter. Or maybe not—we don't know.

States of Matter

Matter exists in four states: *solid, liquid, gaseous,* and *plasma.* In all states the atoms are perpetually moving. In solid state the atoms and molecules vibrate about fixed positions. If the rate of molecular vibration is increased sufficiently, molecules will shake apart and wander throughout the material, vibrating in nonfixed positions. The shape of the material is no longer fixed but takes the shape of its container. This is the liquid state. If more energy is put into the material and the molecules vibrate at even greater rates, they may break away from one another and assume the gaseous state. H_2O is a common example of this changing of states. When solid, it is ice. If we heat the ice, the increased molecular motion jiggles the molecules out of their fixed positions, and we have water. If we heat the water, we can reach a stage where continued molecular vibration results in a separation between water molecules, and we have steam. Continued heating causes the molecules to separate into atoms; if we heat the steam to temperatures exceeding 2000°C, the atoms themselves will be shaken apart, making a gas of free electrons and bare nuclei called **plasma**.

Although the plasma state is less common to our everyday experience, it may be considered the normal state of matter in the universe; the sun and other stars as well as much of the intergalactic matter are in the plasma state. All substances can be transformed from any state to another.

The following three chapters treat these states of matter in turn. We will return to atoms in more detail in Chapter 30.

Summary of Terms

Atom   The smallest particle of an element that has all of its chemical properties.

Brownian motion   The haphazard movement of tiny particles suspended in a gas or liquid resulting from bombardment by the fast-moving molecules of the gas or liquid.

Molecule   The smallest particle of any substance that has all its chemical properties; atoms combine to form molecules.

Chemical reaction   A process in which rearrangement of atoms from one molecule to another occurs.

Avogadro's principle   Equal volumes of all gases at the same temperature and pressure contain the same number of molecules.

Atomic mass unit (amu)   The standard unit of atomic mass, which is equal to one-twelfth the mass of the common atom of carbon, arbitrarily given the value of exactly 12.

Atomic number   The number of protons in the atomic nucleus; also the number of electrons surrounding the nucleus. The atomic number of an atom signifies its numerical position in the periodic table of the elements.

Antimatter   Matter composed of atoms with negative nuclei and positive electrons.

Suggested Reading

Asimov, I. *A Short History of Chemistry.* Chaps. 5 and 8. Science Studies Series. Garden City, N.Y.: Doubleday (Anchor), 1965.

Feynman, R. P., R. B. Leighton, and M. Sands. *The Feynman Lectures on Physics.* Vol. I, Chap. 1. Reading, Mass.: Addison-Wesley, 1963.

Review Questions

1. Distinguish between an atom and a molecule.

2. How many elements compose a water molecule? How many individual atoms compose a water molecule?

3. What is Brownian motion?

4. What is the principal variable that distinguishes one element from another?

5. Which atom is chosen as the standard for atomic mass measurements?

6. Distinguish between an element, a compound, and a mixture.

7. Distinguish between a nucleon, a neutron, a proton, and an electron.

8. As an atom is electrically neutral, how do the number of orbital electrons compare with the number of protons in the nucleus?

9. What kind of force prevents atoms from oozing into one another?

10. What is meant by the *atomic number* of an element?

11. How does antimatter differ from matter?

12. What are the four common states of matter, and what role does temperature play in them?

Home Projects

1. A candle will burn only if oxygen is present. Will a candle burn twice as long in an inverted liter jar as it will in an inverted half-liter jar? Try it and see.

2. By the electrolysis of water, collect samples of oxygen and hydrogen gas. To do this, fasten a couple of pieces of insulated wire, 30 centimeters or thereabouts, bared at the ends, to a couple of pieces of carbon rod. (One source of carbon rod is the inside of a flashlight cell. After breaking open the cell and removing the rod, file it in the middle and snap it into two pieces.) Wrap the end of each wire tightly around the rods and cover the exposed wire with plastic tape so that the wire metal won't interact with the conducting solution. Fasten the other ends of the wires, also bared, to the terminals of a 6- or 12-volt storage battery. Smaller $1\frac{1}{2}$-volt flashlight cells will do if three or four are connected in series (plus to minus, plus to minus, plus to minus—Chapter 21). One electrode is negative because it is connected to the negative terminal of the battery. This is the *cathode.* The other electrode is positive and is called the *anode.*

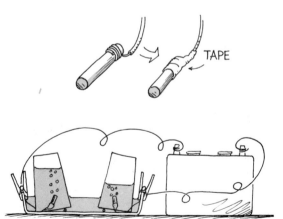

Dissolve a large handful of baking soda in about a liter of water. The baking soda in water will make an electrically conducting solution (salt will also do). Half-fill a shallow cooking dish with the solution. Fill two small drinking glasses with the solution, cover them with a piece of glass (pocket mirror) or cardboard, and invert them in the cooking dish. They should be completely full—without air pockets. Place pieces of wood from a match under the mouths of the glasses so they are slightly raised from the bottom of the dish. Do this while carefully placing the pieces of carbon underneath each glass. A pair of clothespins will help to keep the wires steady.

You'll notice that bubbles start to rise from the carbon electrodes. The details are briefly as follows: H_2O is a polarized molecule. Dissociation occurs in the aqueous solution as the positive side of the molecule is attracted to the cathode and the negative side is attracted to the anode. (We will study later the fundamental rule of electricity: like charges repel and opposite charges attract.) This results in an electric current of positive and negative ions that flows between the carbon electrodes. The positively charged ions are hydrogen nuclei (protons), which gather at the cathode to pick up electrons to become hydrogen atoms; these pair off to become hydrogen molecules and form the hydrogen gas that you see bubbling to the surface. You will note that twice as much of this gas is collected as oxygen gas, which is similarly collected in the other glass. After both glasses are full of gas, slip the sheet of glass under the one that filled first and remove it. Hold a lighted match close to it, and you'll see that the gas in the glass will burn with a quiet, almost invisible flame, or it may produce a loud bang. This is hydrogen. Remove the other glass; light a long wooden splinter, blow out the flame, and then insert the glowing splinter into the glass. The splinter will burst into flame and burn furiously. Or heat a piece of iron or steel wool on the stove until it begins to glow and then place it in the oxygen and watch the iron burn.

Exercises

1. A cat strolls across your backyard. An hour later a dog with his nose to the ground follows the trail of the cat. Explain this occurrence from a molecular point of view.

2. If all the atoms of a body remained intact, would the body have any odor?

3. The average speed of a perfume vapor molecule at room temperature is about 650 meters per second. The speed at which the scent of perfume travels, however, is much less than this. Why?

4. Why is Brownian motion apparent only for microscopic particles?

5. Why are atoms visible with electron microscopes, yet invisible with even ideal optical microscopes?

6. What is the principal variable that determines whether atoms form a solid or liquid or gaseous or plasma state?

7. If the elements were arranged in increasing order by atomic mass rather than by atomic number, would the sequence be the same? (To answer this, check the periodic table inside the front cover.)

8. In what way do the number of protons in the atomic nucleus determine the chemical properties of an element?

9. Which of the following are pure elements: H_2, H_2O, He, Na, NaCl, H_2SO_4, U?

10. If two protons and two neutrons are removed from the nucleus of an oxygen atom, what nucleus remains?

11. You could swallow a capsule of germanium without ill effects. But if a proton were added to each of the germanium nuclei, you would not want to swallow the capsule. Why? (Consult the periodic table of the elements.)

12. Which of the following elements would you predict to have properties most like those of silicon (Si): aluminum (Al), phosphorus (P), or germanium (Ge)? (Consult the periodic table of the elements.)

13. Which would be the more valued result: that of taking protons from or adding protons to the nuclei of gold atoms? Explain.

14. How many grams of oxygen are there in 18 grams of water?

15. How many grams of hydrogen are there in 16 grams of methane gas?

16. If 50 cubic centimeters of alcohol is mixed with 50 cubic centimeters of water, the volume of the mixture is only 98 cubic centimeters. Can you offer an explanation for this?

17. A teaspoon of an organic oil dropped on the surface of a quiet pond spreads out to cover almost an acre. The oil film has a thickness equal to the size of a molecule. If you know the volume of the oil, and measure the film area, how can you calculate the thickness, or the size of the molecule?

18. From an atomic recycling point of view, discuss the practice of cremation as opposed to confining one's remains in a sealed tomb.

19. In what sense can you truthfully say that you are a part of every person around you?

20. What are the chances that at least one of the atoms exhaled by your very first breath will be in your next breath?

21. There are approximately 10^{23} H_2O molecules in a thimbleful of water, and 10^{46} H_2O molecules in the earth's oceans. Suppose that Columbus had thrown a thimbleful of water into the ocean and that the water molecules have by now mixed uniformly with all the water molecules in the oceans. Can you show that if you dip a sample thimbleful of water from anywhere in the ocean, you'll have at least one of the molecules that was in Columbus's thimble? (*Hint:* The ratio of the number of molecules in a thimble to the number of molecules in the ocean will equal the ratio of the number of molecules in question to the number of molecules the thimble will hold.)

22. There are approximately 10^{22} molecules in a single medium-sized breath of air and approximately 10^{44} molecules in the atmosphere of the whole world. Now the number 10^{22} squared is equal to 10^{44}. So how many breaths of air are there in the world's atmosphere? How does this number compare with the number of molecules in a single breath? If all the molecules from Julius Caesar's last dying breath are now thoroughly mixed in the atmosphere, how many of these on the average do we inhale with each single breath?

23. Assume that the present world population of about 4×10^9 people is about one-thirtieth the number of people who ever lived on earth. Then about how many more molecules are there in a single breath compared to the total number of people who have ever lived? Can you show that the number of molecules in a single breath compared to the total number of people who ever lived is about the same as the number of people who ever lived compared to one person?

24. Somebody told your friend that if an antimatter alien ever set foot upon the earth, the whole world would explode into pure energy. Your friend looks to you for verification or refutation of this claim. What do you say?

10 Solids

The classification of solids began when the earliest humans distinguished between rocks for shelter and rocks for tools. Since then the list of solids has grown without pause. Yet only within the last 75 years has the actual structure of solids been confirmed beyond doubt.

Prior to this it was thought that the content of a solid was what determined its characteristics—what made diamonds hard, lead soft, iron magnetic, and copper electrically conducting. It was believed that to change a substance, one merely varied the contents. We have since found that the characteristics of solids have to do with *structure*—that is, the particular order and bonding of atoms that make up the material.

This reemphasis from study of the content to study of the structure of solid materials has changed the role of investigators from being finders and assemblers of materials to actual makers of materials. In the laboratories of today, people daily extend the long list of synthetic materials.

A striking example of the structure of solid matter is shown on the opposite page. The picture is a micrograph produced in 1958 by Erwin Müller. Dr. Müller took an extremely fine platinum needle, ending in a hemisphere forty-millionths of a centimeter in diameter. He enclosed it in a tube of rarefied helium and applied to it a very high positive voltage (25,000 volts). Under such electrical pressure the helium atoms "settling" on the atoms of the needle tip were ionized and torn off as positive ions, streaming away from the tip in a direction almost perpendicular to the surface at every point. They then struck a fluorescent screen, producing the picture of the needle tip, which is enlarged approximately 750,000 times. Clearly, the platinum is crystalline, the atoms arranged like oranges in a grocer's display. Although the photograph is not of the atoms themselves, it shows the positions of the atoms and reveals the microarchitecture of the solids that make up our world.

How Atoms Are Ordered in a Solid

It was realized, near the turn of the nineteenth century, that many solids are composed of crystals. As the atomic theory of matter was being established, crystals themselves were thought to be made up of a particular arrangement of atoms. This was confirmed in 1912 with the use of X rays. Atoms in a crystal are very close together, most between 30 and 100 nanometers apart,[1] which is about the same as the wavelength of X rays. The German physicist Max von Laue discovered that a beam of X rays directed upon a crystal is diffracted into a pattern characteristic

---

[1]For your information, 1 nanometer (nm) = 10^{-9} m.

165

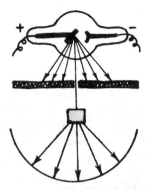

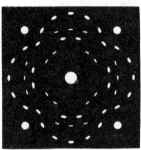

Fig. 10.1 X-ray determination of crystal structure. The photograph of salt is a product of X-ray diffraction. Rays from the X-ray tube are blocked by a lead screen except for a narrow beam that hits the crystal of sodium chloride (common table salt). The radiation that penetrates the crystal to the photographic film makes the pattern shown. The white spot in the center is the main unscattered beam of X rays. The size and arrangement of the other spots result from the latticework structure of sodium and chlorine atoms in the crystal. Every crystalline structure has its own unique X-ray diffraction picture. A crystal of sodium chloride will always produce this same design. (Adapted from *Matter,* Life Science Library, 1965.)

of the crystal, called a *diffraction pattern* (Figure 10.1). The diffraction patterns made by X rays on photographic film reveal the crystal to be a neat mosaic of atoms sited on a regular lattice, like a three-dimensional chessboard or a child's jungle gym. Metals such as iron, copper, and gold have relatively simple crystal structures. Tin and cobalt are only a little more complex. And all are made up of interlocking crystals, each almost perfect, each with the same regular lattice but at a different inclination to the crystal nearby. These metal crystals can be seen when a metal surface is cleaned (etched) with acid (Figure 10.2). You can see them on the surface of galvanized iron that has been exposed to the weather, or on brass doorknobs that have been etched by the perspiration of hands.

Fig. 10.2 A deeply etched surface of single-crystal antimony.

Von Laue's photographs of the X-ray diffraction patterns fascinated a father-and-son team of English scientists, William Henry Bragg and his son William Lawrence Bragg. Through their own experimentation, the Braggs developed a mathematical formula that showed just how X rays should rebound from the various regularly spaced atomic layers in a crystal. With this formula and an analysis of the pattern of spots in a diffraction pattern, they could determine the atomic layer thicknesses and therefore the distances between the atoms in a crystal.

How Atoms Are Held Together in a Solid

A solid can be amorphous or crystalline. In the *amorphous* state, the atoms and molecules exist in a random arrangement, vibrating about fixed positions. In the more common crystalline state, the atoms are in a three-dimensional orderly arrangement—a crystalline latticework of atoms (Figure 10.3). Each atom vibrates about its own fixed position and is unable to move through the lattice. Atoms are tied together in this lattice by an electrical *bonding* force. There are four principal types of **atomic bonding** in solids.

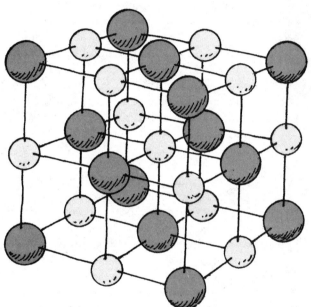

Fig. 10.3 Sodium chloride crystal: the large spheres represent chlorine atoms.

In **ionic bonding** an electron of one element is lost to another element, creating charged atoms called *ions*. The electrical attraction between oppositely charged neighboring ions produces an extremely strong bond. The atoms of sodium and chlorine in the crystals of common table salt are held together by ionic bonding. A sodium atom gives up its single outer electron to a chlorine atom that has a single "open space" in its outer shell, and this results in a strong attraction between the oppositely charged ions. High temperatures are required to shake this bond apart and melt crystals of table salt.

Covalent bonding also consists of the sharing of one or more electrons of one atom with neighboring atoms. But instead of circling one atom, the outer electrons circle both atoms and bind them into one package.

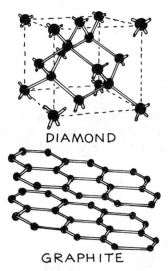

DIAMOND

GRAPHITE

Fig. 10.4 Both diamond and graphite are pure carbon. In a diamond, all four outer electrons are localized in covalent bonds, whereas in graphite only three are.

Covalent bonds are very strong, and they can produce substances with quite different physical properties. The diamond, a form of pure carbon, is extremely hard. But graphite, another form of pure carbon also covalently bonded, is so soft and slippery that it is useful as a lubricant—in door locks, for example. The difference in the physical properties comes about because carbon atoms in a diamond are bonded to neighboring carbon atoms in all directions, whereas in graphite the bonding takes place in layers (Figure 10.4). The crystalline layers in graphite are relatively far apart, which results in a very weak bond between the atoms of adjacent layers. This is why the layers freely slip over one another to provide the excellent lubricating property of graphite.

Metallic bonding is like covalent bonding, but rather than having a situation in which particular electrons are shared by particular atoms, the outer electrons of atoms in metals are free to migrate throughout the atomic lattice. They make up a moving cloud, loosely coupling atoms they drift between. This mobile bonding accounts for the conductivity of electricity and heat in metals. When an electric voltage is applied, these "conduction electrons" drift through the atomic lattice and make up an electric current; or when heat is applied to a metal solid, the conduction electrons increase their speed and move randomly, colliding with the atomic lattice. This is how a metal conducts heat so easily.

The weakest bond in solid crystals is provided by **van der Waals' forces**. These forces are the outcome of tiny imbalances in the electrical charge distributions in matter. Electrons occasionally group together on one side of a molecule, producing for a brief time a negative charge there. At the same time, a positive charge of the same size occurs on the opposite side because of the absence of electrons there. A molecule with this distortion is said to be **polarized**; that is, it has opposite charge distributions on opposite sides. When two polarized molecules are close to each other, the oppositely charged sides produce a weak bond between them. Molecules bonded in this way can easily separate from the crystal. An example of this is the sublimation of dry ice; solid carbon dioxide passes directly from the solid to the gaseous state, bypassing altogether the liquid state.

It is possible to have more than one kind of bond in a crystal. The bonds—whether ionic, covalent, metallic, or van der Waals—are not rigid but are elastic enough to permit vibrations of the atoms about their positions in the lattice. If heat is applied, the vibrations increase until the bonds break. At this point, the crystal melts.

Question   What kind of force binds atoms together to form molecules and underlies all the varieties of interatomic bonding?*

---

*__Answer__   Electrical force.

Liquid Crystals

Fig. 10.5 Liquid crystals form the digits on many digital watches.

The melting temperature is precise for pure crystalline substances and is useful for identification purposes. In certain cases the crystalline properties of the solid state carry over to the liquid state, through a sort of in-between state where the liquid shows degrees of molecular orderliness. This is known as the *liquid crystal* state. There are many compounds that exist in this state at ordinary temperatures. Depending on their viscosities, liquid crystals can be poured like water, easily assuming the shape of their containers. Due to selective reflection, white light striking a film of these crystals causes a different wavelength to be reflected for different angles of incidence, resulting in iridescent colors. Since the crystalline structure is inherently weak, any force that causes this structure to shift to realign itself changes its optical properties. Materials made with liquid crystals reflect colors that can be varied and controlled by thermal, acoustic, electrical, magnetic, or even mechanical means. The list of applications is growing rapidly. Wristwatches and clock faces that dissolve from one display to another are made with liquid crystals. So are windowpanes that can be dimmed in sunlight. An exciting application is full-color, flat TV screens that can be hung on a wall like a painting. Liquid crystals are presently the object of much interest and research.

Alloys

A mixture of copper and tin in a molten state will cool to form a harder, stronger, and more durable solid than either copper itself or tin itself. This solid is bronze, historically the first of the materials we call **alloys**. Brass, a mixture of copper and zinc, is another alloy. Steels are alloys of iron with carbon, often with the addition of other elements for special purposes, like chromium for rustlessness and silicon for high permeability. Gallium arsenide is a newer alloy that is a semiconductor used in electronics for such solid-state devices as the light-emitting diodes in calculator readouts. All alloys are made by basically the same process, by mixing two or more molten metals in varying proportions and letting the mixture cool and solidify.

Of all the metals in the periodic table, only seven are primary alloying metals: iron, copper, zinc, lead, tin, aluminum, and magnesium. These and fifteen other metals have been successfully combined with forty other elements to form alloys. Although the list of possible alloys is enormous, not all elements will combine. An alloy of bismuth and arsenic, for example, would provide some very interesting electronic properties, but attempts on earth to form this alloy have failed. A major problem is that bismuth is a very heavy element and settles to the bottom of any potential mixture with lighter arsenic. If one could "turn off" gravity long enough, the chances for forming this alloy and others would be greatly enhanced.

Metallurgists are therefore looking more and more to the prospects of orbiting laboratories in space where zero-*g* conditions enable the formation of alloys that cannot be made on earth. Only very recently have new "wonder metals" and other unique materials emerged from experiments in space alloying technology. These unique materials promise to add greatly to our technological alternatives.

Density

The masses of atoms and the spacing between atoms determine the **density** of materials. We think of density as the "lightness" or "heaviness" of materials. It is a measure of the compactness of matter, of how much mass is squeezed into a given space; it is the amount of matter per unit volume:[2]

$$\text{Density} = \frac{\text{mass}}{\text{volume}}$$

The densities of a few materials are given in Table 10.1. Mass is measured in grams or kilograms and volume in cubic centimeters (cc) or cubic meters (m^3). A gram of material is defined as equal to the mass of 1 cubic centimeter of water at a temperature of 4°C. Therefore, water has a density of 1 gram per cubic centimeter.[3] Gold, which has a density of 19.3 grams per cubic centimeter, is therefore 19.3 times more massive than an equal volume of water. Osmium, a hard, bluish-white metallic element, is the densest substance on earth. Although the individual osmium atom is less massive than individual atoms of gold, mercury, lead, and uranium, the close spacing of osmium atoms in crystalline form gives it the greatest density. More atoms of osmium fit into a cubic centimeter than other more massive and more widely spaced atoms.

Table 10.1   Densities of a few substances

Material	Grams/cubic centimeter	Kilograms/cubic meter
Ice	0.92	920
Water	1.0	1,000
Aluminum	2.7	2,650
Tin	7.3	7,290
Iron	7.8	7,800
Brass	8.6	8,600
Copper	8.9	8,930
Silver	10.5	10,500
Lead	11.3	11,370
Mercury	13.6	13,600
Gold	19.3	19,320
Platinum	21.4	21,400
Osmium	22.48	22,480

[2]Density can also be expressed by the amount of weight a body has, compared to its volume:

$$\text{Density} = \frac{\text{weight}}{\text{volume}}$$

Weight density is common to British units, where 1 cubic foot of fresh water (almost $7\frac{1}{2}$ gallons) weighs 62.4 pounds. So we say fresh water has a weight density of 62.4 lb/ft³. Salt water is a bit denser, 64 lb/ft³.

[3]A cubic meter is a sizable volume and contains a million cubic centimeters, so there are a million grams of water or, equivalently, a thousand kilograms of water in a cubic meter. Hence, 1 g/cm³ = 1000 kg/m³.

Questions
1. What is the density of 100 kilograms of water?*

2. Which has the greater density—10 kilograms of lead or 100 kilograms of aluminum?†

3. Which has the greater density—a single uranium atom or the whole world?‡

Elasticity

When an object is subjected to external forces, it usually undergoes changes in size or shape or both. These changes depend on the arrangement and bonding of the atoms making up the material.

A weight hanging on a spring causes the spring to stretch. Additional weight stretches the spring still more. The stretch is directly proportional

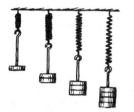

Fig. 10.6 The stretch of the spring is directly proportional to the applied force. If the weight is doubled, the spring stretches twice as much.

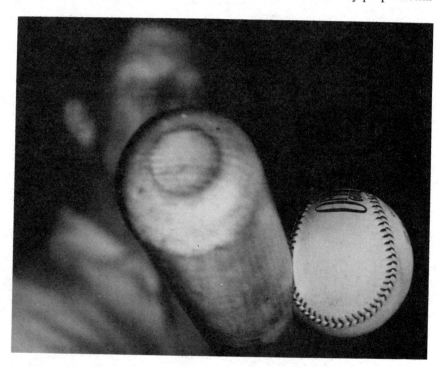

Fig. 10.7 A baseball is elastic.

*Answer   The density of *any* amount of water is 1000 kg/m³, which means that the mass of water that would exactly fill a cubic-meter tank would be 1000 kg. Can you see that the *volume* of 100 kg of water must be 1/10 m³? And that $\frac{100 \text{ kg}}{0.1 \text{ m}^3} = \frac{1000 \text{ kg}}{1 \text{ m}^3}$?

†Answer   Density is a *ratio* of weight or mass per volume, and this ratio is greater for any amount of lead than for any amount of aluminum—see Table 10.1.

‡Answer   A single uranium atom is more dense. If the world were composed entirely of uranium atoms, except for the effects of interatomic spacing and closer packing in the compressed core, the densities of each would be the same.

to the applied force (Figure 10.6).[4] When we remove the weights, the spring returns to its original length. When a batter hits a baseball, he temporarily changes its shape. When an archer shoots an arrow, she first bends the bow, which springs back to its original form when the arrow is released. The spring, the baseball, and the bow are examples of *elastic* objects. **Elasticity** is that property of a body by which it experiences a change in shape when a deforming force acts on it and by which it returns to its original shape when the deforming force is removed.

Materials that do not resume their original shape after being distorted are said to be *inelastic*. Clay, putty, and dough are elastic materials. Lead is relatively inelastic, since it is easy to distort permanently.

Steel is an elastic material. It can be stretched and it can be compressed. Because of its strength and elastic properties, it is used to make not only springs but construction girders as well. Vertical girders of steel used in the construction of tall buildings undergo only slight compression. A typical 25-meter-long vertical girder used in high-rise construction is compressed only a millimeter when it carries a 10-ton load. Most deformation occurs when girders are used horizontally, where the tendency is to sag under heavy loads.

Horizontal beams undergo both compression and stretching. To understand this, first consider the heavy horizontal branches of the tree shown in Figure 10.8. The branches tend to bend downward due to gravity, which results in enormous tensions in the wood fibers. Wood fibers in the top part of the branches are being stretched, and fibers in the lower part are being compressed. Between, in the middle, fibers make up a neutral layer. Fibers there are neither stretched nor compressed.

Fig. 10.8 The upper half of the branches is being stretched by its weight, while the lower half is being compressed. In what location is the wood neither compressed nor stretched?

---

[4]This relationship, which applies only within the limits of permanent distortion of the material being stretched, was noted by Robert Hooke in the mid-seventeenth century and is called Hooke's law, $F \sim d$.

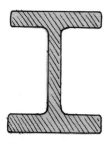

Fig. 10.9 Most of the steel in an I-beam is in the upper and lower flanges. Why?

Back to steel beams: Have you ever wondered why the cross section of steel girders has the form of an I (Figure 10.9)? Most of the material in these I-beams is concentrated in large flanges at the top and bottom, whereas the web joining the flanges is of relatively thin construction. Why is this the shape? If the beam is horizontal and supported at both ends, the situation is just the opposite of the tree branches in Figure 10.8; it tends to sag. The top flange tends to be squeezed shorter, while the lower flange is stretched (Figure 10.10). So the tension in the top flange tends to resist compression, while the bottom flange tends to resist stretching. Between the top and bottom flanges is a tension-free region that acts principally to connect the top and bottom flanges together. This is the neutral layer, where comparatively little material is needed. An I-beam is nearly as strong as if it were a solid bar, and it is considerably lighter.

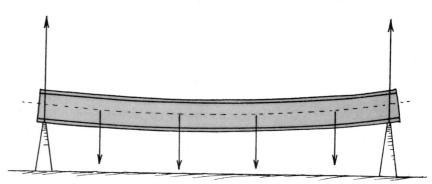

Fig. 10.10 Above the neutral layer (dotted line) the beam is compressed, and below it the beam is stretched.

Question   If you had to make a hole through one of the tree branches shown in Figure 10.8 in a location that would weaken it the least, would you bore it through the top, the middle, or the bottom?*

Scaling†   Did you ever notice how strong an ant is for its size? An ant can easily lift much more than its own weight. What if an ant could somehow increase, in its natural proportion, to the size of a human? Would this "super ant" be many times stronger than a human? Could it lift loads heavier than

---

*****Answer**   Through the neutral layer in the middle. If you drilled above or below this layer, you would considerably weaken the branch. Do you see why?

†Material in this section is based on two delightful and informative essays. "On Being the Right Size," from *Possible Worlds* by J. B. S. Haldane. Copyright 1928 by Harper & Row, Publishers, Inc. Renewed 1956 by J. B. S. Haldane. By permission of Harper & Row, Publishers, Inc. and Chatto and Windus Ltd. "On Magnitude," from *On Growth and Form* by Sir D'Arcy Wentworth Thompson (New York: Cambridge University Press). Used with permission.

those that an elephant can? The answer is no. Such an ant would not be able to lift its own weight off the ground. Its legs would probably break under its own weight. Let's consider a simpler example. How would a typical football player fare if he were somehow doubled in size—that is, if he were a giant twice as tall, twice as wide, his bones twice as thick, and every linear dimension increased by a factor of 2?[5] Could he lift a barbell equal to his new weight over his head? And what would this weight be? In doubling his height, is his weight doubled? The answer is no. Both his mass and weight are increased by a factor of 8. If his mass were initially 100 kilograms, his new mass would be 800 kilograms! He would have trouble doing push-ups, let alone trying to lift his own weight. To see why this is so, consider the simplest case of a solid cube of matter, 1 centimeter on a side. To make the numbers easy, we'll assume the cube has a density of 1 gram per cubic centimeter.

A 1-cubic-centimeter cube would have a cross section of 1 square centimeter. That is, if we sliced through the cube parallel to one of its faces, the sliced area would be 1 square centimeter. Compare this cube to a cube that has double the linear dimensions, a cube 2 centimeters on each side. Its cross-sectional area will be 2 × 2 (or 4) square centimeters and its volume 2 × 2 × 2 (or 8) cubic centimeters. At the same density of 1 gram per cubic centimeter, its mass would be 8 grams. Investigation of Figure 10.11 shows that the area grows as the square of the linear dimensions increases; and the volume, mass, and weight grow as the cube of the linear dimensions increases.

Fig. 10.11 As the linear dimensions of an object change by some factor, the area changes as the square of this factor, and the volume (and hence the weight) changes as the cube of this factor. We see that when the linear dimensions are doubled (factor = 2), the area grows by $2^2 = 4$, and the volume grows by $2^3 = 8$.

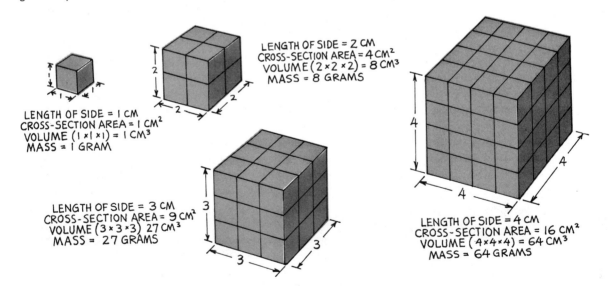

LENGTH OF SIDE = 1 CM
CROSS-SECTION AREA = 1 CM²
VOLUME (1 × 1 × 1) = 1 CM³
MASS = 1 GRAM

LENGTH OF SIDE = 2 CM
CROSS-SECTION AREA = 4 CM²
VOLUME (2 × 2 × 2) = 8 CM³
MASS = 8 GRAMS

LENGTH OF SIDE = 3 CM
CROSS-SECTION AREA = 9 CM²
VOLUME (3 × 3 × 3) 27 CM³
MASS = 27 GRAMS

LENGTH OF SIDE = 4 CM
CROSS-SECTION AREA = 16 CM²
VOLUME (4 × 4 × 4) = 64 CM³
MASS = 64 GRAMS

[5]It is interesting to note that humans are already giants—one of the largest of all the animals. The names of all animals larger than humans can be written on a single sheet of paper, while the names of those that are smaller would fill volumes.

The volume multiplies much more than the corresponding increase of cross-sectional area. Although we use the simple example of a cube in the figure, the principle applies to an object of any shape. Double the size of any object and you multiply its area four times, but its volume (and hence mass and weight) increases eight times. For our football player, the cross section of his bones increases fourfold when his height and other linear dimensions double, but his weight increases eight times. The fibers in his bones are then subjected to twice the stress. Unless his legs were increased out of proportion in thickness or constructed with a material stronger than bone, he would have to walk very carefully and avoid jumping. Otherwise, his legs would be in splints much of the time. If the ant's height, width, and length were increased by 100, the cross section of its legs would go up as the square of this increase $(100)^2$ or 10,000 times, but its volume and weight would increase by $(100)^3$ or 1,000,000. Its legs would be too thin to support its weight.

We find in nature that large animals have disproportionally thick legs compared to small animals. The smaller the size of a creature, the thinner its legs are proportionally. Consider the different legs of an elephant and a daddy longlegs.

It is interesting to compare the overall surface of an object with its volume. A study of Figure 10.12 shows that as the size of an object is increased, the volume increases disproportionally to the surface area. The increase in volume is greater than the corresponding increase in surface area. Conversely, as an object becomes smaller, its surface area increases when compared to its volume, mass, or weight. This principle applies to

Fig. 10.12 As the size of an object increases, there is a greater increase in volume than in surface area; as a result, the ratio of surface area to volume decreases.

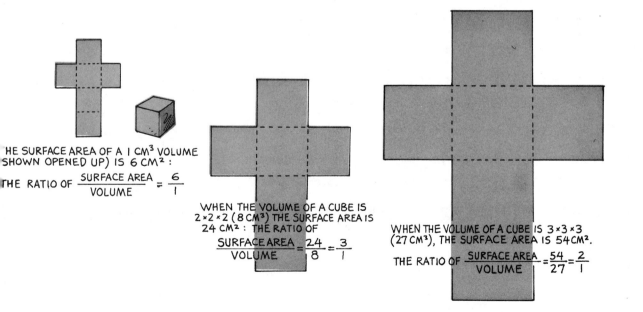

THE SURFACE AREA OF A 1 CM³ VOLUME (SHOWN OPENED UP) IS 6 CM² :

THE RATIO OF $\dfrac{\text{SURFACE AREA}}{\text{VOLUME}} = \dfrac{6}{1}$

WHEN THE VOLUME OF A CUBE IS 2×2×2 (8 CM³) THE SURFACE AREA IS 24 CM² : THE RATIO OF $\dfrac{\text{SURFACE AREA}}{\text{VOLUME}} = \dfrac{24}{8} = \dfrac{3}{1}$

WHEN THE VOLUME OF A CUBE IS 3×3×3 (27 CM³), THE SURFACE AREA IS 54 CM². THE RATIO OF $\dfrac{\text{SURFACE AREA}}{\text{VOLUME}} = \dfrac{54}{27} = \dfrac{2}{1}$

objects of any shape. A good cook knows that more skin results when peeling 5 kilograms of small potatoes than when peeling 5 kilograms of large potatoes. Smaller objects have more surface area per kilogram. The surface of ice cools the beverage in which it is immersed; therefore, an ice cube will cool the beverage much faster if it is crushed because the surface area is multiplied many times. The rusting of iron is a surface phenomenon. Iron rusts when exposed to moist air, but it rusts much faster, and is soon eaten away, if it is first reduced to a heap of small filings.

The fact that smaller things have proportionally more surface area is especially important to living organisms. The cells of living organisms obtain nourishment by diffusion and osmosis through the cell membranes. Small cells have a large surface area compared to their size, and this is crucial to their existence. As cells increase in size, area increases, but not in proportion to volume. The surface area per volume decreases. For example, if the surface area doubles, the volume has increased almost threefold; if the surface area increases fourfold, the corresponding volume and mass have increased by eight times. So we see the cell cannot adequately nourish itself because the diffusion rate increases only in proportion to the increase in surface area and not in proportion to the more important and greater increase in volume.

Consider falling bodies. Air resistance to movement in the air is proportional to the surface of the moving object. Small bodies therefore fall slower than large bodies because of their greater surface area per weight. An insect can fall from a high ceiling to the floor without harm. Because of its relatively great surface area, its terminal velocity is quite modest. An insect can fall unharmed from heights that would be fatal to larger creatures. Whereas a human needs a parachute that increases surface area in order to fall safely, an insect in a sense *is* its own parachute. If a creature's length, breadth, and height are each divided by 10, its weight is reduced a thousandth, but its surface only a hundredth. This means that there is ten times as much resistance compared to the pull of gravity for the smaller creature.

Although in the case of falling the relatively large surface area of an insect is an asset, it is a serious liability if the insect is wet. A person of average size coming out of a bath carries about $\frac{1}{2}$ kilogram or so of water over the body. A mouse emerging from a bathtub carries about its own weight of water, and a wet fly carries many times its own weight. Surface tension is dangerous for small creatures. When an insect drinks water, it is in danger of falling into the grasp of surface tension. If it gets wet, it may stay wet until it drowns. Some insects, like water beetles, have a waxy surface that resists wetness.

The heat a creature radiates is proportional to its surface area. The ultimate source of this heat is the food the creature eats. The amount of food we human beings eat each day is a very small fraction of our body weight, whereas a mouse eats about a quarter of its own weight in food every day to stay warm. This is because a mouse has a relatively large surface area and loses more heat per body weight than we do. An elephant, on the other hand, has relatively less surface area compared to its weight and consequently consumes only a tiny fraction of its body weight in food each day.

Thin people have relatively more radiating surface area than fat people and, as a result, are the first to feel chilly in cool weather but are usually more comfortable than fat people in hot weather. Mammals adapted to live in the hot desert are usually slender, which results in more radiating surface area per body weight. The fat necessary as food reserves in such animals is carried in localized regions of the body; hence, camels and certain cattle have humps, and desert sheep have fat tails. The long tail of the vervet monkey makes up one-third of its total body mass and is used, in addition to contributing to balance, to regulate body temperature. The ears of the African elephant do likewise. Elephants have very little surface area compared to their great body weight—that is, when their ears are not

Fig. 10.13 The long tail of the monkey not only helps maintain balance but also effectively radiates heat.

taken into account. The large ears of the African elephant greatly increase overall surface area, which facilitates cooling (Figure 10.14). So the next time you visit the zoo, think physics!

Fig. 10.14 The African elephant has less surface area compared to its weight than other animals. It compensates for this with its large ears, which significantly increase its radiating surface area and promote cooling.

Summary of Terms

Atomic bonding The linking together of atoms to form solids. The different kinds of bonding are ionic, covalent, and metallic bonding, and van der Waals' forces.

Alloy A substance composed of two or more metals or of a metal and a nonmetal.

Density The mass of a substance per unit volume:

$$\text{Density} = \frac{\text{mass}}{\text{volume}}$$

Or it can be expressed as weight/volume, in which case it is called *weight density*.

Elasticity The property of a material by which it experiences a change in shape when a deforming force acts on it and by which it returns to its original shape when the deforming force is removed.

Suggested Reading

For more about the relationships among the size, area, and volume of objects, read these essays: "On Being the Right Size," by J. B. S. Haldane, and "On Magnitude," by Sir D'Arcy Wentworth Thompson. Both are in J. R. Newman, ed. *The World of Mathematics*, Vol. II. New York: Simon & Schuster, 1956.

Morrison, P. *Powers of Ten.* New York: W. H. Freeman, 1982.

Review Questions

1. How do the arrangements of atoms differ in a crystalline and noncrystalline substance?

2. What are the four types of atomic bonding?

3. A diamond and graphite are both composed of carbon atoms, yet physically different from each other. Why?

4. Distinguish between iron and steel.

5. What is an electrically *polarized* molecule?

6. Why does a cupful of water have the same density as a lakeful of water?

7. If twice as much force is applied to an elastic spring, by how much will its elongation be increased?

8. Why is the cross section of construction girders in the shape of an I?

9. If the linear dimensions of an object are doubled, by how much does the overall area increase? By how much does the volume increase?

10. As the volume of an object is increased, its surface area is also increased, but the ratio of its surface area *per volume* (or weight) is *decreased*. True or false? Explain.

11. Which will cool a drink quicker—a 10-gram ice cube or 10 grams of crushed ice? Explain.

12. Which has more skin—an elephant or a mouse? Which has more skin *per body weight*—an elephant or a mouse?

Home Projects

1. If you live in a region where snow falls, collect some snowflakes on black cloth and examine them with a magnifying glass. You'll see that all the many shapes are hexagonal crystalline structures, among the most beautiful sights in nature.

2. Simulate atomic close packing with a couple of dozen or so pennies. Arrange them in a square pattern so each penny inside the perimeter makes contact with four others. Then arrange them hexagonally, so contact with six others occurs. Compare the areas taken by the same number of pennies close packed both ways.

Exercises

1. Silicon is the chief ingredient of both glass and semiconductor devices, yet the physical properties of glass are different from those of semiconductor devices. Explain.

2. Distinguish the following as elements or alloys: iron, steel, copper, bronze, brass, chromium, nickel.

3. Is iron necessarily heavier than cork? Explain.

4. What happens to the density of a loaf of bread when you squeeze it?

5. The uranium atom is the heaviest and most massive of the common naturally occurring elements. Why, then, isn't a solid bar of uranium the densest metal?

6. A certain spring stretches 4 additional centimeters when a load of 10 newtons is suspended from it. How much will the spring stretch if an identical spring also supports the load as shown in *a* and *b?* (Neglect the weights of the springs.)

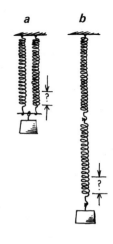

7. Why does a suspended spring stretch more at the top than at the bottom?

8. A thick rope is stronger than a thin rope of the same material. How much stronger is a twice-as-thick rope? How much stronger is a twice-as-long rope?

9. Tensions of compression and of stretching occur in a partially supported horizontal beam when it sags due to gravity or supports a load. Make a simple sketch to show a means of supporting the beam so that stretching occurs at the top part and compression at the bottom. Sketch another case where compression is at the top and stretching occurs at the bottom.

10. Why do you suppose that girders are so often arranged to form triangles in the construction of bridges and other structures? (Compare the stability of three sticks nailed together to form a triangle with four sticks nailed to form a rectangle, or any number of sticks to form similar geometric figures. Try it and see!)

11. Consider two bridges that are exact replicas of each other except that every dimension in the larger is exactly twice that of the other, that is, twice as long, structural elements twice as thick, etc. Which bridge is stronger?

12. A candymaker making taffy apples decides to use 100 kilograms of large apples rather than 100 kilograms of small apples. Will the candymaker need to make more or less taffy?

13. Will a person twice as heavy as another use more or less than twice as much suntan lotion at the beach?

14. If the linear dimensions of a storage tank are reduced to half their former values, by how much does the overall surface area of the tank decrease? By how much does its volume decrease?

15. Why is it easier to start a fire with kindling rather than large sticks and logs of the same kind of wood?

16. Why does a chunk of coal burn when ignited, whereas coal dust explodes?

17. Chain reactions don't build up to appreciable intensities in small pieces of uranium that are not too close together; but if the same pieces are suddenly assembled into a single unit, the result may be a nuclear explosion. Why is this so?

18. Why are the absorption rods in a nuclear reactor long and thin instead of short and fat?

19. Why is there less heat loss in a dwelling if it is dome-shaped?

20. Why is heating more efficient in large apartment buildings than in single-family dwellings?

21. Why do thin french fries cook faster than fat fries?

22. Why do hamburgers cook faster than meatballs of the same weight?

23. If you use a batch of cake batter for cupcakes instead of cake and bake them for the time suggested for baking a cake, what will be the result?

24. Why are mittens warmer than gloves on a cold day?

25. Which parts of the body are most susceptible to frost-bite? Why?

26. Why is the heartbeat of large creatures generally slower than the heartbeat of smaller creatures?

27. Nourishment is obtained from food through the inner surface area of the intestines. Why is it that a small organism, such as a worm, has a simple and relatively straight intestinal tract, while a large organism, such as a human being, has a complex and many-folded intestinal tract?

28. The human lungs have a volume of only about 4 liters, yet an internal surface area of nearly 100 square meters. Why is this important and how is this accomplished?

29. Why are the living cells in a whale about the same size as those in a mouse?

30. Which fall faster—large or small raindrops?

31. Why can a large fish swim faster than a small fish?

32. Do the effects of scaling help or hinder a small swimmer on a racing team?

33. Why doesn't a hummingbird soar like an eagle and an eagle flap its wings like a hummingbird?

34. Consider eight little spears of mercury, each with a diameter of 1 millimeter. When they coalesce to form a single spear, how big will it be? How does its surface area compare to the total surface area of the previous eight little spears?

35. Can you relate the idea of scaling to the governance of small versus large groups of citizens? Explain.

11 Liquids

Unlike the molecules in a solid, molecules in a liquid state are not confined to fixed positions. They easily move from position to position by sliding over one another, and the body of the liquid assumes the shape of its container. Molecules of a liquid are close together and greatly resist compressive forces. Liquids are practically incompressible.

A liquid contained in a vessel exerts forces against the walls of the vessel. To discuss the interaction between the liquid and the walls, it is convenient to introduce the concept of **pressure**. Pressure is defined as force per area on which it acts:[1]

$$\text{Pressure} = \frac{\text{force}}{\text{area}}$$

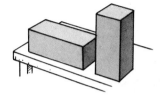

Fig. 11.1 Although the weight of both blocks is the same, the upright block exerts greater pressure against the table.

As an illustration of the distinction between pressure and force, consider the two blocks in Figure 11.1. The blocks are identical, but one stands on its end and the other on its side. Both blocks are of equal weight and therefore exert the same force on the surface, but the upright block exerts a greater pressure against the surface. This is because the force is distributed over a smaller area, which increases the pressure. If the block were tipped up so that it made contact with a single corner, the pressure would be greater still.

Pressure in a Liquid

When you swim under water, you can feel the water pressure acting against your eardrums. The deeper you swim, the greater the pressure. What causes this pressure? It is simply the weight of the water above you pushing against you. If you swim twice as deep, there is twice the weight of water above, and you therefore feel twice the water pressure. Liquid pressure depends on depth. If you were submerged in a liquid more dense than water, the pressure would be proportionally greater. Liquid pressure depends also on the density of the liquid. The pressure in a liquid is equal to the product of weight density and depth:[2]

$$\text{Liquid pressure} = \text{weight density} \times \text{depth}$$

---

[1]Pressure may be measured in any unit of force divided by any unit of area. The standard international (SI) unit of pressure, the newton per square meter, is called the *pascal* (Pa), named after the seventeenth-century theologian and scientist, Blaise Pascal. A pressure of 1 Pa is very small, and approximately equals the pressure exerted by a dollar bill resting flat on a table. Science types more often use kilopascals (1 kPa = 1000 Pa).

[2]This is derived from the definitions of pressure and density. Consider an area at the bottom of a vessel of liquid. The weight of the column of liquid directly above this area produces pressure. From the definition Weight density = weight/volume, we can express this weight of liquid as Weight = weight density × volume, where the volume of the column is simply the area multiplied by the depth. Then we get

$$\text{Pressure} = \frac{\text{force}}{\text{area}} = \frac{\text{weight}}{\text{area}} = \frac{\text{weight density} \times \text{volume}}{\text{area}}$$

$$= \frac{\text{weight density} \times \text{area} \times \text{depth}}{\text{area}} = \text{weight density} \times \text{depth} \quad \textbf{181}$$

Fig. 11.2 The dependence of liquid pressure on depth is not a problem for the giraffe because of its large heart and intricate system of valves and elastic, absorbent blood vessels in the brain. Without these structures, it would faint when suddenly raising its head and would be subject to brain hemorrhaging when lowering it.

Simply put, the pressure that a liquid exerts against the sides and bottom of a container depends only on the density and depth of the liquid. At twice the depth, the liquid pressure against the bottom is twice as great; at three times the depth, the pressure is threefold; and so on. Or, if the liquid is two or three times as dense, liquid pressure is correspondingly two or three times as great for any given depth.[3]

It is important to note that the pressure does not depend on the volume of liquid. For example, you feel the same pressure whether you dunk your head a meter under water in a small pool or a meter under water in the middle of a large lake. The same is true for a fish. Refer to the connecting

---

[3]The density of fresh water is 1000 kg/m^3. Since the weight (mg) of 1000 kg is 1000 × 9.8 = 9800 N, the weight density of water is 9800 newtons per cubic meter (9800 N/m^3). The water pressure beneath the surface of a lake is simply equal to this density multiplied by the depth in meters. For example, water pressure is 9800 N/m^2 at a 1-m depth, and 98,000 N/m^2 at a depth of 10 m. In SI units, pressure is measured in pascals, so this would be 9800 Pa and 98,000 Pa, respectively; or, in kilopascals, 9.8 kPa and 98 kPa, respectively.

vases in Figure 11.3. If we hold a goldfish by its tail and dunk its head a couple of centimeters under the surface, the water pressure against the fish's head will be the same in any of the vases. If we release the fish and it swims a few centimeters deeper, the pressure on the fish will increase with depth but be the same no matter what vase it is in. If the fish swims to the bottom, the pressure will be greater, but it makes no difference what vase it is in. All vases are filled to equal depths, so the water pressure is the same at the bottom of each vase, regardless of its shape or volume. If the water pressure at the bottom of a vase were greater than the water pressure at the bottom of a neighboring, narrower vase of smaller volume, the greater pressure would force water sideways and then up the narrower vase to a higher level until the pressures at the bottom were equalized. But this doesn't happen. Pressure is depth dependent and not volume dependent, so we see there is a reason why water seeks its own level.

Fig. 11.3 Liquid pressure is the same for any given depth below the surface, regardless of the shape of the containing vessel.

The fact that water seeks its own level can be demonstrated by filling a garden hose with water and bringing the two ends together. The levels will be equal. They will also be equal when both ends are far apart. This fact was not known by the Roman engineers who designed elaborate aqueducts with tall arches and roundabout routes to ensure that water would always flow slightly downward from the reservoir to the city. If pipes were laid in the ground and followed the natural relief, in some places the water would have to flow upward, and the Romans were skeptical of this. In any event, convincing experimentation was not yet the mode. Slave labor was plentiful, so elaborate aqueducts were built.

Fig. 11.4 The average water pressure acting against the dam depends on the average depth of the water and not on the volume of water held back. The large, shallow lake exerts only one-half the pressure that the small, deep pond exerts.

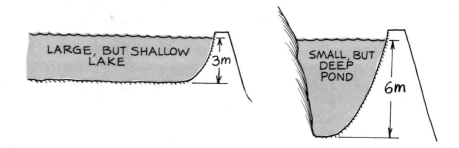

LARGE, BUT SHALLOW LAKE

3m

SMALL, BUT DEEP POND

6m

Fig. 11.5 If Roman engineers had understood that water can be made to flow uphill as well as downhill via a pipe, the Roman aqueducts might never have been built.

Another interesting fact about liquid pressure is that it is exerted equally in all directions. For example, if we are submerged in water, no matter which way we tilt our heads we feel the same amount of water pressure on our ears. Because a liquid can flow, the pressure isn't only downward. We know pressure acts sideways when we see water spurting sideways from a leak in the side of an upright can. We know pressure also acts upward when we try to push a beach ball beneath the surface of the water. The bottom of a boat is certainly pushed upward by water pressure. Pressure in a liquid at any point is exerted in equal amounts in all directions.

Against a submerged surface, the many components of forces that produce pressure combine to give a resulting pressure that is directed perpendicularly to the surface of the submerged body (Figure 11.6).

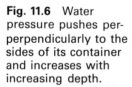

Fig. 11.6 Water pressure pushes per-perpendicularly to the sides of its container and increases with increasing depth.

Buoyancy Anyone who has ever lifted a submerged object out of water is familiar with buoyancy, the apparent loss of weight of objects submerged in a liquid. For example, lifting a large boulder off the bottom of a riverbed is a relatively easy task as long as the boulder is below the surface. When it is lifted above the surface, however, the force required to lift it is considerably more. This is because when the boulder is submerged, the water exerts an upward force on it that is exactly opposite in direction to gravity.

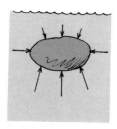

Fig. 11.7 The greater pressure against the bottom of a submerged object produces an upward buoyant force.

This upward force is called the **buoyant force** and is a consequence of pressure increasing with depth. Figure 11.7 shows why the buoyant force acts upward. Pressure is exerted everywhere against the object in a direction perpendicular to its surface. The arrows represent the magnitude and directions of points of pressure. Forces that produce pressures against the sides are at equal depths and cancel one another. Pressure is greatest against the bottom of the boulder because the bottom of the boulder is at a greater depth. Since this upward pressure is greater than the downward pressure acting against the top, the pressures do not cancel, and there is a net pressure upward. This net pressure distributed over the bottom of the boulder produces a net upward force, which is the buoyant force.

If the weight of the submerged object is greater than the buoyant force, the object will sink. If the weight is equal to the buoyant force on the submerged object, it will remain at any level, like a fish. If the buoyant force is greater than the weight of the completely submerged object, it will rise to the surface and float.

Understanding buoyancy requires understanding the meaning of the expression "volume of water displaced." If a stone is placed in a container brimful of water, some water will overflow (Figure 11.8). Water is *displaced* by the stone. A little thought will tell us that the *volume*—that is, the amount of space taken up or the number of cubic centimeters—*of water displaced* is equal to the *volume of the stone*. Place any object into a container partly filled with water, and the level of the surface rises (Figure 11.9). By how much? Exactly the same as if a volume of water were poured in that was equal to the volume of the submerged object. This is a good method for determining the volume of irregularly shaped objects: *A completely submerged object always displaces a volume of liquid equal to its own volume.*

Fig. 11.8 When a stone is submerged, it displaces a volume of water equal to the volume of the stone.

The relationship between buoyancy and displaced liquid was first discovered by the ancient Greek philosopher Archimedes. It is stated as

An immersed body is buoyed up by a force equal to the weight of the fluid it displaces.

This is called **Archimedes' principle**. It is true of liquids and gases, which are both fluids. If an immersed body displaces 1 kilogram of fluid, the buoyant force acting on it is equal to the weight of 1 kilogram.[4] *By immersed, we mean either completely or partially submerged.* If we immerse a sealed 1-liter container halfway into water, it will displace a half-liter of water and be buoyed up by the weight of a half-liter of water. If we immerse it completely (submerge it), it will be buoyed up by the weight of a full liter or 1 kilogram of water. Unless the container is compressed, the buoyant force will equal the weight of 1 kilogram at *any*

WATER DISPLACED

Fig. 11.9 The raised level equals that which would occur if water equal to the stone's volume were poured in.

---

[4] A kilogram is not a unit of force but a unit of mass. So, strictly speaking, the buoyant force is not 1 kilogram, but the *weight* of 1 kilogram, which is 9.8 newtons. We could as well say that the buoyant force is 1 *kilogram weight*, not simply 1 kilogram.

Fig. 11.10   A liter of water occupies a volume of 1000 cm³, has a mass of 1 kg, and weighs 9.8. N. Its density may therefore be expressed as 1 kg/l, and its weight density as 9.8 N/l. (Seawater is slightly denser, 10 N/l.)

depth, as long as it is completely submerged. This is because at any depth it can displace no greater volume of water than its own volume. And the weight of this volume of water (not the weight of the submerged object!) is equal to the buoyant force.

If a 25-kilogram body displaces 20 kilograms of fluid upon immersion, its apparent weight will be equal to the weight of 5 kilograms. Note that in Figure 11.11 the 3-kilogram block has an apparent weight equal to the weight of 1 kilogram when submerged. The apparent weight of a submerged object is its weight in air minus the buoyant force.

Fig. 11.11   A 3-kg block weighs more in air than in water. When submerged, its loss in weight is the buoyant force, which is equal to the weight of water displaced.

Questions

1. A 1-liter container completely filled with lead has a mass of 11.3 kg and is submerged in water. What is the buoyant force acting on it?*

2. A boulder is thrown into a deep lake. As it sinks deeper and deeper into the water, does the buoyant force increase? Decrease?†

3. Since buoyant force is the *net force* that a fluid exerts on a body, and we learned in Chapter 4 that net forces produce accelerations, why doesn't a submerged body accelerate?‡

Answer   The buoyant force is equal to the weight of 1 kg (9.8 N) because the *volume* of water displaced is 1 liter, which has a mass of 1 kg and a weight of 9.8 N. The 11.3 kg of the lead is irrelevant; 1000 cc of anything submerged will displace 1 liter of water and be buoyed upward with a force of 1 kg weight, or 9.8 N.

†*Answer*   Buoyant force does not change as the boulder sinks because the boulder displaces the same volume of water at any depth. Since water is practically incompressible, its density is the same at all depths; hence, the weight of water displaced, or the buoyant force, is the same at all depths.

‡*Answer*   It does accelerate if the buoyant force is not balanced by other forces that act on it—the force of gravity and fluid resistance. The net force on a submerged body is the result of the net force the fluid exerts (buoyant force), the weight of the body, and, if moving, the force of fluid friction.

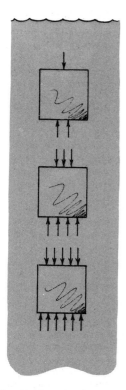

Fig. 11.12 The difference in the upward and downward force acting on the submerged block is the same at any depth.

Perhaps your instructor will show by way of a numerical example that the difference between the upward-acting and downward-acting forces due to the similar pressure differences on a submerged cube is numerically identical to the weight of fluid displaced. It makes no difference how deep the cube is placed, for although the pressures are greater with increasing depths, the *difference* in pressure up against the bottom of the cube and down against the top of the cube is the same at any depth (Figure 11.12). Whatever the shape of the submerged body, the buoyant force is equal to the weight of fluid displaced.

Note again that the buoyant force has to do with the volume of fluid displaced, not the weight of the object displacing the fluid. The buoyant force that acts on a submerged 1000-cubic-centimeter block of lead is identical to the buoyant force acting on 1000 cubic centimeters of aluminum, or 1000 cubic centimeters of any submerged material. The buoyant force depends only on the volume of fluid displaced and is numerically equal to the weight of this displaced volume of fluid.

We can see in this example that if the 1000-cubic-centimeter object has a mass of 1 kilogram, then its density exactly matches that of water. In this case, the buoyant force exactly cancels its weight, and the net force on it when submerged in water is zero. It will neither sink nor float, because there is no net force to move it from its submerged position. From this we can deduce a simple rule relating the density and behavior of objects immersed in a fluid:

1. If an object is more dense than the fluid in which it is immersed, it will sink.
2. If an object is less dense than the fluid in which it is immersed, it will float.
3. If an object's density equals the density of the fluid in which it is immersed, it will neither sink nor float.

An object with a density exceeding that of water will sink in water, and an object with a density less than that of water will float on water. We can see that the overall density of a fish must be the same as that of water. Most people can float, so an average human being must have a density slightly less than that of water. Some people cannot float; they're simply too dense.

The purpose of a life jacket is to decrease one's overall density by increasing volume while adding very little to one's weight. The density of a submarine is controlled by taking water into and out of its ballast tanks; in this way the weight of the submarine is varied to achieve the desired density. A fish regulates its density by expanding and contracting an air sac. The fish can move upward by expanding its air sac and increasing its volume (which displaces more water and increases buoyant force) and downward by contracting its volume (which decreases the buoyant force to a value slightly less than its weight). The overall density of a crocodile increases when it swallows stones. From 4 to 5 kilograms of stones have

Fig. 11.13 A crocodile coming toward you in the water.

A stoned crocodile coming toward you in the water.

been found lodged in the front part of the stomach in large crocodiles. Because of this increased density, the crocodile swims lower in the water, thus exposing less of itself to its prey (Figure 11.13).

Iron is much denser than water and therefore sinks, but an iron ship floats. Why is this so? Consider a solid 1-ton block of iron. Iron is nearly eight times as dense as water, so when it is submerged, it will displace only $\frac{1}{8}$ ton of water, which is hardly enough to keep it from sinking. Suppose we reshape the same iron block into a bowl, as shown in Figure 11.14. It still weighs 1 ton. When we place it in the water, it settles into the water, displacing a greater volume of water than before. The deeper it is immersed, the more water it displaces and the greater the buoyant force acting on it. When the buoyant force equals 1 ton, it will sink no further.

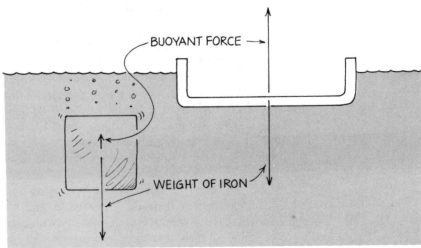

Fig. 11.14 An iron block sinks, while the same block shaped like a bowl floats.

Fig. 11.15 The weight of a floating object equals the weight of the water displaced by the submerged part.

When the iron boat displaces a weight of water equal to its own weight, it floats. This is sometimes called the **principle of flotation**, which states:

A floating object displaces a weight of fluid equal to its own weight.

Fig. 11.16 A floating object displaces a weight of fluid equal to its own weight.

Every ship, every submarine, and every dirigible must be designed to displace a weight of fluid equal to its own weight. Thus, a 10,000-ton ship must be built wide enough to displace 10,000 tons of water before it sinks too deep in the water. The same holds true in air. A dirigible or balloon that weighs 100 tons displaces at least 100 tons of air. If it displaces more, it rises; if it displaces less, it falls. If it displaces exactly its weight, it hovers at constant altitude.

Since the buoyant force upon a body equals the weight of the fluid it displaces, denser fluids will exert a greater buoyant force upon a body than less dense fluids of the same volume: A ship therefore floats higher in salt water than in fresh water because salt water is slightly denser than fresh water. In the same way, a solid chunk of iron will sink in water but float in mercury.

Fig. 11.17 The same ship empty and loaded. How does the weight of its load compare to the weight of extra water displaced?

Questions   Fill in the blanks for these statements:

1. The volume of a submerged body is equal to the ____ of the fluid displaced.*

2. The weight of a floating body is equal to the ____ of the fluid displaced.†

Pascal's Principle

One of the most important facts about fluid pressure is that a change in pressure at one part of the fluid will be transmitted undiminished to other parts. For example, if the pressure of city water is increased at the pumping station—say, by an amount equal to 10 units of pressure—the pressure everywhere in the pipes of the connected system will be increased by the same amount (provided the water is at rest). The pressure will be increased 10 units in the lower parts of the city as well as in the higher parts. The *increase* is the same everywhere. This rule is called **Pascal's principle**:

Changes in pressure at any point in an enclosed fluid at rest are transmitted undiminished to all points in the fluid and act in all directions.

For example, if we fill a U-tube with water and place pistons at each end, as shown in Figure 11.18, pressure exerted against the left piston will be transmitted throughout the liquid and against the bottom of the right piston. The pressure acting down on the left piston will be exactly equal to the pressure acting upward on the right piston. Well, this is really nothing to get excited about. But suppose we make the right part of the tube wider and use a piston of larger area, as in Figure 11.19. Suppose that the small piston has a cross-sectional area of 1 square centimeter and that the large piston has fifty times the area of the small piston. If we place a mass of 10 kilograms on the small piston, the pressure throughout the fluid will

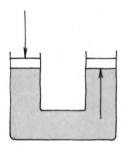

Fig. 11.18 The force exerted on the left piston increases the pressure in the liquid and is transmitted to the right piston.

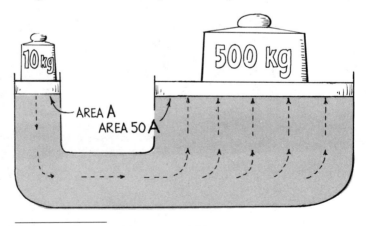

Fig. 11.19 A 10-kg load on the left piston will support 500 kg on the right piston.

AREA A
AREA 50 A

---

*Answer   volume   †**Answer**   weight

increase by the weight of 10 kilograms per square centimeter (98 newtons per square centimeter). This pressure exerted up against the large piston is distributed over *every* square centimeter of its surface. Since there are 50 square centimeters, the overall force exerted on the large piston is fifty times that on the small piston and will raise a 500-kilogram load. This is a remarkable consequence, for we can lift 500 kilograms with the weight of only 10 kilograms, 5000 kilograms with 100 kilograms, and so on. By further increasing the area of the large piston (or reducing the area of the small piston), we can multiply forces to any amount. This is the principle of the hydraulic press, a device that multiplies forces very effectively. This is in keeping with energy conservation, for the multiplication in force is compensated for in distance. When the small piston in the foregoing example is moved downward 10 centimeters, the large piston will be raised only one-fiftieth of this, $\frac{1}{5}$ centimeter. The input force multiplied by the distance it moves is equal to the output force multiplied by the distance it moves.

Pascal's principle applies to all fluids, gases as well as liquids. A typical application of Pascal's principle to gases and liquids is the automobile lift pump seen in gasoline stations (Figure 11.20). Compressed air exerts pressure on the oil in a reservoir. The oil in turn transmits the pressure to a cylinder, which lifts the automobile. The relatively low pressure that exerts the lifting force against the piston is about the same as the air pressure in the tires of the automobile. Even a low pressure exerted over a relatively large area produces a considerable force.

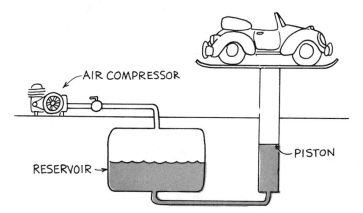

Fig. 11.20 Pascal's principle in a gas station.

Surface Tension

Suppose we suspend a piece of clean wire from a sensitive spiral spring, lower the wire into water, and then raise it. As we attempt to free the wire from the water surface, we see from the stretched spring that the water surface exerts an appreciable force on the wire. The water surface has a greater resistance to being stretched than the spring. The surface of water tends to contract. We note this tendency when a fine-haired paintbrush has been wet. When the brush is under water, the hairs are fluffed pretty much

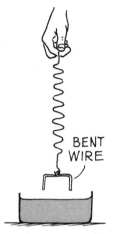

Fig. 11.21 When the bent wire is lowered into the water and then raised, the spring will stretch because of surface tension.

as they are when the brush is dry, but when the brush is lifted out, the surface film of water contracts and pulls the hairs together (Figure 11.22). We call this contractive force of the surface of liquids **surface tension**.

Surface tension accounts for the spherical shape of liquid drops. Raindrops, drops of oil, and falling drops of molten metal are all spherical because their surfaces tend to contract and force each drop into the shape having the *least* surface. This is a sphere, for a sphere is the geometrical figure that has the least surface for a given volume. For this reason the mist and dewdrops on spider webs or on the downy leaves of plants are tiny spheres (Figure 11.23).

Fig. 11.23 Small blobs of water are drawn by surface tension into sphere-like shapes. Can you apply the ideas of Figure 10.11 in the last chapter to an explanation of why the larger drops are more flattened by gravity, while the tinier drops are more spherical?

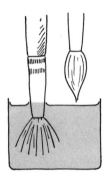

Fig. 11.22 When the brush is taken out of the water, the hairs are held together by surface tension.

Surface tension results from the contraction of liquid surfaces, which in turn is caused by molecular attractions. Beneath the surface, each molecule is attracted in *every* direction by neighboring molecules, with the result that there is no tendency to be pulled in any preferred direction. A molecule on the surface of a liquid, however, is pulled only by neighbors to each side and downward from below; there is no pull upward (Figure 11.24). These molecular attractions thus tend to pull the molecule from the surface into the liquid. This tendency to pull surface molecules into the liquid causes the surface to become as small as possible. The surface behaves as if it were tightened into an elastic film. This is evident when dry steel needles or razor blades seem to float on water; actually they are supported by the surface tension of the water.

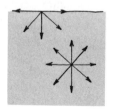

Fig. 11.24 A molecule at the surface is pulled only sideways and downward by neighboring molecules. A molecule beneath the surface is pulled equally in all directions.

The surface tension of water is greater than that of other common liquids. For example, clean water has a stronger surface tension than soapy water. We can see this when a little soap film on the surface of water is pulled out over the entire surface. The same is true of oil or grease floating on water. Oil has less surface tension than water and is drawn out into a film covering the whole surface, except when the water is hot. The surface tension of hot water decreases because the faster-moving molecules are not held as cohesively. This allows the grease or oil in hot soups to float in little bubbles on the surface of the soup. But when the soup cools and the surface tension of water increases, the grease or oil is dragged out over the surface of the soup. The soup becomes "greasy." Hot soup tastes different from cold soup primarily because the surface tension of water changes with temperature.

Capillarity

If the end of a thoroughly clean glass tube having a small inside diameter is dipped into water, the water will wet the inside of the tube and rise in it. In a tube with a bore of about $\frac{1}{2}$ millimeter in diameter, for example, the water will rise slightly higher than 5 centimeters. With a still smaller bore, the water will rise much higher (Figure 11.26).

Water molecules are attracted to glass more than to each other. The attraction between unlike substances is called **adhesion**. The attraction between like substances is called **cohesion**. When a glass tube is dipped into water, the adhesion between the glass and the water causes a thin film of water to be drawn up over the surfaces of the tube (Figure 11.27a). Surface tension causes this film to contract (Figure 11.27b). The film on the inner surface continues to contract, raising water with it until the adhesive force is balanced by the weight of the water lifted (Figure 11.27c). In a small tube, the weight of the water in the tube is small and is lifted higher than if the tube were large.

Fig. 11.25 Tom Noddy blows smoke into an interior soap bubble to reveal its hexagonal prism shape, caused by the surface tension of surrounding bubbles.

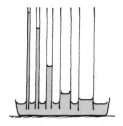

Fig. 11.26 Capillary tubes.

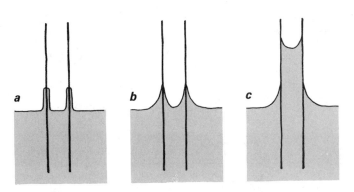

Fig. 11.27 Hypothetical "stages" of capillary action, as seen in a cross-sectional view of a capillary tube.

If a paintbrush is dipped into water, the water will rise up into the narrow spaces between the bristles by capillary action. Hang your hair in the bathtub, and water will seep up to your scalp in the same way. This is how oil soaks upward in a lamp wick and water into a bath towel when one end hangs in water. Dip one end of a lump of sugar in coffee, and the entire lump is quickly wet. The capillary action in soils is important in bringing water to the roots of plants. We see capillary action at work in many phenomena of nature. How nice.

Summary of Terms

Pressure   The ratio of the amount of force per area over which that force is distributed:

$$\text{Pressure} = \frac{\text{force}}{\text{area}}$$

Liquid pressure = weight density × depth

Buoyant force   The net force that a fluid exerts on an immersed object.

Archimedes' principle   An immersed body is buoyed up by a force equal to the weight of the fluid it displaces.

Principle of flotation   A floating object displaces a weight of fluid equal to its own weight.

Pascal's principle   The pressure applied to a fluid confined in a container is transmitted undiminished throughout the fluid and acts in all directions.

Surface tension   The tendency of the surface of a liquid to contract in area and thus behave like a stretched rubber membrane.

Capillarity   The rise or fall of a liquid in a fine, hollow tube or in narrow spaces.

Suggested Reading

Rogers, E. *Physics for the Inquiring Mind.* Princeton, N.J.: Princeton University Press, 1960. Chapter 6 of this oldie but goodie textbook treats surface tension in interesting detail.

Review Questions

1. Distinguish between *force* and *pressure*.

2. Why are reservoirs that supply water to communities elevated?

3. How does the volume of a submerged object compare with the volume of water displaced?

4. If a submerged object displaces 10 newtons of water, what is the buoyant force on the object? What is the buoyant force when the object is submerged to twice its depth?

5. Does the buoyant force on a submerged object depend on the weight of the object itself or on the weight of the fluid displaced? Does it depend on the weight of the object itself or on its volume? Explain.

6. If liquid pressure were the same at all depths, would there be a buoyant force on an object submerged in the liquid? Explain.

7. Why is it that some people, try as they may, simply cannot float?

8. Why will a boat float higher in denser salt water than in less dense fresh water?

9. Why does the surface of a volume of water contract?

10. What geometrical shape has the least surface area?

11. Why is cold soup greasy?

12. Distinguish between *adhesive* and *cohesive* forces.

Home Projects

1. Try to float an egg in water. Then dissolve salt in the water until the egg floats. How does the density of an egg compare to that of tap water? To salt water?

2. Weight the end of a pencil with a tack in the eraser so that it will float upright partly above the water. Mark the pencil at the water line and number it "1," the arbitrary number for water. Put the pencil in kerosene and it will sink deeper. Mark the surface point and number it "0.8" (kerosene has 0.8 the density of water). With these two reference 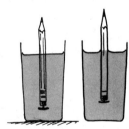 points you can calibrate markings at 0.1 intervals on the pencil. You now have a hydrometer, which can be used for testing the relative densities of different liquids. In a saturated solution of salted water, it should sink to the 1.25 mark. In alcohol, it should read 0.9. If you have antifreeze in your car radiator mixture, the reading should be less than 1.

3. Make a Cartesian diver. Completely fill a large, pliable plastic bottle with water and partially fill a small pill bottle so that it just barely floats when turned upside down and placed in the large bottle (you may have to experiment to get it just right). Once the pill bottle is barely floating, secure the lid or cap on the large bottle so that it is airtight. When you press the sides of the large bottle, the pill bottle sinks; when you release it, the bottle returns to the top. Experiment by squeezing the bottle different ways and getting different results. Can you explain the behavior you see?

4. Punch a couple of holes in the bottom of a water-filled container, and water will spurt out because of water pressure. Now drop the container, and as it freely falls note that the water no longer spurts out! If your friends don't understand this, could you figure it out and then explain it to them?

5. Soap greatly weakens the cohesive forces between water molecules. You can see this by putting some oil in a bottle of water and shaking it so that the oil and water mix. Notice that the oil and water quickly separate as soon as you stop shaking the bottle. Now add some soap to the mixture. Shake the bottle again and you will see that the soap makes a fine film around each little oil bead and that a longer time is required for the oil to gather after you stop shaking the bottle.

This is how soap works in cleaning. It breaks the surface tension around each particle of dirt so that the water can reach the particles and surround them. The dirt is carried away in rinsing. Soap is a good cleaner only in the presence of water.

Exercises

1. How many liters are there in 1 cubic meter?

2. Persons confined to bed are less likely to develop bedsores on their bodies if they use a waterbed rather than an ordinary mattress. Why is this so?

3. You know that a sharp knife cuts better than a dull knife. Do you know *why* this is so?

4. If water faucets upstairs and downstairs are turned fully on, will more water per second come out of the upstairs or downstairs faucet?

5. How does the water pressure at the bottom of a narrow 10-meter tube of water compare to the water pressure 10 meters below the surface of a 50,000-liter reservoir?

6. Which do you suppose exerts more pressure on the ground—an elephant or a lady standing on spike heels? Can you approximate a rough calculation for each?

7. Physics professor Joe Pizzo smashes glass bottles in front of his class and spreads the broken pieces across his lecture table. He then walks barefoot across the broken glass! What physical concepts is he demonstrating, and why is he careful that the broken pieces are small and numerous?

8. Why does your body get more rest when lying than when sitting? And why is blood pressure measured in the upper arm, at the elevation of your heart? Is blood pressure in your legs greater?

9. A block of aluminum having a volume of 10 cubic centimeters is placed in a beaker of water filled to the brim. Water overflows. The same is done in another beaker with a 10-cubic-centimeter block of lead. Does the lead displace more, less, or the same amount of water?

10. A block of aluminum having a mass of 1 kilogram is placed in a beaker of water filled to the brim. Water overflows. The same is done in another beaker with a 1-kilogram block of lead. Does the lead displace more, less, or the same amount of water?

11. A block of aluminum having a weight of 10 newtons is placed in a beaker of water filled to the brim. Water overflows. The same is done in another beaker with a 10-newton block of lead. Does the lead displace more, less, or the same amount of water? (Why are your answers to this exercise and Exercise 10 different from your answer to Exercise 9?

12. There is a legend of a Dutch boy who bravely held back the whole Atlantic Ocean by plugging a hole in a dike with his finger. Is this possible and reasonable?

13. If you've wondered about the flushing of toilets on the upper floors of city skyscrapers, how do you suppose the plumbing is designed so that there is not an enormous impact of sewage arriving at the basement level? (Check your speculations with someone who is into architecture.)

14. How much force is needed to push a nearly weightless but rigid 1-liter carton beneath a surface of water?

15. Why does water seek its own level?

16. Suppose you wish to lay a level foundation for a home on hilly and bushy terrain. How can you use a garden hose filled with water to determine equal elevations for distant points?

17. When you are bathing on a stony beach, why do the stones hurt your feet less when you get in deep water?

18. The density of a rock doesn't change when it is submerged in water, but *your* density changes when you are submerged. Why is this so?

19. When floating, why do you tend to float lower in the water when you exhale?

20. How does the fraction of the submerged part of a floating body relate to its density compared to the density of the liquid in which it floats?

21. The Himalayan Mountains are slightly less dense than the mantle material upon which they "float." Do you suppose that, like floating icebergs, they are deeper than they are high?

22. Consider the plug in your bathtub the next time you take a bath. When the tub is full, is there a buoyant force on the plug when it is in the drain? When it's pulled out and lying submerged on the tub bottom beside the drain? Defend your answer.

23. Why is it inaccurate to say that heavy objects sink?

24. A piece of iron placed on a block of wood makes it float lower in the water. If the iron were instead suspended beneath the wood, would it float as low, lower, or higher? Defend your answer.

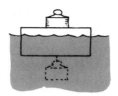

25. Compared to an empty ship, would a ship loaded with a cargo of styrofoam sink deeper into water or rise in water? Defend your answer.

26. If a submarine starts to sink, will it continue to sink to the bottom if no changes are made? Explain.

27. A barge filled with scrap iron is in a canal lock. If the iron is thrown overboard, does the water level at the side of the lock rise, fall, or remain unchanged? Explain.

28. Would the water level in a canal lock go up or down if a battleship in the lock sank?

29. A balloon is weighted so that it is barely able to float in water. If it is pushed beneath the surface, will it come back to the surface, stay at the depth to which it is pushed, or sink? Explain. (*Hint:* What change in density, if any, does the balloon undergo?)

30. In answering the question of why bodies float higher in salt water than fresh water, your friend replies that the reason is that salt water is denser than fresh water. (Does your friend often answer questions by reciting only factual statements that relate to the answers but don't provide any concrete reasons?) How would you answer the same question?

31. The weight of the human brain is about 15 newtons. The buoyant force supplied by fluid around the brain is about 14.5 newtons. Does this mean that there must be at least 14.5 newtons of fluid surrounding the brain? Defend your answer.

32. Alcohol is less dense than water. Do ice cubes float higher or lower in a mixed drink? What can you say about a cocktail in which the ice cubes lie submerged at the bottom of the glass?

33. When an ice cube in a glass of water melts, does the water level in the glass rise, fall, or remain unchanged?

34. Suppose the ice cube in the preceding question has many air bubbles. When it melts, what happens to the water level in the glass? What if the ice cube contains many grains of heavy sand?

35. A half-filled bucket of water is on a spring scale. Will the reading of the scale increase or remain the same if a fish is placed in the bucket? (Will your answer be different if the bucket is initially filled to the brim?)

36. The weight of the container of water is equal to the weight of the stand and suspended solid iron ball as shown in *a*. When the suspended ball is lowered into the water, the balance is upset (*b*). Will the additional weight that must be put on the right side to restore balance be greater than, equal to, or less than the weight of the ball?

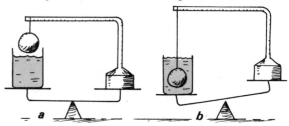

37. Would a fish float to the surface, sink, or stay at the same depth if the gravitational field of the earth increased?

38. What differences would we experience when swimming in water at the $\frac{1}{2}$-g level in an orbiting space habitat? Would we float in the water as we do on earth?

39. If you release a Ping-Pong ball beneath the surface of water, it will be buoyed to the surface. Would it do the same if the vessel of water were located at the hub of an orbiting structure in space? Explain.

40. So you're on a run of bad luck and you fall quietly into the water as hungry crocodiles lurking at the bottom are appreciating Pascal's principle. What does Pascal's principle have to do with their delight?

41. In the hydraulic arrangement shown, the larger piston has an area which is 50 times that of the smaller piston. The strongman hopes to exert enough force on the large piston to raise the 10 kilograms that rest on the small piston. Do you think he will be successful? Defend your answer.

42. In the hydraulic arrangement shown in Figure 11.19, the multiplication of force is equal to the ratio of areas of the large and small pistons. Some people are confused to learn that the arrangement shown in Figure 11.20 is immaterial. What is your explanation to clear up this confusion?

43. Why is it that wet sand is firm under your feet, while dry sand is not?

44. Why will hot water leak more readily through small leaks in a car radiator than cold water?

45. Would it be correct to say that the *reason* water rises in small, hollow tubes is because of capillarity? Explain.

12 Gases and Plasmas

Gases as well as liquids flow; hence, both are called *fluids*. The primary difference between a gas and a liquid is the distance between molecules. In a gas, the molecules are far apart and free from the cohesive forces that dominate their motions when in the liquid and solid state. Their motions are less restricted. When the molecules in a gas collide with one another and with the walls of a container, they rebound without loss of kinetic energy. A gas expands indefinitely and fills all space available to it. Only when the quantity of gas is very large, such as the earth's atmosphere or a star, do the gravitational forces limit the size or determine the shape of the mass of gas.

The Atmosphere

Because molecules in the air are energized by sunlight and are continually in motion, and because of gravity, the earth has an atmosphere. If gas molecules were not constantly moving, our "atmosphere" would be just so much more matter on the ground. It would lie on the surface of the earth just as dormant popcorn lies at the bottom of a popcorn machine. But add heat to the popcorn and the atmospheric gases, and both will

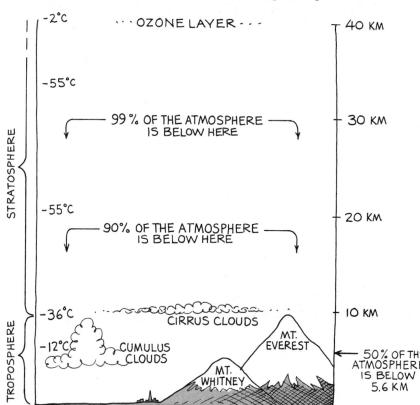

Fig. 12.1 The atmosphere. Air is more compressed at sea level than at higher altitudes. Like a huge pile of feathers, those at the bottom are more squashed than the ones nearer the top.

198

bumble their way up to higher altitudes. Pieces of popcorn in a popper attain speeds of a few kilometers per hour and are able to occupy altitudes up to a meter or two; molecules in the air move at speeds of about 1600 kilometers per hour and bumble up to many kilometers in altitude. If there were no gravity, both the popcorn and the atmospheric gases would fly into outer space. Fortunately there is an energizing sun, and there is gravity, and we have an atmosphere.

The exact height of the atmosphere has no real meaning, for the air gets thinner and thinner the higher one goes and eventually thins out into interplanetary space. Even in the vacuous regions of interplanetary space, however, there is a gas density of about 1 molecule per cubic centimeter. This is primarily hydrogen, the most plentiful element in the universe. About 50 percent of the atmosphere is below an altitude of 5.6 kilometers, 75 percent is below 11 kilometers, 90 percent is below 17.7 kilometers, and 99 percent is below about 30 kilometers (Figure 12.1). A detailed description of the atmosphere can be found in any encyclopedia.

Atmospheric Pressure

We live at the bottom of an ocean of air. The atmosphere, much like the water in a lake, exerts a pressure. One of the most celebrated experiments demonstrating the pressure of the atmosphere was conducted in 1654 by Otto von Gueicke, burgomaster of Magdeburg and inventor of the vacuum pump. Von Gueicke placed together two copper hemispheres about $\frac{1}{2}$ meter in diameter to form a sphere, as shown in Figure 12.2. He set a

Fig. 12.2 The famous "Magdeburg hemispheres" experiment of 1654, demonstrating atmosphere pressure. Two teams of horses couldn't pull the evacuated hemispheres apart. Were the hemispheres sucked together or pushed together? By what?

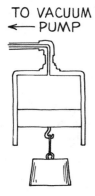

TO VACUUM
← PUMP

Fig. 12.3 Is the piston pulled up or pushed up?

gasket made of a ring of leather soaked in oil and wax between them to make an airtight joint. When he evacuated the sphere with his vacuum pump, two teams of eight horses each were unable to pull the hemispheres apart.

When the air pressure inside a cylinder like that shown in Figure 12.3 is reduced, there is an upward force on the piston. This force is large enough to lift a heavy weight. If the inside diameter of the cylinder is 10 centimeters or greater, a person can be lifted by this force.

What do the experiments of Figures 12.2 and 12.3 demonstrate? Do they show that air exerts pressure or that there is a "force of suction"? If we say there is a force of suction, then we assume that a vacuum can exert a force. But what is a vacuum? It is an absence of matter; it is a condition of nothingness. How can nothing exert a force? The hemispheres are not sucked together, nor is the piston holding the weight sucked upward. The hemispheres and the piston are being pushed against by the pressure of the atmosphere.

Just as water pressure is caused by the weight of water, atmospheric pressure is caused by the weight of air. We have adapted so completely to the invisible air that we sometimes forget it has weight. Perhaps a fish "forgets" about the weight of water in the same way. The reason we don't feel this weight crushing against our bodies is that the pressure inside our bodies equals that of the surrounding air. There is no net force for us to sense.

At sea level, 1 cubic meter of air has a mass of about $1\frac{1}{4}$ kilograms. So the air in your kid sister's small bedroom weighs about as much as she does! The density of air decreases with altitude. At 10 kilometers, for example, 1 cubic meter of air has a mass of about 0.4 kilograms. To compensate for this, airplanes are pressurized; the additional air to fully pressurize a 747 jumbo jet, for example, is more than 1000 kilograms. Air is heavy if you have enough of it.

Question   About how many kilograms of air occupy a classroom that has a 200-m² floor area and a 4-m-high ceiling?*

Consider the mass of air in an upright 30-kilometer-tall bamboo pole having an inside cross-sectional area of 1 square centimeter. If the density of air in the hollow part matched the density of the air outside, the enclosed mass of air would be about 1 kilogram. So the air pressure at the bottom of the bamboo pole would correspond to 1 kilogram per square centimeter (1 kg/cm²). Of course, the same is the case without the bamboo

---

*__Answer__   1000 kg. The volume of air is 200 m² × 4 m = 800 m³; each m³ of air has a mass of about $1\frac{1}{4}$ kg, so 800 m³ × $1\frac{1}{4}$ kg/m³ = 1000 kg.

pole. There are 10,000 square centimeters in 1 square meter, so a column of air 1 square meter in cross section extending up through the atmosphere has a mass of about 10,000 kilograms, which weighs about 100,000 newtons (10^5 N). This produces a pressure of 100,000 newtons per square meter or, equivalently, 100,000 pascals, or 100 kilopascals. To be more exact, the average atmospheric pressure at sea level is 101.3 kilopascals (101.3 kPa). We call 101.3 kilopascals of pressure 1 **atmosphere** (atm).[1]

The pressure of the atmosphere is far from uniform. Besides altitude variations, there are variations in atmospheric pressure at any one locality due to moving air currents and storms. Measurement of changing air pressure is important to meteorologists in predicting weather.

Barometers

Instruments used for measuring atmospheric pressure are called **barometers**. Figure 12.4 illustrates a simple mercury barometer. A glass tube, longer than 76 centimeters, is filled with mercury and inverted into a dish of mercury. The mercury in the tube will run out the submerged open bottom until the level falls to about 76 centimeters above the level in the dish. Since no air was allowed to enter the tube, the empty space above is a vacuum. The vacuum, of course, does not pull the mercury up; the mercury is supported in the tube because the weight of the atmosphere outside pushes on the surface of mercury in the open dish. This pressure is transmitted through the mercury (Pascal's principle) and up into the mouth of the inverted tube. At the level in the bottom of the tube, equal to the free surface level outside, atmospheric pressure is exactly counterbalanced by a downward pressure due to the weight of the column of mercury. Since the downward and upward pressures are equal (since there is no motion), the mercury must extend high enough in the tube to produce a pressure equal to that of the air outside. If the column of mercury were a fat square meter in cross section, it would weigh 10^5 newtons, exactly as much as the weight of a 1-square-meter, 30-kilometer-or-so counterbalancing column of air. The pressure of the column of mercury, 10^5 newtons per square meter, then equals the pressure of the surrounding air. When air pressure is 10_5 newtons per square meter (1 atmosphere), the mercury will stand 76 centimeters, or 760 millimeters, high, whatever the width of the barometer tube.[2] As the atmospheric pressure decreases, the level of mercury in the barometer falls. Increased pressure pushes the mercury to a higher level in the tube.

If water were used in a barometer instead of mercury, would the column be higher or lower than a mercury column? Since water is so much less dense than mercury, we can see that a much higher column would be

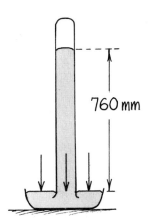

760 mm

Fig. 12.4 A simple mercury barometer.

---

[1]In British units, 1 atm = 14.7 lb/in².

[2]Equivalent units for atmospheric pressure are: 1 atm $\approx 10^5$N/m² = 101.3 kPa = 760 mm of mercury.

needed to push down as hard as denser mercury. In fact, it's easy to find just how high it would be. Since water is only $\frac{1}{13.6}$ as dense as mercury, the column would have to be 13.6 times taller than a 76-centimeter mercury column. This would be about 10.3 meters. A water barometer would be too tall to make measurements in ordinary conditions.

Fig. 12.5 Strictly speaking, they do not suck the soda up the straw. They instead reduce pressure in the straw and allow the weight of the atmosphere to press the liquid up into the straws. Could they drink a soda this way on the moon?

Question   What is the maximum height to which water could be drunk through a straw?*

A small portable instrument that measures atmospheric pressure is the *aneroid barometer* (Figure 12.6). It uses a small metal box that is partially exhausted of air and has a slightly flexible lid that bends in or out with changes in atmospheric pressure. Motion of the lid is indicated on a scale by a mechanical spring-and-lever system. Since the atmospheric pressure decreases with increasing altitude, a barometer can be used to determine the elevation. An aneroid barometer calibrated for altitude is called an *altimeter* (altitude meter). Some of these instruments are sensitive enough to indicate a change in elevation less than a meter.

---

*__Answer__   However strong your lungs may be, or whatever device you use to make a vacuum in the straw, at sea level the water could not be pushed by the atmosphere higher than 10.3 m.

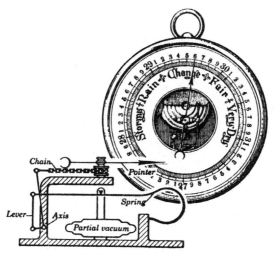

Fig. 12.6 The aneroid barometer.

The pressure (or vacuum) inside a television picture tube is about one ten-thousandth pascal (10^{-4} Pa). At an altitude of about 500 kilometers, artificial satellite territory, gas pressure is about one ten-thousandth of this (10^{-8} Pa). This is a pretty good vacuum by earthbound standards. Still greater vacuums exist in the wakes of satellites orbiting at this distance, and they can reach 10^{-13} pascals. This is a hard vacuum.

Vacuums on the earth are produced by pumps, which work by virtue of a gas tending to fill its container. If a space with less pressure is provided, gas will flow from the region of higher to lower pressure. A vacuum pump simply provides a region of lower pressure into which the normally fast-moving gas molecules randomly fill. The air pressure is repeatedly lowered by piston and valve action (Figure 12.7). The best vacuums attainable with mechanical pumps are about a pascal. Better vacuums, up to 10^{-8} pascals, are attainable with vapor diffusion pumps. Greater vacuums are very difficult to attain. Technologists requiring hard vacuums are looking more to the prospects of orbiting laboratories in space.

Fig. 12.7 A mechanical vacuum pump. When the piston is lifted, the intake valve opens and air moves in to fill the empty space. When the piston is moved downward, the outlet valve opens and the air is pushed out. What changes would you make to convert this pump into an air compressor?

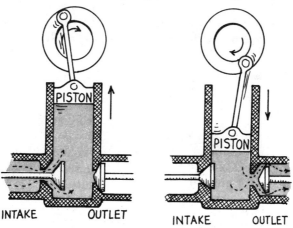

Boyle's Law

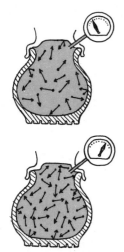

Fig. 12.8 When the density of gas in the tire is increased, pressure is increased.

The air pressure inside the tires of an automobile is considerably greater than the atmospheric pressure outside. This is because a volume of air at atmospheric pressure is compressed to about a third its volume inside the containing tire. The molecules of air behave like tiny tennis balls, perpetually moving helter-skelter and banging against the inner walls. Their impacts on the inner surface of the tire produce a jittery force that appears to our coarse senses as an average push. This pushing force averaged over a unit of area gives us the pressure of the enclosed air.

Suppose we have twice as many molecules in the same volume (Figure 12.8). Then the air density is doubled. If the molecules move at the same average speed, or, equivalently, if they have the same temperature, then, to a close approximation, the number of collisions will be doubled. So we see from this doubling of pressure due to the doubling of density that pressure is proportional to density.

We could instead double the density by simply compressing the air to half its volume. Consider the cylinder with the movable piston in Figure 12.9. If we push the piston downward so that the volume is half its original value, the density of molecules will be doubled, and the pressure will be correspondingly doubled. Decrease the volume to a third its original value, and the pressure will be increased threefold, and so forth. Notice from these examples that the product of pressure and volume is the same; that is, a doubled pressure multiplied by a halved volume gives the same value as a tripled pressure multiplied by a one-third volume. In general, we can say that the product of pressure and volume for a given mass of air is a constant as long as the temperature does not change. The pressure of a quantity of air multiplied by its volume at one period of time is the same as any different pressure multiplied by its correspondingly different volume at any other time. In shorthand notation,

$$P_1V_1 = P_2V_2$$

where P_1 and V_1 represent the original pressure and volume, respectively, and P_2 and V_2 the second, or final, pressure and volume. This relationship is called **Boyle's law**, after Robert Law, the seventeenth-century physicist who is credited with its discovery.[3] Note the similarity of this relationship to the conservation of mechanical energy, where the product of force and distance is constant: $F_1d_1 = F_2d_2$.

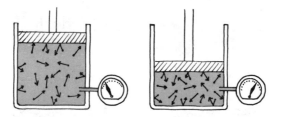

Fig. 12.9 When the volume of gas is decreased, density and therefore pressure are increased.

---

[3] Humor aside, Boyle's law is named after Robert Boyle.

Boyle's law applies to ideal gases. An **ideal gas** is one in which the disturbing effects of intermolecular forces and molecular volume are so small that they are negligible. Air and other gases under normal pressures approach ideal gas conditions.[4]

Question   A scuba diver 10.3 m deep breathes compressed air. If she holds her breath while returning to the surface, by how much does the volume of her lungs tend to increase?*

Buoyancy of Air   A crab lives at the bottom of its ocean of water and looks upward at jellyfish and other lighter-than-water objects floating above it. Similarly, we live at the bottom of our ocean of air and look upward to balloons and other lighter-than-air objects floating above us. A balloon floats in air and a jellyfish "floats" in water for the same reason: each is buoyed upward by a displaced weight of fluid equal to its own weight. In one case the displaced fluid is air, and in the other case it is water. In water, immersed objects are buoyed upward because the pressure acting up against the bottom of the object exceeds the pressure acting down against the top. Likewise, air pressure acting up against an object immersed in air is greater than the pressure above pushing down. The buoyancy in both cases is numerically equal to the weight of fluid displaced. We can state **Archimedes' principle for air** as

An object surrounded by air is buoyed up by a force equal to the weight of the air displaced.

A cubic meter of air at ordinary atmospheric pressure and temperature has a mass of about $1\frac{1}{4}$ kilograms. Therefore, any 1-cubic-meter object in air is buoyed up with a force equal to the weight of $1\frac{1}{4}$ kilograms. If the mass of the object is more than $1\frac{1}{4}$ kilograms, it falls when released. If the object has a mass less than $1\frac{1}{4}$ kilograms, it rises in the air. Any object that has a mass less than the mass of an equal volume of air will rise in air. Gas-filled balloons that are less dense than air rise for this reason.

Gas is used in balloons simply because its presence prevents the atmosphere from collapsing the balloon. Helium is usually used because its mass is small enough that the combined weight of helium, balloon, and

Fig. 12.10 All bodies are buoyed up by a force equal to the weight of air they displace. Why, then, don't all bodies float like this balloon?

*Answer   Unfortunately for her, by *twice*, since at 10.3 m the water pressure—and therefore air pressure—is twice that at the surface! A first lesson in scuba diving is *not* to hold your breath when ascending. To do so can be fatal.

[4]A general law that takes temperature changes into account is $\dfrac{P_1 V_1}{T_1} = \dfrac{P_2 V_2}{T_2}$, where T_1 and T_2 represent the initial and final *absolute* temperatures, measured in kelvins (Chapter 14).

whatever the cargo happens to be is less than the weight of air it displaces.[5] Helium is used in a balloon for the same reason cork is used in a swimmer's life preserver. The cork possesses no strange tendency to be drawn toward the surface of water, and helium possesses no strange tendency to rise. Both are buoyed upward like anything else. They are simply light enough for the buoyancy to be significant.

Unlike water, there is no sharp surface at the "top" of the atmosphere. Furthermore, unlike water, the atmosphere becomes less dense with altitude. Whereas cork will float to the surface of water, a released helium-filled balloon does not rise to any atmospheric surface. Will a lighter-than-air balloon rise indefinitely? How high will a balloon rise? We can state the answer in several ways. A balloon will rise only so long as it displaces a weight of air greater than its own weight. Since air becomes less dense with altitude, a lesser weight of air is displaced per given volume as the balloon rises. When the weight of displaced air equals the total weight of the balloon, upward acceleration of the balloon will cease. We can also say that when the buoyant force on the balloon equals its weight, the balloon will cease rising. Equivalently, when the density of the balloon equals the density of the surrounding air, the balloon will cease rising. Helium-filled toy balloons usually break when released in the air because as the balloon rises to regions of less pressure, the helium in the balloon expands, increasing the volume and stretching the rubber until it ruptures.

| **Question** | Two balloons that have the same weight and volume contain equal amounts of helium. One is rigid, and the other is free to expand as the pressure outside decreases. When released, which balloon will rise higher?* |

Large dirigible airships are designed so that when loaded they will slowly rise in air; that is, their total weight is a little less than the weight of air displaced. When in motion, the ship may be raised or lowered by means of horizontal rudders or "elevators."

Thus far we have treated pressure only as it applies to stationary fluids. Motion produces an additional influence.

| **Bernoulli's Principle** | Mark this statement true or false: Atmospheric pressure increases in a gale, tornado, or hurricane. If you answered true, sorry; the statement is false. High-speed winds may blow the roof off your house, but the pressure within the winds is actually less than for still air of the same density inside |

---

*__Answer__   The balloon that is free to expand will displace more air as it rises than the balloon that is restrained. Hence, the balloon that is free to expand will experience more buoyant force and rise higher.

[5]Hydrogen is the least dense gas but is highly flammable, so it is seldom used.

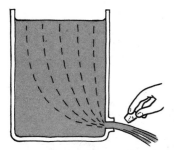

Fig. 12.11 The pressure in the spout reduces when the plug is removed.

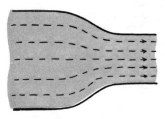

Fig. 12.12 Water speeds up when it flows into the narrower pipe. The constricted streamlines indicate increased speed and decreased internal pressure.

the house. As strange as it may at first seem, when the speed of a fluid increases, the internal pressure decreases proportionally. This is true whether the fluid is a gas or a liquid.

Consider liquid initially at rest in a container. The pressure that the liquid exerts against the walls of the container depends on both the density and the depth of the liquid. The pressure on the sides or bottom is reduced if there is an opening through which the liquid moves (Figure 12.11). This is because the molecules in the liquid no longer push against the container but move into the opening. The pressure at the opening is lessened, and the pressure in the moving liquid is lessened. In fact, the faster the liquid moves, the lower is the pressure in that liquid. Daniel Bernoulli, a Swiss scientist of the eighteenth century, studied the relationship of fluid velocity and pressure. This relationship is called **Bernoulli's principle**. In a loose sense, it states:

The pressure in a fluid decreases with increased velocity of the fluid.

When a fluid flows through a narrow constriction, its velocity increases (Figure 12.12). You can see this easily if you consider the increased flow of water in a brook through the narrow parts. The fluid must speed up in the constricted region if the flow of fluid is to be continuous. How does the fluid pick up this extra speed? Bernoulli reasoned that this extra speed was acquired at the expense of a lowered internal pressure.

When a fluid is moving, energy is present in the fluid in different forms. Some of the energy is stored in the pressure of the fluid, and some is stored as kinetic energy in the motion of the fluid. If the speed of the fluid does not change, each of these forms has its fixed share of the energy. But if the speed of the fluid changes, one of the forms of energy gets a larger share at the expense of the other. If the speed suddenly increases, there is more kinetic energy than there was before. To make up for this increase in kinetic energy, the pressure within the fluid suddenly drops, so that the overall energy is unchanged.

Applying energy conservation to fluids, Bernoulli stated that *in a fluid flow where no energy is added or taken away, the kinetic energy (represented by velocity) plus the potential energy (represented by pressure) plus the gravitational potential energy (represented by elevation) equals constant total energy.* In the strict sense, Bernoulli's principle is the mathematical expression of this statement.

This is illustrated in Figure 12.13. Vertical tubes act as pressure gauges and indicate the pressure of the water flowing through the wide pipe into the narrower pipe and then into the wider pipe again. Since the pipe is level, there are no differences in gravitational potential energy, and the total energy of water in any region along the pipe is the sum of the kinetic and potential energies. We can see that where the water has its greatest speed, its pressure is least, and vice versa. And we can see that this is in accord with the conservation of energy.

Fig. 12.13 Total energy (kinetic + potential) is the same at points A, B, and C. The greater kinetic energy at B is at the expense of potential energy, which is evidenced by the lower level of fluid in the vertical tube.

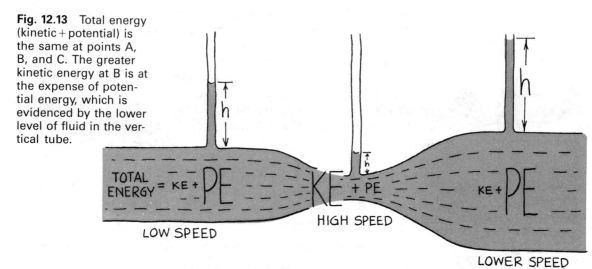

The velocity of fluid is best represented with *streamlines*. These are the smooth paths of steady flow shaped by the boundaries of flow. Streamlines are indicated in the figures by broken lines. The lines are crowded in the narrower regions, indicating a greater flow velocity. Where streamlines are close together, the pressure within the fluid is less. If the velocity of flow is too great, the streamlines may become turbulent and curl into *eddies* or *vortices*. The proportional relationship between fluid flow and pressure then breaks down. Bernoulli's principle applies only to steady flow and does not apply to turbulent flow.

Applications of Bernoulli's Principle

Fig. 12.14 The paper rises when air is blown across its top surface.

Hold a sheet of paper in front of your mouth as shown in Figure 12.14. When you blow across the top surface, the paper rises. This is because the moving air pushes against the top surface of the paper with less pressure than the air at rest, which pushes against the lower surface. The greater pressure beneath pushes the paper upward into the region of reduced pressure.

We began our discussion of Bernoulli's principle by stating that atmospheric pressure decreases in a strong wind, tornado, or hurricane. As it turns out, an unvented building with airtight closed windows is in more danger of losing its roof than a well-vented building. This is because the air pressure inside may be appreciably greater than the atmospheric pressure outside, and the roof is more likely to be pushed off by the relatively compressed air in the building than blown off by the wind. When the wind is directed over a peaked roof as shown in Figure 12.15, the effect is even more pronounced. The crowding of the streamlines shows this. The difference in outside and inside pressures need not really be very much. A small pressure applied over a large area can produce a formidable force.

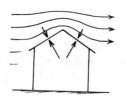

Fig. 12.15 Air pressure above the roof is less than air pressure beneath the roof.

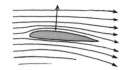

Fig. 12.16 Air pressure is less above the wing than below the wing.

So if you're ever caught in an *unvented* building in a tornado or a hurricane, consider opening the windows a bit so that the pressures inside and outside are more nearly equal.[6]

If we think of the blown-off roof as an airplane wing, we can better understand the lifting force that supports a heavy airliner. In both cases a greater pressure below pushes the roof and wing into a region of lesser pressure above. A cambered roof is more apt to be blown off than a flat roof. Similarly, a wing with more curvature on the top surface has greater lift than a wing with flat surfaces (like the wings of those balsa-wood gliders you used to play with). Whether it has flat or curved wings, an airplane will fly by virtue of air impact against the lower surface of wings that are tilted back slightly so they will deflect oncoming air downward (as discussed briefly in Chapter 4). The airfoil of a curved wing, however, adds considerably to lift and results in a greater difference in pressure against the lower surface than against the upper surface. This net upward pressure multiplied by the surface area of the wing gives the net lifting force. The lift is greater when there is a large wing area and when the plane is traveling fast. Gliders have a very large wing area so that they do not have to be going very fast for sufficient lift. At the other extreme, fighter planes designed for high speed have very small wing areas. Consequently, they must take off and land at relatively high speeds.

Fig. 12.17 Where is air pressure greater—on the top or bottom surface of the hang glider?

---

[6]Opening windows is not generally recommended, for most structures have venting adequate to tolerate the sudden pressure drop without the need to open windows. Opening the wrong window can actually increase damage.

Bernoulli's principle accounts for the fact that passing ships run the risk of a sideways collision. Water flowing between the ships travels faster than water flowing past the outer sides. The streamlines are more compressed between the ships than outside. Water pressure acting against the hulls is therefore reduced between the ships. Unless the ships are steered to compensate for this, the greater pressure against the outer sides of the ships then forces them together. Figure 12.18 shows how this can be demonstrated in your kitchen sink or bathtub.

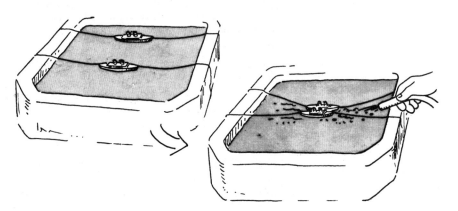

Fig. 12.18 Try this in your sink. Loosely moor a pair of toy boats side by side. Then direct a stream of water between them. The boats will draw together and collide. Why?

We all know that a baseball pitcher can throw a ball in such a way that it will curve off to one side in its trajectory. Any ball will follow a curved path if it is hit or thrown in such a way that it has both a high speed and a large spin. A thin layer of air is dragged by friction around the ball as it spins through the air. The threads on the baseball provide added friction, and the overall result is a crowding of streamlines on one side (Figure 12.19). Note that the streamlines are more crowded at B than at A for the direction of spin shown. Air pressure is greater at A, and the ball curves in the direction shown.

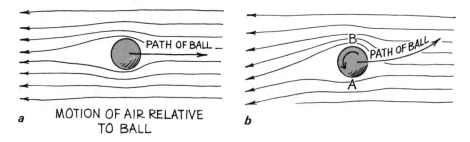

a   MOTION OF AIR RELATIVE TO BALL   b

Fig. 12.19 (*a*) The streamlines are the same on either side of a nonspinning baseball. (*b*) A spinning ball produces a crowding of streamlines and curves as shown.

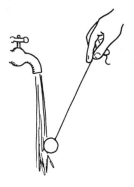

Fig. 12.20 Pressure is greater in the stationary fluid (air) than in the moving fluid (water stream). The atmosphere pushes the ball into the region of reduced pressure.

An interesting example of Bernoulli's principle is shown in Figure 12.20. Allow a Ping-Pong ball to swing from a string into a stream of running water, and it will remain in the stream even when tugged slightly to the side as shown. This is because the pressure of the stationary atmosphere is greater than the pressure in the moving water. The ball is *pushed* into the region of reduced pressure by the atmosphere.

The same thing happens to a bathroom shower curtain when the shower water is turned on full blast. The pressure in the shower stall is reduced, and the relatively greater pressure outside pushes the curtain inward. The next time you're taking a shower and the curtain swings in against your legs, think of Daniel Bernoulli!

Fig. 12.21 The curved shape of an umbrella can be disadvantageous on a windy day.

Question Will the mercury level in a barometer be affected on a windy day?*

Plasma In addition to solids, liquids, and gases, there is a fourth state of matter, the least common state in our everyday environment—**plasma** (not to be confused with the clear, formless liquid part of blood, also called *plasma*). A plasma looks and behaves like a high-temperature gas but with an important difference: it conducts electricity. The atoms and molecules that make it up are *ionized*, stripped of one or more electrons as a result of the more violent collisions at high temperatures. Recall that a neutral atom has as many positive protons inside the nucleus as it has negative electrons outside the nucleus. When one or more of these electrons is stripped from the atom, the atom has more positive charge than negative charge and is called a positive **ion** (under some conditions, it may have extra electrons, in which case it is called a negative ion). In an ideal plasma, all atoms are completely stripped of electrons, and the mixture is composed entirely of free electrons and bare nuclei. Although the electrons and ions are themselves electrically charged, the plasma as a whole is electrically neutral

****Answer** Atmospheric pressure is reduced on a windy day, so the barometer level will fall from its normal position.

because there are still equal numbers of positive and negative charges. The free electrons and ions make up an electrically conducting medium. The configuration of electrons and ions can be shaped, molded, and moved by electric and magnetic fields.

The major difference between a neutral gas and a plasma is that the particles in a plasma are ionized and can exert electromagnetic forces on one another.

Plasma in the Everyday World

If you happen to be reading this by the light emitted by a fluorescent lamp, you don't have to look far to see plasma in action. Within the glowing tube of the lamp is a plasma consisting of low-pressure mercury vapor. It is ionized by voltage across electrodes at the ends of the tube and conducts an electric current, which causes the plasma to radiate, which in turn causes the phosphor coating on the inner surface of the tube to glow.

The neon gas in an advertising sign similarly becomes a plasma when it is ionized by a high voltage and made to conduct electricity. The different colors seen in these signs correspond to different kinds of atoms glowing in the plasma state. Vapor lamps used in street lighting also emit the light of glowing plasmas (Figure 12.22). The bluish-green white light from some of these lamps is emitted from mercury atoms in the plasma state, and the yellow light from others is from a plasma made up of sodium atoms.

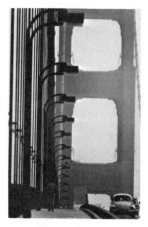

Fig. 12.22 Streets are illuminated at night by the glowing plasma of vapor lamps.

The light from a laser is emitted from a glowing plasma. Strictly speaking, a gas laser is a plasma laser once the high voltage is applied. We will treat lasers in detail in Chapter 28.

The aurora borealis (sometimes called the northern lights) is emitted from glowing plasma in the upper atmosphere. Layers of low-temperature plasma encircle the whole earth. Occasionally, showerings of electrons from outer space and from radiation belts enter the "magnetic windows" at the earth's poles, crash into the layers of plasma, and produce light.

These layers, which extend upward from about 55 kilometers, make up the ionosphere and act as mirrors to low-frequency radio waves. Higher-frequency radio and TV waves pass through the ionosphere. This is why you can pick up radio stations from long distances on your lower-frequency AM radios, but you have to be in the "line of sight" of broadcasting or relay antennas to pick up higher-frequency FM and TV signals. Ever notice that at nighttime you can pick up very distant stations on your AM set? This is because the plasma layers settle closer together in the absence of the energizing sunlight and act as better radio reflectors.

Plasma Power

A higher-temperature plasma is the exhaust of a rocket engine. It is a weakly ionized plasma, but when small amounts of potassium salts or cesium metal are added, it becomes a very good conductor, and when it is directed into a magnet, electricity is generated! This is MHD power, the **m**agneto**h**ydro**d**ynamic interaction between a plasma and a magnetic field. (We will treat the mechanics of *how* electricity is generated in this way in Chapter 23.) Low-pollution MHD power is now in the developmental stage and is in operation at a few places in the world already. We can expect to see more plasma power with MHD.

A more promising achievement will be plasma power of a different kind—the controlled fusion of atomic nuclei. We will treat the physics of fusion in Chapter 32. Even more important than the physics of thermonuclear fusion is the social impact of its controlled use. We have already seen the early effects of uncontrolled thermonuclear fusion, the hydrogen bomb. Like all technology, fusion can be applied for humanity's benefit as well as for its destruction. The benefits of controlled fusion may well be more far-reaching than the harnessing of electrical power in the last century. Fusion plants should ultimately not only make electrical energy abundant but provide the energy and means to recycle and even synthesize elements as well. The control of fusion may provide the setting for a new age. For the first time in our evolution, we could build a civilization having a base of abundance. This civilization should be very different from those that have had to deal with the scarcity of two important fundamentals—energy and material.

We have come a long way with our mastery of the first three states of matter. Our mastery of the fourth state should bring us ever so much farther.

Summary of Terms

Atmospheric pressure   The pressure, exerted against bodies immersed in the atmosphere, which results from the weight and motion of molecules of atmospheric gases. At sea level, atmospheric pressure is about 101 kilopascals.

Barometer   Any device that measures atmospheric pressure.

Boyle's law   The product of pressure and volume is a constant for a given mass of confined gas regardless of changes in either pressure or volume individually, so long as temperature remains unchanged:

$$P_1V_1 = P_2V_2$$

Archimedes' principle for air   An object surrounded by air is buoyed up with a force equal to the weight of displaced air.

Bernoulli's principle   The pressure in a fluid decreases with an increase in fluid velocity.

Plasma   Hot matter beyond the gaseous state composed of electrically charged particles. Most of the matter in the universe is in the plasma state.

Review Questions

1. What is the energy source for the motion of atmospheric molecules? What prevents them from escaping to outer space?

2. How high above sea level would you have to rise for half of the atmosphere to be below you?

3. What is the cause of atmospheric pressure?

4. What is the mass of a cubic meter of air at sea level? Of 1000 cubic meters (the size of a large classroom)?

5. When drunk through a straw, is the liquid sucked up or pushed up? Explain.

6. Why is it necessary for scuba divers who make emergency ascents to continuously exhale as they ascend?

7. If you compress an inflated balloon to half its volume, by how much will the gas pressure increase? (Disregard temperature changes.)

8. How high will a helium-filled balloon rise in the atmosphere?

9. Does atmospheric pressure increase or decrease on a windy day?

10. What does Bernoulli's principle have to do with the conservation of energy?

11. Why is it that the shower curtain draws toward you when you are taking a shower with the water on full-blast?

12. How does a plasma differ from a gas?

Home Projects

1. Try this in the bathtub or when you're washing dishes. Lower a drinking glass, mouth downward, over a small floating object. What do you observe? How deep will the glass have to be pushed in order to compress the enclosed air to half its volume? (You won't be able to do this in your bathtub unless it's 10.3 meters deep!)

2. You ordinarily pour water from a full glass to an empty glass by simply placing the full glass above the empty glass and tipping. Have you ever poured air from one glass to the other? The procedure is similar. Lower two glasses in water, mouths downward. Let one fill with water by tilting its mouth upward. Then hold the water-filled glass mouth downward above the air-filled glass. Slowly tilt the lower glass and let the air escape, filling the upper glass. You will be pouring air from one glass to another!

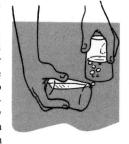

3. Raise a filled glass of water above the water line, but with its mouth beneath the surface. Why does the water not run out? How tall would a glass have to be before water began to run out? (You won't be able to do this indoors unless you have a 10.3-meter ceiling.)

4. Place a card over the open top of a glass filled to the brim with water, and invert it. Why does the card stay intact? Try it sideways.

 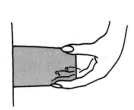

5. Invert a water-filled pop bottle or small-necked jar. Notice that the water doesn't simply fall out, but gurgles out of the container—a result of the weight of the many kilometers of air above pushing down and forcing air up into the bottle and forcing the water out. How would a water-filled, inverted bottle empty if this were done on the moon?

6. Pour about a half cup of water in a 5-or-so-liter metal can that has a screw top. Place the can *open* on a stove and heat until the water boils and steam comes out of the opening. Quickly remove the can and screw the cap on tightly. Allow the can to stand and observe the results. The effect can be hastened by cooling the can with a dousing of cold water. Explain your observations.

7. Heat a small amount of water to boiling in an aluminum soda-pop can and invert it quickly into a dish of cold water. Surprisingly dramatic!

8. Make a small hole near the bottom of an open tin can. Fill it with water, which proceeds to spurt from the hole. Cover the top of the can firmly with the palm of your hand and the flow stops. Explain.

9. Lower a narrow glass tube or drinking straw in water and place your finger over the top of the tube. Lift the tube from the water and then lift your finger from the top of the tube. What happens? (You'll do this often if you enroll in a chemistry lab.)

10. Fold the ends of a filing card down so that you make a little bridge. Stand it on the table and blow through the arch as shown. No matter how hard you blow, you will not succeed in blowing the card off the table (unless you blow against the side of it). Try this with your nonphysics friends. Then explain it to them!

11. Push a pin through a small card sheet and place it in the hole of a thread spool. Try to blow the card from the spool by blowing through the hole. Try it in all directions.

12. Check an encyclopedia to see how Bernoulli's principle applies to an automobile carburetor. In this and similar applications, it is called the *Venturi effect*.

13. Hold a spoon in a stream of water as shown and feel the effect of the differences in pressure.

Exercises

1. Why is there no atmosphere on the moon?

2. What is the purpose of the ridges that prevent the funnel from fitting tightly in the mouth of a bottle?

3. How would the density of air in a deep mine compare to the air density at the earth's surface?

4. Why do bubbles of gas rising in a liquid become larger as they approach the surface?

5. Why do your ears "pop" when you ascend to higher altitudes?

6. Two teams of eight horses each were unable to pull the Magdeburg hemispheres apart (Figure 12.2). Why? Suppose two teams of nine horses each *could* pull them apart. Then would one team of nine horses succeed if the other team were replaced with a strong tree? Defend your answer.

7. Before boarding an airplane, you buy a roll of camera film or any item packaged in an airtight foil package, and while in flight notice that it is puffed up. Explain why this happens.

8. Why should a flight attendant avoid wearing an inflatable bra when in flight?

9. Why do you suppose that airplane windows are small compared to bus windows?

10. A half cup or so of water is poured into a 5-liter can, which is put over a source of heat until most of the water has boiled away. Then the top of the can is screwed on tightly and the can is removed from the source of heat and allowed to cool. What happens to the can, and why?

11. Will breaking a TV picture tube cause it to implode or to explode? Explain.

12. We can understand how pressure in water depends on depth by considering a stack of bricks. The pressure at the bottom of the first brick corresponds to the weight of the entire stack. Halfway up the stack, the pressure is half because the weight of the bricks above is half. To explain atmospheric pressure, we should consider compressible bricks, like foam rubber. Why is this so?

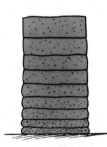

13. The "pump" in a vacuum cleaner is merely a high-speed fan. Would a vacuum cleaner pick up dust from a rug on the moon? Explain.

14. From how deep a well can water be pumped with a perfect vacuum pump?

15. If a liquid only half as dense as mercury were used in a barometer, how high would its level be on a day of normal atmospheric pressure?

16. Why does the size of the cross-sectional area of a barometer not affect the height of the enclosed fluid?

17. From how deep a vessel could mercury be drawn with a siphon?

18. Would it be slightly more difficult to draw soda through a straw at sea level or on top of a very high mountain? Explain.

19. The pressure exerted against the ground by an elephant's weight distributed evenly over its four feet is *less* than 1 atmosphere. Why, then, would you be crushed beneath the foot of an elephant, while you're unharmed by the pressure of the atmosphere?

20. Why is it so difficult to breathe when snorkeling at a depth of 1 meter, and practically impossible at a 2-meter depth? Why can't a diver simply breathe through a hose that extends to the surface?

21. Why does the weight of an object in air differ from its weight in a vacuum? Cite an example where this would be an important consideration.

22. Estimate the buoyant force that acts on you. (To do this, you can estimate your volume by knowing your weight and by assuming that your weight density is a bit less than that of water.)

23. A little girl sits in a car at a traffic light holding a helium-filled balloon. The windows are up and the car is relatively airtight. When the light turns green and the car accelerates forward, her head pitches backward but the balloon pitches forward. Explain why.

24. Would a bottle of helium gas weigh more or less than an identical bottle filled with air at the same pressure? Than an identical bottle with the air pumped out?

25. A balloon filled with air falls to the ground, but a balloon filled with helium rises. Why is this so?

26. The gas pressure inside an inflated rubber balloon is always greater than the air pressure outside. Why is this so?

27. Two identical balloons filled with air to the same volumes are suspended on the ends of a stick that is horizontally balanced. One of the balloons is then punctured. Is the balance of the stick upset? If so, which way does it tip?

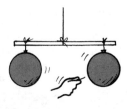

28. Two balloons that have the same weight and volume are filled with equal amounts of helium. One is rigid and the other is free to expand as the pressure outside decreases. When released, which will rise higher? Explain.

29. A weather balloon released in the atmosphere grows as it rises. As it grows, does the buoyant force acting on it increase, decrease, or remain the same?

30. Imagine a huge space colony that consists of a rotating air-filled cylinder. How would the density of air at "ground level" compare to the air densities "above"?

31. Would a helium-filled balloon "rise" in the atmosphere of a rotating space habitat? Defend your answer.

32. Cite some differences in the sport of ballooning in the atmosphere of a cylindrical or spherical space colony. Do the same for hang gliding.

33. The force of the atmosphere at sea level against the outside of a 10-square-meter store window is about a million newtons. Why does this not shatter the window?

34. Estimate the *force* that the atmosphere exerts on you, and then compare this force with the weight of something familiar in your environment. To do this, estimate your surface area from the area of your clothes, and multiply by 10^5 newtons/meter^2. Why is your answer to this exercise so different from your answer to exercise 22?

35. Nitrogen and oxygen in their liquid states have densities only 0.8 and 0.9 that of water. Atmospheric pressure is due primarily to the weight of nitrogen and oxygen gas in the air. If the atmosphere liquefied, would its depth be greater or less than 10.3 meters?

36. Why are industrial chimneys tall?

37. Why is the draft in a fireplace better on a windy day?

38. What provides the lift to keep a Frisbee in flight?

39. Why is it that when passing an oncoming truck on the highway, your car tends to lurch toward the truck?

40. Why is it that the canvas roof of a convertible automobile bulges upward when the car is traveling at high speeds?

41. Why is it that the windows of older trains break when a high-speed train passes by on the next track?

42. A steady wind blows over the waves of an ocean. Why does the wind increase the humps and troughs of the waves?

43. In answering the question of why a flag flaps in the wind, your friend replies that it flaps because of Bernoulli's principle. (Not really a convincing explanation, is it?) How would you answer the same question?

44. Wharves are made with pilings that permit the free passage of water. Why would a solid-walled wharf be disadvantageous to ships attempting to pull alongside?

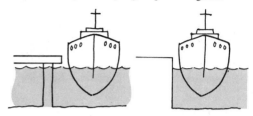

45. Under what conditions are the mercury atoms in a fluorescent lamp in a gaseous state? In a plasma state?

PART 3 Heat

13 Temperature, Heat, and Expansion

All matter is composed of continually jiggling atoms and molecules. Whether the atoms and molecules combine to form solids, liquids, gases, or plasmas depends on the rate of molecular vibrations. In this and the following two chapters we are going to investigate more closely the effects of the chaotic and haphazard motion of atoms and molecules that we call *thermal motion*. We begin by considering that which a body has by virtue of this energetic motion: **thermal energy**.

Thermal energy is involved in all aspects of our everyday lives, from cooking our food to warming our feet. The molecules of substances are constantly jiggling in some sort of back-and-forth vibratory motion. The greater this random molecular kinetic energy in a substance, the hotter the substance is. When we strike a piece of metal with a hammer, for example, the metal becomes warm. This is because the hammer's blow causes the molecules in the metal to jostle faster. Later, as the molecules in the substance slow down by giving some of their energy to the air or surrounding medium, the metal becomes cool again.

As recently as the eighteenth century, people had no way to measure accurately the hotness or coldness of a body. A physician judged the extent of a patient's fever by feeling the forehead; a baker estimated the hotness of his oven by the color of the glowing coals. The coldness of winter was determined by the thickness of ice in frozen ponds. There was a need for an accurate way of describing hotness and coldness.

Fig. 13.1 Can we trust our sense of hot and cold? Will both fingers feel the same temperature when they are put in the warm water?

Temperature

When we touch a hot stove, thermal energy enters the hand because the stove is warmer than the hand. When we touch a piece of ice, however, thermal energy passes out of the hand and into the colder ice. In such instances the direction of energy transfer is always from a warmer body to a neighboring cooler body. The quantity that tells how warm or cold something is with respect to a standard body is called **temperature**. We say temperature is a measure of the random translational motion of atoms and molecules in a body; more specifically, it is a measure of the *average kinetic energy* of atoms and molecules in a body. We know, for example, that there is twice the thermal energy in 2 liters of boiling water as in 1 liter of boiling water, because 2 liters of boiling water will melt twice as much ice as 1 liter. But the temperatures of both amounts of water are the same because the average kinetic energy of molecules in each is the same. So we see there is a difference between thermal energy, which is measured in joules, and temperature, which we measure in degrees.

Question   Suppose you add thermal energy to a half cup of tea and its temperature rises 2°. If you add the same amount of thermal energy to a full cup of tea, by how much will its temperature rise?*

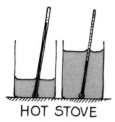

HOT STOVE

Fig. 13.2 Although the same quantity of heat is supplied to both containers, the container with the smaller amount of water has the higher temperature.

Measuring Temperature

Generally, when the temperature of a piece of matter changes, several things may happen to it. Its size and its electrical, magnetic, or optical properties may become different, and any such change can be used to detect and measure its change in temperature. The simplest to use, in most cases, is the change in size. Nearly all materials expand when their temperature is raised and shrink when it is lowered.

A thermometer is a common instrument that measures temperature by means of the expansion and contraction of a liquid, usually mercury or colored alcohol. To fix a scale for a thermometer, the number 0 is assigned to the temperature at which water freezes, and the number 100 to the temperature at which water boils (at standard atmospheric pressure). The space between is divided into 100 equal parts, called degrees; hence, a thermometer so calibrated has been called a *centigrade* thermometer. It is now called a *Celsius* thermometer in honor of the Swedish astronomer Anders Celsius, who first suggested the scale.

In the United States the number 32 is assigned to the temperature at which water freezes, and the number 212 is assigned to the temperature at which water boils. Such a scale makes up a *Fahrenheit* thermometer, named after its illustrious originator, the German physicist G. D. Thermometer. This scale is becoming obsolete, and will be if and when the United States goes metric.[1]

---

*Answer   1°, because there are twice as many molecules in the full cup, and each molecule will receive only half as much thermal energy on the average. So the average kinetic energy, and thus the temperature, will increase half as much.

[1]The conversion to Celsius will put the United States in step with the rest of the world, where the Celsius scale is the standard. Americans are slow to convert, perhaps because the Fahrenheit scale seems much better suited to everyday use. For example, its degrees are smaller (1 F° = $\frac{5}{9}$ C°), which gives greater accuracy when reporting the weather in whole-number temperature readings. Then, too, people somehow attribute a special significance to numbers increasing by an extra digit, so that when the temperature of a hot day is reported to reach 100°F, the idea of heat is conveyed more dramatically than by saying it is 37.7°C. Like so much of the British system of measure, the Fahrenheit scale is geared to human beings. (And, for the record, the Fahrenheit scale is named after Gabriel Daniel Fahrenheit.)

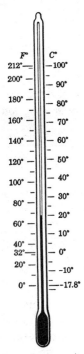

F°   C°
212° —|— -100°
200° —   — 90°
180° —   — 80°
160° —   — 70°
140° —   — 60°
120° —   — 50°
100° —   — 40°
80° —   — 30°
          — 20°
60° —   — 10°
40° —
32°—|—   — 0°
20° —   — -10°
0° —   —|—-17.8°

Fig. 13.3 Fahrenheit and Celsius scales on a thermometer.

Fig. 13.4 Just as water in the pipes seeks a common level (where the pressures at the bottom are the same), the thermometer and its immediate surroundings reach a common temperature (where the average molecular KE for both is the same).

Arithmetic formulas are used for converting from one temperature scale to the other, an exercise popular in classroom exams. The probability of your doing this task elsewhere does not merit our concern with it here. Besides, the conversion can be very closely approximated simply by reading the corresponding temperature from the side-by-side scales in Figure 13.3.

The temperature scale used by scientists is the *Kelvin* scale, in which the number 0 is assigned to the lowest possible temperature, where a substance has absolutely no thermal energy to give up—**absolute zero**. On the Celsius scale, absolute zero corresponds to $-273°C$. Degrees on the Kelvin scale are the same size as those on the Celsius scale, so the temperature of melting ice is $+273$ Kelvin degrees. There are no negative numbers on the Kelvin scale. We will discuss this scale, often called the *thermodynamic* scale, when we study thermodynamics in Chapter 16.

Interestingly enough, a thermometer registers its own temperature. When a thermometer is in thermal contact with something the temperature of which we wish to know, thermal energy will flow between the two until their temperatures are equal and *thermal equilibrium* is established. If we then know the temperature of the thermometer, we also know the temperature of the something. A thermometer should be small enough that it doesn't appreciably alter the temperature of the something being measured. If you are measuring the room air temperature, then your thermometer is small indeed. But if you are measuring a drop of water, its temperature after thermal contact may be quite different from its initial temperature.

Heat It is common to think of thermal energy as another word for *heat*. A body contains thermal energy, which is the overall energy of motion of its molecules, so we commonly say a body contains heat. The physicist, however, makes a distinction between thermal energy and heat and defines **heat** as *the thermal energy that is transferred from one body to another because of a temperature difference between the bodies.* Heat flows, while thermal energy may or may not. Once heat has been transferred to a body or substance, it ceases to be heat and instead becomes thermal energy.

For bodies or substances in thermal contact, heat will flow from the substance at a higher temperature into the substance at a lower temperature, but it will not necessarily flow from a substance with more thermal energy into a substance with less thermal energy. There is more thermal energy in a bowl of warm water than there is in a red-hot thumbtack; if the tack is immersed in the water, heat will not flow from the warm water to the tack. Instead, heat will flow from the hot tack to the relatively cooler water. Heat never flows of itself from a cooler body into a hotter body.

How much heat flows depends not only on the temperature difference between substances but on the amount of material as well. For example, a barrelful of hot water will transfer more heat to a cooler substance than a cupful of water at the same temperature. There is more thermal energy in the larger amount of water.

In addition to the thermal energy of jostling molecules in a substance, there is energy in other forms, mainly the potential energy within and between molecules, which affects the thermal properties of substances. The grand total of all energies internal to a substance is called **internal energy**. We will see that internal energy is thermal energy and then some when we study thermodynamics in Chapter 16.

Quantity of Heat

A body does not contain heat. A body contains internal energy. Heat is the thermal energy, or more correctly, the internal energy transferred from one body to another by virtue of a temperature difference. The quantity of heat involved in such a transfer is measured by some change that accompanies the process. In determining the energy value in food, for example, the amount of internal energy that is released as heat is measured by burning. Fuels are rated on how much energy a certain amount of fuel will produce in burning. The unit of heat is defined as the heat necessary to produce some standard, agreed-on change. The most commonly used unit for heat is the **calorie**:[2]

The calorie is the amount of heat required to change the temperature of 1 gram of water by 1 Celsius degree.

Another common unit of heat is the kilocalorie, which is 1000 calories (the energy required to heat 1 kilogram of water 1°C). The Calorie (most often written with a capital C) used in rating foods is actually a kilocalorie. The calorie and the Calorie are units of energy. These names are historical carry-overs from the early idea that heat was an invisible fluid called *caloric*. This view persisted almost to the nineteenth century. We

---

[2]Another common unit of heat quantity is the British thermal unit (Btu). The Btu is defined as the amount of heat required to change the temperature of 1 lb of water 1°F.

Fig. 13.5 The filling of hot apple pie may be too hot to eat, whereas the crust is not.

now know that heat is a form of energy, and we are presently in a transition period to the International System of units (SI), where the quantity of heat is measured in joules,[3] the SI unit for all forms of energy.

Specific Heat

Have you ever noticed that some foods remain hotter much longer than others? Boiled onions and squash on a hot dish, for example, are often too hot to eat when mashed potatoes may be eaten comfortably. The filling of hot apple pie can burn your tongue while the crust will not, even when the pie has just been taken out of the oven. And the aluminum covering on a frozen dinner can be peeled off with your bare fingers as soon as it is removed from the oven. A piece of toast may be comfortably eaten a few seconds after coming from the hot toaster, but we must wait several minutes before eating soup from a stove no hotter than the toaster. Evidently, different substances have different capacities for storing internal energy. If we heat a bucket of water on a stove, we might find that it requires 15 minutes to raise it from room temperature to its boiling temperature. But if we put an equal mass of iron on the same flame, we would find that it would rise through the temperature range in only about 2 minutes. For silver, the time would be less than a minute. We find that different materials require different quantities of heat to raise the temperature of a given mass of the material through a specified number of degrees. Different materials absorb energy in different ways. The energy may increase the back-and-forth vibrational motion, which will increase the temperature, or it may go into potential energy, which increases the internal rotational and vibrational states within the molecules without increasing temperature. Generally, there is a combination of both.

A gram of water requires 1 calorie of energy to raise the temperature 1°C. It takes only about one-eighth as much energy to raise the temperature of 1 gram of iron by the same amount. Water absorbs more heat than iron. We say iron has a lower **specific heat** (sometimes called *specific heat capacity*).

The specific heat of any substance is defined as the quantity of heat required to raise the temperature of a unit mass of the substance by 1 degree of temperature.

Question   Which has a higher specific heat—water or sand?*

---

*__Answer__   Water. The temperature of water increases less than the temperature of sand in the same sunlight. Sand's low specific heat, as evidenced by how quickly the surface warms in the morning sun and how quickly it cools at night, affects local climate.

[3]For your information, 4.186 J = 1 calorie.

Fig. 13.6 Water has a high specific heat and is transparent, so it takes more energy to heat up than land. Solar energy incident upon land is concentrated at the surface, but upon water it extends and dilutes beneath the surface.

Water has a much higher capacity for storing energy than all but a few uncommon materials. A relatively small amount of water absorbs a great deal of heat for a correspondingly small temperature rise. Because of this, water is a very useful cooling agent, as evidenced by the fact that it is used in automobile cooling systems. If a liquid of lower specific heat were used in the cooling system, its temperature would rise higher for a comparable absorption of heat. (Of course, if the temperature of the liquid increases to the temperature of the engine, no cooling takes place.)

Water also takes a long time to cool, a fact that not only adds to the usefulness of the hot-water bottle but also appreciably improves the climate in many places. The next time you are looking at a world globe, notice the high latitude of Europe. If water did not have a high specific heat, the countries of Europe would be as cold as the northeastern regions of Canada, for both Europe and Canada get about the same amount of sunlight per square kilometer. The Gulf Stream holds much of its internal energy long enough to reach the North Atlantic off the coast of Europe, where it then cools. The energy released, a calorie per degree for each gram of water that cools, is carried by the westerly winds over the European continent. A similar effect occurs in the United States.

The winds in the latitude of the United States are westerly. Air moves from the Pacific Ocean landward. On the east coast, air moves from land seaward. Cooling of the Pacific Ocean releases internal energy and warms the air moving over the land, raising the temperature of coastal regions. As a result, San Francisco is much warmer in the winter than Washington, D.C., which is at about the same latitude. In summer months the energy flow is reversed as the warm air heats the relatively cooler water. Then the presence of water cools the air.

Islands and peninsulas that are more or less surrounded by water do not have the same extremes of temperatures that are observed in the interior of a continent. The high summer and low winter temperatures common in the central United States, for example, are largely due to the absence of large bodies of water. Europeans, islanders, and people living in coastal regions should be glad that water has such a high specific heat. San Franciscans are!

Expansion   When the temperature of a substance is increased, its molecules are made to jiggle faster. The more energetic collisions between molecules force them to move farther apart, resulting in an expansion of the substance. All forms of matter—solids, liquids, gases, and plasmas—generally expand when they are heated and contract when they are cooled.

In many cases the changes in the size of objects are not very noticeable, but careful observation will usually detect them. Telephone wires are longer and sag more on a hot summer day than they do on a cold winter

Fig. 13.7 Place a dented Ping-Pong ball in boiling water, and you'll remove the dent. Why?

day. Metal lids to glass fruit jars can often be loosened by heating them under hot water. If one part of a piece of glass is heated or cooled more rapidly than adjacent parts, the expansion or contraction that results may break the glass. This is especially true with thick glass. Pyrex glass is specially formulated to expand very little with increasing temperature.

Liquids expand appreciably with increases in temperature. When you get your gasoline tank filled at a filling station and then park the car, the gasoline often overflows after it sits in the gas tank for a while. This is because the gasoline is cold when it comes from the underground gas tanks; as it sits in the gas tank of your car, it warms up to the temperature of your car. As the gasoline warms, it expands; its volume increases and overflows the gas tank. Similarly, an automobile radiator filled to the brim with cold water overflows when heated.

Gases expand even more when heated. Hold an air-filled balloon over a hot stove and notice that the size of the balloon increases; this is because the air inside expands with increasing temperature.

Different substances expand at different rates. In most cases the expansion of liquids is greater than the expansion of solids. The gasoline overflowing a gas tank on a hot day is evidence of this. If the tank expanded at the same rate, both tank and contents would expand together, and no overflow would occur. Similarly, if the expansion of the glass of a thermometer were as great as the expansion of the mercury, the mercury would not rise with increasing temperature. The mercury in a thermometer rises with increasing temperature because the expansion of liquid mercury is greater than the expansion of glass.

Fig. 13.9 A thermostat. When the coil expands, the mercury rolls away from the electrical contacts and breaks the circuit. When the coil contracts, the mercury rolls against the contacts and completes the electrical circuit.

ROOM
TEMPERATURE

  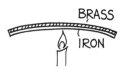

Fig. 13.8 A bimetallic strip. Brass expands (or contracts) more when heated (or cooled) than iron does, so the strip bends as shown.

When two strips of differents metals—one of brass, say, and the other of iron—are welded or riveted together, the greater expansion of one metal is demonstrated (Figure 13.8). Such a compound bar is called a *bimetallic strip*. The difference in the amounts of expansion of brass and iron shows up easily because the double strip bends into a curve when its temperature changes. The movement of the strip may be used to turn a pointer, to regulate a valve, or to close a switch.

A practical application of this is the thermostat (Figure 13.9). The back-and-forth bending of the bimetallic strip opens and closes an electrical circuit. When the room becomes too cold, the strip bends toward the brass side, and in so doing activates an electrical switch that turns on the heat. When the room becomes too warm, the strip bends toward the iron side, which activates an electric contact that turns off the heating unit. Elec-

Fig. 13.10 This gap is called an *expansion joint;* it allows the bridge to expand and contract.

trical refrigerators are equipped with special thermostats to prevent them from becoming either too warm or too cold. Bimetallic strips are used in oven thermometers, electric toasters, automatic chokes on carburetors, and various other devices.

The expansion of substances must be considered in the construction of structures and devices of all kinds. A dentist uses fillings that have the same rate of expansion as teeth; the aluminum pistons of an automobile engine are made enough smaller in diameter than the steel cylinders to allow for the much greater expansion rate of aluminum. A civil engineer uses reinforcing steel that has the same expansion rate as concrete. Long steel bridges are provided with rollers or rockers at the ends, and the roadway itself is segmented with tongue-and-groove gaps called *expansion joints* (Figure 13.10). Similarly, concrete roadways and sidewalks are intersected by gaps, sometimes filled with tar, so that the concrete can expand freely in summer and contract in winter.

Question   How would a thermometer be different if glass expanded more with increasing temperature than mercury?*

Expansion of Water

Increase the temperature of any common liquid and it will expand. But not water at temperatures near the freezing point: ice-cold water does just the opposite! Water of the temperature of melting ice, 0°C or 32°F, *contracts* when the temperature is increased. This is most unusual. As the water is heated and its temperature rises, it continues to *contract* until it reaches a temperature of 4°C or 39.2°F. With further increase in temperature, the water then begins to *expand;* the expansion continues all the way to the boiling point, 100°C. The result of this odd behavior is that water has its smallest volume and greatest density at 4°C. This is shown graphically in Figure 13.11. A given amount of water has its smallest volume and greatest

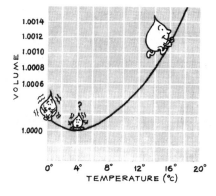

Fig. 13.11 The expansion of water with increasing temperature.

*****Answer**   The scale would have to be upside down. Do you know why?

density at 4°C, and its largest volume and smallest density (neglecting steam) in its solid form, ice. The volume of ice is not shown in Figure 13.11, for if it were plotted to the same exaggerated scale, the graph would extend far beyond the top of the page. After water has turned to ice, further cooling causes it to contract.

Ice has a crystalline structure. For most crystals, the orderly arrangement of the molecules in the solid state results in a smaller volume than the same molecules in the liquid state; in the case of ice, however, the angular shape of the water molecules, and the fact that the binding forces are strongest at certain angles, result in open-structured crystals (Figure 13.12) that occupy a greater volume in the solid than in the liquid state. Consequently, ice is less dense than water and floats on water.

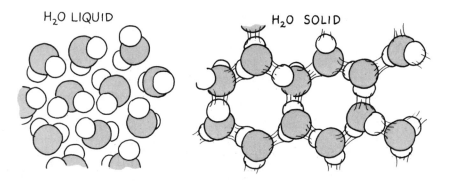

Fig. 13.12 Water molecules in their crystal form have an open-structured hexagonal arrangement that results in the expansion of water upon freezing. Ice therefore is less dense than water.

The reason for the dip in the curve of Figure 13.11 is that two types of volume changes are taking place. The open-structured crystals that make up solid ice are also present, to a much smaller extent, in ice-cold water. They exist as a sort of microscopic slush. These crystals gradually collapse as the temperature is increased and thus reduce the volume of the water. At about 10°C, all the ice crystals have dissolved. Figure 13.13a indicates how volume changes due only to the collapsing of the microscopic ice crystals. Figure 13.13b, on the other hand, shows how volume changes as a result of the increased molecular motion and the consequent expansion. Whether ice crystals are in the water or not, increased vibra-

Fig. 13.13 The collapsing of ice crystals plus increased molecular motion with increasing temperature produce the overall effect of water being most dense at 4°C.

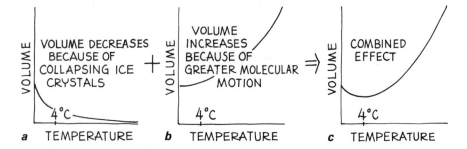

tional motion of the molecules and crystals increases the volume of the water. When we combine these two effects, one of contraction and the other of expansion, the curve looks like Figure 13.13c, or Figure 13.11.

This behavior of water is of great importance in nature. If the greatest density of water were at its freezing point, as is true of most liquids, then the coldest water would settle to the bottom and ponds would freeze from the bottom up. Marine life would be destroyed in winter months. But this doesn't happen because the densest water that settles at the bottom of a pond is 4 degrees above the freezing temperature. Water at the freezing point, 0°C, is less dense, so ice forms at the surface.

Let us examine this in more detail. Most of the cooling in a pond takes place at its surface, when the surface air is colder than the water. As the surface water is cooled, it becomes denser, and sinks to the bottom. Water will "float" at the surface for further cooling only if it is equal to or less dense than the water below. Consider a body of water that initially is at, say, 10°C. It cannot possibly be cooled to 0°C without first being cooled to 4°C. And water at 4°C cannot remain at the surface for further cooling unless all the water below has at least an equal density—that is, unless all the water below is 4°C. If the water below the surface is any temperature other than 4°C, any surface water at 4°C will be denser and sink before it can be further cooled. So before any ice can form, *all* the water in a pond must be cooled to 4°C. Only when this condition is met can the surface water be cooled to 3°, 2°, 1°, and 0°C without sinking. Then ice can form. We can see that the water at the surface is first to freeze. Continued cooling of the pond results in the freezing of the uppermost water against the ice, so a pond freezes from the surface downward. In a cold winter the ice will be thicker than in a milder winter.

Fig. 13.14 As water is cooled, it sinks until the entire pond is 4°C. Only then can surface cooling to the freezing point take place without further sinking.

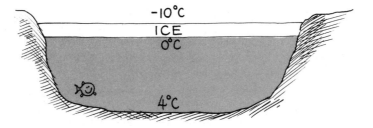

Because all the water in a lake must be cooled to 4°C before lower temperatures can be reached, very deep bodies of water are not ice-covered even in the coldest of winters. This is because the winter is not long enough for all the water to be cooled to 4°C. If only some of the water is 4°C, it lies on the bottom. Because of water's high specific heat and poor ability to conduct heat, the bottom of deep lakes in cold regions is a constant 4°C the year round. Fish should be glad that this is so.

Summary of Terms

Thermal energy The energy in a substance due to the random motion of its molecules.

Temperature A measure of the average kinetic energy per molecule in a body, measured in degrees Celsius, Fahrenheit, or Kelvin.

Absolute zero The lowest possible temperature that a substance may have—the temperature at which molecules of a substance have their minimum kinetic energy.

Heat The thermal energy that flows from a body of higher temperature to a body of lower temperature, commonly measured in calories or joules.

Internal energy The total of all molecular energies, kinetic energy plus potential energy, internal to a substance.

Specific heat The quantity of heat per unit mass required to raise the temperature of a substance by 1°C.

Review Questions

1. Why does a piece of metal become warm when you strike it with a hammer?

2. Distinguish between *thermal energy* and *heat* and between *temperature* and *heat*.

3. What is the lowest possible temperature in Celsius degrees? Fahrenheit degrees? Kelvins?

4. What is a calorie? A Calorie?

5. Why do some substances increase in temperature more than others for the same absorption of heat? (Why does the statement "Because different substances have different specific heats" *not* answer the question?)

6. How does the specific heat of water compare with the specific heat of other common materials?

7. Why do material things generally expand when their temperature is increased?

8. Which expand more for increases in temperature—solids or liquids?

9. Why does ice water *contract* as its temperature is increased?

10. At what temperature is water densest?

11. Why is ice less dense than water?

12. Why do deep lakes in cold climates remain unfrozen in winter?

Home Project

Place a tray of cool water and a tray of hot water in a freezer. Periodically check to see which freezes first. Which does, and why?

Exercises

1. Why can't you establish whether you are running a high temperature by touching your own forehead?

2. Which has the greater amount of thermal energy—an iceberg or a cup of hot coffee? Explain.

3. Would a common mercury thermometer be feasible if glass and mercury expanded at the same rates for changes in temperature? Explain.

4. Does it make sense to talk about the temperature of a vacuum?

5. Explain what is meant by stating that a thermometer measures its own temperature.

6. If you drop a hot rock into a pail of water, the temperature of the rock and water will change until both are equal. The rock will cool and the water will warm. Does this hold true if the hot rock is dropped into the Atlantic Ocean? Explain.

7. Would you expect the temperature of water at the bottom of Niagara Falls to be slightly higher than the temperature at the top of the falls? Why?

8. Why does the pressure of a gas enclosed in a rigid container increase as the temperature increases?

9. Bermuda is close to North Carolina, but unlike North Carolina it has a tropical climate year round. Why is this so?

10. If the winds at the latitude of San Francisco and Washington, D.C., were from the east rather than from the west, why might San Francisco be able to grow cherry trees and Washington, D.C., palm trees?

11. San Francisco is warmer in winter than Washington, D.C. Why isn't it warmer in the summer?

12. In addition to the to-and-fro vibrations of the center of mass of a molecule that are associated with temperature, some molecules can absorb large amounts of energy in the form of internal vibrations and rotations of the molecule itself. Would you expect materials composed of such molecules to have a high or a low specific heat? Explain.

13. The desert sand is very hot in the day and very cool at night. What does this tell you about its specific heat?

14. Would you or the gas company gain by having gas warmed before it passed through your gas meter?

15. A metal ball is just able to pass through a metal ring. When the ball is heated, however, it will not pass through the ring. What would happen if the ring, rather than the ball, were heated? Does the size of the hole increase, stay the same, or decrease?

16. After a machinist very quickly slips a hot, snugly fitting iron ring over a very cold brass cylinder, there is no way that the two can be separated intact. Can you explain why this is so?

HOT IRON RING

COLD BRASS

17. Suppose you cut a small gap in a metal ring. If you heat the ring, will the gap become wider or narrower?

18. When a mercury thermometer is warmed, the mercury level momentarily goes down before it rises. Can you give an explanation for this?

19. One of the reasons the first light bulbs were expensive was that the electrical lead wires into the bulb were made of platinum, which expands at about the same rate as glass when heated. Why is it important that the metal leads and the glass have the same "coefficient of expansion"?

20. If you measure a lot of land with a steel tape on a hot day, will your measurements of the lot be larger or smaller than they actually are?

21. What was the precise temperature at the bottom of Lake Superior at 12:01 A.M. on October 31, 1894?

22. Suppose that water is used in a thermometer instead of mercury. If the temperature is at 4°C and then changes, why can't the thermometer indicate whether the temperature is rising or falling?

23. How does the combined volume of the billions and billions of hexagonal open spaces in the structures of ice crystals in a piece of ice compare to the portion of ice that floats above the water line?

24. How would the shape of the curve in Figure 13.11 differ if *density* were plotted against temperature, instead of volume? Make a rough sketch.

25. Why is it important to protect water pipes so they don't freeze?

26. If cooling occurred at the bottom of a pond instead of at the surface, would a lake freeze from the bottom up? Explain.

27. If water had a lower specific heat, would ponds be more likely to freeze or less likely to freeze?

28. Suppose a bar 1 meter long expands $\frac{1}{2}$ centimeter when heated. By how much will a bar 100 meters long of the same material expand when similarly heated?

29. Steel expands 1 part in 100,000 (10^{-5}) for each Celsius degree increase in temperature. Suppose the 1.3-kilometer main span of the Golden Gate Bridge had no expansion joints. How much longer would it be for an increase in temperature of 10 Celsuis degrees?

30. Consider a 40,000-kilometer steel pipe that forms a ring to fit snugly all around the circumference of the world. Suppose people along its length breathe on it so as to raise its temperature 1 Celsius degree. The pipe gets longer. It also is no longer snug. How high does it stand above ground level? (To simplify, consider only the expansion of its radial distance from the center of the earth, and apply the geometry formula that relates circumference C and radius r, $C = 2\pi r$. The result is surprising!)

14 Heat Transfer

Heat always tends to pass from warmer to cooler bodies. If several bodies of different temperatures are in thermal contact, those that are warm become cooler and those that are cool become warmer. They tend to reach a common temperature. This equalizing of temperature is brought about in three ways: by *conduction,* by *convection,* and by *radiation.*

Conduction

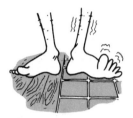

Fig. 14.1 The tile floor feels colder than the wooden floor even though both floor materials are the same temperature. This is because tile is a better conductor than wood, and heat is more readily conducted from the foot that makes contact with the tile.

Hold one end of an iron nail in a flame. It will quickly become too hot to hold. The heat enters the metal nail at the end kept in the flame and is transmitted along its whole length. The transmission of heat in this manner is called **conduction**. The fire causes the molecules at the heated end of the nail to move more rapidly. Because of this increased motion, these molecules and free electrons collide with their neighbors and cause them to move faster. These in turn collide with their neighbors, and so on. This process continues until the increased motion has been transmitted to all the molecules and the entire body has become hot. The conduction of heat is accomplished by electron and molecular collisions.

How well an object conducts heat depends on the electrical bonding of the molecular structure. Solids whose molecules have a "loose" outer electron conduct heat (and electricity) well. Metals have the "loosest" outer electrons and are the best conductors of heat and electricity for this reason. Silver is the best, copper is next, and, among the common metals, aluminum and then iron are next in order. Wool, wood, straw, paper, cork, and styrofoam are poor conductors of heat. Poor conductors are called **insulators**.

Liquids and gases, in general, are poor conductors. Air is a very poor conductor. Porous substances that have many small air spaces are poor conductors and good insulators. The good insulating properties of such things as wool, fur, and feathers are largely due to the air spaces they contain. Be glad that air is a poor conductor; if it weren't, you'd feel quite chilly on a 25°C (77°F) day!

Snow is a poor conductor and hence is popularly said to keep the earth warm. Its flakes are formed of crystals, which collect into feathery masses, imprisoning air and thereby interfering with the escape of heat from the earth's surface. The winter dwellings of the Eskimos are shielded from the cold by their snow covering. Animals in the forest find shelter from the cold in snowbanks and in holes in the snow. The snow doesn't provide them with heat; it simply prevents the heat they generate from escaping.

Fig. 14.2 Snow patterns on the roof of a house reveal the conduction, or lack of conduction, of heat through the roof.

Heat is transmitted from a higher to a lower temperature. We often hear people say they wish to "keep the cold out" of their homes. A better way to put this is to say that they want to prevent the heat from escaping. There is no "cold" that flows into a warm home. If the home becomes colder, it is because heat flows out. Homes are insulated with rock wool or spun glass to prevent heat from escaping rather than to prevent cold from entering. Interestingly enough, insulation of whatever kind does not actually prevent heat from getting through it; it simply slows the rate at which heat penetrates. Even a well-insulated, warm home in winter will gradually cool. Insulation delays the transfer of heat.

Question   In desert regions that are hot in the daytime and cold at nighttime, the walls of houses are usually made of mud. Why is it important that the mud walls be thick?*

Convection   Liquids and gases transmit heat mainly by **convection**, which is transmission by means of currents. For example, air in contact with a hot surface is heated and then ascends and is replaced by cooler air. After the warm column rises, it cools and then descends, whereupon the process is repeated. To understand convection currents in the atmosphere, we must first understand why warm air rises—then why it cools.

Why Warm Air Rises
We all know that warm air rises. From our study of buoyancy we can understand why this is so. Warm air expands and becomes less dense than

Fig. 14.3 Convection currents in air.

*Answer   A wall of optimum thickness provides a lag in thermal conductivity that will produce maximum interior temperature during the sleeping hours when it is cold outside and minimum interior temperature during midday when it is hot outside.

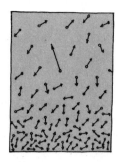

Fig. 14.4 A fast-moving molecule tends to migrate toward the region of least obstruction—upward. Warm air therefore rises.

the surrounding air and is buoyed upward like a balloon. The buoyancy is in an upward direction because the air pressure below a region of warmed air is greater than the air pressure above. And the warmed air rises because the buoyant force is greater than its weight.

We can understand the rising of warm air from a different point of view—by considering the motion of individual molecules. Consider a fairly large region of identical gas molecules. Because of gravity we would find more molecules near the bottom of our region than near the top; the gas would be slightly denser toward the ground. Suppose the region is of uniform temperature; then each molecule, on the average, has the same kinetic energy and the same average velocity. Each molecule, therefore, has the same tendency to migrate throughout the region. Suppose now that we introduce a faster-moving molecule—a "hot" one. Until it gives up its excess energy to slower-moving molecules, it will migrate farther and more rapidly than any of its neighbors. If our sample molecule is placed in the middle of our region, it will bump into and rebound from molecules in all directions. It will rebound, however, from a greater number of molecules whenever it happens to be moving downward rather than upward. This is because the density of molecules is greater below; there is more opposition to a downward migration than to an upward migration where the air is less dense. Furthermore, when our "hot" molecule moves in an upward direction, it travels farther before making a collision than when it travels downward. We say it has a longer "mean-free path" when moving upward. So our faster-moving molecule will tend to bumble upward in its random jostling.

We have simplified the idea of rising warm air by considering the behavior of a single molecule. A single fast-moving molecule would, of course, soon share its excess energy and momentum with its less energetic neighbors and would not rise very high.[1] However, if we start with a large cluster of energetic molecules, many of these will rise to appreciable heights before their energy and momentum dissipate.

Why Expanding Air Cools

Because warm air rises, we might expect the atmosphere to be warmer with increasing altitude; we might expect the mountaintops to be warm and

[1]Interestingly enough, a single helium atom will indeed rise in the atmosphere for exactly the reasons stated in the previous paragraph, because even with no excess kinetic energy its average speed will always be considerably greater than the speeds of neighboring heavier molecules of nitrogen and oxygen. Can you see that in a mixture of gaseous particles of various masses at the same temperature, the particles of least mass have the greatest speeds? (How else would they have the same temperature, that is, the same kinetic energy?) And can you see that lightweight helium with its greater speed will migrate more than its heavier neighbors, bumble its way to the top of the atmosphere, and escape into outer space? Can you see why helium, which in fact is the seventh most common gas in the earth's atmosphere, doesn't normally exist in the lower atmosphere?

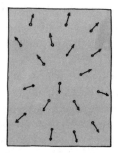

Fig. 14.5 A molecule in a region of expanding air collides primarily with receding molecules, not approaching ones. Its velocity of rebound therefore lessens with each collision and results in a cooling of expanding air.

green and the valleys below to be cold and snow-covered. But this is not the case; the atmosphere is cooler with increasing altitude. As strange as it may seem, a primary reason for the cooling at higher altitudes has very much to do with the fact that warm air rises. Warm air rises from a region of greater atmospheric pressure at the ground to a region of lesser pressure above. And while it is rising to regions of lesser pressure, it *expands*. The rapid expansion of air is a cooling process. To see why the air cools, consider a molecule in the midst of a region of expanding air. When a molecule collides with another that is approaching at a greater speed, its rebound velocity increases; when a molecule collides with another that is receding, its rebound velocity decreases. (This is easy to see: a Ping-Pong ball rebounds faster from an approaching paddle than from a receding paddle.) If the molecule is in a region of air that is expanding, then, on the average, it will collide with a greater proportion of molecules that are receding than approaching. Its velocity therefore decreases. The effect of many molecules slowing down results in a lowering of temperature (Figure 14.5). Where does the energy go in this case? It actually goes into the work done on the surrounding air as the volume of expanding air pushes outward. In this way internal energy is diluted. So we see that the energy per volume and the temperature is less.[2]

A common misconception about temperature and molecular motion is that heat is produced by the collisions of molecules against one another—that the more frequently molecules collide, the higher the temperature of the gas should be. This is not true. A pair of molecules bouncing off one another have the same total energy and momentum before and after a collision. Temperature is a measure of their kinetic energies rather than a measure of their collision rates. The temperature of gas would be no different if all the molecules were able to move without colliding with one another. Let's put this another way: when a gas is heated, the molecules collide more often. We can say they collide more often *because* they're heated. But we can *not* say that they are heated because they collide more often; cause produces the effect and not the other way around.

Compression heats air; expansion cools it. But once compressed or expanded, it soon comes to a temperature equilibrium with its surroundings. Put your hand on a tank of compressed air and you will find that it has the same temperature as its surroundings. Put your hand in the path of the same air as it escapes and expands from a nozzle and you will find a considerably lower temperature. A more dramatic example occurs with steam that expands through the nozzle of a pressure cooker (Figure 14.6). The cooling effect of both expansion and rapid mixing with cooler air allows one to comfortably hold one's hand in the jet of condensed vapor.

Fig. 14.6 The hot steam expands from the pressure cooker and is cool to Millie's touch.

_____

[2]When a gas expands without heat input, the process is called *adiabatic expansion*. A gas that expands adiabatically does external work, and its internal energy and hence its temperature decrease. Conversely, temperature increases with adiabatic compression of air—the heating of a bicycle pump, for example. More about this in Chapter 16.

Fig. 14.7 Convection currents in liquid.

Convection is a means of heat transmission in all fluids, whether liquids or gases. Whether we heat water in a pan or warm air in a room, the process is the same (Figure 14.7). If the fluid is heated from below, its molecules increase in speed and rise, permitting cooler fluid to come to the bottom. In this way, convection currents keep the fluid stirred up as it heats.

Convection currents stirring the atmosphere result in winds. Some parts of the earth's surface absorb heat from the sun more readily than others, and as a result the air near the surface is heated unevenly and convection currents form. This is most evident at the seashore. In the daytime the shore warms more easily than the water; air over the shore rises and cooler air from above the water takes its place. The result is a sea breeze. At night the process reverses because the shore cools off more quickly than the water, and then the warmer air is over the sea (Figure 14.8). Build a fire on the beach and you'll notice that the smoke sweeps inward during the day and seaward at night.

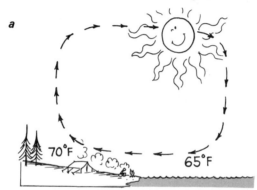
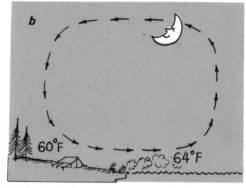

Fig. 14.8 Convection currents produced by unequal heating. (*a*) Warmed air rises and cools as it expands. It then sinks and flows toward the warmed area to replace the air that has risen. The land is warmer than the water in the day (lower specific heat) and (*b*) cooler than the water at night, so the direction of air flow reverses.

Radiation

Implied by the preceding discussion of wind is the fact that heat from the sun somehow passes through the atmosphere and warms the earth's surface. This heat does not pass through the atmosphere by conduction, for air is one of the poorest conductors. Nor does it pass through by convection, for convection begins only after the earth is warmed. We also know that neither convection nor conduction is possible in the empty space between our atmosphere and the sun. We can see that heat must be transmitted by another process; we call this process **radiation**.

It is common to associate the word *radiation* with *radioactivity*, reactions involving the atomic nucleus and characteristic of nuclear power plants and the like. But radiation is not confined to nuclear reactions of this kind and includes energy transfer associated with a wide variety of common phenomena including radio, television, medical X rays, and, most familiarly, sunlight. *Radiation* can be generally defined as the transfer of energy by the rapid oscillations of electromagnetic fields in space. These oscillations travel in the form of waves, called *electromagnetic waves,* with a characteristic *wavelength* (distance between successive crests or troughs) and *frequency* (number of wave crests passing a point per time, or the number of oscillations per time). We will discuss the general properties of waves in Chapter 17.

So radio waves, television waves, X rays, and sunlight are basically the same phenomenon—electromagnetic waves that can transmit energy through a vacuum. The basic differences among these forms of radiation are frequency and wavelength. All travel at the same speed, 3×10^8 meters per second, equal to the speed of light. These waves originate from vibrating or accelerating electrons. Since all matter contains vibrating electrons, all matter in the universe, whether hot or cold, radiates energy. The universe is filled with a montage of electromagnetic waves. We will discuss electromagnetic radiation more fully in Chapter 24.

Strictly speaking, heat is not transmitted in the radiation process. The thermal energy of a radiating body is transformed, at the instant of radiation, into *radiant energy*. Thermal energy from the sun, for example, is transformed into radiant energy and travels in that form in electromagnetic waves through space; it is *retransformed into thermal energy when it strikes an object*. Throughout the space between the radiating and the receiving objects, radiation is a form of energy entirely distinct from thermal energy or molecular motion. Strictly speaking, radiation is the transmission of radiant energy, not heat. For the sake of brevity, we speak of *heat radiation*.

Temperature Dependence of Radiation

The radiations bodies emit are not all alike. The length of the waves depends on the temperature of the source. As the temperature of a radiating object increases, the vibratory motion of electrons in the atoms and molecules of the source increases and vibrations of shorter wavelength are emitted. You can picture this by imagining you are holding the end of a long rope attached to a distant wall (Figure 14.9). When you shake the rope up and down slowly (low temperature), long, lazy waves travel along the rope. But when you shake it briskly (high temperature), the waves are shorter and more numerous along the rope. So bodies at room temperature emit longer waves, while a hot iron, for example, emits waves that are shorter (but not short enough to affect our vision). As the temperature of a radiating body increases, it emits, on the average, shorter and shorter waves. At a temperature of about 500°C, a piece of metal begins to emit waves so short that they stimulate vision. These are the longest waves we can see, red light. At about 800°C, the metal gives off still shorter waves

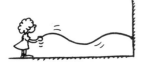

Fig. 14.9 Shorter wavelengths are produced when the rope is shaken more vigorously.

and is a bright cherry color. At 1100°C, enough even shorter waves are added to make it yellowish. At about 1200°C, it emits all the different waves to which the eye is sensitive and becomes "white hot." At the same time, however, it continues to emit the long waves, thus radiating a variety of wavelengths. When any kind of matter absorbs these waves, the waves set the electrons in the matter into vibration, and temperature increases. If radiation in the wave range of infrared falls on our skin, it may excite the sensation of warmth. In this way, heat is transmitted from the radiating object to the receiving object.

Emission, Absorption, and Reflection of Radiation

All bodies continually radiate energy. Why, then, doesn't the temperature of all bodies continually decrease? The answer is that all bodies also continually absorb radiant energy. If a body is radiating more energy than it is absorbing, its temperature does decrease; but if a body is absorbing more energy than it is emitting, its temperature increases. A body that is warmer than its surroundings emits more energy than it receives and therefore cools; a body colder than its surroundings is a net gainer of energy and its temperature therefore increases. A body whose temperature is constant, then, emits as much radiant energy as it receives. If it receives none, it will radiate away all its available energy, and its temperature will approach absolute zero.

The rate at which a body radiates or absorbs radiant energy depends on the nature of the body and the difference between its temperature and the surrounding temperature. Emission and absorption occur at the surface of a body. A rough surface is therefore a better absorber and emitter than a smooth surface since microscopically it has a greater surface area. This is true both for long-wavelength heat radiation and shorter-wavelength light radiation. So we can see that a very good absorber of radiation reflects very little light and therefore appears black. Things that reflect little visible light appear black to us. A perfect absorber reflects no radiation and

Fig. 14.10 The hole looks perfectly black and indicates a black interior, when in fact the interior has been painted a bright white.

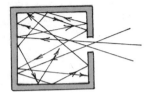

Fig. 14.11 Radiation that enters the cavity has little chance of leaving before it is completely absorbed.

Fig. 14.12 When the containers are filled with hot (or cold) water, the blackened one cools (or warms) faster.

appears perfectly black. The pupil of the eye, for example, allows radiation to enter with almost none exiting and therefore appears perfectly black (except when illuminated directly, with a flash camera, for example, in which case it appears pink).

Holes and cavities appear black because the radiation that enters is reflected from the inside walls many times and is partly absorbed at each reflection until none remains (Figure 14.11). So radiation is absorbed by the hole. If the walls of the hole or cavity are heated, however, the radiation given off in all directions is more than the incoming radiation, and there is a net escape of radiation from the hole. This total absorption or emission is called **blackbody radiation**.

A very poor absorber of radiation reflects most of the incident radiation and appears to mirror the illuminating light. That's why a polished or mirrorlike surface is a poor absorber.

Good absorbers are also good emitters. And poor absorbers are poor emitters. For example, a radio antenna constructed to be a good emitter of radio waves will also by its very design be a good receiver of radio waves. And a poorly designed transmitting antenna will also be a poor receiver. Likewise with the atomic and molecular construction of matter.

This can be illustrated by placing thermometers in a pair of metal containers of the same size and shape, one having a brightly polished surface and the other a blackened surface (Figure 14.12). If they are filled with hot water, we will find that the container with the blackened surface cools faster. The blackened surface is a better radiator. Coffee will stay hot longer in a polished pot than in a blackened one. If we repeat the same experiment, but this time fill each container with ice water, we will find that the container with the blackened surface warms up faster. The blackened surface is also a better absorber of radiant energy.

Whether a surface plays the role of radiator or absorber depends on whether its temperature is above or below the temperature of its surroundings. If the surface is hotter than the surrounding air, for example, it will be a net radiator and cool. If the surface is colder than the surrounding air, it will be a net absorber and become warmer.

The heat radiation that falls on a body and is absorbed obviously cannot also be reflected; likewise, the heat radiation that falls on a body and is reflected cannot be absorbed. Hence, a good absorber is a poor reflector of radiation, and a good reflector is a poor absorber.

Clean snow is a good reflector and therefore does not melt rapidly in sunlight. If the snow is dirty, it absorbs radiant energy from the sun and melts faster. Dropping black soot by aircraft on snowed-in mountainsides is a technique sometimes used in flood control. Controlled melting at favorable times rather than a sudden runoff of melted snow is accomplished.

Questions

1. Why is black a good absorber?*
2. Why is a good absorber black?†

Cooling at Night by Radiation

Bodies that radiate more energy than they receive become cooler. This happens at night when solar radiation is absent. Objects out in the open radiate energy into the night and, because of the absence of warmer bodies, may receive very little energy in return. They give out more energy than they receive and become cooler. If the object is a good conductor of heat—like metal, stone, or concrete—heat from the ground will be conducted to it, somewhat stabilizing its temperature. But materials such as wood, straw, and grass are poor conductors, and little heat is conducted into them from the ground. These insulating materials radiate without recompensation and *get colder than the air*. It is common for frost to form on these kinds of materials even when the temperature of the air does not go down to freezing. Have you ever seen a frost-covered lawn or field on a chilly but above-freezing morning before the sun is up? The next time you see this, notice that the frost forms only on the grass, straw, or other poor conductors, while none forms on cement, stone, or other good conductors.

Snow itself is a case in point. During the day the snow gains very little heat energy from the sun because of its reflectivity, and at night it loses energy rapidly by radiating infrared radiation to space. Because of snow's poor conductivity, the ground conducts very little heat to it, and the surface of the snow becomes cooler than the surrounding air. Snow therefore has a significant cooling effect on the earth and atmosphere. The bottom of the snow remains at about ground temperature. That's why Midwestern wheat farmers like deep snows in winter, to protect their wheat fields from the harsh, cold weather.

*****Answer** This question is misleading because it is based on the mistaken notion that because a surface is black it will be a good absorber. Strictly speaking, it is the other way around: a surface that is a good absorber will by consequence of that absorption appear black. The pupil of the eye, for example, absorbs light and is *therefore* black. It is incorrect to say that because the pupil of the eye is black it absorbs light. Its blackness is evidence that it absorbs light but not the reason it absorbs light. The question must be turned around so that it makes sense—as in Question 2.

†**Answer** A surface cannot be both a good absorber and a good reflector at the same time. If it is a good absorber, then, by definition, it must be a poor reflector. If a surface absorbs all the radiant energy incident upon it and reflects none, then it appears dark—black.

Greenhouse Effect

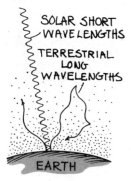

Fig. 14.13 The greenhouse effect of the earth's atmosphere. Carbon dioxide in the atmosphere acts to absorb and retain heat that would otherwise be radiated from earth into space.

Radiation from the sun consists of very short wavelengths, most of which are transmitted rather freely through the atmosphere to be absorbed at the earth's surface. The earth in turn reradiates energy (terrestrial radiation) back into the atmosphere, but at relatively long wavelengths since the earth is so much cooler than the sun. This radiation is not transmitted freely through the atmosphere, however, and is partially absorbed and reemitted by carbon dioxide, water vapor, and other atmospheric components. This secondary radiation is emitted in all directions, some upward and some downward. The downward radiation plays a part in warming the atmosphere near the earth's surface. The upward part that leaves the atmosphere produces cooling that is essential to the heat balance of the earth-atmosphere system.

Because florists' greenhouses retain terrestrial energy in a similar manner, the overall process is known as the **greenhouse effect** (Figures 14.13 and 14.14). Glass has the property of being transparent to short waves and opaque to longer waves, so the glass acts as a one-way valve, allowing short-wavelength solar radiation to enter and preventing long-wavelength radiation from leaving. This warms the greenhouse interior. Most of the warmth in a greenhouse, however, is due to the role the enclosure plays in preventing air warmed by contact with the ground from escaping. Outside the greenhouse, the ground conducts heat to the air, which is continually being replaced by convection currents. Inside the greenhouse, the ground warms only the air limited by the enclosure. So, ironically, the earth is significantly warmed by the greenhouse effect while greenhouses are not.

Fig. 14.14 Glass is transparent to short-wavelength radiation but opaque to long-wavelength radiation due to the sun's high temperature. Reradiated energy from the plant is long wavelength because the plant has a relatively low temperature.

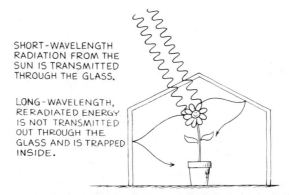

SHORT-WAVELENGTH RADIATION FROM THE SUN IS TRANSMITTED THROUGH THE GLASS.

LONG-WAVELENGTH, RERADIATED ENERGY IS NOT TRANSMITTED OUT THROUGH THE GLASS AND IS TRAPPED INSIDE.

Newton's Law of Cooling

A body at a different temperature from its surroundings will ultimately come to a common temperature with its surroundings. A hot body will cool as it warms the surroundings, and a body cooler than its surroundings will warm up as the surroundings cool. The *rate* at which heat transfer takes place—whether by conduction, convection, or radiation—is roughly proportional to ΔT, the temperature difference between the body and its surroundings; that is, how rapidly a body hotter than its surroundings loses heat is proportional to the difference between its temperature and that of the surroundings. This is known as **Newton's law of cooling**. (Guess who was the first to establish this.) A cup of coffee at 80 degrees above its surroundings cools degree by degree twice as rapidly as when it is 40 degrees above its surroundings. It cools faster if placed in a freezer because ΔT is greater. The same law holds for heating: if a body is cooler than its surroundings, its rate of warming is also proportional to ΔT.

The rate of cooling we experience on a cold day can be further increased by the added convection of wind. We speak of this in terms of *wind chill*. A wind chill of $-20°C$—when the temperature is $0°C$, for example—means we are losing heat at the same rate as if the temperature were $-20°C$. The extra heat loss is carried away by the wind.

Question   Since a hot body loses heat more rapidly than a cooler body, can we say that a hot cup of coffee will cool to its surroundings *before* a not-so-hot cup of coffee?*

Solar Power   Step from the shade into the sunshine and you're noticeably warmed. The warmth you feel isn't so much because the sun is hot, for its surface temperature of 6000°C is no hotter than the flames of some welding torches; you are warmed principally because the sun is so *big*. As a result, it emits enormous amounts of radiant energy, less than one part in a billion of which reaches the planet Earth. The amount of radiant energy received each second over each square meter that is at right angles to the sun's rays at the top of the atmosphere is 1400 joules (1.4 kJ) (Figure 14.15). This amount of energy is called the **solar constant**. Expressed in terms of power, this is 1.4 kilowatts per square meter (1.4 kW/m²). The amount of solar power that reaches the ground is attenuated by the atmosphere and reduced by nonperpendicular elevation angles of the sun. As a result, the United States averages over the four seasons about 13 percent of the solar constant, which is 0.18 kilowatts per square meter. This rate of heating corresponds to about twice the energy needed per day to heat or cool the average American house. This is part of the reason we see more and more homes using solar power for domestic heating and cooling.

Fig. 14.15 Over each square meter that is perpendicular to the sun's rays at the top of the atmosphere, the sun pours 1400 J of radiant energy each second. Hence the solar constant is 1.4 kJ/s/m², or 1.4 kW/m².

*Answer   No! Although the rate of cooling is greater for the hotter cup, it has farther to cool to reach thermal equilibrium. The extra time is equal to the time it takes to cool to the initial temperature of the not-so-hot cup of coffee. Cooling rate and cooling time are not the same thing.

Fig. 14.16 Solar water heaters are covered with glass to provide a greenhouse effect, which further heats the water. Why are the collectors painted black?

Solar heating utilizes some sort of distribution system to move solar energy from the collector to the storage or living space. When the distribution system requires external energy to operate fans or pumps, we have an *active system*. When the distribution is by natural means (conduction, convection, or radiation), we have a *passive system*. Even in the northern states, solar homes with either active or passive systems are essentially problem-free and economical.

On a larger scale, the problems of utilizing solar power are greater. First, there is the fact that no energy arrives at night. This calls for supplemental sources of energy or efficient solar-energy storage devices. Variations in the weather, particularly cloud cover, produce a variable energy supply from day to day and from season to season. Even in clear daylight hours, the sun is high in the sky only part of the day. Solar-energy collecting and concentration systems, whether by arrays of mirrors or photovoltaic cells, at this writing are not yet competitive with the costs of electrical power generation by conventional power sources. Projections indicate the story may be different before the turn of the century.

Of considerable interest is the idea of harvesting solar energy at the solar constant 24 hours per day via solar power plants in orbit, where weightlessness and the absence of wind and rain would permit the building of huge structures of low mass. These power plants could consist of giant banks of photovoltaic cells in geosynchronous orbits (similar to the orbits of communications satellites), which would render them always stationary with respect to any desired location on earth. These cells could collect solar energy continuously and convert it to electricity that would be fed to microwave generators aboard the satellites. The microwaves would be beamed to earth and picked up by antennae that could be located just about anywhere—on land, at sea, or in the desert.

The receiving antenna for such an arrangement would be a network of wires with about 80 percent open space. Sunlight would pass through it, but the microwaves would be absorbed by the antenna, making it safe for occupants beneath it. If we ever have cities covered by huge geodesic domes of the type proposed by Buckminster Fuller, these domes could be effective microwave-receiving antennae, since they would be transparent to sunlight and opaque to microwaves.

Whether large-scale solar plants are better located on earth or in space is debatable. The technological proficiency is now available for doing either, however, and is beyond debate. The essential considerations involve economic priorities rather than technology.

Fig. 14.17 A Boeing conception of a photovoltaic power satellite being constructed in low earth orbit. A space-shuttle orbiter (upper right) docks at the facility's assembly bay. To the left, an upper stage of a massive-lift launch vehicle approaches the facility to discharge its cargo of construction materials. The weightlessness in orbit allows the use of large weblike structures of a sort that would be crushed if used on earth. The satellite would be deployed in geosynchronous orbit after completion. The electricity produced would be converted to microwaves and beamed to earth.

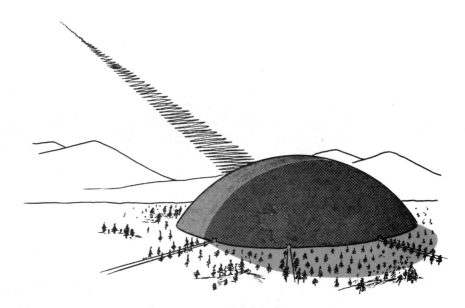

Fig. 14.18 A climate-controlled city beneath a microwave-receiving antenna.

Excess Heat Problem

A radiative equilibrium exists that balances the amount of radiation the earth receives from the sun with the radiation the earth emits into space. This equilibrium results in an average temperature that supports the development of life as we know it. The power generation in the present century is a new factor in the radiative equilibrium. We are all familiar with the problem of thermal pollution that is a by-product of power generation plants. The earth's environment is the heat sink for wasted energy. Power plants in orbit will not solve the problem. The amount of heat put into the earth's environment has to do not only with power production but also with energy *consumption*.

Suppose, for example, that enough power to supply all our needs were beamed to earth by orbiting solar power stations, and even suppose that all thermally polluting facilities from steel mills to manufacturing plants were also located in space. Then the heat sink for these facilities would be outer space rather than the planet Earth. This would significantly cut down the buildup of heat on earth that would otherwise match continued growth, but as long as more and more energy is *consumed* on earth, more and more heat is the end product. When you make your toast in the morning, the extra heat your kitchen or breakfast nook receives has little to do with the nature of the power plant that produced the electricity. All the energy we consume, whether from making toast, operating a television set, or running a power saw, ultimately becomes heat.

However you look at it, increased energy consumption on earth results in increased heating of the planet Earth. If an increase in terrestrial radiation into space does not compensate for the increase in heating on earth, the temperature of the earth increases and climate changes. Warmer

Fig. 14.19 All the energy we consume ultimately becomes heat.

$$g \propto \frac{A}{L} \Delta T \qquad g = -K A \Delta T$$
$$\underset{L}{}$$

oceans result in increased evaporation and increased snowfalls in polar regions that in turn result in increased glacial growth. The fraction of the earth that is presently beneath glaciers is about equal to the total area used for farmlands. Glacial growth and the corresponding larger areas of white snow reflect more solar radiation, which may well lead to a significant drop in global temperature. So overheating the earth might well trigger the next ice age! Or it might not. We don't know.

$K = THERMAL\ CONDUCTIVITY$

We can speculate about the long-term effects of worldwide climate changes. On one hand, it may turn out that overall changes will be tolerable and that civilizations will adapt and continue to function. On the other hand, the effects may be intolerable and be the final push to the colonization of space, where communities in orbit could easily control their temperatures by radiating excess energy into the enormous heat sink of space. Would space colonies thermally pollute the solar system in time? No, the energy used in a space facility is intercepted from a tiny fraction of the energy radiated from the sun, and whatever is discharged partially fills in what was taken away. It simply goes back to where it came from in the first place. Regulating the temperature of a colony in the vacuum of space should be an achievable task; doing the same on the planet Earth surrounded by its insulating atmosphere is a different story.

Back here on earth we must seriously question the idea of continued growth. (Please take time to read Appendix V, "Exponential Growth and Doubling Time"—very important stuff.)

Summary of Terms

Conduction The transfer and distribution of thermal energy that moves from molecule to molecule within a body.

Convection The transfer of thermal energy in a gas or liquid by means of currents in the heated fluid. The fluid moves, carrying energy with it.

Radiation The transfer of energy at the speed of light by means of electromagnetic waves.

Blackbody radiation The broad range of radiant energy that emanates from a perfect emitter of radiation. A perfect emitter is likewise a perfect absorber; this makes it appear black at low temperatures.

Greenhouse effect The heating effect of a medium such as glass or the earth's atmosphere that is transparent to the short-wavelength radiation of sunlight but opaque to long-wavelength terrestrial radiation. Energy of sunlight that enters the glass of a florist's greenhouse or the atmosphere of the earth is absorbed and reradiated at a longer wavelength that is consequently trapped, which produces heating.

Newton's law of cooling The rate of loss of heat from a body is proportional to the excess temperature of the body over the temperature of its surroundings.

Solar power Energy per unit time derived from the sun.

Solar constant 1400 joules per meter^2 received from the sun each second at the top of the earth's atmosphere; expressed in terms of power, 1.4 kilowatts per meter^2.

Review Questions

1. A row of dominoes is placed upright, one next to the other. When one of the dominoes on the end is tipped over, it knocks against its neighbor, which does the same in cascade fashion until the whole row collapses. Which of the three means of heat transmission is this most analogous to?

2. Why will a covering of snow protect crops from damage on a suddenly cold day?

3. Why do molecules of warm air migrate upward? (Why not sideways or downward?)

4. Do collisions between molecules in a gas increase the temperature of the gas? Explain.

5. Why does the direction of coastal winds change from day to night?

6. In what form is energy transmitted from the sun to the earth? Why do we not say that heat is transmitted directly from the sun to the earth?

7. When we speak about the temperature dependence of radiation, are we talking about the temperature of the sources of radiation or the temperature of radiation itself?

8. Since all bodies are absorbing energy from their surroundings, why doesn't the temperature of all bodies continually increase?

9. If a good absorber of radiation were a poor emitter, how would its temperature compare to the temperature of its surroundings?

10. If a poor absorber of radiation were a good emitter, how would its temperature compare to the temperature of its surroundings?

11. Why do good absorbers of heat appear black?

12. How are materials that are poor heat conductors able to cool to temperatures less than the surrounding air at night?

13. By what two means is the temperature within a florist's greenhouse greater than the air temperature outside?

14. Does Newton's law of cooling apply to cold bodies being warmed in a hot environment?

15. How does worldwide energy consumption relate to the average temperature of the world?

Home Projects

1. Hold the bottom end of a test tube full of cold water in your hand. Heat the top part in a flame until it boils. The fact that you can still hold the bottom shows that water is a poor conductor of heat. This is even more dramatic if you wedge chunks of ice at the bottom; then the water above can be brought to a boil without melting the ice. Try it and see.

BOILING WATER

STEEL WOOL

ICE

2. If you live where there is snow, do as Benjamin Franklin did nearly two centuries ago and lay samples of light and dark cloth on the snow. Note the difference in the rate of melting beneath the cloths.

Exercises

1. Why is it difficult to estimate the temperature of things by touching them?

2. At what common temperature will a block of wood and a block of metal both feel neither hot nor cold to the touch?

3. If you hold one end of a metal nail against a piece of ice, the end in your hand soon becomes cold. Does cold flow from the ice to your hand? Explain.

4. Silver is a very good conductor of heat. Is this quality favorable or unfavorable for silverware? Explain.

5. Why do restaurants serve baked potatoes wrapped in aluminum foil?

6. Many tongues have been injured by licking a piece of metal on a very cold day. Why would no harm result if a piece of wood were licked on the same day?

7. Wood is a better insulator than glass. Yet fiberglass is commonly used as an insulator in wooden buildings. Explain.

8. Why do double windows (Thermopane) keep a house warmer in winter?

9. If you were caught in freezing weather with only a candle for a heat source, would you be warmer in an Eskimo igloo or a wooden shack?

10. When it is daytime on the moon, the moon-rock surface is hot enough to melt solder. Why is this not so on earth?

11. If you wish to cool something by placing it in contact with ice, should you put it on top of the ice block or put the ice block on top of it?

12. You can bring water in a paper cup to a boil by placing it in a hot flame. Why doesn't the paper cup burn?

13. Why can you comfortably hold your fingers close beside a candle flame, but not very close above the flame?

14. In a still room, smoke from a cigarette will sometimes rise and then settle in the air before reaching the ceiling. Explain why.

15. Why would you expect a single helium molecule to continually rise in an atmosphere of nitrogen and oxygen? Why doesn't it "settle off" like the smoke in the preceding question?

16. In a mixture of hydrogen and oxygen gases at the same temperature, which molecules move faster? Why?

17. One container is filled with argon gas and the other with krypton gas. If both gases have the same temperature, in which container are the atoms moving faster?

18. If we warm a volume of air, it expands. Does it then follow that if we expand a volume of air, it warms? Explain.

19. How would the drawing in Figure 14.5 differ to illustrate the heating of air when it is compressed? Make a sketch of this case.

20. A snow-making machine used for ski areas consists of a mixture of compressed air and water blown through a nozzle. The temperature of the mixture may initially be well above the freezing temperature of water, yet crystals of snow are formed as the mixture is ejected from the nozzle. Explain how this happens.

21. What does the high specific heat of water have to do with convection currents in the air at the seashore?

22. What would be the most efficient color for steam radiators?

23. Why does a good *emitter* of heat radiation appear black at room temperature?

24. A number of bodies at different temperatures placed in a closed room will ultimately come to the same temperature. Would this thermal equilibrium be possible if good absorbers were poor emitters and if poor absorbers were good emitters? Explain.

25. From the rules that a good absorber of radiation is a good radiator and a good reflector is a poor absorber, state a rule relating the reflecting and radiating properties of a surface.

26. Suppose at a restaurant you are served coffee before you are ready to drink it. In order that it be hottest when you are ready for it, would you be wiser to add cream to it right away or when you are ready to drink it?

27. Even though metal is a good conductor, frost can be seen on parked cars in the early morning even when the air temperature is above freezing. Can you explain this?

28. Why is whitewash sometimes applied to florist greenhouses in the summer?

29. On a very cold sunny day you wear a black coat and a transparent plastic coat. Which should be worn on the outside for maximum warmth?

30. If the composition of the upper atmosphere were changed so that it permitted a greater amount of terrestrial radiation to escape, what effect would this have on the earth's climate? How about if the atmosphere reduced the escape of terrestrial radiation?

31. Is it important to convert temperatures to the Kelvin scale when we use Newton's law of cooling? Why or why not?

32. Some people think a can of soft drink will cool just as fast if put into the regular part of the refrigerator as it will if put into the colder freezer. What do you think? What physical law do you bring into your thinking?

33. If you wish to save fuel and you're going to leave your warm house for a half hour or so on a very cold day, should you turn your thermostat down a few degrees, turn it off altogether, or let it remain at the room temperature you desire?

34. If you wish to save fuel and you're going to leave your cool house for a half hour or so on a very hot day, should you turn your air conditioning thermostat up a bit, turn it off altogether, or let it remain at the room temperature you desire?

35. As more energy is consumed on earth, the overall temperature of the earth tends to rise. Regardless of the increase in energy, however, the temperature does not rise indefinitely. By what process is an indefinite rise prevented? Explain your answer.

15 Change of State

The matter in our environment exists in four common states. Ice, for example, is the **solid** state of H_2O. Add energy, and you add motion to the rigid molecular structure, which breaks down to form H_2O in the **liquid** state, water. Add more energy, and the liquid changes to the **gaseous** state, vapor, or, if hot enough, steam. Add still more energy, and the molecules break into ions, giving the **plasma** state. The state of matter depends on its temperature and the pressure that is exerted on it. Changes of state require changes in temperature or pressure or both—or, more generally, a transfer of energy.

Evaporation and Condensation

Molecules in the liquid state move haphazardly, sometimes colliding head-on, sometimes glancing off one another, hitting sideways and every which way, resulting in many different speeds. The temperature of the liquid is associated with the average of these speeds, or, more specifically, with the average molecular kinetic energy. There are about as many molecules moving faster than the average speed as there are molecules moving at speeds less than the average. If the very fast molecules happen to be at or near the surface, and are moving in the right direction, they will overcome the surface-tension forces, fly into the space above the liquid, and become molecules of a gas. This is the process of **evaporation**—a change of state from liquid to gas. Since the faster or more energetic molecules evaporate, the average kinetic energy of remaining molecules in the liquid is lowered. Hence, the temperature of the liquid is lowered. We say that evaporation is a cooling process.

The cooling effect of evaporation is strikingly demonstrated when rubbing alcohol is poured on your back. The alcohol evaporates very rapidly, cooling the surface of the body quickly. The more rapid the evaporation, the faster the cooling.

The converse process takes place also. Gas molecules near the surface of a liquid are attracted to the liquid and strike the surface with increased kinetic energy, which is absorbed by the liquid. This process is called **condensation**. The kinetic energy of the absorbed gas molecules is given to the liquid, which increases the liquid's temperature. We say that condensation is a warming process.

If a dish of water undergoes no apparent evaporation and no change in temperature during some period of time, we might conclude that "nothing is happening." But in this case both evaporation and condensation are taking place at the same rate. The number of molecules and the amount of energy leaving the liquid's surface by evaporation are counteracted by as many molecules and as much energy returning by condensation. We say that the liquid is in equilibrium, since evaporation and condensation have canceling effects.

Fig. 15.1 The cloth covering on the sides of the canteen promotes cooling when it is wet. As the fastest-moving molecules evaporate from the wet cloth, the temperature of the cloth decreases and cools the metal, which in turn cools the water within. In this way the water in the canteen is cooled appreciably below air temperature.

When we emerge from a shower and step into a dry room, we feel cold. This is because evaporation is taking place at a rate that produces considerable cooling. If we remain in the shower, even with the water off, we do not feel as chilly. This is because we are in a moist environment, and moisture from the air condenses on our skin, producing a warming effect that counteracts the cooling effect of evaporation. If as much moisture condenses as evaporates, we feel no change in body temperature. We can dry ourselves with a towel much more comfortably if we remain in the shower area; to dry ourselves thoroughly, we can finish the job in a less moist area.

Both evaporation and condensation take place at the surface of liquids. If the rate of evaporation exceeds the rate of condensation, the liquid is cooled. If the rate of condensation exceeds the rate of evaporation, the liquid is warmed. This is the reason we feel uncomfortably warm on a humid or muggy day. We will discuss humidity shortly.

Boiling   Under the right conditions, evaporation can take place beneath the surface of a liquid, forming bubbles of gas that are buoyed to the surface where they escape. The speed of the molecules forming gas must be great enough to exert as much pressure within the bubble as is exerted by the atmosphere and water above (Figure 15.2). We therefore find that boiling depends not only on temperature but on pressure as well. Unless the molecules in the liquid are moving fast enough, atmospheric pressure acting on the surface of the liquid will collapse any bubbles that tend to form. As the atmospheric pressure is increased, the molecules in the liquid must move faster to exert a vapor pressure within the bubble equal to the impressed atmospheric pressure. So increasing the pressure on the surface of a liquid raises its boiling point.

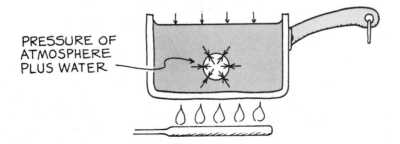

Fig. 15.2 The motion of molecules in the bubble of steam (much enlarged) creates an outward vapor pressure that counterbalances the atmospheric and water pressure pushing inward on the bubble.

PRESSURE OF ATMOSPHERE PLUS WATER

A pressure cooker works by means of this fact. As the evaporating vapor builds up under the lid of the pressure cooker, pressure on the surface of the liquid is increased, which prevents boiling. This raises the boiling point. The increased temperature of the boiling water cooks the food faster. Conversely, if the pressure is reduced, the temperature at

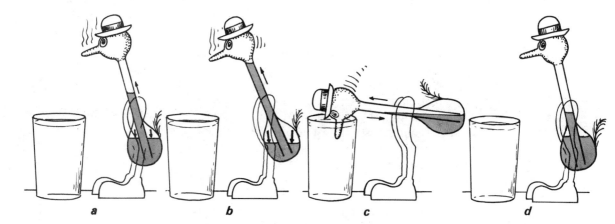

a   b   c   d

Fig. 15.3 The toy drinking bird operates by the evaporation of ether inside its body and by the evaporation of water from the outer surface of its head. The lower body contains liquid ether, which evaporates rapidly at room temperature. As it (a) vaporizes, it (b) (inside arrows) creates pressure, which pushes ether up the tube. Ether in the upper part does not vaporize because the head is cooled by the evaporation of water from the outer felt-covered beak and head. When the weight of ether in the head is sufficient, the bird (c) pivots forward, permitting the ether to run back to the body. Each pivot wets the felt surface of the beak and head, and the cycle is repeated.

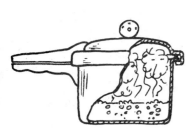

Fig. 15.4 A pressure cooker.

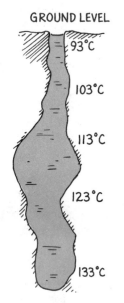

Fig. 15.5 An Old Faithful type of geyser.

which boiling takes place is reduced. At atmospheric pressure, water boils at 100°C, or 212°F. In regions of reduced atmospheric pressure—the mountains, for example—boiling takes place at temperatures lower than 100°C. In Denver, Colorado, the "mile-high city," water boils at 95°C. As a result, food cooked in boiling water requires more time. (A 3-minute egg in Denver is "runny.") When atmospheric pressure is absent, water boils away even at freezing temperatures. That's why no ponds or lakes can exist on the moon.

Geysers

A geyser is a periodically erupting pressure cooker. It consists of a long, narrow, vertical hole into which underground streams seep (Figure 15.5). The column of water is heated by volcanic heat below to temperatures exceeding 100°C. This is because the vertical column of water exerts pressure on the deeper water, thereby increasing the boiling point. The narrowness of the shaft shuts off convection currents, which allows the deeper portions to become considerably hotter than the water surface. Water at the surface, of course, will boil at 100°C. The water is heated from below, so a temperature high enough to permit boiling is reached near the bottom before it is at the top. Boiling therefore begins near the bottom, the rising bubbles push out the column of water above, and the eruption starts. As the water gushes out, the pressure in the remaining water is reduced. It then rapidly boils and erupts with great force.

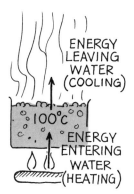

Fig. 15.6 Heating warms the water, and boiling cools it.

Boiling Is a Cooling Process

Evaporation is a cooling process. So is boiling. At first thought, this may seem surprising—perhaps because we always associate boiling with hot water or because we don't make a distinction between heating and boiling. But on second thought, the fact that the temperature of water is not increased above the boiling temperature when heated supports the idea that cooling is taking place. At 100°C, water at sea level is in thermal equilibrium. It is being cooled by boiling as fast as it is being heated by energy from the stove (Figure 15.6). If cooling did not take place, continued application of heat to a pot of boiling water would result in a continued increase in temperature. A pressure cooker reaches higher temperatures because it prevents cooling by boiling.

Question   Since boiling is a cooling process, would it be a good idea to cool your hot and sticky hands by dipping them into boiling water?*

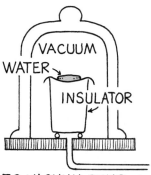

Fig. 15.7 Apparatus to demonstrate that water will freeze and boil at the same time in a vacuum. A gram or two of water is placed on a sheet of plastic wrap over the open end of an insulating styrofoam cup.

Boiling and Freezing at the Same Time

You can demonstrate these ideas dramatically by placing some water at room temperature in a vacuum jar (Figure 15.7). The pressure in the jar is slowly reduced by a vacuum pump until the water starts to boil. The vaporization of water takes heat away from the water left in the dish, which cools to a lower temperature. As the pressure is further reduced, more and more of the slower-moving molecules boil away. Continued boiling results in a lowering of temperature until the freezing point of approximately 0°C is finally reached. Continued cooling by evaporation results in the formation of ice over the surface of the boiling water. Boiling and freezing are taking place at the same time! This must be witnessed to be appreciated. Frozen bubbles of boiling water are a remarkable sight.

Spray some drops of coffee into a vacuum chamber and they, too, will boil until they freeze. Even after they are frozen, the water molecules will continue to evaporate into the vacuum until little crystals of coffee solids are left. These would be properly called *freeze-dried coffee*. And this is actually the way freeze-dried coffee is manufactured! These coffee solids reconstitute easily with little or no detergent or ''wetting agent'' added, and the low temperature of the process tends to keep the chemical structure of the coffee solids from changing. Boiling really is a cooling process!

---

*Answer   No, no, no! When we say boiling is a cooling process, we mean that the *water* (not your hands!) is being cooled relative to the higher temperature it would attain otherwise. Because of cooling, it remains at 100°C instead of getting hotter. A dip in 100°C water would be most uncomfortable for your hands!

Melting and Freezing

Suppose you held hands with someone, and each of you started jumping around randomly. The more violently you jumped, the more difficult keeping hold would be. If you jumped violently enough, keeping hold would be impossible. Something like this happens to the molecules of a solid when it is heated. As heat is absorbed, the molecules vibrate more and more violently. If enough heat is absorbed, the attractive forces between the molecules will no longer be able to hold them together. The solid melts.

Freezing is the converse of this process. As energy is withdrawn from a liquid, molecular motion diminishes until finally the molecules, on the average, are moving slowly enough so that the attractive forces between them are able to cause cohesion. The molecules then vibrate about fixed positions and form a solid.

Regelation

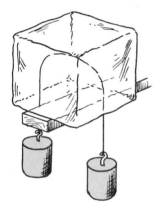

Fig. 15.8 Regelation. The wire will gradually pass through the ice without cutting it in half.

Because H_2O molecules form open structures in the solid state (Figure 13.12 in Chapter 13), the application of pressure lowers the melting point of ice. The crystals are simply crushed to the liquid state. (The temperature of the melting point is lowered only slightly, being 0.007°C for each additional atmosphere of pressure added.) When the pressure is removed, refreezing occurs. This phenomenon of melting under pressure and freezing again when the pressure is reduced is called **regelation**. It is one of the properties of water that make it different from other materials.

Regelation is neatly illustrated in Figure 15.8. A fine wire with weights attached to its ends is hung over a block of ice. The wire will slowly cut its way through the ice, but its track will be left full of ice. So the wire and weights will fall to the floor, leaving the ice in a solid block.

The making of snowballs is a good example of regelation. When we compress the snow with our hands, we cause a slight melting of the ice crystals; when pressure is removed, refreezing occurs and binds the snow together. Making snowballs is difficult in very cold weather because the pressure we can apply is not enough to melt the snow.

A skater skates on a thin film of water between the blade and the ice, which is produced by blade pressure and friction. As soon as the pressure is released, the water refreezes.

Freezing Point of Solutions

Dissolving sugar in water lowers the water's freezing temperature. This is because the sugar molecules do not enter into the hexagonal ice-crystal structure. As a result, molecules of sugar get in the way of water molecules that ordinarily would join together. As ice crystals do form, the hindrance is intensified; the ratio of sugar to water molecules increases, and connections become more and more difficult. Only when the water molecules move slowly enough for attractive forces to play an unusually large part in the process can freezing be completed. In general, adding anything to water has this result; antifreeze is a practical application of this process.

Energy of Changes of State

If we continually add heat to a solid or liquid, the solid or liquid will eventually change state. A solid will liquefy, and a liquid will vaporize. Energy is required for both liquefaction of a solid and vaporization of a liquid. The general behavior of many substances can be illustrated with a description of the changes that occur with water. To make the numbers easy to deal with, suppose we have a 1-gram cube of ice at a temperature of −50°C in a closed container, and we put it on a stove to heat. A thermometer in the container reveals a slow increase in temperature up to 0°C. At 0°C, the temperature stops rising, yet heat is continually being supplied. This heat melts the ice. To melt the whole gram, 80 calories of heat are absorbed by the ice.[1] Not until all the ice melts does the temperature again begin to rise. Each additional calorie absorbed by the water increases its temperature by 1°C until it reaches the boiling temperature, 100°C. Again, as heat is added, the temperature remains constant, while more and more of the gram of water is boiled away and becomes steam. The water must absorb 540 calories of heat to vaporize the whole gram.[2] Finally, when all the water has become steam at 100°C, the temperature begins to rise once more. It will continue to rise as long as we continue to add heat. This process is graphed in Figure 15.9.

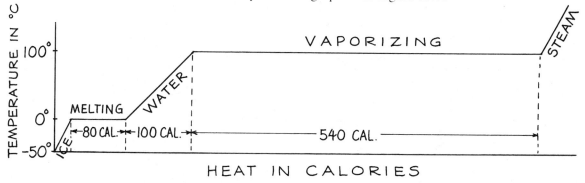

Fig. 15.9 A graph showing the energy involved in the heating and the change of state of 1 g of H_2O.

Note that the 540-calorie section of the graph shows a relatively large amount of energy, much more than would be required to bring a gram of

---

[1]The process is reversible. If the molecules in a gram of ice water fuse to become solid ice, 80 cal of energy are released by the water; we say the heat of *fusion* of water is 80 cal/g or, in SI units, 334880 J/kg.

[2]Similarly, we say the heat of *vaporization* of water is 540 cal/g or 2.26 MJ/kg.

ice at absolute zero to boiling water! Although steam and boiling water at 100°C both have the same molecular kinetic energy, more potential energy exists in the vibrational and rotational states of the H_2O molecules in steam. Steam contains a vast amount of internal energy.

Water can change to vapor by boiling or by evaporation at lower temperatures. At temperatures lower than 100°C, vaporization requires even more energy. So whether by boiling or by evaporation, at least 540 calories of heat are needed to convert each gram of liquid water into vapor. When perspiration evaporates from your skin, it draws much of this heat out of your body.

ENERGY IS ABSORBED WHEN CHANGE OF STATE
IS IN THIS DIRECTION.

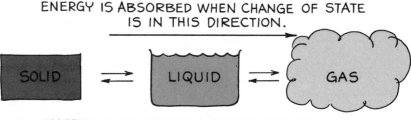

ENERGY IS RELEASED WHEN CHANGE OF STATE
IS IN THIS DIRECTION.

Fig. 15.10 Change of state and energy changes.

A refrigerator is cooled in a similar way. Pipes in the refrigerator contain a special liquid with a low boiling temperature. As this liquid turns into a gas in the cooling unit, it draws heat from the things stored in the food compartment. The gas is then directed outside the cooling unit to condensation coils located in the back, where the heat is given off to the air as the gas condenses into the liquid state. A motor pumps the fluid through the system, where it is made to undergo the cyclic processes of vaporization and condensation. The next time you're near a refrigerator, place your hand near the condensation coils in the back and you will feel the heat that has been extracted from within the cooling unit.

An air conditioner employs the same principles and simply pumps heat from one part of the unit to another. When the roles of vaporization and condensation are reversed, the air conditioner becomes a heater.

Question   How does the formation of snow or rain affect the temperature of the air?*

---

*__Answer__   The formation of snow or rain from water vapor is accompanied by an increase in atmospheric temperature. This is because gas, in this case the water vapor, releases energy in transforming to the liquid and/or solid state. So it is always a little warmer when it rains or snows than it would be otherwise!

You dare not touch your dry finger to a hot skillet on a hot stove, but you can certainly do so without harm if you first wet your finger and touch the skillet momentarily. You can even touch it a few times in succession as long as your finger remains wet. This is because energy that ordinarily would go into burning your finger goes instead into changing the state of the moisture on your finger. The energy converts the moisture to a vapor, which then provides an insulating layer between your finger and the skillet. Similarly, you are able to judge the hotness of a hot clothes iron. And in the same way, one is able to walk barefoot on red-hot coals, as physics professor Jearl Walker demonstrates in front of his physics class (Figure 15.11). This is accomplished via the perspiration on the soles of his feet—perspiration facilitated by nervousness. On one occasion, the confidence he gained from his successful prior fire walk was accompanied by less nervousness and hence less perspiration: He burned his feet. Jearl called this a "no-sweat situation."

Briefly summarizing, we see that a solid must absorb energy to melt; a liquid must absorb energy to vaporize. Conversely, a gas must release energy to liquefy; a liquid must release energy to solidify.

Fig. 15.11 Physicist Jearl Walker demonstrates change of state by walking barefoot on red-hot coals. His feet are protected only by natural perspiration, which in vaporizing absorbs otherwise harmful energy and creates an insulating layer of water vapor between the coals and his skin.

Humidity

The air always contains some water vapor; at any given temperature, however, there is a limit to the amount that the air will support. When this limit is reached, we say the air is *saturated*. More water is required to saturate the atmosphere when the temperature is high than when it is low. This is because water vapor molecules, which would coalesce and liquefy upon low-speed collisions, bounce apart and remain in the gaseous state upon high-speed collisions. This is similar to a fly making a grazing contact with flypaper. At low speed it would surely get stuck, whereas at high speed it stands a greater chance of rebounding into the air. So the faster that water vapor molecules move, the less chance there is that they will stick to one another and form droplets. Warmer air therefore will support a greater number of water molecules in the vapor state than cooler air.

The ability of air to hold water increases rapidly with increasing temperature. An approximate rule is that an increase in temperature of 10°C allows the air to hold twice as much water vapor. That's why hot days are often humid.

Usually we speak of humidity in one of two ways. Sometimes we speak of the percentage of water vapor in a unit volume of air; this is called the **absolute humidity**. Thus, we might say the absolute humidity of some

Fig. 15.12

FAST-MOVING H_2O MOLECULES REBOUND UPON COLLISION

SLOW-MOVING H_2O MOLECULES COALESCE UPON COLLISION

sample of air is 25 percent, meaning that 25 percent of our sample is water vapor. More often, however, we speak of the amount of water vapor present as compared to the amount of vapor present in saturated air at the same temperature; this is called **relative humidity**. Thus, if the relative humidity is 50 percent, the air contains half the amount it would contain when saturated at the same temperature.

Our bodily comfort depends on the humidity as well as on the temperature. For the average person, conditions are ideal when the temperature is about 20°C and the relative humidity is about 50 to 60 percent. When the relative humidity is high, condensation of moist air counteracts evaporation of perspiration, and we say the weather is "muggy."

Body temperature in a rising-temperature environment is regulated by the evaporation of sweat. In dry desert air, with a well-ventilated hat and adequate salt replacement, a person can work with comfort when the air temperature is as high as 52°C. When the air is very dry, sweat is hardly noticed as it evaporates rapidly.

As the humidity rises toward the saturation point, it becomes harder and harder to evaporate sweat. Net evaporation declines as condensation of moisture from the air increases. When the air is saturated, sweat rolls off a person in streams without evaporating; vital salt (NaCl) is lost, and body

Fig. 15.13 Pigs have no sweat glands and therefore cannot cool by the evaporation of perspiration. Instead, they wallow in the mud to cool themselves.

temperature begins to rise. This is one form of "sunstroke" (not due to the sun necessarily) where victims may be found unconscious with a temperature of 41.6°C and shock due to low blood salt.[3]

Fog and Clouds

Fig. 15.14 Why is it common for clouds to form where there are updrafts of warm air?

Warm air rises. As it rises, it expands. As it expands, it chills. As it chills, water vapor molecules begin coalescing rather than bouncing off one another upon glancing collisions. Condensation takes place, and we have a cloud. The presence of larger and slower-moving particles or ions facilitates this process as water molecules adhere to them. We can stimulate cloud formation by "seeding" the air with appropriate particles or ions.

Warm breezes blow over the ocean. When the moist air moves from warmer to cooler waters or from warm water to cool land, it chills. As it chills, water vapor molecules begin coalescing rather than bouncing off one another upon glancing collisions. Condensation takes place, and we have fog.

All fogs and clouds are produced by the cooling of moist air. Fast-moving water molecules in the vapor state coalesce into tiny droplets upon chilling. If the process takes place overhead in the atmosphere, we call the result a cloud. If it takes place near the ground, we call the result fog. Flying through a cloud is much like driving through a fog.

---

[3]The other form of "sunstroke" is usually, though not always, due to the sun and is not related to humidity. For instance, hatless exposure to a "hot" sun may result in the brain temperature rising faster than it can be cooled by the blood, and at about 40°C to 40.5°C (104°F to 105°F), the heat-regulating center in the hypothalamus fails, and sweating stops. In this case a person may be unconscious, with a temperature of perhaps 41.6°C (107°F) but not in shock. Blood is normal, and sweating does not occur.

Snowflakes are formed when water vapor solidifies in the air at a temperature below freezing. With an adequate amount of vapor solidification, the snowflakes become large enough to fall. Tiny flakes may have such a slow rate of fall that they appear to float in the air. High cirrus and cirrostratus clouds consist of snowflakes or minute ice crystals. A haze of minute ice crystals high up in the atmosphere sometimes produces a large halo around the sun and moon.

Hail results from raindrops being blown upward into freezing temperatures where they solidify. Sometimes they are blown upward again and again, becoming very heavy before they finally fall to earth.

Summary of Terms

Evaporation   The change of state at the surface of a liquid as it passes to vapor. This results from the random motion of molecules that occasionally escape from the liquid surface. Cooling of the liquid results.

Condensation   The change of state from vapor to liquid; the opposite of evaporation. Warming of the liquid results.

Boiling   A rapid state of evaporation takes place within the liquid as well as at its surface. As with evaporation, cooling of the liquid results.

Melting   The change of state from the solid to the liquid form. Energy is absorbed by the substance that is melting.

Freezing   The change of state from the liquid to the solid form: the opposite of melting. Energy is released by the substance undergoing freezing.

Regelation   The process of melting under pressure and the subsequent refreezing when the pressure is removed.

Absolute humidity   A measure of the amount of water vapor per unit volume in a sample of air.

Relative humidity   The ratio of the amount of water vapor in a sample of air to the amount of water vapor the sample of air is capable of supporting at a given temperature. A relative humidity of 100 percent means that the air is saturated.

Review Questions

1. Why is evaporation a cooling process and condensation a warming process?

2. Why do you feel less chilly if you dry yourself in a damp region after taking a shower?

3. Why do you feel uncomfortably warm on a humid or muggy day?

4. Can you heat a substance without raising its temperature? Give an example.

5. What is the difference between evaporation and boiling?

6. Why does the temperature at which a liquid boils depend on atmospheric pressure?

7. Why would a pressure cooker be more desirable when cooking in the mountains than when cooking at sea level?

8. Boiling is a cooling process. Does this mean that if you wish to cool your hands, you should place them in a pot of boiling water? Why or why not?

9. Why does adding antifreeze to water lower the temperature at which freezing takes place?

10. When water freezes, is energy absorbed or released by the water? How about when ice melts?

11. Is the food compartment in a refrigerator cooled by vaporization or by condensation of the refrigerating liquid?

12. Why does warmer air support a greater amount of water vapor?

13. Why does water vapor in the air condense when the air is chilled?

14. On cold days the windows in your home sometimes get wet on the inside. Why is this so?

15. Why is your finger unharmed by momentarily touching a hot iron if you first wet your finger?

Home Projects

1. Place a Pyrex funnel mouth down in a saucepan full of water so the straight tube of the funnel sticks above the water. Rest a part of the funnel on a nail or coin so water can get under it. Place the pan on a stove and watch the  water as it begins to boil. Where do the bubbles form first? Why? As the bubbles rise, they expand rapidly and push water ahead of them. The funnel confines the water, which is forced up the tube and driven out at the top. Now do you know how a geyser and a coffee percolator work?

2. Watch the spout of a teakettle of boiling water. Notice that you cannot see the steam that issues from the spout. The cloud that you see farther away from the spout is not steam, but condensed water droplets. Now hold a candle in the cloud of condensed steam. Can you explain your observations?

3. You can make rain in your kitchen. Put a cup of water in a Pyrex saucepan or Silex coffeemaker and heat it slowly over a low flame. When the water is warm, place on top of it a saucer filled with ice cubes. As the
water below is heated, droplets form at the bottom of the cold saucer and combine until they are large enough to fall, producing a steady "rainfall" as the water below is gently heated. How does this resemble and how does it differ from the way in which natural rain is formed?

4. Compare the temperature of boiling water with the temperature of a boiling solution of salt and water.

5. Suspend a heavy weight by wire over an ice cube. In a matter of minutes the wire will be pulled through the ice. The ice will melt beneath the wire and refreeze above it, leaving a visible path if the ice is clear.

6. Place a tray of nearly boiling water and a tray of water with a temperature between 60°C and boiling in a freezer. See that the hotter water will freeze first! Then figure out why (taking the high heat of vaporization of water into account).

7. If you suspend a container of water in a larger container of boiling water, the water in the inner container will reach 100°C but will not boil. Can you figure out why this is so?

Exercises

1. Wind direction can be determined by wetting your finger and holding it up in the air. Explain.

2. Can you give two reasons why blowing over hot soup cools the soup?

3. Can you give two reasons why pouring a cup of hot coffee into the saucer results in faster cooling?

4. A covered glass of water sits for days with no drop in water level. Strictly speaking, can you say that nothing has happened, that no evaporation or condensation has taken place? Explain.

5. Would evaporation be a cooling process if all the molecules in a liquid had the same speed? Explain.

6. Where does the energy come from that keeps the dunking bird in Figure 15.3 operating?

7. An inventor claims to have developed a new perfume that lasts a long time because it doesn't evaporate. Comment on this claim.

8. Porous canvas bags filled with water are used by travelers in hot weather. When slung on the outside of a fast-moving car, the water inside is cooled considerably. Explain.

9. At a picnic, why will wrapping a bottle in a wet cloth be a better method of cooling than placing the bottle in a bucket of cold water?

10. Why does the temperature of boiling water remain the same as long as the heating and boiling continue?

11. Water will boil spontaneously in a vacuum—on the moon, for example. Could you cook an egg in this boiling water? Explain.

12. Our inventor friend proposes a design of cookware that will allow boiling to take place at a temperature less than 100°C so that food can be cooked with less energy. Comment on this idea.

13. Your instructor hands you a closed flask of water. When you hold it in your bare hands, the water begins to boil. Quite impressive! How is this accomplished?

14. Why does putting a lid over a pot of water on a stove shorten the time it takes for the water to come to a boil, whereas after the water is boiling, use of the lid only slightly shortens the cooking time?

15. In a nuclear submarine power plant, the temperature of the water in the reactor is above 100°C. How is this possible?

16. Explain why the eruptions of many geysers repeat with notable regularity.

17. Would regelation occur if ice crystals were not open-structured? Explain.

18. Why is it important that the wire that cuts through the ice in Figure 15.8 be a good thermal conductor? (Copper wire of high thermal conductivity will cut through much faster than an iron wire that has lower conductivity. A thin string of very low conductivity will not pass through at all.)

19. You'll slip on a smoothly polished floor much more readily than on an unpolished floor, but you'll slip more readily on bumpy ice than on smooth ice. Why is this so? (*Hint:* Why is ice slippery?)

20. People who live where snowfall is common will attest to the fact that air temperatures are higher on snowy days than on clear days. Some misinterpret this by stating that snowfall can't occur on very cold days. Explain this misinterpretation.

21. How does melting ice change the temperature of the surrounding air?

22. Some old-timers found that when they wrapped newspaper around the ice in their iceboxes, melting was inhibited. Discuss the advisability of this.

23. Why is it that in cold winters a tub of water placed in a farmer's canning cellar helps to prevent canned food from freezing?

24. Why will spraying fruit trees with water before a frost help to protect the fruit from freezing?

25. The human body can maintain its customary temperature of 37°C on a day when the temperature is above 40°C. How is this done?

26. For comparable temperatures, which sample of air has more water vapor—a sample at 50 percent absolute humidity or a sample at 50 percent relative humidity?

27. Why are icebergs often surrounded by fog?

28. What is the relationship of Figure 15.12 to the moisture that forms on the inside of car windows when you're parking on a cool night?

29. You know that the windows in your warm house get wet on a cold day. But can moisture form on the windows if the inside of your house is cold on a hot day? How is this different?

30. Air-conditioning units contain no water whatever, yet it is common to see water dripping from them when they're running on a hot day. Explain.

31. Why do clouds often form above mountain peaks? (*Hint:* Consider the updrafts.)

32. Why will clouds tend to form above either a flat or a mountainous island in the middle of the ocean? (*Hint:* Compare the specific heat of the land with that of the water and the subsequent convection currents in the air.)

33. There are several theories for the advent of an ice age. Consider the following one: If the temperature of the world increases, evaporation of the oceans increases, precipitation increases, and snowfall increases. This results in more snow left in cooler places at the end of each summer, with successively increasing depths each winter that become hard-packed from the bottom up to become ice. All the while, the ice reflects more solar radiation that would otherwise be absorbed. This, in turn, cools the earth further and allows for more freezing. Can you continue this sequence and see how the process reverses itself and how the ice age withdraws?

34. It so happens that drops of water placed in a hot skillet will quickly vaporize away, but drops in a very hot skillet will "dance around" for a longer time before vaporizing away. Can you think of a reason for this?

35. Why does a hot dog pant?

16 Thermodynamics

The study of heat and its transformation to mechanical energy is called **thermodynamics** (which stems from Greek words meaning "movement of heat"). The science of thermodynamics was developed at the beginning of the last century, before the atomic and molecular theory of matter was understood. Whereas in previous chapters we described heat in terms of the microscopic behavior of jiggling atoms and molecules, in this chapter we invoke only macroscopic notions such as mechanical work, pressure, and temperature and their roles in energy transformations. Thermodynamics is a powerful theoretical science that bypasses the molecular details of a system altogether. Its foundation is the conservation of energy and the fact that heat flows from hot to cold and not the other way around. It provides the basic theory of heat engines, from steam turbines to fusion reactors, and the basic theory of refrigerators and heat pumps. We begin our study of thermodynamics with a look at one of its early concepts—a lowest limit of temperature.

Absolute Zero In principle, there is no upper limit of temperature. As thermal motion increases, a solid body first melts and then evaporates; as the temperature is further increased, molecules break up into atoms, and atoms lose some or all of their electrons, thereby forming a cloud of electrically charged particles—a plasma. This situation exists in stars, where the temperature is many millions of degrees Celsius.

In contrast, there is a definite limit at the other end of the temperature scale. This was suggested in experiments with gases in the nineteenth century. Gases expand when heated and contract when cooled. It was found that all gases, regardless of their initial pressures or volumes, change by $\frac{1}{273}$ of their volume at 0°C for each degree Celsius change in temperature, provided the pressure is held constant. So if a gas at 0°C were cooled down by 273°C, it would contract $\frac{273}{273}$ volumes and be reduced to zero volume. Clearly, we cannot have a substance with zero volume. It was also found that the pressure of any gas in any container of *fixed* volume would change by $\frac{1}{273}$ for each degree Celsius change in temperature. So a gas in a container of fixed volume cooled 273°C below zero would have no pressure whatsoever. In practice, every gas liquefies before it gets this cold. Nevertheless, these decreases by $\frac{1}{273}$ increments suggested the idea of a lowest temperature: −273°C. So there is a limit to coldness. When atoms and molecules lose all available kinetic energy, they reach the **absolute zero** of temperature.[1] At absolute zero, no more energy can be

---

[1]Even at absolute zero, molecules still have a small kinetic energy, called the *zero-point energy*. Helium, for example, has enough motion at absolute zero to keep it from freezing. The explanation for this involves modern quantum theory.

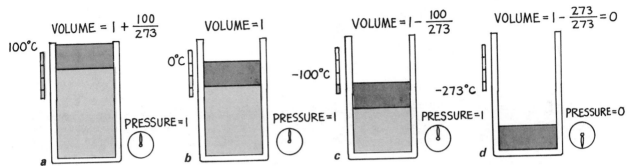

Fig. 16.1 The volume of a gas changes by $\frac{1}{273}$ its volume at 0°C with each degree Celsius change in temperature, when the pressure is held constant. (*a*) At 100°C the volume is $\frac{100}{273}$ greater than it is at (*b*) 0°C. (*c*) When the temperature is reduced to −100°C, the volume is reduced by $\frac{100}{273}$. (*d*) At −273°C the volume of the gas would be reduced by $\frac{273}{273}$ and be zero.

extracted from a substance, and no further lowering of its temperature is possible. This limiting temperature is actually 273.15° *below* zero on the Celsius scale (and 459.69° *below* zero on the Fahrenheit scale).

On the absolute temperature scale, called the Kelvin scale,[2] absolute zero is 0K (rather than saying "0 degrees Kelvin," we simply say "0 Kelvin," written 0K). On this scale, absolute zero is 0K. There are no minus numbers in the Kelvin scale. Degrees on the Kelvin scale are calibrated with the same-sized divisions as on the Celsius scale. The melting point of ice thus is 273.15K, and the boiling point of water is 373.15K.

Question A piece of metal has a temperature of 0°C. A second, identical piece of metal is twice as hot (has twice the thermal energy). What is its temperature?*

Internal Energy We all know that there is a vast amount of energy locked in all materials—this book, for example. The pages of this book are composed of molecules that are in constant motion. They have thermal energy. Due to interactions with neighboring molecules, they also have potential energies. The pages can be easily burned, so we know they store chemical energy, which is really electric potential energy at the molecular level. We know there are vast amounts of energy associated with atomic nuclei. Then there is the "energy of being," described by the celebrated equation $E = mc^2$ (mass energy). Energy within a substance is found in these and other forms,

---

*__Answer__ The piece of metal with twice the thermal energy has twice the absolute temperature, or two times 273K. This would be 546K, or 273°C.

[2]Named after the famous British physicist Lord Kelvin, who coined the word *thermodynamics* and was the first to suggest this thermodynamic temperature scale.

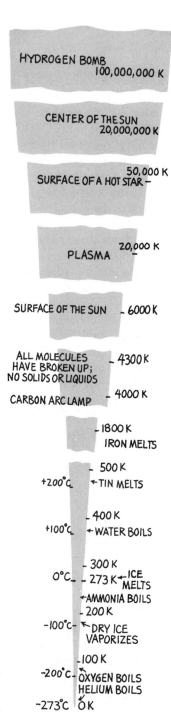

HYDROGEN BOMB
100,000,000 K

CENTER OF THE SUN
20,000,000 K

50,000 K
SURFACE OF A HOT STAR –

PLASMA   20,000 K
–

SURFACE OF THE SUN – 6000 K

ALL MOLECULES   – 4300 K
HAVE BROKEN UP;
NO SOLIDS OR LIQUIDS

CARBON ARC LAMP   – 4000 K

– 1800 K
IRON MELTS

– 500 K
+200°C   ←TIN MELTS

– 400 K
+100°C   ←WATER BOILS

– 300 K
0°C – 273 K ←ICE
MELTS
←AMMONIA BOILS
– 200 K
–100°C – ←DRY ICE
VAPORIZES
–100 K
–200°C – OXYGEN BOILS
HELIUM BOILS
–273°C 0 K

Fig. 16.2 Some absolute temperatures.

which, when taken together, are called **internal energy**. The internal energy in even the simplest substance can be quite complex. But in our study of heat changes and heat flow, we will be concerned only with the *changes* in the internal energy of a substance. Changes in temperature will indicate these changes in internal energy.

If this book were poised at the edge of a table and ready to fall, it would possess gravitational potential energy; if it were tossed into the air, it would possess kinetic energy. But these are *not* examples of internal energy, because they involve more than just the elements of which the book is composed. They include gravitational interactions with the earth and motion with respect to the earth. If we wish to include these forms, we must do so in terms of a larger "system"—a system enlarged to include both the book and the earth. They are not part of the internal energy of the book itself.

First Law of Thermodynamics

More than a hundred years ago heat was thought to be an invisible fluid called *caloric*, which flowed like water from hot bodies to cold bodies. Caloric was conserved in its interactions, and this discovery was the forerunner of the law of conservation of energy. In the mid-1880s, it became apparent that the flow of heat was nothing more than the flow of energy itself. The caloric theory of heat was gradually abandoned.[3] Today we view heat as a form of energy. Energy cannot be either created or destroyed.

When the law of energy conservation is applied to thermal systems, we call it the **first law of thermodynamics**. We state it generally in the following form:

Whenever heat is added to a system, it transforms to an equal amount of some other form of energy.

By *system*, we mean a well-defined group of atoms, molecules, particles, or objects. The system may be the steam in a steam engine; it may be the body of a living creature; or it may be the whole earth's atmosphere. The important point is that we must be able to define what is contained within the system and what is outside it. If we add heat energy to the steam in a steam engine, to the body of a living creature, or to the earth's atmosphere, we will increase the internal energies of these bodies. And because of this added energy, these bodies may do work on external bodies. This added energy does one or both of two things: (1) it increases the internal energy of the system if it remains in the system, or (2) it does work on bodies

---

[3]Popular ideas, when proved wrong, are seldom suddenly discarded. People tend to identify with the ideas that characterize their time; hence, it is often the young who are more prone to discover and accept new ideas and push the human adventure forward.

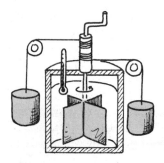

Fig. 16.3 Paddle-wheel apparatus first used to compare thermal energy with mechanical energy. As the weights fall, they give up potential energy and warm the water accordingly. This was first demonstrated by James Joule, for whom the unit of energy is named.

external to the system if it leaves the system. So, more specifically, the first law states:

Heat added = increase in internal energy + external work done by the system.

Note that the first law is an overall principle that is not concerned with the inner workings of the system itself. Whatever the details of molecular behavior in the system, the added heat energy serves only two functions: it increases the internal energy of the system, or it enables the system to do external work (or both). Our ability to describe and predict the behavior of systems that may be too complicated to analyze in terms of atomic and molecular processes is one of the beauties of thermodynamics. Thermodynamics bridges the microscopic and macroscopic worlds.

Put an airtight can of air on a hot stove and heat it up. Since the can is of fixed size, no work is done on it, and all the heat that goes into the can increases the internal energy of the enclosed air. Its temperature rises. If the can is fitted with a movable piston, then the heated air can do work as it expands and pushes the piston outward. Can you see that the temperature of the enclosed air will be less than if no work were done on the piston? If heat is added to a system that does no external work, then the amount of heat added is equal to the increase in the internal energy of the system. If the system does external work, then the increase in internal energy is correspondingly less. The first law of thermodynamics is simply the thermal version of the conservation of energy.

Consider a given amount of heat supplied to a steam engine. The amount supplied will be evident in the increase in the internal energy of the steam and in the mechanical work done. The sum of the increase in internal energy and the work done will equal the heat input. In no way can energy output in any form exceed energy input.

Questions

1. If 100 J of energy are added to a system that does no external work, by how much is the internal energy of that system raised?*

2. If 100 J of energy are added to a system that does 40 J of external work, by how much is the internal energy of that system raised?†

Adding heat to a system so that it can do mechanical work is only one application of the first law of thermodynamics. If instead of adding heat to a system, we do mechanical work on it, the first law tells us what we can expect: an increase in internal energy. A bicycle pump provides a good example. When we pump on the handle, the pump becomes hot. Why? Because we are putting mechanical work into the system and rais-

---

Answer   100 J

†**Answer**   60 J: we see from the first law that 100 J = 60 J + 40 J.

Fig. 16.4 Do work on the pump by pressing down on the piston and you compress the air inside. What happens to the temperature of the enclosed air? What happens to its temperature if the air expands and pushes the piston outward?

ing its internal energy. If the process happens quickly enough so that very little heat is conducted from the system during compression, then most of the work input will go into increasing the internal energy, and the system becomes hot.

Adiabatic Processes

A process wherein no heat enters or leaves a system is said to be **adiabatic**. Adiabatic processes can be achieved either by thermally insulating a system from its surroundings (with styrofoam, for example) or by performing the process rapidly. If we set the "heat added" part of the first law to zero, we see that changes in internal energy are equal to the work done on or by the system.[4] If work is done on a system—compressing it, for example—the internal energy increases. We raise its temperature. If work is done by the system—expanding against its surroundings, for example—the internal energy decreases. It cools.

A common example of an adiabatic process is the compression and expansion of the gases in the cylinders of an automobile engine (Figure 16.5). Compression and expansion occur in only hundredths of a second, which is too short a time for any appreciable amount of energy to leave the combustion chamber. For very high compressions, like those which occur in diesel engines, the temperatures achieved are sufficient to ignite a fuel mixture without the use of spark plugs. Diesel engines have no spark plugs.

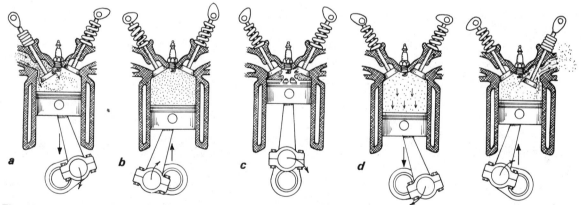

Fig. 16.5 A four-cycle internal-combusion engine. (a) A fuel-air mixture from the carburetor fills the cylinder as the piston moves down. (b) The piston moves upward and compresses the mixture—adiabatically, since no heat transfer occurs. (c) The spark plug fires, ignites the mixture, and raises it to a high temperature. (d) Adiabatic expansion pushes the piston downward, the power stroke. (e) The burned gases are pushed out the exhaust pipe. Then the intake valve opens and the cycle repeats.

[4]Δ Heat $= \Delta$ internal energy $+$ work
$0 = \Delta$ internal energy $+$ work

Then we can say

$-$ Work $= \Delta$ internal energy

Meteorology and the First Law

Meteorologists use the first law of thermodynamics extensively in analyzing weather. Since changes in internal energy are proportional to changes in temperature, and the work done on or by a gas is proportional to pressure changes in the gas, meteorologists express the first law in the following form:

Change in temperature ~ heat added (or subtracted)
+ pressure change

The temperature of the air may be changed by adding or subtracting heat, changing the pressure of the air, or both. Heat is added by solar radiation, by longwave earth radiation, or by contact with the warm ground. An increase in air temperature results. The atmosphere may lose heat by radiation to space, by evaporation of rain falling through dry air, or by contact with cold surfaces. The result is a drop in air temperature.

There are many atmospheric processes, usually involving time scales of a day or less, in which the amount of heat added or subtracted is very small—small enough so that the process is nearly adiabatic. Then we have the adiabatic form of the first law:

Change in temperature ~ pressure change

Adiabatic processes in the atmosphere are characteristic of large volumes of air, which we'll call *blobs*, that have dimensions on the order of kilometers. Because of their large sizes, mixing of different temperatures or pressures of air at the edges does not appreciably alter the overall composition of the blob, which behaves as if it were enclosed in a giant, tissue-light garment bag. As the blob of air flows up the side of a mountain, its pressure lessens. Reduced pressure results in reduced temperature.[5] Measurements show that the temperature of a blob of dry air will decrease by 10°C for a decrease in pressure that corresponds to an increase in altitude of 1 kilometer. So dry air cools 10°C for each kilometer it rises (Figure 16.6). Air flowing over tall mountains or rising in thunderstorms or cyclones may change elevation by several kilometers. Thus, if a blob of dry air at ground level with a comfortable temperature of 25°C were lifted to 6 kilometers, the temperature would be a frigid − 35°C. On the other hand, if air at a typical temperature of − 20°C at an altitude of 6 kilometers sank to the ground, its temperature would be a whopping 40°C. These changes in temperature are produced without any heat input or output whatsoever. The effect of expansion or compression on gases is quite impressive.[6]

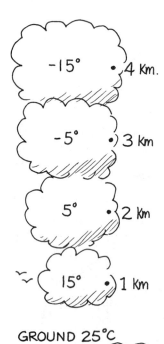

Fig. 16.6 The temperature of a blob of dry air that expands adiabatically changes by about 10°C for each kilometer of elevation.

[5] Recall that in Chapter 13 we treated the cooling of rising air at the microscopic level by considering the behavior of colliding molecules. With thermodynamics, we consider only the macroscopic measurements of temperature and pressure to get the same results. It's nice to analyze things from more than one point of view.

[6] Interestingly enough, when you're flying at high altitudes where outside air temperature is typically − 35°C, you are quite comfortable in your warm cabin—but *not* because of heaters. The process of compressing outside air to near–sea-level cabin pressure would normally heat the air to a roasting 55°C (131°F). So air conditioners must be used to extract heat from the pressurized air.

Fig. 16.7 A thunderhead is the result of the rapid adiabatic cooling of a rising mass of moist air.

Fig. 16.8 The layer of campfire smoke over the lake indicates a temperature inversion. The air above the smoke is warmer than the smoke, and the air below is cooler.

A rising blob cools as it expands. But the surrounding air is cooler with increasing elevation also. The blob will continue to rise so long as it is warmer (less dense) than the surrounding air. If it gets cooler (denser) than its surroundings, it will sink toward its original level. Under some conditions, large blobs of cold air sink and remain at a low level, with the result that the air above is warmer. When the upper regions of the atmosphere are warmer than the lower regions, we have a **temperature inversion**. Unless any rising warm air is less dense than this upper layer of warm air, it will rise no farther. It is common to see evidence of this over a cold lake where visible gases and particles, such as smoke, spread out in a flat layer above the lake rather than rise and dissipate higher in the atmosphere (Figure 16.8). Temperature inversions trap smog and other thermal pollutants. The smog of Los Angeles is trapped by such an inversion,

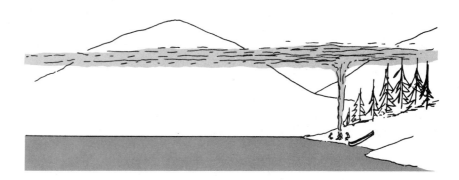

caused by low-level cold air from the ocean over which is a layer of hot air that moves over the mountains from the hotter Mojave Desert. The mountains aid in holding the trapped air (Figure 16.9). The mountains on the edge of Denver play a similar role in trapping smog beneath a temperature inversion.

Fig. 16.9 Smog in Los Angeles is trapped by the mountains and a temperature inversion caused by warm air from the Mojave Desert overlying cool air from the Pacific Ocean.

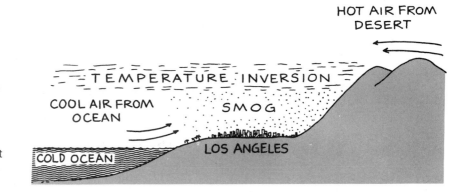

Adiabatic blobs are not restricted to the atmosphere, and changes in them do not necessarily happen quickly. Convection of some deep ocean currents takes thousands of years for circulation. The water masses are so huge and conductivities are so low that no appreciable quantities of heat are transferred to or from these blobs over these long periods of time. They are warmed or cooled adiabatically by changes in internal pressure. Changes in adiabatic ocean convection, as evidenced by El Niño in recent years, have a great effect on the earth's climate. Ocean convection is influenced by the temperature of the ocean floor, which in turn is influenced by convection currents in the molten material that lies beneath the earth's crust (Figure 16.10). Knowledge of the behavior of molten material in the earth's mantle is harder to come by. Once a blob of hot liquid material deep within the mantle begins to rise, will it continue to rise to the earth's crust? Or will its rate of adiabatic cooling find it cooler and denser than its surroundings, at which point it will sink? Is convection self-perpetuating? Geophysical types are currently pondering these questions.

Fig. 16.10 Do convection currents in the earth's mantle drive the continents as they drift across the global surface? Do rising blobs of molten material cool faster or slower than the surrounding material? Do sinking blobs heat to temperatures above or below those of the surroundings? The answers to these questions are not known at this writing.

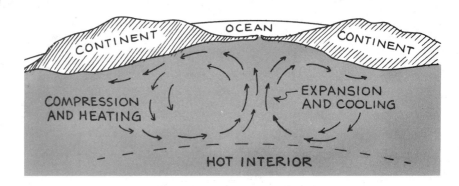

Questions

1. If a blob of air initially 0°C expands adiabatically while flowing upward alongside a mountain a vertical distance of 1 km, what will its temperature be? When it has risen 10 km?*

2. Imagine a giant dry-cleaner's garment bag full of air at a temperature of −10°C floating like a balloon with a string hanging from it 5 km above the ground. If you were able to yank it suddenly to the ground, what would its approximate temperature be?†

*****Answer**   At 1-km elevation, its temperature will be −10°C; at 10 km, −100°C.

†**Answer**   If it is pulled down so quickly that heat conduction is negligible, it would be adiabatically compressed by the atmosphere to a piping-hot 50°C (122°F), just as air is heated by compression in a bicycle pump.

Second Law of Thermodynamics

Suppose we place a hot brick next to a cold brick in a thermally insulated region. A certain quantity of thermal energy is associated with both bricks. We know that the hot brick will cool as it gives heat to the colder brick, which will warm. They will arrive at a common temperature: thermal equilibrium. No energy will be lost, in accordance with the first law of thermodynamics. But suppose the hot brick extracts heat from the cold brick and becomes hotter. Would this violate the first law of thermodynamics? Not if the cold brick becomes correspondingly colder so that the thermal energy of both bricks taken together remains the same. If this were to happen, it would not violate the first law. But it would violate the **second law of thermodynamics**. The second law distinguishes the direction of energy transformation in natural processes. The second law of thermodynamics can be stated in many ways, but most simply it is

Heat will never of itself flow from a cold body to a hot body.

Heat flows one way, downhill from hot to cold. Heat can be made to flow the other way, but only by imposing external effort. There is an enormous quantity of thermal energy in the ocean, but without external effort all this energy cannot be used to light a single flashlight lamp. Gas expands adiabatically and cools. Once cooled, it will never spontaneously contract to its original volume and warm up to its original temperature. Outside work would be necessary. We could squeeze it back, but it would not do so of its own accord. To do so would violate the second law.

The thermal energy converted to useful work in heat engines flows from a hot region to a cold region. Every heat engine—whether it be a steam engine, gasoline or diesel engine, gas turbine, or jet engine—will (1) absorb thermal energy from a reservoir of higher temperature, (2) convert some of this energy into mechanical work, and (3) discard the remaining energy to some lower-temperature reservoir, usually called a *sink* (Figure 16.11). In a gasoline engine, for example, the burning fuel in the combustion chamber is the high-temperature reservoir, mechanical work is done on the piston, and the discarded thermal energy goes out the exhaust. The second law tells us that no heat engine can convert all the thermal energy supplied into mechanical energy. Some heat is always exhausted in the process. Applied to heat engines, the second law may be stated:

When work is done by a heat engine running between two temperatures, T_{hot} and T_{cold}, only some of the input energy at T_{hot} can be converted to work, and the rest is rejected as heat at T_{cold}.

Always there is exhaust, which may be desirable or undesirable. Hot steam discharged in a laundry on a cold winter day may be quite desirable, while the same steam on a hot summer day is something else. When exhausted heat is undesirable, we call it *thermal pollution*.

Before the second law was understood, it was thought that for a very low friction heat engine, nearly all the input energy could be converted to useful work. But not so. It turns out that the fraction of thermal energy

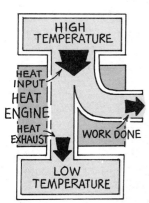

Fig. 16.11 When heat flows in any heat engine from a high-temperature place to a low-temperature place, part of this heat can be turned into work. (If work is put *in*to a heat engine, the flow of energy may go from a low-temperature to a high-temperature place, as in a refrigerator or air conditioner.)

that can be converted to useful work even under ideal conditions depends on the temperature difference between the hot reservoir and the cold reservoir or sink. The French physicist Sadi Carnot discovered this in 1824. His equation is

$$\text{Ideal efficiency} = \frac{T_{hot} - T_{cold}}{T_{hot}}$$

where T_{hot} is the temperature of the hot reservoir and T_{cold} the temperature of the cold.[7] Whenever ratios of temperatures are involved, the absolute temperature scale must be used. So T_{hot} and T_{cold} are expressed in kelvins. As an example, suppose the hot reservoir in a steam turbine has a temperature of 400K (127°C), and the cold reservoir 300K (27°C). The ideal efficiency is

$$\frac{400 - 300}{400} = \frac{1}{4}$$

This means that even under *ideal* conditions, only 25 percent of the internal energy of the steam can be converted into work, while the remaining 75 percent is waste. Carnot's equation is by no means restricted to steam engines; it holds for heat engines of all kinds. In practice, friction is ever-present in all engines, and efficiency is further reduced.

Questions

1. What would be the ideal efficiency of an engine if its hot reservoir and exhaust were the same temperature—say, 400K?*

2. What would be the ideal efficiency of a machine having a hot reservoir at 400K and a cold reservoir at absolute zero, 0K?†

The first law of thermodynamics states that energy can be neither created nor destroyed. The second law qualifies this by adding that the form energy takes in transformations "deteriorates" to less useful forms. Energy becomes more diffuse. The energy of gasoline burned in an automobile engine, for example, is not destroyed. It dissipates. Part of it goes into moving the car; part heats the engine and surroundings; part may turn a generator and produce light energy for the headlights or supply energy for the radio; part goes out the exhaust. All these parts add up to the

---

*__Answer__   Zero efficiency; $\frac{400 - 400}{400} = 0$. So no work output is possible for any heat engine unless a temperature difference exists between the reservoir and the sink.

†__Answer__ $\frac{400 - 0}{400} = 1$; only in this idealized case is an ideal efficiency of 100 percent possible.

[7]For a derivation of this equation, see almost any introductory-level thermodynamics textbook. A good one is Keith Stowe, *Introduction to Statistical Mechanics and Thermodynamics* (New York: Wiley, 1984).

Fuel energy in = cooling water losses + engine energy output + exhaust losses
100%   36%   26%   38%

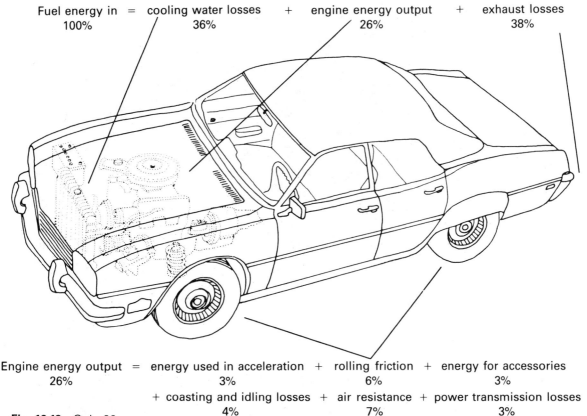

Engine energy output = energy used in acceleration + rolling friction + energy for accessories
26%   3%   6%   3%

   + coasting and idling losses + air resistance + power transmission losses
                          4%   7%   3%

Fig. 16.12 Only 26 percent of the heat energy produced in burning the gasoline in an automobile becomes useful mechanical energy, and most of that is lost in friction and in overcoming air resistance. The losses shown here are for a typical American automobile and are averaged over different driving situations. (Courtesy of the Exploratorium, San Francisco.)

energy given up by the gasoline on combustion (Figure 16.12). All the energy exists, and none is really lost. But it is dissipated and unavailable for driving another car. Like the Big Bang spreading through the universe, the energy from the gasoline will not exist in so concentrated a form again—at least not in this cycle of universal expansion.

Organized energy degenerates into disorganized energy. The organized energy in the form of electricity that goes to the electric lights in homes and office buildings degenerates into thermal energy. This is a principal source of heating in many office buildings where the climate is moderate, such as the Transamerica Pyramid in San Francisco. Every bit of the electrical energy in the lamps, even the part that briefly exists in the form of light, is turned into thermal energy, which is used to heat the buildings. This explains why the lights are on most of the time. The thermal energy so generated has no further use. All the thermal energy in the buildings cannot be reused to light a single lamp without some outside organizational effort. Heat is the graveyard of useful energy.

Fig. 16.13 The Transamerica Pyramid and some other buildings are heated by electric lighting, which is why the lights are on most of the time.

Fig. 16.14 Push a heavy crate across a rough floor and all your work will go into heating the floor and crate. Work against friction turns into thermal energy, which cannot in turn do work on the crate to even begin to push it back to its starting place.

Fig. 16.15 Molecules of gas go from the bottle to the air and not vice versa.

We see that the "quality" of energy is lowered with each transformation. Energy of an organized form tends to disorganized forms. In this broader regard, the second law can be stated:

Natural systems tend to proceed toward a state of greater disorder.

You finally straighten up your room. Everything is dust-free and in its proper place. A week later it is cluttered again. You stack a pile of pennies on your table, all heads up. Somebody walks by and accidentally bumps into the table and the pennies topple to the floor below, certainly not all heads up. You're on the job at a large company and a co-worker puts up a joke sign that reads, "If you thing thinks are confused now, just wait." These are examples of the tendency of the universe and all that is in it to become disordered. In the broadest sense, that is the message of the second law of thermodynamics. By *disordered*, we mean more random. For example, molecules of gas all moving in harmony make up an orderly state—and also an unlikely state. Molecules of gas moving in haphazard directions and speeds make up a disorderly state—a more probable state. If you remove the lid of a bottle containing some gas, the gas molecules escape into the room and make up a more disorderly state (Figure 16.15). Relative order becomes disorder. You would not expect the reverse to happen; that is, you would not expect the molecules to spontaneously order themselves back into the bottle and thereby return to the more ordered containment. Such processes in which disorder goes to order are simply not observed to happen.

Disordered energy can be changed to ordered energy only at the expense of some organizational effort or work input. For example, in a refrigerator, water freezes and becomes more ordered because work is put into the refrigeration cycle; or gas can be ordered into a small region if work input is supplied to a compressor. But without some imposed work input, processes in which the net effect is an increase in order are not observed in nature.

Entropy

The idea of lowering the "quality" of energy is embodied in the idea of **entropy**.[8] Entropy is the measure of the *amount of disorder*. The second law states that, in the long run, entropy always increases. Gas molecules escaping from a bottle move from a relatively orderly state to a disorderly state. Disorder increases; entropy increases. Whenever a physical system is allowed to distribute its energy freely, it always does so in a manner such that entropy increases while the available energy of the system for doing work decreases.

Consider the old riddle, "How do you unscramble an egg?" The answer is simple: "Feed it to a chicken." But even then you won't get all your original egg back—egg making has its inefficiencies too. All living organisms, from bacteria to trees to human beings, extract energy from their surroundings and use it to increase their own organization. In living organisms, entropy decreases. But the order in life forms is maintained by increasing entropy elsewhere; life forms plus their waste products have a net increase in entropy.[9] Energy must be transformed into the living system to support life. When it isn't, the organism soon dies and tends toward disorder.

Fig. 16.16 Why is the motto of this contractor—"Increasing entropy is our business"—so appropriate?

---

[8] Entropy can be expressed as a mathematical equation, which states that the increase in entropy ΔS in an ideal thermodynamic system is equal to the amount of heat added ΔQ to a system divided by the temperature T of the system: $\Delta S = \Delta Q/T$.

[9] Interestingly enough, the American writer Ralph Waldo Emerson, who lived during the time the second law of thermodynamics was the new science topic of the day, philosophically speculated that not everything becomes more disordered with time and cited the example of human thought. Ideas about the nature of things grow increasingly refined and better organized as they pass through the minds of succeeding generations. Human thought is evolving toward more order.

The first law of thermodynamics is a universal law of nature for which no exceptions have been observed. The second law, however, is a probabilistic statement. Given enough time, even the most improbable states may occur; entropy may sometimes decrease. For example, the haphazard motions of air molecules could momentarily become harmonious in a corner of the room, just as a barrelful of pennies dumped on the floor could all come up heads. These situations are possible—but not probable. The second law tells us the most probable course of events, not the only possible one.

The laws of thermodynamics are often stated this way: You can't win (because you can't get any more energy out of a system than you put into it), you can't break even (because you can't even get as much energy out as you put in), and you can't get out of the game (entropy in the universe is always increasing).

Summary of Terms

Thermodynamics   The study of heat and its transformation to other forms of energy.

Absolute zero   The lowest possible temperature that a substance may have; the temperature at which molecules of a substance have their minimum kinetic energy.

Internal energy   The total of all molecular energies, kinetic energy and potential energy, internal to a substance. *Changes* in internal energy are of principal concern in thermodynamics.

First law of thermodynamics   A restatement of the law of energy conservation, usually as it applies to systems involving changes in temperature: The heat added to a system equals an increase in internal energy plus external work done by the system.

Adiabatic process   A process, usually of expansion or compression, wherein no heat enters or leaves a system.

Temperature inversion   The condition wherein the upper regions of the atmosphere are warmer than the lower regions.

Second law of thermodynamics   Heat will never spontaneously flow from a cold body to a hot body. Also, no machine can be completely efficient in converting energy to work; some input energy is dissipated as heat. And finally, all systems tend to become more and more disordered as time goes by.

Entropy   A measure of the disorder of a system. Whenever energy freely transforms from one form to another, the direction of transformation is toward a state of greater disorder and therefore of greater entropy.

Suggested Reading

Stepp, R. D. *Making Theories to Explain the Weather.* 1983. This different and delightful little book is published by the author at Humboldt State University, Arcata, Calif. 95521.

Review Questions

1. What is the lowest possible temperature on the Celsius scale? On the Kelvin scale?

2. Distinguish between *thermal energy* and *internal energy*.

3. How does the law of the conservation of energy relate to the first law of thermodynamics?

4. What happens to the internal energy of a system when mechanical work is done on it? What happens to its temperature?

5. What happens to the internal energy of a system if no heat is added and work is done by the system on something else? What happens to its temperature?

6. What is an adiabatic process? What condition is necessary for adiabatic expansion of air in the atmosphere?

7. What is a temperature inversion?

8. How does the second law of thermodynamics relate to the direction of heat flow?

9. What exactly is thermal pollution?

10. If all friction could be removed from a heat engine, would it be 100 percent efficient? Explain.

11. What is the physicist's term for *measure of messiness?*

12. What relationship exists between the second law of thermodynamics and entropy?

Exercises

1. Consider a piece of metal that has a temperature of 10°C. If it is heated until it is twice as hot, what will be its temperature? Why is your answer *not* 20°C?

2. On a chilly 10°C day your friend who likes cold weather says she wishes it were twice as cold. Taken literally, what temperature would this be?

3. If you shake a can of liquid back and forth, will the temperature of the liquid increase? (Try it and see.)

4. When air is compressed, why does its temperature increase?

5. When you pump a tire with a bicycle pump, the cylinder of the pump becomes hot. Give two reasons why this is so.

6. Is it possible to wholly convert a given amount of mechanical energy into heat? Is it possible to wholly convert a given amount of heat into mechanical energy? Cite examples to illustrate your answers.

7. Why do diesel engines need no spark plugs?

8. Everybody knows that warm air rises. So it might seem that the air temperature should be higher at the top of mountains than down below. But the opposite is most often the case. Why?

9. Imagine a giant dry-cleaner's bag full of air at a temperature of − 35°C floating like a balloon with a string hanging from it 10 kilometers above the ground. Estimate its temperature if you were able to suddenly yank it to the earth's surface.

10. We say that energy "deteriorates" in energy transformations. Does this contradict the first law of thermodynamics? Defend your answer.

11. The combined molecular kinetic energies of molecules in a very large container of cold water are greater than the combined molecular kinetic energies in a cup of hot tea. Pretend you partially immerse the teacup in the cold water and that the tea absorbs 10 units of energy from the water and becomes hotter, while the water that gives up 10 units of energy becomes cooler. Would this energy transfer violate the first law of thermodynamics? The second law of thermodynamics? Explain.

12. Why is *thermal pollution* a relative term?

13. Is it possible to construct a heat engine that produces no thermal pollution? Defend your answer.

14. Why is it advantageous to use steam as hot as possible in a steam-driven turbine?

15. How does the ideal efficiency of an automobile relate to the temperature of the engine and the temperature of the environment in which it runs? Be specific.

16. What happens to the efficiency of a heat engine when the temperature is lowered in the reservoir into which heat energy is rejected?

17. To increase the efficiency of a heat engine, would it be better to increase the temperature of the reservoir while holding the temperature of the sink constant or decrease the temperature of the sink while holding the temperature of the reservoir constant? Explain.

18. What is the ideal efficiency of an automobile engine in which gasoline is heated to 2700K and the outdoor air is 300K?

19. Under what conditions would a heat engine be 100 percent efficient?

20. Suppose one wishes to cool a kitchen by leaving the refrigerator door open and closing the kitchen door and windows. What will happen to the room temperature? Why?

21. Why will a refrigerator with a fixed amount of food consume more energy in a warm room than in a cold room?

22. In buildings that are being electrically heated, is it at all wasteful to turn all the lights on? Is turning all the lights on wasteful if the building is being cooled by air conditioning?

23. Using both the first and the second law of thermodynamics, defend the statement that 100 percent of the electrical energy that goes into lighting a lamp is converted to thermal energy.

24. Is the heat energy of the universe becoming more unavailable with time? Explain.

25. Water evaporates from a salt solution and leaves behind salt crystals that have a higher degree of molecular order than the more randomly moving molecules in the salt water. Has the entropy principle been violated? Why or why not?

26. Water put into a freezer compartment in your refrigerator goes to a state of less molecular disorder when it freezes. Is this an exception to the entropy principle? Explain.

PART 4 Sound

17 Vibrations and Waves

Many things in nature wiggle and jiggle. We call a wiggle in time a *vibration* and a wiggle in space and time a *wave*. A wave cannot exist in one place but must extend from one place to another. Similarly, a vibration cannot exist in one instant but needs time to move to and fro. Light and sound are both vibrations that propagate through space as waves, but of two very different kinds. Sound is the propagation of vibrations through a material medium—a solid, liquid, or gas. If there is no medium to vibrate, then no sound is possible. On the other hand, light is a vibration of nonmaterial electric and magnetic fields, a vibration of pure energy. The propagation through space of these vibrating fields makes up light waves. Although light waves can pass through many material media, they need none; light can propagate through a vacuum. Sound and light are fundamentally different. But the source of all waves—whether sound, light, or whatever—is something that is vibrating. For sound it can be the vibrating prongs of a tuning fork, and for light the vibrations of electrons in an atom. We shall begin our study of vibrations and waves by considering the motion of a pendulum.

Pendulums

If we suspend a small, massive object at the end of a piece of string, we have a simple pendulum. Pendulums swing to and fro with such regularity that they have long been used to control the motion of clocks. Galileo discovered that the time a pendulum takes to swing to and fro through small distances depends only on the *length of the pendulum and the acceleration of gravity*.[1] The time of a to-and-fro swing, called the *period*, does not depend on the mass or on the size of the arc through which it swings.[2]

In addition to controlling timekeeping devices, pendulums can be used to measure the acceleration due to gravity. Oil and mineral prospectors use very sensitive pendulums to detect slight differences in this acceleration, which is affected by the densities of underlying formations.

A long pendulum swings to and fro more slowly than a short pendulum. When walking, we allow our legs to swing with the help of gravity, like a pendulum. Just as a long pendulum takes a long time to swing to and fro,

---

[1]The exact relationship for the period of a simple pendulum is $T = 2\pi\sqrt{l/g}$, where T is the period, l is the length of the pendulum, and g is the acceleration of gravity.

[2]If you release two rocks of different masses, they fall together with the same g; and if you let them slide without friction down the same incline, they slide together with the same fraction of g. So it is reasonable to expect that two pendulum bobs of different masses at the ends of strings of the same length will similarly swing to and fro in unison. Try it and see!

a person with long legs tends to walk with slower strides than a person with short legs. This is most noticeable in long-legged animals—like giraffes, horses, and ostriches—that run with a slower gait than do shorter-legged animals—like cats, dogs, and mice.

The motion of a pendulum bob swinging to and fro along a single plane in a small arc is called **simple harmonic motion**. If a pendulum bob leaks sand while undergoing simple harmonic motion, it will trace out and retrace a short straight line (Figure 17.1a). Suppose, however, that we swing such a pendulum above a conveyor belt that moves in a perpendicular direction to the plane of the swinging pendulum (Figure 17.1b). The trace it now makes is called a **sine curve**. This curve is not itself a wave but is a pictorial representation of a wave and lends itself to a description of some of the technical terms used in describing wave motion.

a   b

Fig. 17.1 Frank Oppenheimer at the Exploratorium demonstrates (a) a straight line traced by a swinging pendulum bob that leaks sand on the stationary conveyor belt. (b) When the conveyor belt is uniformly moving, a sine curve is traced.

Some Technical Terms

The terms *crest* and *trough* refer to the highest and lowest parts of the curve, or the wave it represents (Figure 17.2). The term **amplitude** refers to the distance from the midpoint (straight line) to the crest (or to the trough) of the wave. Thus, amplitude is equal to the maximum displacement of the pendulum from its position of rest. The distance from the top of one crest to the top of the next one is equal to the **wavelength**. Or the wavelength is the distance between successive identical parts of a wave.

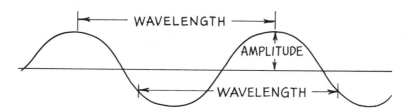

Fig. 17.2 A sine curve.

Another important term we will use to describe vibrations and waves is **frequency**. In the case of a vibrating pendulum, we can speak of its frequency in terms of how frequently it swings to and fro or, more specifically, by the number of vibrations per second that it undergoes. We call a complete to-and-fro swing one vibration (or cycle). If the pendulum swings to and fro at the rate of two complete vibrations each second, we say the frequency is two vibrations per second, or two cycles per second. The unit of frequency is called the **hertz** (Hz), after Heinrich Hertz, who demonstrated radio waves in 1886. So two cycles per second is 2 hertz. Higher frequencies are measured in kilohertz (kHz), and still higher, in megahertz (MHz). In the case of radio waves, the Federal Communications Commission licenses a radio station to broadcast at a special frequency all its own. For example, 94 kHz on your AM radio dial corresponds to radio waves that have a frequency of vibration of 94,000 hertz. And 101 MHz on your FM dial corresponds to vibrations of 101,000,000 hertz. These radio-wave frequencies are the frequencies that electrons are forced to oscillate in the antenna of a radio station's transmitting tower. The frequency of a wave—that is, the number of times its crests (or troughs) pass a set point during 1 second—is the same as the frequency of the vibrating source.

Wave Motion

Fig. 17.3 Water waves.

Most information about our surroundings comes to us in some form of waves. It is through wave motion that sounds come to our ears, light to our eyes, and electromagnetic signals to our radios and television sets. By *wave motion*, we mean the transfer of energy from a source to a distant receiver without the transfer of matter between the two points.

Wave motion can be most easily understood by first considering its simplest case in a horizontally stretched string. If one end of such a string is shaken up and down, a rhythmic disturbance travels along the string. Each particle of the string moves up and down, while at the same time the disturbance moves along the length of the string. The medium returns to its initial condition after the disturbance has passed.

Perhaps a more familiar example of wave motion is a water wave. If a stone is dropped into a quiet pond, waves will travel outward in expanding circles, the centers of which are at the source of the disturbance. In this case we might think that water is being transported with the waves, since water is splashed onto previously dry ground when the waves meet the shore. We should realize, however, that barring obstacles the water will run back into the pond, and things will be much as they were in the beginning: the surface of the water will have been disturbed, but the water itself will have gone nowhere. Again, the medium returns to its initial condition after the disturbance has passed.

Let us consider another example of a wave to illustrate that what is transported from one place to another is a disturbance in a medium, not the medium itself. If you view a wheat field from an elevated position on a gusty day, you will see waves travel across the wheat. The individual stems do not leave their places; instead, they swing to and fro. Furthermore, if you stand in a narrow footpath, the wheat that blows over the edge of the path, brushing against your legs, is very much like the water that doused the shore in our earlier example. While wave motion continues, the wheat swings back and forth, vibrating between definite limits but going nowhere. When the wave motion stops, the wheat returns to its initial position. It is characteristic of wave motion that the medium carrying the wave returns to its initial condition after the disturbance has passed.

Wave Velocity

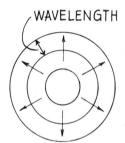

WAVELENGTH

Fig. 17.4 A top view of water waves.

The velocity of wave motion is related to the frequency and wavelength of the waves. We can understand this by considering the simple case of water waves (Figure 17.4). Imagine that we fix our eyes at a stationary point on the surface of a body of water and observe the waves passing by this point. If we count the number of crests of water passing this point each second (the frequency) and also observe the distance between crests (the wavelength), we can then calculate the horizontal distance that a particular crest travels each second. We will then know how frequently a distance equal to the wavelength is traveled—that is, the **wave velocity**.

For example, if two crests pass a stationary point each second, and if the wavelength is 10 meters, then 2 × 10 meters of waves pass by in 1 second. The waves therefore travel at 20 meters per second. We can say

$$\text{Wave velocity} = \text{frequency} \times \text{wavelength}$$

This relationship holds true for all kinds of waves, whether they are water waves, sound waves, or light waves.

Questions

1. If a train of freight cars, each 10 m long, rolls by you at the rate of 3 cars each second, what is the speed of the train?*

2. If a water wave oscillates up and down 3 times each second and the distance between wave crests is 2 m, what is its frequency? Its wavelength? Its wave velocity?†

---

*Answer   30 m/s; we can see this two ways, the Chapter 2 way and the Chapter 17 way: from $v = \frac{d}{t} = \frac{3 \times 10\,\text{m}}{1\,\text{s}} = 30\,\frac{\text{m}}{\text{s}}$, where d is the distance of train that passes you in t seconds. If we compare our train to wave motion, where wavelength corresponds to 10 m, and frequency is 3 Hz, then Velocity = frequency × wavelength = 3 × 10 = 30 m/s.

†Answer   The frequency of the wave is 3 Hz; its wavelength is 2 m; and its Wave velocity = frequency × wavelength = 3 × 2 = 6 m/s. It is customary to express this as the equation $v = f\lambda$, where v is wave velocity, f is wave frequency, and λ (Greek letter lambda) is wavelength.

Transverse Waves

Suppose we fasten one end of a long rope to a wall and hold the other end in our hand. If we suddenly jerk the end we are holding, a pulse will travel along the rope and back (Figure 17.5). In this case the motion of the rope (up and down arrows) is at right angles to the direction of wave velocity. The right-angled, or sideways, motion is called *transverse*. Since the motion of the medium is transverse to the direction the wave travels, this type of wave is called a **transverse wave**.

WAVE VELOCITY ⟶

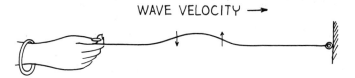

Fig. 17.5   A transverse wave pulse.

Waves in the stretched strings of musical instruments and upon the surfaces of liquids are transverse; so also are electromagnetic waves (radio waves, visible light, X rays, and so forth).

Longitudinal Waves

A WAVE OF STRETCH TRAVELS UP THE SPRING

Fig. 17.6   A longitudinal wave.

Not all waves are transverse. In some cases the particles of the medium move back and forth along the same direction the wave travels. When this happens, we call it a **longitudinal wave**. A simple demonstration of longitudinal waves can be made with a vertically suspended, long spiral spring (Figure 17.6). If a part of the spring is compressed, a wave of compression will travel up and down the spring. We call this region of compression a *condensation*. If a part of the spring is stretched and then released, a wave of stretch will travel up and down the spring. We call this region of stretch a *rarefaction*. A little thought will make it clear that when the compressed part of the spring is released, it will compress the part farther up, and so on. Each part of the spring is successively pushed forward when the compression reaches it. Similarly, when a stretched part of the spring is released, it will stretch the part farther up, and so on. Each part of the spring is successively pulled back when the rarefaction reaches it. And when a part of the spring is repeatedly compressed and stretched in a vibratory motion, a train of waves consisting of a series of condensations and rarefactions following one behind the other travels along the spring. Each part of the spring vibrates to and fro.

The wavelength of a longitudinal wave is the distance between successive regions of condensation or, equivalently, the distance between successive rarefactions.

In the case of longitudinal waves, the medium vibrates to and fro along the direction in which the wave is traveling. The most common example of longitudinal waves is sound in air. Each air molecule vibrates back and forth about some equilibrium position as the waves move by.

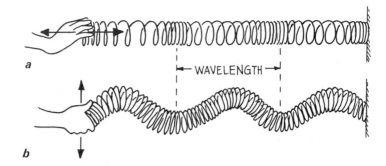

Fig. 17.7 (a) Back-and-forth motion produces a longitudinal wave in the coiled spring. (b) And up-and-down motion produces a transverse wave.

Interference

Whereas a material body like a rock will not share its space with another rock, more than one vibration or wave can exist at the same time in the same space. If you drop two rocks in water, the waves each produces can overlap and form an **interference pattern**. Within the pattern, wave

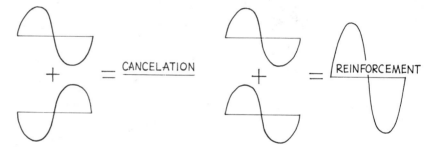

Fig. 17.8 Destructive and constructive interference in a transverse wave.

effects may be increased, decreased, or neutralized. When the crest of one wave coincides with the crest of another, their individual effects add together, and a crest of increased amplitude results. This is called *constructive interference* (Figure 17.8). When the crest of one wave coincides with the trough of another, their individual effects are reduced. This is called *destructive interference*. Figure 17.9 shows the interference pattern

Fig. 17.9 Two sets of overlapping water waves interfere.

of identical water waves. You can see the regions where a crest of one wave overlaps the trough of another and produces a region of zero amplitude. At points along these regions, the waves arrive "out of step." We say they are *out of phase* with one another.

Interference is characteristic of all wave motion, whether the waves are water waves, light waves, or sound waves. It is characteristic of both transverse and longitudinal waves.

Standing Waves

If we shake a rope up and down, a train of waves will be sent out and be reflected from the fixed end. By shaking the rope just right, we can cause the incident and reflected waves to form a **standing wave**, where parts of the rope, called the **nodes**, are stationary.

Standing waves are a result of interference. When two sets of waves of equal amplitude and wavelength pass through each other in opposite directions, the waves are steadily in and out of phase and produce stable regions of constructive and destructive interference (Figure 17.10).

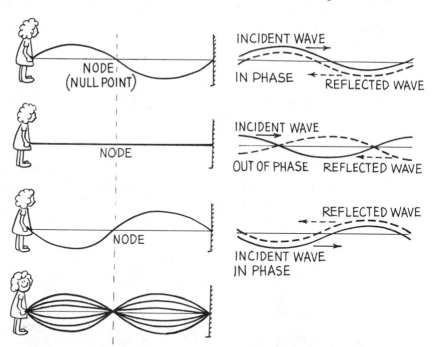

Fig. 17.10 The incident and reflected waves interfere to produce a standing wave.

It is easy to make standing waves yourself. Tie a rope or, better, a rubber tube between two firm supports. Shake the tube from side to side with your hand near one of the supports. If you shake the tube with the right frequency, you will set up a standing wave as shown in Figure 17.11. Shake the tube with twice the frequency, and a standing wave of half the previous

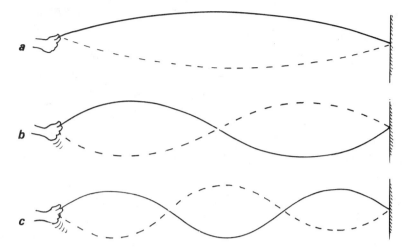

Fig. 17.11 (a) Shake the rope until you set up a standing wave of one segment ($\frac{1}{2}$ wavelength). (b) Shake it with twice the frequency and you'll produce a standing wave with two segments (1 wavelength). (c) Shake with three times the frequency and you'll produce three segments ($1\frac{1}{2}$ wavelengths).

wavelength, two loops, will result. (The distance between successive nodes is a half wavelength; two loops comprise a full wavelength.) Triple the frequency, and a standing wave with one-third the original wavelength, three loops, results, and so forth. Unless you find the right frequency in shaking the tube, the waves reflecting up and down the tube will not interfere to form a standing wave. But when you find a proper frequency, a very small motion of your hand will produce a standing wave of large amplitude.

Standing waves are set up in a guitar string when it is plucked, in a violin string when it is bowed, and in a piano string when it is struck. They are set up in the air in an organ pipe and in the air of a soda-pop bottle when air is blown over the top. Standing waves can be set up in a tub of water or a cup of coffee by sloshing it back and forth with the right frequency.

In transverse waves, the medium (rope, rubber tube, water, or whatever) vibrates to and fro in a direction at right angles (transverse) to the direction in which the wave travels. In the case of standing longitudinal waves—a column of air vibrating in a musical wind instrument, for example—the medium vibrates to and fro along the direction in which the wave is traveling.

Doppler Effect

A pattern of water waves produced by a bug jiggling its legs and bobbing up and down in still water is shown in Figure 17.12. The bug is not going anywhere but is merely treading water in a fixed position. Note that the waves are concentric circles, the larger circles being those produced first. The circles represent wave crests and travel at a constant speed. They are actually expanding, so we can consider that the figure is a snapshot of the moving waves. If the bug bobs in the water at a constant frequency, the distance between wave crests (the wavelength) is the same for all suc-

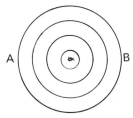

Fig. 17.12 Water waves made by a stationary bug jiggling in still water.

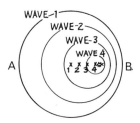

Fig. 17.13 Water waves made by a bug swimming in still water.

cessive waves. Waves would encounter point A as frequently as they would point B; that is, the frequency of wave motion is the same at points A and B or anywhere in the vicinity of the bobbing bug. And the wave frequency is the same as the bobbing frequency of the bug.

Suppose that the jiggling bug moves across the water at a speed less than the wave velocity. In effect, it chases the waves it has produced. The wave pattern is distorted and no longer concentric (Figure 17.13). Wave 1 was produced first when the bug was in position 1; wave 2 was produced when the bug was in position 2; and so forth. An interesting thing to note here is that even though the bug maintains the same bobbing frequency as before in Figure 17.12, an observer at point B would encounter a *higher* frequency. This is because each successive wave has a shorter distance to travel and therefore arrives at point B more frequently than it would if the bug were not moving toward point B. An observer at A encounters a *lower* frequency because the time between wave-crest arrivals increases as each successive wave must travel farther. The waves arrive less frequently, which is to say, at a lower frequency. This change in frequency due to the motion of the source (or receiver) is called the **Doppler effect**.

Water waves spread over the flat surface of the water. Sound and light waves, on the other hand, travel in all directions like an expanding balloon. Just as circular waves are closer together in front of the swimming bug, spherical sound or light waves ahead of a moving source are closer together and encounter a receiver more frequently.

In the case of sound, a listener perceives a higher frequency as a higher pitch—as a higher note on a musical scale. When the source (or listener) moves away, a lower frequency is perceived as a lower pitch, or lower tone. The Doppler effect is evident to a listener standing by the side of a highway. A higher-pitched sound is heard from cars approaching the listener, and a lower-pitched sound is heard when the cars recede.

And the same for light. When a light source approaches, there is an increase in its measured frequency, and when it recedes, there is a decrease in its frequency. An increase in light frequency is called a *blue shift,* because the increase is toward the high-frequency, or blue, end of the color spectrum. A decrease in frequency is called a *red shift,* referring to the lower-frequency, or red, end of the spectrum. The galaxies, for example, show a red shift in the light they emit. A measurement of this shift enables a calculation of their speed of recession. A rapidly spinning star shows a red shift on the side turning away from us and a relative blue shift on the side turning toward us, which enables a calculation of the star's spin rate.

The most dramatic example of the Doppler effect is the presence of the low-frequency microwave radiation that seems to permeate the universe. It is believed that this is the high-frequency fireball radiation emitted during the earliest years of the expanding universe, which has been absorbed and reradiated countless times by receding matter—and now appears as a faint relic of the past.

Question   Would you measure an increase in the speed of a wave if its source were moving toward you and a decrease in wave speed if it were moving away?*

Wave Barriers

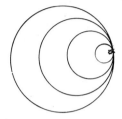

Fig. 17.14 Wave pattern made by a bug swimming as fast as the wave velocity.

Suppose the bug in our earlier example swims as fast as the wave velocity. Then it "keeps up" with the waves it produces. Instead of the waves moving ahead of it, they pile up or superimpose on one another directly in front of it (Figure 17.14). The bug encounters a "wave barrier." Considerable effort is required for the bug to swim over this barrier before it can swim faster than the wave velocity. An airplane traveling at the speed of sound similarly encounters a barrier of superimposed waves—wave upon wave of sound creating a barrier of compressed air along the leading edges of the aircraft. Considerable thrust is required for the aircraft to push through this barrier. But once through, the aircraft is *supersonic*, and no barrier exists to interrupt further acceleration. Once the bug gets over its wave barrier, the water in front of it is relatively smooth and undisturbed.

Bow Waves

When the bug swims faster than the wave velocity, ideally it produces a wave pattern as shown in Figure 17.15. It outraces the waves it produces, which seem to be dragging behind it. But most important, the waves overlap where the X's are shown. When one wave crest overlaps another wave crest, a greater wave crest is produced. Only a few waves are shown

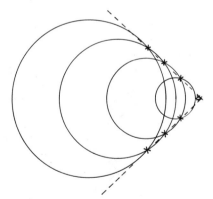

Fig. 17.15 Wave pattern made by a bug swimming faster than the wave velocity.

*Answer   No! The speed of a wave is not affected by motion of the source, but depends on the nature of the medium through which the wave travels. You would measure an increase or a decrease in wave *frequency* but not in speed. However, if you the receiver move toward the source (instead of the source moving toward you), then you would observe an increase in speed as well as frequency. (We will see later that this increase in speed does *not* apply to electromagnetic waves.) Be clear about the distinction between frequency and speed. How frequently a wave vibrates is altogether different from how fast it moves from one place to another.

in the figure, and therefore only a few positions of overlapping waves are indicated. If more waves were indicated, there would be a greater number of X's. In practice, the bug produces many waves, so many that the X's form a solid V shape called a **bow wave**. The familiar bow wave generated by a speedboat knifing through the water is actually the superposition of many circular waves!

Figure 17.16 summarizes some wave patterns made by sources of various velocities. Note that after the velocity of the wave source exceeds the wave velocity, increased velocity produces a narrower V shape.[3]

Fig. 17.16 Summary of idealized wave patterns.

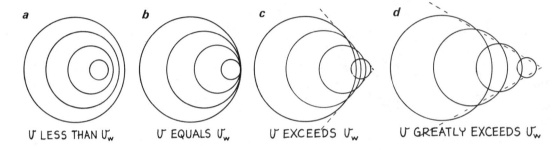

a   b   c   d

v LESS THAN v_w v EQUALS v_w v EXCEEDS v_w v GREATLY EXCEEDS v_w

Shock Waves

Just as overlapping circular waves tend to form a V, overlapping spherical waves produced by a supersonic aircraft form a cone. And just as the bow wave of a speedboat spreads until it reaches the shore of a lake, the conical wake generated by a supersonic aircraft spreads until it reaches the ground (Figure 17.17). This cone of sound is called a **shock wave**. It is a thin,

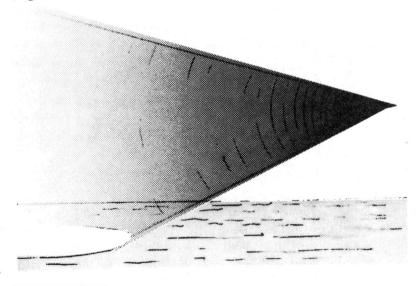

Fig. 17.17 A shock wave.

---

[3]Bow waves generated by boats in water are more complex than is indicated here. Our idealized treatment serves as an analogy for the production of the less complex shock waves in air.

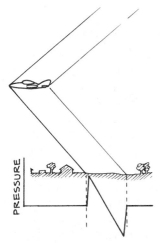

PRESSURE

Fig. 17.18 The shock wave is actually a high-pressure cone the apex of which is at the bow and a low-pressure cone with the apex at the tail. A graph of the air pressure at ground level between the cones takes the shape of the letter N.

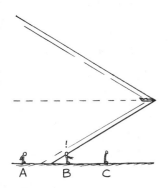

A   B   C

Fig. 17.19 The shock wave has not yet encountered listener A, is now encountering listener B, and has already encountered listener C.

conical shell of compressed air that sweeps behind the aircraft just as a bow wave of water is swept behind a swimming bug or speeding boat. Suppose you are floating in tranquil water and a speedboat happens by. After the boat passes, its bow wave soon reaches you. You might describe your dousing as being hit by a "water boom." In the same way, when the shock wave of a supersonic aircraft is incident on a listener, a sharp crack is heard, which we call the **sonic boom**.

The sound produced by a slower-than-sound, or subsonic, aircraft reaches our ears one wave at a time and is perceived as a continuous tone. Encountering a shock wave, however, is receiving many sound waves in a single burst. The sudden increase in pressure is much the same in effect as the sudden expansion of air produced by an explosion. Both processes direct a burst of high-pressure air to the listener. Pressure from an explosion results from rapidly expanding air, while for the sonic boom the pressure results from the superposition of many waves. The ear is hard put to tell the difference between the two!

A water skier is familiar with the fact that next to the high hump of the V-shaped bow wave is a V-shaped depression. The same is true of a shock wave, which actually consists of two cones: a high-pressure cone generated at the bow of the supersonic aircraft and a low-pressure cone that follows at the tail of the craft. So between these two cones the air pressure rises sharply to above atmospheric pressure, then falls *below* atmospheric pressure before sharply returning to normal beyond the inner tail cone (Figure 17.18). This overpressure suddenly followed by underpressure intensifies the sonic boom.

A common misconception is that sonic booms are produced *when* an aircraft breaks through the sound barrier—that is, just as the aircraft surpasses the speed of sound. This would be equivalent to saying that a boat produces a bow wave *when* it overtakes its own waves. This is not so. The fact is that a shock wave and its resulting sonic boom are swept continuously behind an aircraft traveling *faster* than sound, just as a bow wave is swept continuously behind a speedboat. In Figure 17.19, listener B is in the process of hearing a sonic boom, while listener A has already heard it and listener C will hear it shortly. The aircraft generating this shock wave may have broken through the sound barrier hours ago!

It is not necessary that the moving object emit sound to produce shock waves. Once an object is moving faster than the speed of sound, it will *make* sound. A bullet passing overhead produces a cracking sound, which is a small sonic boom. If the bullet were larger and disturbed more air in its path, the crack would be more boomlike. When the lion tamer cracks his circus whip, the cracking sound is actually a sonic boom produced by the tip of the whip, which travels faster than the speed of sound! Both the bullet and the whip are not in themselves sound sources; but when traveling at supersonic speeds, they produce their own sound as waves of air are generated to the sides of the moving objects.

Summary of Terms

Simple harmonic motion   A vibratory or periodic motion, like that of a pendulum, in which the force acting on the vibrating body is proportional to its displacement from its central equilibrium position and acts toward that position.

Sine curve   A wave form traced by simple harmonic motion that is uniformly moving in a perpendicular direction, like the wavelike path traced on a moving conveyor belt by a pendulum swinging at right angles above the moving belt.

Amplitude   For a wave or vibration, the maximum displacement on either side of the equilibrium position.

Wavelength   The distance between successive crests, troughs, or identical parts of a wave.

Frequency   For a body undergoing simple harmonic motion, the number of vibrations it makes per unit time. For a wave, the frequency of a series of waves is the number of waves that pass a particular point per unit time.

Period   The time required for a vibration or a wave to make a complete cycle; equal to 1/frequency.

Wave velocity   The speed with which waves pass by a particular point:

$$\text{Wave velocity} = \text{frequency} \times \text{wavelength}$$

Transverse wave   A wave in which the individual particles of a medium vibrate from side to side perpendicular (transverse) to the direction in which the wave travels.

Longitudinal wave   A wave in which the individual particles of a medium vibrate back and forth in the direction in which the wave travels. Sound consists of longitudinal waves.

Interference   The superposition of different sets of waves that produces mutual reinforcement in some places and cancelation in others.

Standing wave   A stationary wave pattern formed in a medium when two sets of identical waves pass through the medium in opposite directions.

Doppler effect   The change in frequency of wave motion resulting from motion of the sender or receiver.

Bow wave   The V-shaped wave made by an object moving across a liquid surface at a speed greater than the wave velocity.

Shock wave   The cone-shaped wave made by an object moving at supersonic speed through a fluid.

Sonic boom   The loud sound resulting from the incidence of a shock wave.

Review Questions

1. What is the difference between a *vibration* and a *wave?*

2. What is required as a source for all wave motion?

3. Why do short-legged animals run with quicker strides than long-legged animals?

4. What is meant by the *amplitude* of a wave?

5. How does the frequency of a wave compare to the frequency of its vibrating source?

6. If a wave oscillates up and down 3 times each second and travels an average distance of 30 meters each second, what is its frequency? Its wave velocity?

7. What is the relationship of wave frequency to the period of a wave?

8. If a wave vibrates 2 times each second, and its wavelength is 4 meters, what is its frequency? Its period? Its speed?

9. Clearly distinguish between *wave frequency* and *wave velocity.*

10. If the frequency of a wave is 1 vibration each second and its wavelength is 4 meters, what is its frequency? Its period? Its speed?

11. What is a standing wave? How does interference play a role in the production of a standing wave?

12. How do the vibrations of a transverse wave differ from the vibrations of a longitudinal wave?

13. Does a longitudinal wave have a frequency, a wavelength, and an amplitude?

14. Is it physically possible for one wave to be canceled by another?

15. What is meant by the expression *out of phase?*

16. Would it be correct to say that the Doppler effect is the apparent change in the speed of a wave by virtue of motion of the source or receiver? Why or why not?

17. What is the difference between a *subsonic* aircraft and a *supersonic* aircraft?

Home Project

Tie a rubber tube, a spring, or a rope to a fixed support and produce standing waves. See how many nodes you can produce.

Exercises

1. Will a pendulum clock that is accurate at sea level gain time or lose time when located high in the mountains?

2. If a pendulum is shortened, what happens to its frequency? Its period?

3. You swing an empty suitcase to and fro at its natural frequency. If the case were filled with books, would the natural frequency be less, greater, or the same as before?

4. Is the time required to swing to and fro (the period) on a playground swing more or less when you stand rather than sit? Explain.

5. Why do short people tend to walk with quicker strides than tall people?

6. What kind of motion should you impart to the nozzle of a garden hose so that the resulting stream of water approximates a sine curve?

7. What kind of motion should you impart to a stretched, coiled spring (or Slinky) to provide a transverse wave? A longitudinal wave?

8. If a gas tap is turned on for a few seconds, someone a couple of meters away will hear the gas escaping long before she smells it. What does this indicate about the way in which sound waves travel?

9. If we double the frequency of a vibrating object, what happens to its period?

10. What is the frequency of the second hand of a clock? The minute hand? The hour hand?

11. You successively dip your finger into a puddle of water and make waves. What happens to the wavelength if you dip your finger more frequently?

12. How does the frequency of vibration of a small object floating in water compare to the number of waves passing it each second?

13. How far does a wave travel during one period?

14. Does the light intensity alternate 60 times each second or 120 times each second for a fluorescent lamp that operates on a 60-hertz source?

15. A rock is dropped in water, and waves spread over the flat surface of the water. What becomes of the energy in these waves when they die out?

16. The wave patterns seen in Figure 17.4 are composed of circles. What does this tell you about the wave speeds in different directions?

17. Why is lightning seen before thunder is heard?

18. Would there be a Doppler effect if the source of sound were stationary and the listener in motion? Why or why not? In which direction should the listener move to hear a higher frequency? A lower frequency?

19. When you blow your horn while driving toward a stationary listener, an increase in frequency of the horn is heard by the listener. Would the listener hear an increase in horn frequency if he were also in a car traveling at the same speed in the same direction? Explain.

20. Is there a Doppler effect when motion of the source is at right angles to a listener? Explain.

21. How does the Doppler effect aid police in detecting speeding motorists?

22. How does the phenomenon of interference play a role in the production of bow or shock waves?

23. Would the conical angle of a shock wave open wider, narrow down, or remain constant for faster-moving supersonic aircraft?

24. If the sound of an airplane does not come from the part of the sky where the plane is seen, does this imply that the airplane is traveling faster than the speed of sound? Explain.

25. Does a sonic boom occur at the moment when an aircraft exceeds the speed of sound? Explain.

26. Why is it that a subsonic aircraft, no matter how loud it may be, cannot produce a sonic boom?

27. Imagine a super-fast fish that is able to swim faster than the speed of sound in water. Would such a fish produce a "sonic boom"?

28. A skipper on a boat notices wave crests passing his anchor chain every 5 seconds. He estimates the distance between wave crests to be 15 meters. He also correctly estimates the speed of the waves. What is this speed?

29. A weight suspended from a spring is seen to bob up and down over a distance of 20 centimeters twice each second. What is its frequency? Its period? Its amplitude?

30. Radio waves travel at the speed of light—300,000 kilometers per second. What is the wavelength of radio waves received at 100 megahertz on your radio dial?

18 Sound

If a tree fell in the middle of a deep forest hundreds of kilometers away from any living being, would there be a sound? Different people will answer this question in different ways. "No," some will say, "sound is subjective and requires a listener. If there is no listener, there will be no sound." "Yes," others will say, "a sound is not something in a listener's head. A sound is an objective thing." Discussions of questions like this one often are beyond agreement because the participants fail to realize that they are arguing not about the nature of sound but about the definition of the word. Either side is right, depending on which definition is taken, but investigation can proceed only when a definition has been agreed on. The physicist usually takes the objective position and defines sound as a form of energy that exists whether or not it is heard and goes on from there to investigate its nature.

Origin of Sound

All sounds are waves produced by the vibrations of material objects. In pianos and violins, the sound is produced by the vibrating strings; in a clarinet, by a vibrating reed. The human voice results from the vibration of the vocal cords. In each of these cases a vibrating source sends a disturbance through the surrounding medium, usually air, in the form of longitudinal waves. The *loudness* of sound depends on the amplitude of these waves, that is, on how much air is set into motion. The **pitch** of sound is directly related to the frequency of the sound waves, which is identical to the frequency of the vibrating source. The pitches produced by lower frequencies are heard as low bass tones, and higher pitches are produced by higher frequencies. (We will treat pitch and loudness further in the next chapter.) The human ear can normally hear sounds made by vibrating bodies of frequencies in the range between 16 and 20,000 hertz. Sound waves with frequencies below 16 hertz are called **infrasonic**, and those with frequencies above 20,000 hertz are called **ultrasonic**. We cannot hear infrasonic and ultrasonic sound waves.

Nature of Sound in Air

When we clap our hands, we produce a wave pulse that travels out in all directions. Being longitudinal, this wave pulse vibrates the air in the same way that a similar pulse would vibrate a coiled spring. Each particle moves back and forth along the direction of the expanding wave.

For a clearer picture of this process, consider a long corridor as in Figure 18.1a. At one end is an open window with a curtain over it. At the other end is a door. When we open the door, we can imagine the door pushing the adjacent molecules of air away from their initial positions. These molecules strike their neighbors, the neighbors are driven into *their* neighbors, and so on, until the curtain flaps out the window. A pulse of compressed air has moved from the door to the curtain. This pulse of

Fig. 18.1 (*a*) When the door is opened, a condensation travels across the room. (*b*) When the door is closed, a rarefaction travels across the room. (Adapted from A. V. Baez, *The New College Physics: A Spiral Approach.* W. H. Freeman and Company. Copyright © 1967.)

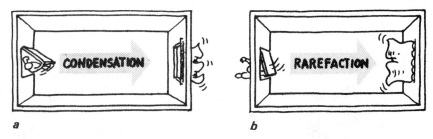

compressed air is what we call a **condensation**. Individual air molecules have tended to move first away from the door, then, after pushing against their neighbors, back toward the door. They have vibrated like points on a spiral spring.

When we close the door (Figure 18.1*b*), the door pushes adjacent air molecules out of the room. This produces an area of low pressure behind the door. Neighboring molecules then move into it, leaving a zone of lessened pressure behind them. Others, which are still farther from the door, in turn move into these rarefied regions, and once again a disturbance moves across the room, as evidenced by the curtain, which flaps inward. This time the disturbance is a **rarefaction**. It is important to note that it is not the air itself but a *pulse* that travels across the room, and in both cases it travels in a direction from the door to the curtain. We know this because in both cases the curtain moves *after* the door is opened or closed. If we continually swing the door open and closed in periodic fashion, we can set up a wave of periodic condensations and rarefactions in the room that will make the curtain swing back and forth. On a much smaller but more rapid scale, this is what happens when a tuning fork is struck. The vibrations of the tuning fork and the waves it produces are considerably higher in frequency and lower in amplitude than in the case of the swinging door. We don't notice the effect of these waves on the curtain, but we are well aware of them when they meet our sensitive eardrums!

Consider sound waves in the long tube shown in Figure 18.2. For simplicity, we represent only the waves that travel through the air in the tube placed near the source, a vibrating tuning fork. When the prong of the fork next to the tube moves to the right, a condensation is started through

Fig. 18.2 Condensations and rarefactions traveling from the tuning fork through the tube.

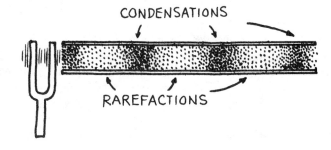

Fig. 18.3 Waves of compressed and rarefied air, produced by the vibrating cone of the loudspeaker, make up the pleasing sound of music.

the tube; when it swings back to the left, a rarefaction follows the condensation. It is like a Ping-Pong paddle moving back and forth in a room packed full of Ping-Pong balls. The paddle strikes the neighboring balls, which in turn strike their neighbors, which in turn do the same, resulting in the transmission of a pulse through the Ping-Pong balls. As the fork vibrates back and forth, a series of compressions and rarefactions of air molecules travels through the tube.

Pause to reflect on the physics of sound while you are quietly listening to your radio sometime. The radio loudspeaker is a paper cone that vibrates in rhythm with an electrical signal. Air molecules that constantly impinge on the vibrating cone of the speaker are themselves set into vibration. These in turn vibrate against neighboring molecules, which in turn do the same, and so on. As a result, rhythmic patterns of compressed and rarefied air emanate from the loudspeaker, showering the whole room with undulating motions. The resulting vibrating air sets your eardrum into vibration, which in turn sends cascades of rhythmic electrical impulses along the cochlear nerve canal and into the brain. And you listen to the sound of music.

Media That Transmit Sound

Most sounds that we hear are transmitted through the air. However, any elastic substance—whether solid, liquid, gas, or plasma—can transmit sound.[1] Compared to solids and liquids, air is a relatively poor conductor of sound. The sound of a distant train can be heard more clearly if the ear is placed against the rail. Similarly, a watch placed on a table beyond hearing distance can be heard by placing your ear to the table. American Indians often placed their ears to the ground to hear sounds that were inaudible in the air. The next time you go swimming, have a friend at some distance click two rocks together beneath the surface of the water while you are submerged. You will find that liquid is an excellent conductor of sound. If you and your friend ever happen to be in a vacuum, perform your last experiment and have her click two rocks together. You won't hear a thing! Sound will not travel in a vacuum. The transmission of sound requires a medium; if there is nothing to compress and expand, there can be no sound.

Speed of Sound

If we watch a person at a distance chopping wood or hammering, we can easily see that the blow takes place an appreciable time before its sound reaches our ears. Thunder is heard after a flash of lightning. These common experiences show that sound requires a recognizable time to travel from one place to another. The speed of sound depends on wind conditions, temperature, and humidity. It does not depend on the loudness or the frequency of the sound; all sounds travel at the same speed. The speed of sound in dry air at 0°C is about 330 meters per second, nearly 1200

---

[1]Recall from Chapter 10 that an elastic substance like steel (in contrast to putty, which is an inelastic substance) has resilience and can transmit energy with little loss.

kilometers per hour. Water vapor in the air increases this speed slightly. Sound travels faster through warm air than through cold air. This is to be expected because the faster-moving molecules in warm air bump into each other more often and therefore can transmit a pulse in less time.[2] For each degree rise in temperature above 0°C, the speed of sound in air increases by 0.6 meter per second. In water, sound travels about four times as fast as it does in air; while in steel, the speed of sound is about fifteen times as great as in air.

| **Question** | What is the approximate distance of a thunderstorm when you note a 3-second delay between the flash of lightning and the sound of thunder?* |

Refraction of Sound

Sound waves bend when parts of the wave fronts travel at different speeds. This happens in uneven winds or when sound is traveling through air of uneven temperatures. This bending of sound is called **refraction**. On a warm day, the air near the ground may be appreciably warmer than the rest of the air, so the speed of sound near the ground increases. Sound waves therefore tend to bend away from the ground, resulting in sound that does not appear to carry well.

On a cold day, or at night when the layer of air near the ground is colder than the air above, the speed of sound near the ground is reduced. The higher speed of the wave fronts above causes a bending of the sound toward the earth, resulting in sound that can be heard over considerably longer distances (Figure 18.4).

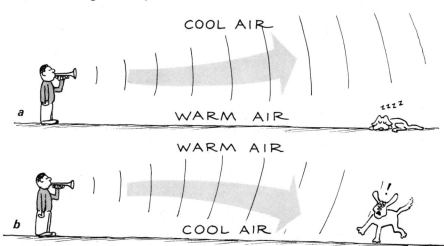

Fig. 18.4 The wave fronts of sound are bent in air of uneven temperatures.

---

*Answer   At normal temperatures (15° to 17°C), the speed of sound in air is about 340 m/s, and in 3 s will travel (340 × 3) 1020 m. We can assume no time delay for the light, so the storm is slightly more than 1 km away.

[2]The speed of sound in a gas is about $\frac{3}{4}$ the average speed of the molecules.

Reflection of Sound

We call the reflection of sound by distant walls, buildings, and the like an echo. If we are in a favorable location and make a sound, in a short time that sound may be reflected back. The time required, of course, depends on the distance to and from the reflecting surface. If the reflector is 170 meters away, the sound must travel 170 meters and back, or 340 meters. Since at normal temperatures sound travels 340 meters in about 1 second, the echo will be heard about 1 second after the sound is produced. In fogs, seafaring men once used this principle in estimating distance to obstacles by timing the echo of their ship's whistle. Today's sailors still use the principle; only instead of using sound, they use the high-frequency radio waves of radar.

Sometimes there is more than one reflecting surface, and we hear a series of echoes. Thus, when a gun is fired in a canyon, the sound bounces first from one surface and then from another, reflecting many times and reaching our ears by many paths, so that it is drawn out into a long peal. We are all familiar with the same effect in the rumble of distant thunder. This prolonging of sound by multiple reflections is called **reverberation**.

The multiple reflections of ultrasonic waves are used by physicians in a technique for harmlessly exploring the body without the use of X rays. When ultrasound enters the body, it is reflected more strongly from the outside of organs than from their interior, and a picture of the outline of the organs can be obtained. When ultrasound is incident upon a moving object, the reflected sound has a slightly different frequency. Using this Doppler effect, a physician can "hear" the beating heart of a fetus as early as 11 weeks (Figure 18.5).

Fig. 18.5 A 5-month-old fetus displayed on a viewing screen by ultrasound.

Question

An oceanic depth-sounding vessel surveys the ocean bottom with ultrasonic sound that travels 1530 meters per second in seawater. How deep is the water if the time delay of the echo from the ocean floor is 2 seconds?*

Energy in Sound Waves

Wave motion of all kinds possesses energy of varying degrees. For electromagnetic waves we find a great amount of energy in X rays, somewhat less in ultraviolet radiation, and less still in visible sunlight. By comparison, the energy in sound is extremely small. That's because producing sound requires only a small amount of energy. For example, 10,000,000 people talking at the same time would produce a sound energy equal only to the energy needed to light a common flashlight. Hearing is possible only because of our remarkably sensitive ears. Very few microphones can detect sound softer than we can hear.

Sound energy is dissipated in the form of heat in air. For waves that vibrate more often in 1 second, the sound energy is transformed into heat

---

*__Answer__   1530 m

more rapidly than for waves of lower frequencies. As a result, sound of low frequencies will travel farther through air than sound of higher frequencies. That's why the foghorns of ships are of a low frequency.

Forced Vibrations

If we strike an unmounted tuning fork, the sound from it may be rather faint. If we hold the same fork against a table and strike it with the same force, the sound is louder. This is because the table is forced to vibrate, and with its larger surface it will set more air in motion. The table will be forced into vibration by a fork of any frequency. This is a case of **forced vibration**. The vibration of a factory floor caused by the running of heavy machinery is an example of forced vibration. A more pleasing example is given by the sounding boards of stringed instruments.

Resonance

When someone drops a wrench on a concrete floor, you are not likely to mistake its sound for that of a baseball bat. This is because both objects vibrate at characteristic frequencies when struck. Any object composed of an elastic material will vibrate at its own special frequency when disturbed. This frequency depends on factors such as the elasticity and shape of the object. We express this fact when we speak of the body's *natural frequency*. This is the frequency at which the least amount of work is required to produce forced vibrations.

When the frequency of forced vibrations on a body matches the body's natural frequency, a dramatic increase in amplitude occurs. This phenomenon is called **resonance**. A common experience illustrating resonance is pushing a swing. As a child, you probably learned how to make someone on a swing move back and forth in a greater and greater arc. This didn't require strength nearly as much as a sense of timing; even small pushes, if delivered in rhythm with the frequency of the swinging motion, ultimately produced a large amplitude. Similarly, a tuning fork adjusted to the frequency of a second fork will be set in vibration by the sound waves from the second fork even when 8 or 10 meters away. When a series of sound waves impinges on the fork, each condensation gives the fork prong a tiny push. Since the frequency of these pushes corresponds to the natural frequency of the fork, the pushes will successively increase the amplitude of vibration. This is because the pushes are repeatedly in the same direction as the instantaneous motion of the fork; they act in the same way as a child's pushes on a swing. This is true for the matched-frequency tuning forks shown in Figure 18.6 and is true for all elastic bodies. Submultiples

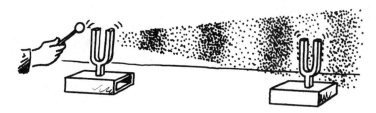

Fig. 18.6   Resonance.

($\frac{1}{2}$, $\frac{1}{3}$, $\frac{1}{4}$, etc.) of the natural frequency will also produce resonance. Resonance occurs when successive impulses are applied to a vibrating object in time with its natural frequency.

Cavalry troops marching across a footbridge near Manchester, England, in 1831 inadvertently caused the bridge to collapse when they marched in rhythm with the bridge's natural frequency. Ever since, troops have been ordered to "break step" when crossing such bridges. A more recent bridge disaster was caused by wind-generated resonance (Figure 18.7).[3]

Fig. 18.7 In November 1940, four months after being completed, the Tacoma Narrows Bridge in the state of Washington was destroyed by wind-generated resonance. The mild gale produced a fluctuating force in resonance with the natural frequency of the bridge, steadily increasing the amplitude until the bridge was destroyed.

A more subtle example of resonance may be the effect of music on plants. The rate at which rhythms of air molecules impinge upon the leaves of plants has been shown to affect their growth rate. However, this may not be simply a case of resonance.

Interference

Sound waves, like any waves, can be made to exhibit interference, a phenomenon discussed in Chapter 17. A comparison of interference in the cases of transverse waves and longitudinal waves is shown in Figure 18.8. An interesting case of sound interference is illustrated in Figure 18.9. If you are an equal distance from two sound speakers both of which emit an identical unvarying tone, the intensity of sound you receive is increased because the condensations and rarefactions from each speaker arrive in phase. However, if you move to the side so that the paths from the speakers to you differ by a half wavelength, then the rarefactions from one speaker

[3]There is a movie of this disaster, taken on the spot in 1940. Most physics departments have this film or have access to it. Ask your instructor to share it with you.

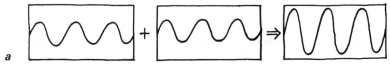

a

The superposition of two identical transverse waves in phase produces a wave of increased amplitude.

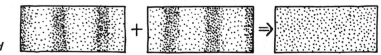

b

The superposition of two identical longitudinal waves in phase produces a wave of increased intensity.

c

Two identical transverse waves that are out of phase destroy each other when they are superimposed.

d

Two identical longitudinal waves that are out of phase destroy each other when they are superimposed.

Fig. 18.8 Constructive and destructive interference.

will be filled in with the condensations from the other speaker, producing destructive interference (like the crest of one water wave filling in the trough of another). If the region were devoid of any reflecting surfaces, little or no sound would be heard.

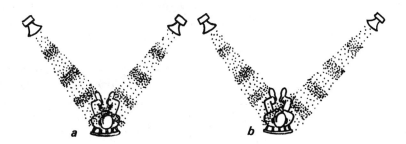

a   *b*

Fig. 18.9 (*a*) Waves of sound constructively interfere (condensation overlaps condensation) when the distances from the speakers to the listener are the same. (*b*) Destructive interference (condensation overlaps rarefaction) occurs when one speaker is farther from the listener than the other by a distance of a half wavelength ($\frac{3}{2}$, $\frac{5}{2}$, etc.).

Beats

When two tones of slightly different frequencies are sounded together, a fluctuation in the loudness of the combined sounds is heard; the sound is loud, then faint, then loud, then faint, and so on. This periodic variation in the loudness of the sound is called **beats** and is due to interference (Figure 18.10). Suppose, for example, that the sources of interfering sounds are two tuning forks, each of a slightly different frequency. Because one fork vibrates with more frequency than the other, the vibrations of the forks will momentarily be in step, then out, then in again, and so on. When the combined waves reaching our ears are in step—say, when a condensation from one fork reaches us at the same time as a condensation from the other fork—the resulting sound is a maximum. A moment later, when the forks are out of step, a condensation from one fork will be met with a rarefaction from the other, resulting in a minimum. The sound reaching our ears will throb between maximum and minimum loudness, producing a tremolo effect.

Fig. 18.10 The interference of two sound waves having different frequencies produces beats.

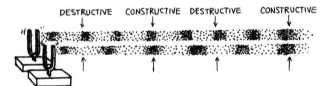

We can understand beats by considering the analogous case of two people walking side by side with different strides. At some moment they will be in step, a little later out of step, then in step again, and so on. Imagine that one person, perhaps with longer legs, takes exactly seventy steps in one minute, while the shorter person takes seventy-two steps. The shorter person then gains two steps per minute on the taller person. A little thought will show that they will momentarily be in step twice each minute. We can generalize this to the following: If two people with different strides walk together, the number of times they are in step each minute is equal to the difference in the frequencies of the steps. We can also apply this to a pair of tuning forks. If one fork undergoes 254 vibrations each second, and the other 256, they will be in step twice each second, and two beats will be heard each second. The overall tone will correspond to the average frequency, 255 hertz.

Question What is the beat frequency when a 256-Hz and a 260-Hz tuning fork are sounded together? A 270-Hz and a 260-Hz fork?*

*****Answer:** For the 256- and 260-Hz forks, the ear will hear 258 Hz, which will beat at (260 − 256) 4 Hz. For the 270- and 260-Hz forks, 265 Hz will be heard, and some people will hear it throb ten times each second. Higher beat frequencies are too rapid to be heard normally.

Fig. 18.11 The unequal spacings of the two combs produce a moiré pattern that is similar to the pattern of beats.

If you overlap two combs of different teeth spacings, you'll see a moiré pattern that is related to beats. The number of beats per unit length will equal the difference in the number of teeth per unit length for the two combs (Figure 18.11).

Beats are used in tuning the notes of any two sources to the same frequency. If two piano strings differ in frequency, they will produce beats when sounded together. When no beats are heard, the frequencies are matched.

The familiar vibrato effect of organ pipes is actually an example of beats. Two pipes tuned to different frequencies are used for each "single" note.

Summary of Terms

Condensation   Region of compression of the medium through which a longitudinal wave travels.

Rarefaction   Rarefied region, or region of lessened pressure, of the medium through which a longitudinal wave travels.

Refraction   The bending of a wave caused by differences in wave velocity through a nonuniform medium.

Reverberation   Reechoed sound.

Forced vibration   The setting up of vibrations in a body by a vibrating force.

Resonance   The result of forced vibrations on a body when the applied frequency matches the natural frequency of the body.

Beats   A series of alternate reinforcements and cancelations produced by the interference of two sets of superimposed waves of different frequencies; heard as a throbbing effect in sound waves.

Review Questions

1. What is the frequency in cycles per second of a 1000-hertz wave?

2. What is the average frequency range of human hearing?

3. Distinguish between *infrasonic* and *ultrasonic* sound waves.

4. Do condensations and rarefactions in a sound wave travel in the same direction or in opposite directions from one another?

5. Approximately how fast does sound travel in dry air? In a vacuum?

6. Does sound travel faster in warm air or in cool air?

7. Would the refraction of sound be possible if the speed of sound were unaffected by wind, temperature, and other conditions?

8. What is an echo? A reverberation?

9. Why are foghorns low in frequency?

10. What is the relationship between resonance and forced vibration?

11. Is it possible for one sound wave to cancel another? Explain.

12. How are beats produced?

Home Projects

1. Suspend the wire grille from a refrigerator or an oven from a string, the ends of which you hold to your ears. Let a friend gently stroke the grille with pieces of broom straw and other objects. The effect is best appreciated in a relaxed condition with your eyes closed. Be sure to try this!

2. In the bathtub, submerge your head and listen to the sound you make when clicking your fingernails together or tapping the tub beneath the water surface. Compare the sound with that you make when both the source and your ears are above water. At the risk of getting the floor wet, slide back and forth in the tub at different frequencies and see how the amplitude of the sloshing waves quickly builds up when you slide in rhythm with the waves. (The latter of these projects is most effective when you are alone in the tub.)

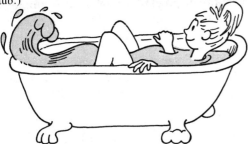

3. Stretch a piece of balloon rubber not too tightly over a radio loudspeaker. Glue a small, very lightweight piece of mirror, aluminum foil, or polished metal near one edge. Project a narrow beam of light on the mirror while your favorite music is playing and observe the beautiful patterns that are reflected on a screen or wall.

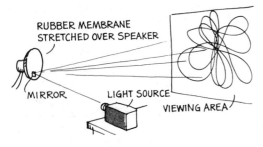

RUBBER MEMBRANE
STRETCHED OVER SPEAKER

MIRROR   LIGHT SOURCE
                VIEWING AREA

Exercises

1. Why do flying bees buzz?

2. When a sound wave propagates past a point in air, are there changes in the density of air at this point? Explain.

3. At the instant that a high-pressure region is created just outside the prongs of a vibrating tuning fork, what is being created inside between the prongs?

4. Why is it so quiet after a snowfall?

5. If a bell is ringing inside a bell jar, we can no longer hear it if the air is pumped out, but we can still see it. What differences in the properties of sound and light does this indicate?

6. Why is the moon described as a "silent planet"?

7. If the speed of sound were frequency-dependent, how would distant music sound?

8. If the frequency of sound is doubled, what change will occur in its speed? In its wavelength?

9. Why does sound travel faster in warm air?

10. Why does sound travel faster in moist air? (*Hint:* At the same temperature, water vapor molecules have the same average kinetic energy as the heavier nitrogen and oxygen molecules in the air. How, then, do the average speeds of H_2O molecules compare with N_2 and O_2 molecules?)

11. Why can the tremor of the ground from a distant explosion be felt before the sound of the explosion can be heard?

12. What kinds of wind conditions would make sound more easily heard at long distances? Less easily heard at long distances?

13. What is the wavelength of a 340-hertz tone in air? What is the wavelength of a 34,000-hertz ultrasonic wave in air?

14. Ultrasonic waves have many applications in technology and medicine. One advantage is that large intensities can be used without danger to the ear. Cite another advantage of their short wavelength. (*Hint:* Why do microscopists use blue light rather than white light to see fine detail?)

15. A bat flying in a cave emits a sound and receives its echo in 1 second. How far away is the cave wall?

16. An oceanic depth-sounding vessel surveys the ocean bottom with ultrasonic sound that travels 1530 meters per second in seawater. How deep is the water if the time delay of the echo from the ocean floor is 6 seconds?

17. Why is an echo weaker than the original sound?

18. How can you estimate the distance of a distant thunderstorm?

19. A rule of thumb for estimating the distance in kilometers between an observer and a lightning stroke is to divide the number of seconds in the interval between the flash and the sound by 3. Is this rule correct?

20. You watch a distant lady driving nails into her front porch at a regular rate of one stroke per second. You hear the sound of the blows exactly synchronized with the blows you see. And then you hear one more blow after you see her stop hammering. How far away is she?

21. If a single disturbance some unknown distance away sends out both transverse and longitudinal waves that travel with distinctly different speeds in the medium, such as in the ground during earthquakes, how could the origin of the disturbance be located? (*Hint:* Consider triangulation from two testing sites.)

22. Apartment dwellers will testify that bass notes are more distinctly heard from music played in nearby apartments. Why do you suppose lower-frequency sounds get through walls, floors, and ceilings more easily?

23. Why will marchers at the end of a long parade following a band be out of step with marchers nearer the band?

24. Why do soldiers break step in marching over a bridge?

25. Why does a tuning fork eventually stop vibrating?

26. If the handle of a tuning fork is held solidly against a table, the sound from the tuning fork becomes louder. Why? How will this affect the length of time the fork keeps vibrating? Explain.

27. Why are you not troubled by sound interference as shown in Figure 18.9 when you listen to stereo or quadraphonic music?

28. An object resonates when a vibrating force either matches its natural frequency or is a submultiple of its natural frequency. Why will it not resonate to multiples of its natural frequency? (Think of pushing a child in a swing.)

29. What beat frequencies are possible with tuning forks of frequencies 256, 259, and 261 hertz, respectively?

30. Suppose a piano tuner hears 3 beats per second when listening to the combined sound from her tuning fork and the piano note being tuned. After slightly tightening the string, she hears 5 beats per second. Should she loosen or further tighten the string?

19 Musical Sounds

Noises and Musical Sounds

Most of the sounds we hear are noises. The impact of a falling object, the slamming of a door, the roaring of a motorcycle, and most of the sounds from traffic in city streets are noises. Noise corresponds to an irregular vibration of the eardrum produced by the irregular vibration of some nearby object. If we make a diagram to indicate the pressure of the air on the eardrum as it varies with time, the graph corresponding to a noise might look like that shown in Figure 19.1a. The sound of music has a different character, having more or less sustained tones—or musical "notes." (Musical instruments can make noise as well!) The graph representing a musical sound has a shape that repeats itself over and over again (Figure 19.1b). Such graphs can be displayed on the screen of an oscilloscope when the electrical signal from a microphone is fed into the input terminal of this useful device.

The line that separates music and noise is thin and subjective. To some contemporary composers, it is nonexistent. Some people consider contemporary music and music from other cultures to be noise. Differentiating these types of music from noise becomes a problem of aesthetics. However, differentiating traditional music—that is, classical music and most types of popular music—from noise presents no problem. A deaf person could distinguish between these by using an oscilloscope.

Musicians usually speak of a musical tone in terms of three characteristics—pitch, loudness, and quality.

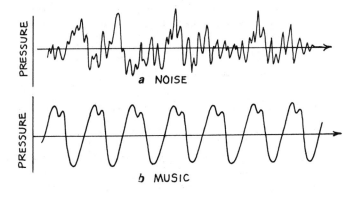

Fig. 19.1 Graphical representations of noise and music.

Pitch

The **pitch** of a sound corresponds to frequency. Rapid vibrations of the sound source produce a shrill high note, whereas slow vibrations produce a deep low note. We speak of the pitch of a sound in terms of its position in the musical scale. When middle C is struck on a piano, a hammer strikes two or three strings, each of which vibrates 264 times in 1 second. The pitch of middle C corresponds to 264 hertz.

Different musical notes are obtained by changing the frequency of the vibrating sound source. This is usually done by altering the size, the tightness, or the mass of the vibrating object. A guitarist or violinist, for example, adjusts the tightness, or tension, of the strings of the instrument when tuning them. Then different notes can be played by altering the length of each string by "stopping" it with the fingers. In wind instruments, the length of the vibrating air column is altered to change the pitch of the note produced.

High-pitched sounds used in music are most often less than 4000 vibrations per second, but the average human ear can hear sounds up to 20,000 vibrations per second. Some people can hear tones of higher pitch than this, and most dogs can. In general, the upper limit of hearing in people gets lower as they grow older. A high-pitched sound is often inaudible to an older person and yet may be clearly heard by a younger one. So by the time you can really afford that high-fidelity music system, you may not be able to hear the difference.

Loudness

Fig. 19.2 James displays a sound signal on a cathode-ray oscilloscope.

The intensity of sound depends on pressure variations within the sound wave; it depends on the amplitude. (More specifically, as with any type of wave, intensity is proportional to the square of the amplitude.) Sound intensity is a purely objective and physical attribute of a wave and can be measured by various acoustical instruments (such as the oscilloscope in Figure 19.2). **Loudness**, on the other hand, is a physiological sensation; it depends on intensity but in a complicated way. If you turned a radio up until it seems about twice as loud as before, you would have to increase the power output and, therefore, the intensity by approximately eight times. Although the pitch of a sound can be judged very accurately, our ears are not very good at judging loudness. The loudest sounds we can tolerate have intensities a million million times greater than the faintest sounds. The difference in loudness, however, is much less than this amount.

The relative loudness of a sound the ear hears is called the *noise level* and is measured in *decibels,* a unit named after Alexander Graham Bell and abbreviated db. Some common sounds and their noise levels are compared in Table 19.1.

Decibel ratings are logarithmic; a sound of 10 decibels is 10 times as loud as 0 decibels, the threshold of hearing; but 20 decibels is not twice as loud again, but 10 times again louder or 100 times as loud as the threshold of hearing. Accordingly, 30 decibels is 1000 times the threshold of hearing, and 40 decibels is 10,000 times. So 60 decibels represents sound intensity a million times greater than 0 decibels; 80 decibels represents sound 100 times as intense as 60 decibels. Physiological hearing damage begins at exposure to 85 decibels, the degree depending on the length of exposure and on frequency characteristics. Damage from loud sounds can be temporary or permanent, depending on whether the organs of Corti, the

Table 19.1 Common sounds and noise levels

Source of sound	Noise level (db)
Jet airplane, 30 meters away	140
Air-raid siren, nearby	125
Disco music, amplified	115
Riveter	95
Busy street traffic	70
Conversation in home	65
Quiet radio in home	40
Whisper	20
Rustle of leaves	10
Threshold of hearing	0

receptor organs in the inner ear, are impaired or destroyed. A single burst of sound can produce vibrations in the organs intense enough to tear them apart. Less intense, but severe, noise can interfere with cellular processes in the organs that cause their eventual breakdown. Unfortunately, the cells of these organs do not regenerate.

From the earliest stages of human evolution, we have been subjected to a wide range of light intensities, and—except for looking at the sun—our eyes are now well adapted to the light we encounter in today's world. But not so with sound. Hearing loud and sustained sounds is something new for humans to contend with. Except for occasional bursts of thunder and the like, our ancestors were not exposed to loud sounds, and as a result we did not develop an ability to adapt to the noise pollution we experience today. You know that you'll ruin your sense of sight if you stare into a source of light as bright as the sun. Please don't ruin your sense of hearing and blow the fine tuning in your ears by subjecting yourself to loud sounds.

Question   Is hearing permanently impaired when attending concerts, discotheques, or functions that feature very loud music?*

Quality   We have no trouble distinguishing between the tone from a piano and a like-pitched tone from a clarinet. Each of these tones has a characteristic sound that differs in quality—or timbre. Most musical sounds consist

---

*__Answer__   Yes, depending on how loud, how long, and how often. Some music groups have emphasized loudness over quality. Tragically, as hearing becomes more and more impaired, members of the group (and their fans) require louder and louder sounds for stimulation. Hearing loss caused by loud sounds is particularly common in the frequency range of 2000 to 5000 Hz, the very range in which speech and music normally occur.

of a fundamental tone and an assortment of higher-pitched and generally fainter tones called **overtones**. Most musical sounds are composed of a superposition of many frequencies. The lowest frequency, called the *fundamental*, determines the pitch of the note. The higher frequencies, the overtones and **harmonics**, give the characteristic quality (Figure 19.3). The frequencies of the overtones are not haphazard but are whole multiples of the lowest frequency.

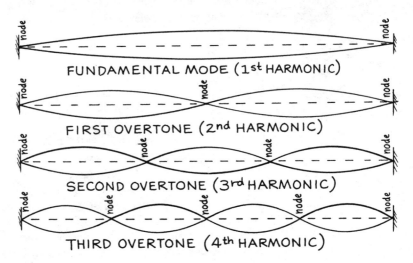

Fig. 19.3 Modes of vibration of a guitar string.

Thus, if we strike middle C on the piano, we produce a fundamental tone with a pitch of about 264 vibrations per second and also a blending of overtones of two, three, four, five, and so on, times the frequency of middle C. The number and relative loudness of the overtones produce the quality of sound associated with the piano. Sound from practically every musical instrument consists of a fundamental and overtones. Pure tones, those having only one frequency, can be produced electronically. The electronic Moog synthesizer, for example, produces pure tones and mixtures of these to give a vast variety of musical sounds.

The quality of a tone is determined by the presence and relative intensity of the various overtones. The sound produced by a certain tone from the piano and that produced by one of the same pitch from a clarinet have different qualities the ear recognizes because their overtones are different.

Fig. 19.4 A composite vibration of the fundamental mode and second overtone (or, equivalently, of the first and third harmonic).

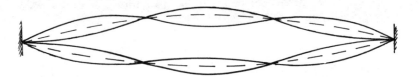

A pair of tones of the same pitch with different qualities have either different overtones or a difference in the relative intensity of the overtones.

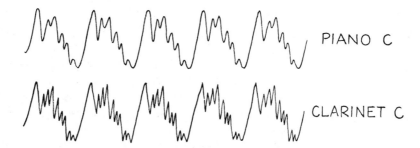

PIANO C

CLARINET C

Fig. 19.5 Sounds from the piano and clarinet differ in quality.

Musical Instruments

Conventional musical instruments can be grouped into one of three classes: those in which the sound is produced by vibrating strings, those in which the sound is produced by vibrating air columns, and those in which the sound is produced by percussion—the vibrating of a two-dimensional surface.

In a stringed instrument, the vibration of the strings is transferred to a sounding board and then to the air, with a considerable dissipation of energy. Stringed instruments are low-efficiency producers of sound, so to compensate for this, we find the string sections in orchestras relatively large. A smaller number of the higher-efficiency wind instruments sufficiently balances a much larger number of violins.

In a wind instrument, sound is emitted directly by the vibrations of air columns in the instrument. There are various ways to set the air columns into vibration. In brass instruments such as trumpets, French horns, and trombones, vibrations of the player's lips interact with standing waves that are set up by acoustic energy reflected within the instrument by the flaring bell. The lengths of the vibrating air columns are manipulated by valves that add or subtract extra segments. In woodwinds such as clarinets, oboes, and saxophones, a stream of air produced by the musician sets a reed vibrating; whereas in fifes, flutes, and piccolos, the musician blows air against the edge of a hole to produce a fluttering stream that sets the air column into vibration.

In percussion instruments, drums, cymbals, and the like, a two-dimensional membrane or elastic surface is struck to produce sound. The fundamental tone produced depends on the geometry, the elasticity, and, in some cases, the tension of the surface. Changes in pitch result from changing the tension in the vibrating surface; depressing the edge of a drum membrane with the hand is one way of accomplishing this. Different modes of vibration can be set up by striking the surface in different places. As in all musical sounds, the quality depends on the number and relative loudness of the overtones.

Electronic musical instruments differ markedly from conventional musical instruments. Instead of strings that must be bowed, plucked, or struck, or reeds over which air must be blown, or diaphragms that must be tapped to produce sounds, electronic instruments use electrons to generate the signals that make up musical sounds. In appearance, electronic musical instruments resemble computers more than "musical" instruments. They are often as complex as a computer as well, and they demand of the composer and player an expertise beyond the knowledge of musicology. Electronic music is a different kind of music, emphasizing pure tones in addition to themes, randomness with imposed order, and looseness as a complement to restricted thought. People who look down on "music by machines" fail to realize that electronic music is not meant to replace musicians; rather, it brings a powerful new tool to the hands of the musician.

Musical Scales

As early as 530 B.C. Pythagoras found that notes played together on stringed instruments were pleasing to the ear when the ratios of the frequencies of these notes were the ratios of whole numbers, and he found them displeasing when the ratios were not (due in part to the beats produced). A succession of notes of frequencies that are in simple ratios to one another is a **musical scale**. Many scales exist that are the outcome of many cultures throughout various ages. The simplest scale in Western culture is the *diatonic major scale,* the familiar "do-re-mi-fa-sol-la-ti-do," where each note is named by the letters A, B, C, D, E, F, and G. In this scale the pitch or frequency of successive notes goes up in the whole-number ratios given in Table 19.2. The ratio in the frequencies of two sounds is called the *musical interval* of the two notes. The ratios in the table are those of the respective intervals relative to C, called *middle C*. We see that for C' above middle C the ratio is 2, or the frequency is double that of note C. This interval is called an *octave*, the Latin word for the "eight" notes in this interval. Middle C is taken to be 264 hertz on this scale, and C' is twice this—528 hertz. As we progress to notes higher than those in the table, each succeeding C has twice the frequency of the preceding C; each D has twice the frequency of the preceding D; and so on. The frequency of each note doubles with each octave (Figure 19.6).

Fig. 19.6 Piano keyboard. Low C (C''') is 33 Hz, and successive overtones of C keep doubling in frequency. The pitch that corresponds to 264 Hz (there are no "black notes" on the piano between these keys) is called *middle C*.

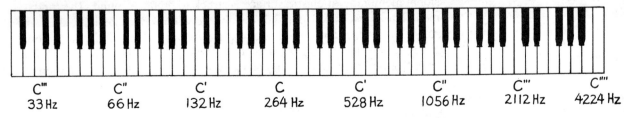

C'''	C''	C'	C	C'	C''	C'''	C''''
33 Hz	66 Hz	132 Hz	264 Hz	528 Hz	1056 Hz	2112 Hz	4224 Hz

Table 19.2 Diatonic C Major scale

Note	Letter name	Frequency (hertz)	Frequency ratio	Interval
do	C	264		
			$\frac{9}{8}$	Whole
re	D	297		
			$\frac{10}{9}$	Whole
mi	E	330		
			$\frac{16}{15}$	Half
fa	F	352		
			$\frac{9}{8}$	Whole
sol	G	396		
			$\frac{10}{9}$	Whole
la	A	440		
			$\frac{9}{8}$	Whole
ti	B	495		
			$\frac{16}{15}$	Half
do	C'	528		

The intervals between notes in the diatonic scale are not all the same. The difference in pitch between the notes "mi" and "fa" and between "ti" and "do" are about half that of the other intervals. These are called *half intervals* (there are no "black notes" on the piano between these keys). To obtain these half intervals when music is played in keys other than C, extra notes raised a half interval called *sharps* (♯) or lowered a half interval called *flats* (♭) are required. For the key of C these are the black keys on a piano keyboard.

Because of the variation in musical intervals, when a piano has been tuned for a diatonic scale in any one key, the frequencies are somewhat off for other keys. For this reason, pianos are not usually tuned to a diatonic scale, but to an *equitempered scale*. On this scale there are thirteen notes

Table 19.3 Equitempered chromatic scale

Note	Frequency (hertz)	Frequency ratio	Interval
C	262	$\sqrt[12]{2}$	
C♯ or D♭	277	$\sqrt[12]{2}$	Half
D	294	$\sqrt[12]{2}$	Half
D♯ or E♭	311	$\sqrt[12]{2}$	Half
E	330	$\sqrt[12]{2}$	Half
F	349	$\sqrt[12]{2}$	Half
F♯ or G♭	370	$\sqrt[12]{2}$	Half
G	392	$\sqrt[12]{2}$	Half
G♯ or A♭	415	$\sqrt[12]{2}$	Half
A	440	$\sqrt[12]{2}$	Half
A♯ or B♭	466	$\sqrt[12]{2}$	Half
B	494	$\sqrt[12]{2}$	Half
C'	524	$\sqrt[12]{2}$	Half

and twelve intervals in one octave, and the ratio between all successive notes is exactly the same (the twelfth root of 2, or 1.05946). This results in the same frequency for C♯ and D♭, D♯ and E♭, and so on, which is not generally the case for the diatonic scale. The standard frequency for the equitempered scale is A = 440 hertz, the only note in the octave that agrees with its counterpart on the diatonic scale. All other notes differ slightly between the two scales, as you can see by comparing Tables 19.2 and 19.3. The equitempered scale is now used for most music written in the Western world.

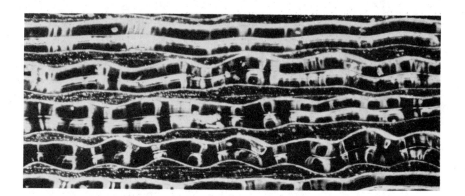

Fig. 19.7 A microscopic view of the grooves in a phonograph record.

Fourier Analysis

Did you ever look closely at the grooves in a phonograph record? And did you notice the variations in the width of the grooves—variations that cause the phonograph needle that rides in the groove to vibrate? And did you ever wonder how all the distinct vibrations made by the various pieces of an orchestra are captured by the single-wave groove of the record? The sound of an oboe when captured by the groove of a phonograph record and displayed on an oscilloscope screen looks like Figure 19.8a. The high points on the wave correspond to the greatest amplitude of the electronic signal produced by the vibrating needle. They also correspond to the amplified signal that activates the loudspeaker of the sound system and to the amplitude of air vibrating against the eardrum. Figure 19.8b shows the wave appearance of a clarinet, and Figure 19.8c shows the wave form when oboe and clarinet are sounded together.

The shape of the wave in Figure 19.8c is the net result of shapes a and b interfering with each other. If we know a and b, it is a simple thing to create c. But it is a far different problem to discern in c the shapes of a and b that make it up. Looking only at shape c, we cannot unscramble the oboe from the clarinet.

But play the record on the phonograph, and our ears will at once know what instruments are being played, what notes they are playing, and what their relative loudness is. Our ears break the overall signal into its component parts automatically.

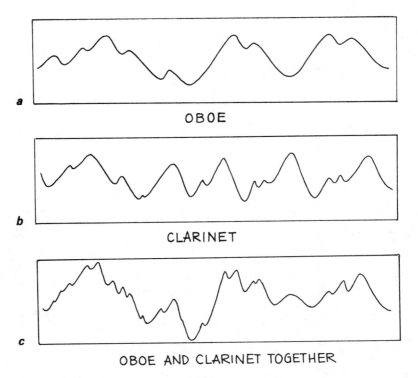

a

OBOE

b

CLARINET

c

OBOE AND CLARINET TOGETHER

Fig. 19.8 Wave forms of (*a*) an oboe, (*b*) a clarinet, and (*c*) the oboe and clarinet sounded together.

Fig. 19.9 A sine wave.

In 1822 the French mathematician Joseph Fourier discovered a mathematical regularity to the component parts of wave motion. He found that even the most complex wave motion could be broken down into simple sine waves. A sine wave is the simplest of waves, having a single frequency (Figure 19.9). Fourier found that all rhythmic waves may be broken down into constituent sine waves of different amplitudes and frequencies. The mathematical operation for doing this is called **Fourier analysis**. We will not explain the mathematics here but simply point out that by such analysis one can find the pure sine tones that compose the tone of, say, a violin. When these pure tones are sounded together, as by striking a number of tuning forks or by selecting the proper keys on an electric organ, they combine to give the tone of the violin. The lowest-frequency sine wave is the fundamental and determines the pitch of the note. The higher-frequency sine waves are the overtones that give the characteristic quality. Thus, the wave form of any musical sound is no more than a sum of simple sine waves.

Since the wave form of music is a multitude of various sine waves, to duplicate sound accurately by radio, record player, or tape recorder, we should have as large a range of frequencies as possible. The notes of a piano keyboard range in the hundreds of cycles per second, but to duplicate the music of a piano composition accurately, the sound system must have a range of frequencies into the thousands of cycles per second. The

Fig. 19.10 The fundamental and its harmonics combine to produce a composite wave.

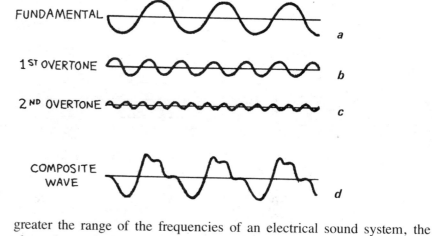

greater the range of the frequencies of an electrical sound system, the closer the musical output approximates the original sound, hence the wide range of frequencies in a high-fidelity sound system.

Our ear performs a sort of Fourier analysis automatically. It sorts the complex jumble of air pulsations reaching it into pure notes. And we recombine various combinations of these pure notes when we listen. What combinations of notes we have learned to focus our attention on determines what we hear when we listen to a concert. We can direct our attention to the sounds of the various instruments and discern the faintest notes from the loudest; we can delight in the intricate interplay of instruments and still detect the extraneous noises of others around us. This is a most incredible feat.

Fig. 19.11 Does each hear the same music?

Laser Discs

You can experience the rich, full sounds of a string quartet or a symphony orchestra in your own room by means of a remarkable sound recording and reproduction technique called *digital audio*. Instead of an LP record and a conventional stylus, the digital player utilizes a laser beam, which is directed onto a plastic reflective disc. The player's pickup is a light sensor rather than a stylus.

On a conventional phonograph record, the stylus is made to vibrate when it rides in the squiggly phonograph groove. The output is a signal like those shown in Figure 19.8. This type of continuous wave form is called an *analog* signal. The shape of the wave can be described by the numeric value of its amplitude during each split second (Figure 19.12). This numeric value can be expressed in a number system that is convenient for computers, called *binary*. In the binary code, any number can be expressed as a succession of ones and zeros; for example, the number 1 is 1, 2 is 10, 3 is 11, 4 is 100, 5 is 101, 17 is 10001, etc. So the shape of the analog wave form can be expressed as a series of "on" and "off" pulses that corresponds to a series of ones and zeros in binary code. That's where the laser disc comes in.

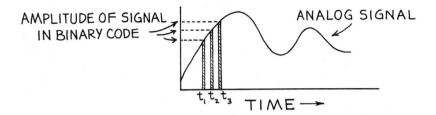

Fig. 19.12 The amplitude of the analog wave form over successive split seconds is recorded in binary code on the reflective surface of the laser disc.

Instead of a squiggly phonograph groove, a laser audio disc has a series of microscopic pits about 30 times thinner than a human hair. When the laser beam falls on a flat portion of the reflective surface, it is reflected directly into the player's optical system; this gives an "on" pulse. When the beam is incident upon a passing pit, very little of the laser beam returns to the optical sensor; this gives an "off" pulse. A flickering of "on" and "off" pulses generates the "one" and "zero" digits of the binary code.

The rate at which these tiny pits on the disc are sampled is 44,100 times per second. A single audio disc is $\frac{1}{16}$ the size of a conventional LP and contains billions of bits of information. All this information is encoded on the reflective surface, which is covered with a protective layer of clear plastic. Since the laser beam is focused onto the signal surface below, it is impervious to dust, scratches, and fingerprints. No more snap, crackle, and pop, so characteristic of the familiar LP. And since the laser beam does not touch the disc, it nevers wears out, no matter how many times you play it.

The most extraordinary feature of the laser disc, however, is the quality of the sound. You can *hear* the difference.

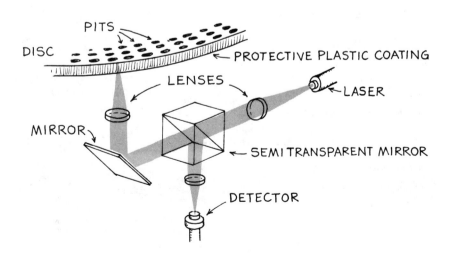

Fig. 19.13 A tightly focused laser beam reads digital information represented by a series of pits on the laser disc.

Summary of Terms

Pitch The "highness" or "lowness" of a tone, as on a musical scale, which is governed by frequency. A high-frequency vibrating source produces a sound of high pitch; a low-frequency vibrating source produces a sound of low pitch.

Loudness The physiological sensation directly related to sound intensity or volume. Relative loudness, or noise level, is measured in decibels.

Quality The characteristic timbre of a musical sound, governed by the number and relative intensities of the overtones or harmonics.

Overtones The sequence of integral multiple tones above the fundamental, so that the first overtone is the second harmonic, the second overtone is the third harmonic, and so on.

Harmonics Modes of vibrations that begin with a lowest or fundamental vibrating frequency and continue as a sequence of tones that are integral multiples of the fundamental frequency.

Musical scale A succession of notes of frequencies that are in simple ratios to one another. In Western culture the principal scales are the diatonic major scale and the equitempered chromatic scale.

Fourier analysis A mathematical method that will resolve any wave form into a series of simple sine waves.

Review Questions

1. Distinguish between *noise* and *music*.

2. Is the pitch of sound directly proportional or inversely proportional to frequency?

3. Which of the three characteristics of sound has most to do with the amplitude of a sound wave?

4. What is a decibel?

5. Is a sound of 60 decibels 60 times greater than the threshold of hearing or 10^6 (a million) times greater?

6. What is meant by musical "quality"?

7. Why do identical notes plucked on a banjo and on a guitar have distinctly different sounds?

8. Do all musical notes that have the same fundamental frequency and the same number of overtones sound the same? What, besides the number of these overtones, would have to be the same for identical sounds?

9. What is an octave?

10. What musical note is common to both the diatonic and equitempered musical scales?

11. What did Fourier discover about complex wave patterns?

12. A high-fidelity sound system may have a frequency range that extends beyond the range of human hearing. Of what use is this extended range?

Home Projects

1. Blow air across the top of a pop bottle, and a puff of air (compression wave) travels downward, bounces from the bottom, and travels back to the opening. When it arrives (less than a thousandths of a second later), it disturbs the flow of air that you are still producing across the top. This causes a slightly bigger puff of air to start again on its way down the bottle. This happens repeatedly until a very large (and loud) vibration is built up and you hear it as sound. The pitch of the sound depends on the time taken for the back-and-forth trip, which depends on the depth of the bottle. If the bottle is empty, a long wave is reinforced and a relatively low tone is produced. With liquid in the bottle, the bottom is closer to the top and the pitch is higher. With a series of bottles properly filled, you can make your own music.

2. Test to see which ear has the better hearing by covering one ear and finding how far away your open ear can hear the ticking of a clock; repeat for the other ear.

3. With a strong magnifying glass, examine the grooves in phonograph records. If you have an old 78-RPM disk, compare the grooves with those of a $33\frac{1}{3}$-RPM disk.

4. Set a record into rotation on a record player. Place the edge of your fingernail in the groove and listen carefully. Notice that you hear only high-frequency sound.

5. Make the lowest-pitched sound you are capable of; then keep doubling the pitch to see how many octaves your voice can span.

6. On a sheet of graph paper, construct the composite wave of Figure 19.10 by superposing various vertical displacements of the fundamental and first two overtones. Your instructor can show you how this is done. Then find the composite waves of overtone selections of your own choosing.

Exercises

1. Explain how you can produce a low-pitched note on a guitar by altering (*a*) the length of the string, (*b*) the tension of the string, and (*c*) the thickness or the mass of the string.

2. Why is the mass per unit length greater for the bass strings of a guitar?

3. Would a plucked guitar string vibrate for a longer or a shorter time if it had no sounding board? Why?

4. If you very lightly touch a guitar string at its midpoint, you can hear a pure tone that is one octave above the fundamental for that string. Explain.

5. If a guitar string vibrates in two segments, where can a tiny piece of folded paper be supported without flying off? How many pieces of folded paper could similarly be supported if the wave form were of three segments?

6. If the wavelength of a vibrating string is reduced, what effect does this have on the frequency of vibration and on the pitch?

7. The amplitude of a transverse wave in a stretched string is the maximum displacement of the string from its equilibrium position. What does the amplitude of a longitudinal sound wave in air correspond to?

8. Which of the two musical notes displayed one at a time on an oscilloscope screen has the higher pitch? Which is the louder?

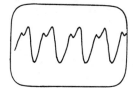 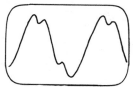

9. In a hi-fi speaker system, why is the woofer (low-frequency speaker) larger than the tweeter (high-frequency speaker)?

10. One person has a threshold of hearing of 5 decibels, and another of 10 decibels. Which person has the more acute hearing?

11. How much more intense is a sound of 10 decibels than the threshold of hearing? 30 decibels? 60 decibels?

12. How much more intense is a sound of 40 decibels compared to a sound of 30 decibels?

13. How is an organ able to produce the sounds made by various musical instruments?

14. When a person talks after inhaling helium gas, the voice is high-pitched. This is principally because the helium gas molecules move faster past the vocal chords than molecules of air. Why do helium molecules move faster?

15. Why does your voice sound fuller in a shower?

16. The frequency range for a telephone is between 500 and 4000 hertz. Why is a telephone inadequate for transmitting music?

17. A certain note has a frequency of 1000 hertz. What is the frequency of a note one octave above it? Two octaves above it? One octave below it? Two octaves below it?

18. How many octaves does normal human hearing span? How many octaves are on a common piano keyboard?

19. The author's range of hearing is from 20 hertz to 14,000 hertz (too many rock concerts in the mid-60s to early 70s). How many octaves can I hear?

20. If the fundamental frequency of a violin string is 440 hertz, what is the frequency of the first overtone? The third?

21. If the fundamental frequency of a violin string is 440 hertz, what is the frequency of the first harmonic? The third harmonic?

22. At an outdoor concert, will the pitch of musical tones be affected on a windy day? Explain.

23. When water is poured into a glass, does the pitch of the sound of the filling glass increase or decrease? (Think about this; *then* try it and hear for yourself.)

24. As you pour water into a glass, repeatedly tap the glass with a spoon. Does the pitch of the tapped glass as it is being filled increase or decrease? (Think about this one also and then try it before confirming your thoughts and/or surprising yourself.)

25. Make a sketch of the resulting wave form when the sine wave and the square wave of the same amplitude and frequency are superposed (*a*) in phase and (*b*) out of phase.

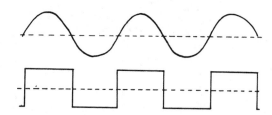

26. Do all the people in a group hear the same music when they listen to it attentively? (Do all see the same sight when looking at a painting? Do all taste the same flavor when sipping the same wine? Do all perceive the same aroma when smelling the same perfume? Do all feel the same texture when touching the same fabric? Do all come to the same conclusion when listening to a logical presentation of ideas?)

PART 5 Electricity and Magnetism

20 Electrostatics

In today's culture we find ourselves surrounded by electrical devices of all kinds. They range from lamps, battery-operated watches, motors, and stereo sets, to computers and so much more. To understand the devices that have become so integral to our everday lives, we must first understand electricity. We begin with this chapter on electricity at rest—**electrostatics**—and first consider the forces that electrical particles exert on one another.

Electrical Forces

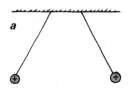

Fig. 20.1 (*a*) Like charges repel. (*b*) Unlike charges attract.

Consider a universal force like gravity, which varies inversely as the square of the distance but which is billions upon billions of times stronger than gravity. If there were such a force and if it were attractive like gravity, the universe would be pulled together into a tight sphere with all the matter in the universe pulled as close together as possible. But suppose this force were a repelling force, where every particle repelled every other particle. What then? The universe would be an ever-expanding gaseous cloud. Suppose, however, that the universe consisted of both attractive and repulsive particles, say, positives and negatives. Suppose that positives repelled positives, but attracted negatives; and that negatives repelled negatives, but attracted positives. Like kinds repel and unlike kinds attract (Figure 20.1). And suppose that there were equal numbers of each. What would the universe be like? The answer is simple: it would be like the one we are living in. For there is such a force. We call it *electrical force*.

Clusters of positives and negatives have been pulled together by the enormous attraction of the electrical force. The result is that the huge forces have balanced themselves out almost perfectly by forming compact and evenly mixed numbers of positive and negatives. Furthermore, between two separate bunches of such mixtures, there is practically no electrical attraction or repulsion at all. Any electrical forces between the earth and moon, for example, have been balanced out. In this way the much weaker gravitational force, which only attracts, is the predominant force between these two bodies.

The terms "positive" and "negative" refer to electric *charge*, that fundamental quantity that underlies all electrical phenomena. Protons are positively charged and electrons are negatively charged. The attractive force between protons and electrons holds atoms together. Between neighboring atoms the negative electrons of one atom may at times be closer to the positive protons of another atom so that the attractive force between these charges is greater than the repulsive force, and the atoms combine to form a molecule. In fact, all the chemical bonding forces that hold atoms together to form molecules are electrical forces acting in small regions where the balance of attractive and repelling forces is not perfect. So anyone planning to study chemistry should first know something about electricity, and before studying electricity should know something about atoms. Here are some important facts about atoms:

1. Every atom is composed of a positively charged nucleus, around which are distributed a number of negatively charged electrons.

2. The electrons of all atoms are identical. Each has the same quantity of negative charge and the same mass.

3. Protons and neutrons compose the nucleus. (The common form of hydrogen that has no neutrons is the only exception.) Protons are almost 2000 times more massive than electrons but carry an amount of positive charge equal to the negative charge of electrons. Neutrons are slightly more massive than protons and have no net charge.

4. All normal atoms have exactly as many electrons surrounding the nucleus as there are protons within the nucleus. Thus, a normal atom has no net charge.

Why don't protons pull the oppositely charged electrons into the nucleus? Sixty-five years ago the answer offered was that the electrons are not pulled into the nucleus by electrical force for the same reason the earth is not pulled into the sun by gravitational force: the electrons are held in orbit by the pull of the protons. This oversimplified explanation serves as a starter for understanding the electrical nature of the atom but must be modified on further study. The answer today has to do with the wave nature of matter, the quantum effects, which we will discuss in Chapter 29.

Why don't the protons in the nucleus mutually repel and fly apart? What holds the nucleus together? The answer is that in addition to electrical forces in the nucleus, there are even greater nonelectrical nuclear forces that are able to hold the protons together in spite of the electrical repulsion. We will treat this further in Chapter 32.

Coulomb's Law The electrical force, like gravitational force, decreases inversely as the square of the distance between charges. This relationship was discovered by Charles Coulomb in the eighteenth century and is called **Coulomb's law**. It states that for charged objects that are much smaller than the distance between them, the force between two charges varies directly as the product of the charges and inversely as the square of the separation. The force is along a straight line from one charge to the other. Coulomb's law can be expressed as

$$F \sim \frac{qq'}{d^2}$$

where q represents the quantity of charge of one particle, q' the quantity of charge of the other particle, and d is the distance between them.[1] The force between like charges is repulsive, and that between unlike charges is attractive. Notice the similarity of Coulomb's law to Newton's law of gravitation ($F \sim mm'/d^2$). The obvious difference between gravitational and electrical forces is that whereas gravity only attracts, electrical forces may either attract or repel.

Questions

1. If a proton at a particular distance from a charged particle is repelled with a given force, by how much will the force decrease when the proton is three times as distant from the particle? Five times distant?*

2. What is the sign of charge of the particle in this case?†

Electrical Shielding

Another difference between electrical and gravitational forces is that electrical forces can be shielded by various materials while gravitational forces cannot. The amount of shielding is characteristic of the medium between the electrical charges. For example, air diminishes the force between a pair of charges slightly more than a vacuum, while oil placed between the charges diminishes the force by about one-eightieth. Metal will completely shield electrical forces.

Consider, for example, an electrical charge on a spherical metal shell. Because of mutual repulsion the charges will spread out uniformly over the outer surface of the shell. It is not difficult to see that the electrical force of a sample test charge placed inside the shell at the exact center would be zero because opposing forces would balance in every direction. But, interestingly enough, the forces would balance out anywhere inside the shell. The reasoning for this is similar to the reasoning we discussed in Chapter 8 when we treated the zero gravitational field inside a hollow planet (recall Figure 8.17). We won't treat the electrical case here, but maybe your instructor will in class. In any event, the conductor need not be a spherical shell for complete internal shielding; it can be any shape—even that shown in Figure 20.2.

Conductors and Insulators

Metals are good electrical shields because electrons in the metal are freely conducted under the influence of electrical forces. All substances can be arranged in order of their ability to conduct electrical charges. Those at the top of the list are called **conductors**, and those at the bottom **insulators**. But the ends of the list are very far apart. Nearly all metals are

---

*__Answer__ $\frac{1}{9}; \frac{1}{25}$

†__Answer__   Positive.

[1]The common unit of charge is called the *coulomb*. It is the charge associated with $6\frac{1}{4} \times 10^{18}$ electrons.

Fig. 20.2 The car is struck by lightning, but the man inside is shielded from the electricity. Electrons from the lightning bolt are mutually repelled to the outer metal surface of the car. They then arc to the ground near the front end.

good conductors, and most nonmetals are poor ones. The conductivity of a metal can be more than a million-trillion times greater than the conductivity of an insulator such as glass. In a power line, charge flows much more easily through hundreds of kilometers of metal wire than through the few centimeters of insulating material separating the wire from its supporting tower. In a common appliance cord, a charge flows through several meters of wire to the appliance, then through the electrical network in the appliance, and then back through the return wire rather than flowing directly across from one wire to the other through a small fraction of a centimeter of rubber insulation.

Fig. 20.3 It is easier for electricity to flow through hundreds of kilometers of metal wire than through a few centimeters of insulating material.

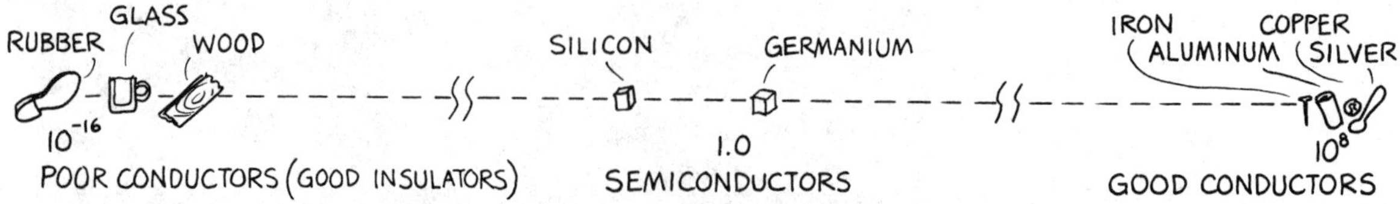

GLASS
RUBBER   WOOD   SILICON   GERMANIUM   IRON   COPPER
                                                                                    ALUMINUM SILVER

10^{-16} 1.0 10^8
POOR CONDUCTORS (GOOD INSULATORS)   SEMICONDUCTORS   GOOD CONDUCTORS

Fig. 20.4 Conductivity of some materials.

Whether a substance is classified as a conductor or an insulator depends on its interatomic bonding and on how tightly the atoms of the substance hold their electrons. Each atom in a metal contributes one or more electrons to a general "sea" of electrons attached to no particular atom. These *conduction electrons* are free to wander long distances throughout the material. On the other hand, all the electrons in an insulating material are attached to particular atoms. Small differences in the bonding between atoms account for very large differences in conductivities.

At temperatures near absolute zero, certain metals become superconducting, having near-infinite conductivity (about zero resistance) to the flow of electrical current. Once current is established in a superconductor, it will flow indefinitely. Exactly why certain materials become superconducting at low temperatures has to do with the quantum nature of matter and is still being researched.

Semiconductors

The interatomic bondings in some materials, such as germanium and silicon, are intermediate between ionic and metallic. These materials are called **semiconductors**. They are good insulators in their pure crystalline form but increase tremendously in conductivity when even one atom in ten million is replaced with an impurity that adds or removes an electron from the crystal structure. Thin layers of semiconducting materials sandwiched together compose **transistors**, which are used to control the flow of currents in circuits, to detect and amplify radio signals, and to produce oscillations in transmitters; they also act as digital switches. These tiny solids were the first electrical components in which materials with different electrical characteristics were not interconnected by wires but were physically joined in one structure. They require very little power and in normal use last indefinitely.

A semiconductor will also conduct when light of the proper color shines on it. A pure silicon plate is normally a good insulator, and any electric charge built up on its surface will remain for extended periods in the dark. If the plate is exposed to light, however, the charge would leak away almost immediately. If a charged silicon plate is exposed to a pattern of light, such as the pattern of light and dark that makes up this page, the charge would leak away only from the areas exposed to light. If a black plastic powder were brushed across its surface, the powder would stick only to the charged areas where the plate had not been exposed to light. Now if a piece of paper were put over the plate and an electric charge put

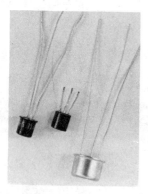

Fig. 20.5 Transistors.

on the back, the black plastic powder would be drawn to the paper to form the same pattern as, say, the one on this page. If the paper were then heated to melt the plastic and fuse it to the paper, you might pay a dime for it and call it a Xerox copy.

Charging

Charging by Friction

We are all familiar with the electrical effects produced by friction: we stroke a cat's fur and hear the crackle of sparks that are produced; or comb our hair, especially in front of a mirror in a dark room, so we can see as well as hear the sparks of electricity; or we scuff our shoes across a rug and produce a spark when we reach for a door knob; or we do the same when sliding across plastic seat covers in an automobile (Figure 20.6). In all these cases electrons are being transferred by the **friction** of contact from one material to another.

Although the innermost electrons in an atom are bound very tightly to the oppositely charged nucleus, the outermost electrons of many atoms are bound very loosely and can be easily dislodged. The force that holds the outer electrons in the atom varies for different substances. The electrons are held more firmly in rubber than in fur, for example, and when a rubber rod is rubbed by a piece of fur, electrons transfer from the fur to the rubber rod. The rubber therefore has an excess of electrons and is negatively charged. The fur, in turn, has a deficiency of electrons and is positively charged. Atoms with missing electrons are **ions**—*positive ions* because their net charge is positive. (If the atom had an extra electron or more, it would be a negative ion.) If we rub a glass or plastic rod with silk, we find that the rod becomes positively charged. Atoms on the surface of the rod have been ionized. The silk has a greater affinity for electrons than the glass or plastic rod. Electrons are rubbed off the rod and onto the silk. We can say

An object that has an imbalance of electrons and protons is electrically charged. If it has more electrons than protons, the object is negatively charged. If it has fewer electrons than protons, it is positively charged.

We see that no electrons are created or destroyed; they are simply transferred from one material to another; so we say that charge is *conserved*. In every event, whether large-scale or at the atomic and nuclear level, the principle of **conservation of charge** has been found to apply. No case of the creation or destruction of net electric charge has ever been found. The conservation of charge ranks with the conservation of energy and momentum in physics.

All the possible charges that occur in nature are a whole multiple of the charge of the electron. An electric charge that is $1\frac{1}{2}$ or $1000\frac{1}{2}$ the charge of the electron has never been found. All charges are some integral number of the charge of the electron (or proton). Charge is "grainy" or made up of

Fig. 20.6 Charging by friction and then by contact.

elementary units called **quanta**. We say that charge is quantized, with the smallest quantum of charge being that of the electron (or proton). No smaller units of charge have ever been found. (High-energy and nuclear physicists currently speculate that a smaller charge in fact exists, one that is one-third the charge of an electron. This charge is associated with a yet-to-be-discovered subatomic particle called a *quark*.)

Question   If you walk across a rug and scuff electrons from your feet, are you negatively or positively charged?*

Charging by Contact

Electrons can be transferred from one material to another by simply touching. A charged rod placed in contact with a neutral body, for example, will transfer charge to the neutral body. If the body is a good conductor, the charge will spread to all parts of its surface because the like charges repel one another. If it is a poor conductor, it may be necessary to touch the rod at several points on the body in order to get a more or less uniform distribution of the charge. This method of charging is called charging by **contact**.

Charging by Induction

We can charge a body by a simple procedure that begins with bringing a charged rod nearby. Consider the noncharged conducting sphere suspended by an insulating string in Figure 20.7. When the negatively charged rod is brought nearby, conduction electrons on the metallic surface of the sphere migrate to the far side of the sphere; as a result, the far side of the sphere becomes negatively charged and the near side positively charged. The sphere swings toward the rod because the force of attraction between the near side of the sphere and the rod is greater (closer distance, stronger force) than the repelling force betwen the far side of the sphere and the rod. We see that there is a net electrical force even though the net charge on the sphere as a whole is zero. Suppose now that we momentarily touch the far side with our finger. Then the electrons can be repelled still farther by the charged rod, some of them being repelled along the

Fig. 20.7 Charging by induction and then by contact.

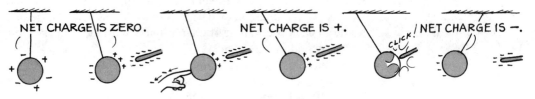

*   NET CHARGE IS ZERO.   NET CHARGE IS +.   NET CHARGE IS −.

---

*__Answer__   You are positively charged (and the rug is negatively charged).

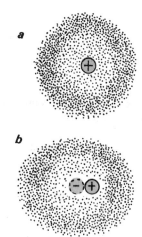

Fig. 20.8 (*a*) The center of the negative electron cloud coincides with the center of the positive nucleus in an atom. (*b*) When an external negative charge is brought nearby to the right, the electron cloud is distorted so the centers of negative and positive charge no longer coincide. The atom is electrically polarized.

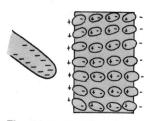

Fig. 20.9 Charge polarization results in surface charges of equal magnitude and opposite sign induced on opposite surfaces of the material.

surface of our moist skin and then conducted to the ground we stand on. The sphere has been "grounded" and now possesses a net positive charge; it has been charged by **induction**. We see that the sphere swings even closer to the rod (why?), and if it touches the rod, there will be a charge by contact as electrons transfer from the rod to the sphere. Then the sphere quickly becomes negatively charged and is repelled by the rod.

Charging by induction is not restricted to conductors but may occur for all materials. In an insulator, there is a reorientation of the charge distribution in the atoms and molecules that make up the material instead of a migration of conduction electrons. Ordinarily, the effective centers of positive charge and negative charge coincide in an atom and in many molecules, but in the presence of a charged body these centers become slightly displaced (Figure 20.8). We say the atoms are **electrically polarized**. The "head-to-tail alignment" of polarized atoms or molecules in an insulating material results in an induced surface charge of equal magnitude and opposite sign on opposite surfaces of the material (Figure 20.9). (Perhaps your instructor will explain how this induced surface charge causes the partial shielding of insulators we discussed earlier.) And we find that there is a slight attraction between the insulator and the charged body, because the induced surface charge on the part of the insulator closest to the charged body is always opposite in sign. This explains why a charged body such as a comb that has been rubbed through hair will attract uncharged bits of paper and other objects. Sometimes bits of paper cling to the charged comb and suddenly fly off. This is because the paper touching the comb acquires a charge by contact, where electrons on the comb repel one another and transfer onto the paper bits. The paper then has the same sign charge as the comb and is repelled.

When we rub an inflated balloon against our hair, we charge the balloon by friction. We rub electrons from our hair onto the balloon, which is then negatively charged. We find that the charged balloon easily sticks to the wall or any insulating surface. This is because the charged balloon induces an opposite surface charge on the surface of the wall. Since the balloon is a good insulator, very little charge transfers from the balloon to the wall.

Charging by induction occurs during thunderstorms. The bottoms of negatively charged clouds induce a positive charge on the surface of the earth. Benjamin Franklin was the first to demonstrate this when he proved during his famous kite-flying experiment that lightning is an electrical phenomenon.[2] Lightning is an electrical discharge between the clouds and oppositely charged ground or between oppositely charged parts of clouds.

---

[2]Franklin was most fortunate that he wasn't electrocuted as were others who attempted to duplicate his experiment. In addition to being a great statesman, Benjamin Franklin was a first-rate scientist. He introduced the terms *positive* and *negative* as they relate to electricity and established the "one-fluid theory" of electricity. He contributed to our understanding of grounding and insulation.

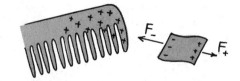

Fig. 20.10 A charged comb attracts an uncharged piece of paper because unbalanced forces act on the induced surface charges.

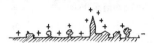

Fig. 20.11 The bottom of the negatively charged cloud induces a positive charge at the surface of the ground below.

Franklin also found that charge leaks off sharp points, and fashioned the first lightning rod. If the rod is placed above a structure connected to the ground, charge induced by the clouds leaks off the rod and prevents a sudden large electrical discharge from occurring. If for any reason sufficient charge does not leak off the rod and lightning strikes anyway, it will be attracted to the rod and directed through the metal path to the ground, thereby sparing the structure.

Electric Field

Fig. 20.12 A gravitational force holds the satellite in orbit about the planet (*a*), and an electric force holds the electron in orbit about the proton (*b*). In both cases there is no contact between the bodies, and the forces are "acting at a distance." We can also say that the orbiting bodies interact with the *force fields* of the planet and proton and are everywhere in contact with these fields. Thus, the force that one electric charge exerts on another can be described as the interaction between one charge and the field set up by the other.

Electrical forces, like gravitational forces, act between bodies that are not in contact with each other as well as between those that are. A conceptual way to describe such forces involves the concept of a *force field*. The properties of space surrounding any mass can be considered to be so altered that another mass introduced to this region will experience a force. The "alteration in space" caused by a mass is called its *gravitational field*. We can think of any other mass as interacting with the field and not directly with the mass producing it. For example, when an apple falls from a tree, we say it is interacting with the mass of the earth, but we can also think of the apple as interacting with the gravitational field of the earth. The field plays an intermediate role in the force between bodies. It is common to think of distant rockets and the like as interacting with gravitational fields rather than with the masses of the earth and other bodies responsible for the fields. Similarly, an electric charge produces an **electric field** around it that interacts with any other charges present.

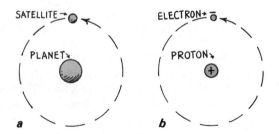

An electric field has both magnitude and direction. The magnitude of the field at any point is simply the force per unit of charge. If a charge q

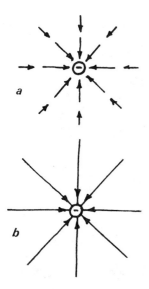

Fig. 20.13 Electric field representations about a negative electric charge. (*a*) A vector representation. (*b*) A lines-of-force representation.

Fig. 20.14 Some electric field configurations. (*a*) Lines of force around a single positive charge. (*b*) Lines of force for a pair of equal but opposite charges. Note that the lines emanate from the positive charge and terminate on the negative charge. (*c*) Uniform lines of force between two oppositely charged parallel plates.

experiences a force F at some point in space, then the electric field at that time is

$$E = \frac{F}{q}$$

The electric field can be depicted with vector arrows as in Figure 20.13*a*. The direction of the field is shown by the vectors and is defined to be the direction in which a small positive test charge at rest would be moved.[3] In the figure we see that all the vectors therefore point to the center of the negatively charged ball. If the ball were positively charged, the vectors would point away from the center because a positive test charge in the vicinity would be repelled.

A more useful way to describe an electric field is with electric lines of force (Figure 20.13*b*). The lines of force shown in the figure represent a small number of the infinitely numerous possible lines that indicate the direction of the field. Where the lines are farther apart, the field is weaker. For an isolated charge, the lines extend to infinity; while for two or more opposite charges, we represent the lines as emanating from a positive charge and terminating on a negative charge. Some electric field configurations are shown in Figure 20.14, and photographs of field patterns are shown in Figure 20.15. The photographs show bits of thread that are suspended in an oil bath surrounding charged conductors. The ends of the bits of thread are charged by induction and tend to line up end to end with the field lines, like iron filings in a magnetic field.

If our only concern regarding electricity were the forces between isolated stationary charges, the electric field concept might be unnecessary. But charges move. This motion is communicated to neighboring charges in the form of a field disturbance that emanates at the speed of light from the accelerating charge. We will learn that the electric field is a storehouse of energy and that energy can be transported over long distances in an electric field, which may be directed through and guided by metal wires or through empty space. We will return to this idea in the next chapter and later when we learn about electromagnetic radiation.

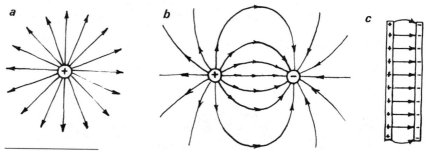

[3]The test charge is *small* so that its electric field is small compared to the electric field already there. If the test charge were large, it might alter the charge distribution of the source of the field we are measuring.

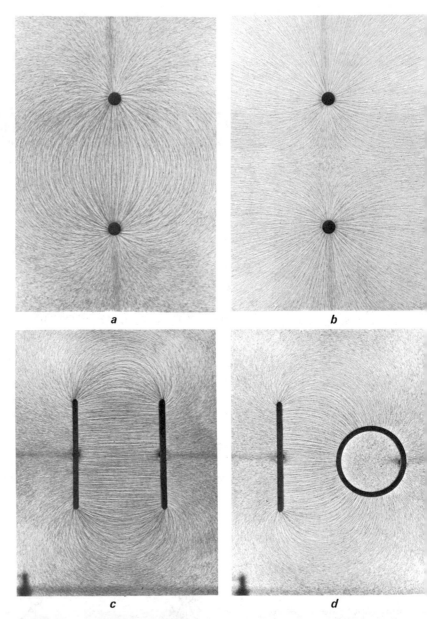

Fig. 20.15 The photographs show bits of fine thread suspended in oil. Charged terminals placed in the oil bath induce opposite charges on the two ends of each bit of thread, which then tend to line up end to end along the direction of the field. (*a*) Equal and opposite charges. (*b*) Equal like charges. (*c*) Oppositely charged plates. Notice that the field is uniform between the plates. (*d*) Oppositely charged cylinder and plate. Notice that the field is shielded inside the cylinder, as indicated by the lack of alignment of the threads.

Electric Potential

When we studied energy in Chapter 6, we learned that an object may have gravitational potential energy because of its location in a gravitational field. Similarly, a charged object can have potential energy by virtue of its location in an electric field. Just as work is required to lift a massive object against the gravitational field of the earth, work is required to push a charged particle against the electric field of a charged body. This work increases the electric potential energy of the charged particle. Consider the small positive charge located at some distance from a positively charged sphere in Figure 20.16a. If you push the small charge closer to the sphere (Figure 20.16b), you will expend energy to overcome electrical repulsion; that is, you will do work in pushing the charge against the electric field of the sphere. This work done in moving the small charge to its new field location is gained by the small charge. We call the energy the charge now possesses by virtue of its location **electric potential energy**. If the charge is released, it accelerates in a direction away from the sphere, and its electric potential energy changes to kinetic energy.

If we push two charges instead, we do twice as much work, and the two charges in the same location have twice the electric potential energy as before; three charges have three times as much potential energy; ten charges ten times the potential energy; and so on. Rather than dealing with the total potential energy of a charged body, it is convenient when working with electricity to consider the electric *potential energy per charge*. We simply divide the amount of energy in any case by the amount of charge. For example, ten charges at a specific location will have ten times as much energy as one, but will also have ten times as much charge; so we see that the energy per charge remains the same at a particular field location whatever the amount of charge. The concept of potential energy per charge is called **electric potential**; that is,

$$\text{Electric potential} = \frac{\text{energy}}{\text{charge}}$$

The unit of measurement for electric potential is the *volt*; hence, electric potential is often called **voltage**.[4] Both terms are common, so we will use *electric potential* and *voltage* interchangeably.

The significance of voltage is that a definite value for it can be assigned to a location whether or not a charge exists at that location. We can speak about the voltages in empty space away from a charged body or the voltages at various locations in an electric circuit. In the next chapter, we'll see that the location of the positive terminal of a 12-volt battery is main-

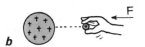

Fig. 20.16 The small charge has more potential energy in (b) than in (a) because work was required to move it to the closer location.

---

[4]For your information, 1 volt equals 1 joule of energy per coulomb of charge. It is common to assign a zero electric potential to places infinitely far away from any charges. Then for a body charged to, say, 10,000 V, it would take 10,000 J of work to assemble the charges by pushing against their mutual repulsions from zero starting points. We will see in the next chapter that for electric circuits the value zero is assigned to the ground.

tained at a voltage 12 volts higher than the location of the negative termi-nal. When a conducting medium connects this voltage difference, charges in the medium will move between these voltage-differing locations.

Question   What does it mean to say that an automobile battery is 12 volts?*

Fig. 20.17 Both the physics enthusiast and the dome of the Van de Graaff generator are charged. Why does her hair stand out? (Courtesy Saunders Publishing Co.)

Fig. 20.18 A simple model of the Van de Graaff generator.

Van de Graaff Generators

A common laboratory device for building up high voltages is the Van de Graaff generator. This is the lightning machine that evil scientists used in old science fiction movies. A simple model of the Van de Graaff gener-ator is shown in Figure 20.18. A large, hollow metal sphere is supported by a cylindrical insulating stand. A motor-driven silk or rubber belt inside the support stand carries charge gathered by friction of the belt rubbing against a glass cylinder or from electrified metal points below to the conducting sphere above. Electrons leak off the belt onto sharp metal points (tiny lightning rods) and are deposited on the inside of the sphere. The electrons repel each other to the outer surface of the conducting sphere. Static charge always lies on the outside surface of any conductor.

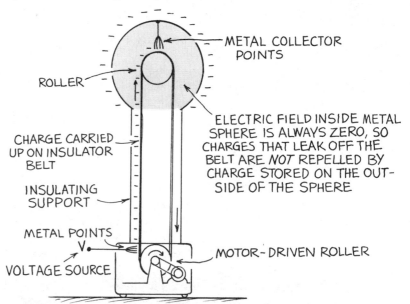

METAL COLLECTOR POINTS

ROLLER

ELECTRIC FIELD INSIDE METAL SPHERE IS ALWAYS ZERO, SO CHARGES THAT LEAK OFF THE BELT ARE *NOT* REPELLED BY CHARGE STORED ON THE OUT-SIDE OF THE SPHERE

CHARGE CARRIED UP ON INSULATOR BELT

INSULATING SUPPORT

METAL POINTS

V

VOLTAGE SOURCE

MOTOR-DRIVEN ROLLER

---

*__Answer__ It means that one of the battery terminals is 12 V higher in potential than the other battery terminal. In the next chapter, we will see that it also means that when a circuit is connected across these terminals, each coulomb of charge that makes up the resulting current will be energized to 12 J of energy.

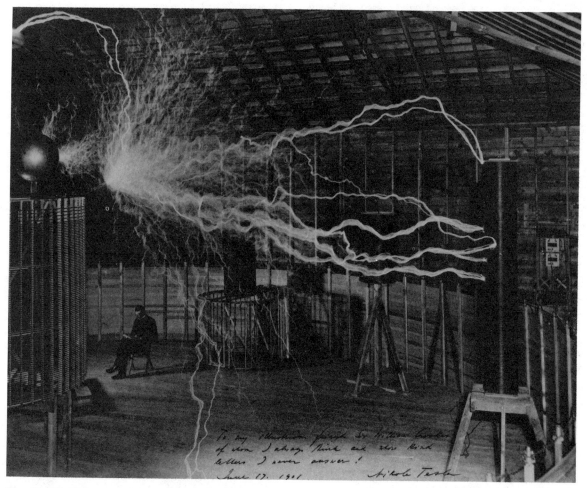

Fig. 20.19 Before electricity became commonplace, it was regarded by most people as a terrifying phenomenon. To help dispel this fear, Nikola Tesla, who was the chief proponent of alternating current when Thomas Edison was promoting direct current, sits indifferently in the midst of a high-voltage sparking demonstration.

This leaves the inside uncharged and able to receive more electrons as they are brought up the belt and, in turn, are continuously transferred to the outside surface of the sphere. In this way a giant charge builds up, establishing a high electic potential—on the order of millions of volts.

The huge voltages generated by practical Van de Graaff generators are useful in producing X rays for medical research. They are also useful in accelerating charged particles used as projectiles for penetrating the nuclei of atoms. Touching one can be a hair-raising experience.

Summary of Terms

Electrostatics   The study of electric charge at relative rest, as opposed to electric currents.

Coulomb's law   The relationship among electric force, charge, and distance:

$$F \sim \frac{qq'}{d^2}$$

If the charges are alike in sign, the force is repelling; if the charges are unlike, the force is attractive.

Conductor   Any material through which charge easily flows when subject to an impressed electrical force.

Insulator   Any material through which charge resists flow when subject to an impressed electrical force.

Semiconductor   A normally insulating material such as crystalline silicon or germanium that becomes conducting when made with certain impurities or when energy is added.

Charging by friction   The transfer of charge between bodies by virtue of rubbing.

Charging by contact   The transfer of charge from one body to another by physical contact between the bodies.

Charging by induction   The redistribution of charge in an object, caused by the electrical influence of a charged body close by, but not in contact.

Electric field   The energetic region of space surrounding a charged body; about a charged point, the field decreases with distance according to the inverse-square law, like the gravitational field. Between oppositely charged parallel plates, the electric field is uniform. A charged object placed in the region of an electric field experiences a force.

Electric potential   The electric potential energy per quantity of charge, measured in volts, and often called *voltage:*

$$\text{Voltage} = \frac{\text{electrical energy}}{\text{charge}}$$

Review Questions

1. The strength of the gravitational force between a pair of electrons is incredibly tiny compared to the electric force, yet gravitation is the predominant force between celestial bodies. Why is this so?

2. What is the fundamental force that underlies all "chemical" forces?

3. How is Coulomb's law similar to Newton's law of gravitation? How is it different?

4. By how much does the electric force between a pair of charged bodies diminish when their separation is doubled? Tripled?

5. Cite three major differences between gravitational and electric forces.

6. Distinguish between a *conductor* and an *insulator* and between a *conductor* and a *semiconductor.*

7. Distinguish between a *positively* and *negatively* charged object.

8. If you rub a glass rod with silk and make the glass positive, what is the sign of charge on the silk?

9. What does it mean to say that in all interactions electrical charge is conserved?

10. Distinguish between *charging by contact* and *charging by induction.*

11. What is charge polarization?

12. Explain how an electrically neutral object can be attracted to a charged body.

13. What is an electric field? How is it similar to a gravitational field?

14. What is the difference between *electric potential* and *electric potential energy?*

15. What is a volt?

Home Projects

1. Demonstrate charging by friction and discharging from points with a friend who stands at the far end of a carpeted room. With leather shoes, scuff your way across the rug until your noses are close together. This can be a delightfully tingling experience, depending on how dry the air is and how pointed your noses are.

2. Bring a comb that has been briskly rubbed on your hair or a woolen garment near a small but smooth stream of running water. Is the stream of water charged?

Exercises

1. We do not feel the gravitational forces between ourselves and the objects around us because these forces are extremely small. Electrical forces, in comparison, are extremely huge. Since we and the objects around us are composed of charged particles, why don't we feel electrical forces?

2. Why do clothes often cling together after tumbling in a clothes dryer?

3. When combing your hair, you scuff electrons from your hair onto the comb. Is your hair then positively or negatively charged? How about the comb?

4. At automobile toll-collecting stations a thin metal wire sticks up from the road and makes contact with cars before they reach the toll collector. What is the purpose of this wire?

5. Why are the tires for trucks carrying gasoline and other flammable fluids manufactured to conduct electricity?

6. An electroscope is a simple device consisting of a metal ball that is attached by a conductor to two fine gold leaves that are protected from air disturbances in a jar as shown. When the ball is touched by a charged body, the leaves that normally hang straight down suddenly spring apart. Why? (Electroscopes are useful not only as charge detectors, but also for measuring the quantity of charge: the more charge transferred to the ball, the more the leaves diverge.)

7. The leaves of a charged electroscope collapse in time. At higher altitudes they collapse more readily. Why is this true? (*Hint:* The existence of cosmic rays was first indicated by this observation.)

8. Would it be necessary for a charged body to actually touch the ball of the electroscope for the leaves to diverge? Defend your answer.

9. Strictly speaking, when an object acquires a positive charge, what happens to its mass? When it acquires a negative charge? (Think small!)

10. How can you charge an object negatively with only the help of a positively charged object?

11. It is relatively easy to strip the outer electrons from a heavy atomic nucleus like that of uranium (which is then a uranium ion) but very difficult to strip the inner electrons. Why do you suppose this is so?

12. If a large enough electric field is applied, even an insulator will conduct an electric current, as is evident in lightning discharges through the air. Explain how this happens, taking into account the opposite charges in an atom and how ionization occurs.

13. If you are caught outdoors in a thunderstorm, why should you not stand under a tree? Can you think of a reason why you should not stand with your legs far apart? Or why lying down can be dangerous? (*Hint:* Consider electric potential *difference*.)

14. Which of the two would be safer—a house with no lightning rod or a house with a lightning rod *not* connected to the ground? Explain.

15. Why is a good conductor of electricity also a good conductor of heat?

16. If you rub an inflated balloon against your hair and place it against a door, by what mechanism does it stick? Explain.

17. How are electrically neutral atoms and molecules able to electrically attract each other?

18. An H_2O molecule is called a "polar" molecule. How does this relate to the fact that charging demonstrations are more difficult to do on a humid day? (*Hint:* How does the presence of polar molecules in the air affect its conductivity?)

19. A gravitational field vector points toward the earth; an electric field vector points toward an electron. Why do electric field vectors point *away* from protons?

20. By what specific means do the bits of fine threads line up in the electric fields shown in Figure 20.15?

21. Suppose a "free" electron and a "free" proton were placed in an electric field. Compare their accelerations and their directions of travel.

22. If you put in 10 joules of work to push a unit charge into an electric field, what will be its voltage with respect to its starting position? When released, what will be its kinetic energy as it flies past its starting position?

23. Can we say that a body with twice the electric potential energy of another has twice the electric potential? Why or why not? (*Hint:* Consider the analogy with thermal energy and temperature.)

24. Can we say that a body with twice the electric potential of another has twice the electric potential energy? Why or why not? (*Hint:* Consider the analogy with temperature and thermal energy.)

25. What safety is offered by staying inside an automobile during a thunderstorm? Defend your answer.

26. Would you feel any electrical effects if you were inside the charged sphere of a Van de Graaff generator? Why or why not?

21 Electric Current

In the last chapter we were introduced to the concept of electric potential, which we measure in volts. In this chapter we will see that this voltage is an "electical pressure" that can produce a flow of charge, *current,* which we measure in amperes; and that the *resistance* that restrains this flow is measured in ohms. We will see that when the flow is unidirectional, we call it *direct current,* dc, and when it flows to and fro we call it *alternating current,* ac. The rate at which energy is transferred by electric current is *power,* which is measured in watts or in thousands of watts, kilowatts. We see here that there are many terms to be sorted out. This is easier to do when we have some understanding of the ideas these terms represent, and these ideas are better understood if we know how they relate to one another. We begin with the flow of electric charge.

Flow of Charge

When the ends of a conductor are placed between two bodies of different temperatures, thermal energy flows from the higher temperature to the lower temperature. Thermal energy flows through the conductor when a difference in temperature exists across its ends. When both ends reach the same temperature, the flow of energy ceases. In a similar way, when an electrical conductor joins two bodies of different electric potential, charge flows from the higher potential to the lower potential. Charge flows when there is a **potential difference** across the ends of a conductor. This flow of charge persists until both bodies reach a common potential. When there is no potential difference, no flow of charge will occur. For example, if one end of a wire were connected to the ground and the other end placed in contact with the sphere of a Van de Graaff generator charged to a high potential, a surge of charge would flow through the wire. The flow of charge would be brief, however, for the sphere would quickly reach a common potential with the ground.

To attain a sustained flow of charge in a conductor, some arrangement must be provided to maintain a difference in potential while charge flows from one end to the other. The situation is analogous to the flow of water from a higher reservoir to a lower one (Figure 21.1*a*). Water will flow in a pipe connecting the reservoirs only as long as a difference in water level exists. The flow of water in the pipe, like a flow of charge in the wire connecting the Van de Graaff generator and the ground, will cease when the pressures at each end are equal (we imply this when we say that water seeks its own level). A continuous flow is possible if the difference in water levels, hence, water pressures, is maintained with the use of a suitable pump (Figure 21.1*b*).

Fig. 21.1 Water flows from the reservoir of higher pressure to the reservoir of lower pressure. (*a*) The flow will cease when the difference in pressure ceases. (*b*) Water continues to flow because a difference in pressure is maintained with the pump.

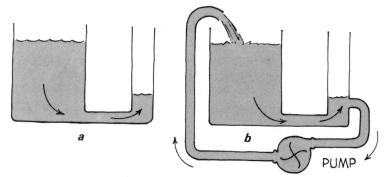

PUMP

Electromotive Force and Current

There are many kinds of electric "pumps," often called "seats" of **electromotive force (emf)**, but more commonly, *voltage sources*. If we charge a rubber rod by rubbing it with fur, we can develop a large voltage between the rod and the fur. Such a combination does not make a good electrical pump because when connected by a conductor, the potentials equalize in a single brief surge of moving charges. Such a combination cannot maintain a sufficient flow of charge to be of practical importance. Chemical batteries or generators, on the other hand, are capable of maintaining a steady flow. Any seat of emf such as a battery or generator is a source of energy. In chemical batteries, the disintegration of zinc or lead in acid results in a separation of electric charge, and the energy stored in the chemical bonds is converted into electrical energy.[1] Both batteries and generators separate charges and move them to opposing terminals—negative on one and positive on the other. The work done in separating the opposite charges is in effect stored at the terminals and constitutes the difference in potential (voltage) that provides the "electrical pressure" that will move electrons through a circuit joined to the terminals. Electromotive force, like potential difference, is measured in volts.

Fig. 21.2 A very simple circuit.

Electric current is simply the flow of electric charge. In solid conductors, it is the electrons that flow through the circuit; while in fluids, ions as well as electrons may compose the flow of electric charge. The *rate* of electrical flow is measured in **amperes**, or, simply, amps. An ampere is the rate of flow of 1 unit of charge per second.[2] In a wire carrying 5 amperes, for example, 5 units of charge pass any cross section in the wire each second. In a wire carrying 10 amperes, twice as many charges pass any cross section each second.

Note that a current-carrying wire is not electrically charged. Negative conduction electrons swarm through an atomic lattice made up of an *equal number* of positively charged atomic nuclei. So the net charge of the wire is normally zero at every moment.

Fig. 21.3 An unusual source of emf. The electric potential energy per charge between the head and tail of the electric eel (*Electrophorus electricus*) can be up to 600 volts.

---

[1] You can find how this is done in almost any chemistry text.

[2] The electric charge of $6\frac{1}{4}$ million-million-million electrons is the standard unit of charge, called the *coulomb*: 1 coulomb of charge passing a point in 1 second is called an *ampere*.

Electrical Resistance

A battery or generator of some kind is the prime mover and source of voltage in an electrical circuit. How much current flows depends not only on the potential difference but also on the resistance that the medium offers to the flow of charge. This is similar to the rate of water flow in a pipe, which depends not only on the pressure behind the water but on the resistance offered by the pipe itself. We have already discussed the fact that different substances conduct electricity differently. It is also a fact that the electrical resistance of a wire depends on its thickness and length. Electrical resistance is less in thick wires. The longer the wire, of course, the greater the resistance. In addition, electrical resistance depends on temperature. The greater the jostling about of atoms within the conductor, the greater resistance the conductor offers to a flow of charge. For most conductors, increased temperature is accompanied by increased resistance.[3] The resistance of some metals approaches zero at very low temperatures. These are the superconductors we discussed briefly in the last chapter.

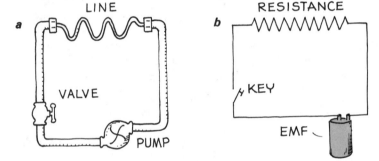

Fig. 21.4 Analogy between a simple hydraulic circuit (a) and an electric circuit (b).

Electrical resistance is measured in units called **ohms**, after Georg Simon Ohm, a German physicist who in 1826 discovered a simple and very important relationship among voltage, current, and resistance.

Ohm's Law

The relationship among voltage, current, and resistance is summarized by a statement called **Ohm's law**. Ohm discovered that the amount of current in a circuit is directly proportional to the voltage (emf) impressed across the circuit and is inversely proportional to the resistance of the circuit. In short,

$$\text{Current} = \frac{\text{voltage}}{\text{resistance}}$$

Or, in units form,

$$\text{Amperes} = \frac{\text{volts}}{\text{ohms}}$$

---

[3]Carbon is an interesting exception. At high temperatures, electrons are shaken from the atom, which increases electric current. Its resistance, in effect, lowers with increasing temperature. This is an important reason for its use in arc lamps.

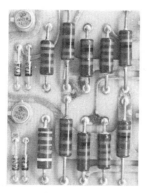

Fig. 21.5 Resistors: the symbol of resistance in an electric circuit is ⌐⌐⌐⌐⌐ .

So, for a given circuit of constant resistance, voltage and current are proportional. For example, if we double the voltage across the ends of a circuit, the current will double. But if we instead double the resistance of a circuit, the current will be reduced by half. The greater the voltage, the more current; the greater the resistance, the less current. We can see that if a potential difference of 1 volt is impressed across a circuit having a resistance of 1 ohm, a current of 1 ampere will be produced. The resistance in a typical lamp cord is much less than 1 ohm, while a typical light bulb has a resistance of about 100 ohms. From Ohm's law, we see that 100 volts impressed upon such a bulb draws about 1 ampere of current. An iron or electric toaster has a resistance of 15 to 20 ohms. The low resistance permits a large current, which produces considerable heat. Inside radio and television receivers, current is regulated by circuit elements called *resistors*, whose resistance may be a few ohms or millions of ohms.

All simple electrical calculations are based on Ohm's law, which holds for metallic conductors over a wide range of temperatures. If two of the three values are known, the third can be calculated. Many conductors such as vacuum tubes and transistors, however, do not obey Ohm's law, and the resistance of these devices changes when different voltages are applied.

Question   What is the resistance of an electric frying pan that draws 11 amperes when connected to a 110-volt circuit?*

Electric Shock and Ohm's Law

What causes electric shock—current or voltage? The damaging effects of shock are the result of current passing through the body. From Ohm's law, we know that this current depends on the voltage applied as well as on the electrical resistance of the human body. The resistance of the body depends on the body's condition and can range from 100 ohms to 500,000 ohms according to circumstances. The resistance of a person touching two electrodes with dry fingers is about 100,000 ohms; with both hands immersed in salt water, the resistance is about 700 ohms. Salt water is rich in ions and therefore is a very good conductor, while ordinary water also contains some ions and is a good conductor. Therefore, the very moist internal flesh conducts electricity well, while the dry outer layers of skin do not. For ordinary voltages, the resistance of the skin is usually enough to limit the current to a safe intensity. With normally dry skin, for example, you cannot even feel 12 volts, and voltage of twice that just barely tingles.

*__Answer__   10 ohms; we rearrange Ohm's law to read

$$\text{Resistance} = \text{voltage/current} = 110 \text{ V}/11 \text{ A} = 10 \ \Omega$$

The Greek letter omega (Ω) is usually used as a symbol for resistance.

Fig. 21.6 The bird can stand harmlessly on one wire of high potential, but it had better not reach over and grab a neighboring wire! Why not?

Fig. 21.7 The third prong connects the body of the appliance directly to ground. Besides grounding possible contact between the live wire and the body of the appliance, it conducts any charge that otherwise builds up to the ground. This prevents shock to the person who handles the appliance.

But if your skin is moist, even such a low voltage can produce an uncomfortable shock.

Many people are killed each year by current from common 110-volt electric circuits. If you touch a faulty 110-volt light fixture with your hand while your feet are on the ground, there is a 100-volt "electric pressure" between your hand and the ground. Resistance to current flow is usually greatest between your feet and the ground, so the current usually isn't great enough to do serious harm. But if your feet and the ground are wet, there is a low-resistance electrical bond between you and the ground. Your overall resistance is so lowered that the 110-volt potential difference between your hand and your feet may produce a current greater than your body can withstand. Currents as low as 0.05 ampere can be fatal, while a current of 0.10 ampere in the heart is nearly always fatal. People are often electrocuted while handling electrical devices while taking a bath or shower or while standing on a wet floor. Although distilled water is a good insulator, the ions in ordinary water greatly reduce the electrical resistance. And if you step into a pan of distilled water with bare feet, material from your feet will dissolve into the water in the form of ions. Distilled water is easily contaminated. Especially with dirty feet.

To receive a shock, there must be a *difference* in electrical potential between one part of your body and another part. Current will pass along a path in your body connecting these two points. Suppose you fell from a bridge and managed to grab onto a high-voltage power line, halting your fall. So long as you touch nothing else of different electric potential, you will receive no shock at all. Even if the wire is a few thousand volts above ground potential and even if you hang by it with two hands, no current will flow from one hand to the other. This is because there is no difference in potential between your hands. If, however, you reach over with one hand and grab onto a wire of different potential . . . *zap!* We have all seen birds perched on high-voltage wires. Every part of their bodies is at the same high potential as the wire, and they feel no ill effects.

Most electric sockets today are wired with three, instead of two, connections. The principal two are for the current-carrying double wire, one part of which is live and the other neutral, while the third contact in the socket is connected directly to earth. Appliances such as irons, stoves, and refrigerators are connected with these three wires. If the live wire accidentally comes in contact with the metal part of the appliance, the current will flow directly to ground rather than causing a shock to anyone handling it.

Electric shock, in addition to upsetting nerves in general, upsets the nerve center that controls breathing. In rescuing victims, the first thing to do is to clear them from the electric supply with a wooden stick or some other nonconducting material so that you don't get electrocuted yourself, and then apply artificial respiration.

Question   What causes electric shock—current or voltage?*

Direct Current and Alternating Current

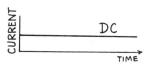

Fig. 21.8 A time graph of dc and ac.

Electric current may be dc or ac. By dc, we mean **direct current**, which refers to the flowing of charges in *one direction*. A battery produces direct current in a circuit because the terminals of the battery always have the same sign of charge. Electrons move from the repelling negative terminal and toward the attracting positive terminal, always moving through the circuit in the same direction. Even if the current moves in unsteady pulses, so long as it moves in one direction only, it is dc.

Alternating current (ac) acts as the name implies. Electrons in the circuit are moved first in one direction and then in the opposite direction, alternating back and forth about relatively fixed positions. This is accomplished by alternating the direction of voltage at the energy source. The overwhelming majority of commercial ac circuits involve voltages and currents that alternate back and forth at a frequency of 60 cycles per second. This is 60-hertz current. In some places, 25-hertz, 30-hertz, or 50-hertz current is used. The popularity of ac arises from the fact that electric energy in the form of ac can be transmitted great distances with easy voltage step-ups that result in lower heat losses in the wires. Why this is so will be discussed in Chapter 23. The primary use of electric current, whether dc or ac, is to transfer energy quietly, flexibly, and conveniently from one place to another.

Speed and Source of Electrons in a Circuit

When you flip on the light switch on your wall and the circuit is completed, the light bulb appears to glow immediately. When you make a long-distance telephone call, the electrical signal carrying your voice travels through the connecting wires at seemingly infinite speed. It is actually electric energy that is transmitted through conductors at a speed approaching the speed of light.

It is *not* the electrons, however, that move at the speed of light in a circuit.[4] Although electrons inside metal at room temperature have an average speed of a few million kilometers per hour, they make up no current because they are moving in all possible directions. There is no net and grab onto a wire of different potential . . . *zap!* We have all seen birds

---

***Answer**   Electric shock *occurs* when current flows in the body, which is *caused* by an impressed voltage.

[4]Much effort and expense are expended in building cyclotrons and the like to accelerate electrons to speeds near that of light. If electrons in a common circuit traveled that fast, one would only have to bend a wire at a sharp angle and electrons traveling through the wire would possess so much momentum that they would fail to make the turn and fly off, providing a beam comparable to that produced by accelerators!

Fig. 21.9 The electric field lines between the terminals of a battery are directed through a conductor, which joins the terminals. A metal bar is shown here, but the conductor is usually an electric circuit. (If you do this, you won't be shocked, but the bar will heat quickly and may burn your hand!)

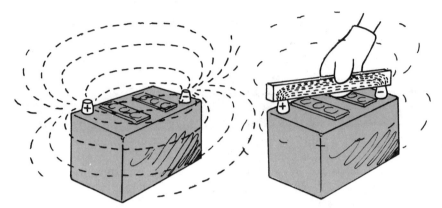

flow in any preferred direction. But when a battery or generator is connected, an electric field is established inside the conductor; the electrons continue their random motions while simultaneously being nudged through the conductor by this field. It is the **electric field** that travels through a circuit at about the speed of light. The conducting wire acts as a guide to the electric field lines established at the voltage source (Figure 21.9). If the source is dc, the electric field lines are maintained in one direction in the wires, and free electrons encountering the field are accelerated along the direction of the field lines. Before they can gain appreciable speed, they encounter the anchored metallic ions in their paths and lose some of their kinetic energy to them, causing the wires to increase in temperature. These collisions continually interrupt the motion of the electrons so that the actual drift velocity of the electron current is extremely slow. In a typical dc circuit—the electric system of an automobile, for example—electrons have a net average drift velocity of about a hundredth-centimeter per second. At this rate, it would take years for an electron to travel across a transcontinental circuit. Long-distance telephone calls would be most arduous.

Fig. 21.10 The solid lines depict a possible random path of an electron bouncing about in an atomic lattice at a speed of about $\frac{1}{200}$ the speed of light. The dashed lines show an exaggerated view of how this path is altered when an electric field is applied. The electron drifts toward the right with a speed much less than a snail's pace.

A common misconception regarding electrical currents is that the current is propagated through the conducting wires by electrons bumping into one another—that an electrical pulse is transmitted in a manner similar to the way the pulse of a tipped domino is transferred along a row of standing, closely spaced dominoes. This simply isn't true! The domino idea is a good model for the way sound is transmitted, but not for the transmission of electrical energy. Electrons that are free to move in a conductor are accelerated by the electric field impressed on them, but not because they bump into each other. True, they do "bump into" each other and other atoms, but this slows them down and offers resistance to their motion. Electrons throughout the entire closed path of a circuit all move simultaneously. In an ac circuit, the conducting electrons don't go anywhere. They oscillate rhythmically back and forth about relatively fixed positions.

Fig. 21.11 The conduction electrons that surge to and fro in the filament of the lamp do not come from the voltage source. They are in the filament. The voltage source simply provides them with surges of energy.

When you talk to someone by a transcontinental telephone call, it is the *pattern* of oscillating motion that is carried across the country at nearly the speed of light. The electrons already along the wires vibrate to the rhythm of the traveling pattern.

This brings up another misconception regarding electricity. Many people think that the electrical outlets in the walls of their homes are the source of electrons—that electrons flow from the power company through the power lines and into the home via these outlets. The fact is that no electrons flow from the power lines or the wall sockets into electrical appliances plugged into these outlets. The outlets in your home are ac. When you plug a lamp into an outlet, *energy* flows from the outlet into the lamp, not electrons. This energy is carried by the electric field and causes vibratory motion of the electrons that already exist in the lamp filament. If 110 volts are impressed on the lamp, then 110 units of energy are given to each unit of charge in the current. Most of this electrical energy is transformed into heat, while some of it takes the form of light.

When you are jolted by an electrical shock, the electrons making up the current in your body originate in your body. Electrons do not come out of the wire and through your body and into the ground. Energy does. The energy simply causes free electrons in your body to vibrate in unison, which can be fatal.

Electric Power

Fig. 21.12 The voltage and power on the hair dryer read "120 volts 1200 watts." How many amperes will flow through it?

When a charge moves in a circuit, it does work. Usually, this results in heating the circuit or in turning a motor. The rate at which work is done—that is, the rate that electric energy is converted into another form like mechanical, heat, or light—is called **power**. Electric power is equal to the product of current and voltage across a circuit:[5]

$$\text{Power} = \text{current} \times \text{voltage}$$

If the voltage is measured in volts and the current in amperes, then the power is expressed in watts. So, in units form,

$$\text{Watts} = \text{amperes} \times \text{volts}$$

If a lamp were rated at 120 watts and operated on a 120-volt line, we see that it would draw a current of 1 ampere (120 watts = 1 ampere × 120 volts). A 60-watt lamp draws $\frac{1}{2}$ ampere on a 120-volt line. This relationship becomes a practical matter when we wish to know the cost of electrical energy, which varies from 1 cent to 10 cents per kilowatt-hour depending

---

[5]Note that the units check:

$$\text{Power} = \frac{\text{charge}}{\text{time}} \times \frac{\text{energy}}{\text{charge}} = \frac{\text{energy}}{\text{time}}$$

on the locality and other circumstances. A *kilowatt* is 1000 watts, and a *kilowatt-hour* represents the amount of energy consumed in 1 hour at the rate of 1 kilowatt.[6] Thus, at a location where electric energy costs 5 cents per kilowatt-hour, a 100-watt electric light bulb can be run for 10 hours at a cost of 5 cents, or a half cent for each hour. A toaster or iron, which draws more current and therefore more energy, costs several times as much to operate.

Questions

1. If a 120-volt line to a socket is limited to 15 amperes by a safety fuse, will it operate a 1200-watt hair dryer?*

2. At 10¢ per kilowatt-hour, what does it cost to operate the 1200-watt hair dryer for 1 hour?†

Electrical Circuits

Any path along which electrons can flow is called a *circuit*. For a current to flow, there must be a complete circuit without gaps. A gap is usually provided by an electric switch that can be opened or closed to either cut off or allow electron flow. Electric appliances and other devices are commonly connected in a circuit in one of two ways: *series* or *parallel*. When connected in series, they form a single pathway for current flow between the terminals of the battery, generator, or wall socket (which is simply an extension of these terminals). When connected in parallel, they form branches, each of which is a separate path for the flow of electrons. Both series and parallel connections have their own distinctive characteristics. We shall briefly treat circuits using these two types of connections.

Series Circuits

A simple **series circuit** is shown in Figure 21.13. Three lamps are connected in series with a dry cell. A current exists almost immediately in all three lamps when the switch is closed. The current does not "pile up" in any lamp but flows *through* each lamp. Electrons making up this current leave the negative terminal of the cell, pass through each of the resistive filaments in the lamps in turn, and then return to the positive terminal of the cell. This is the only path of the electrons through the circuit. A break anywhere in the path results in an open circuit, and the flow of electrons ceases. Burning out of one of the lamp filaments or simply opening the switch could cause such a break.

---

*__Answer__   From the expression Watts = amperes × volts, we see that Amperes = 1200 watts/120 volts = 10 amps; so the hair dryer will operate when connected to the circuit. But two hair dryers on the same plug will blow the fuse.

†__Answer__   12¢ (1200 W = 1.2 kW; 1.2 kW × 1 h × 10¢/1 kWh = 12¢)

[6]Since Power = energy/time, simple rearrangement gives Energy = power × time; hence, energy can be expressed in the unit kilowatt-hours.

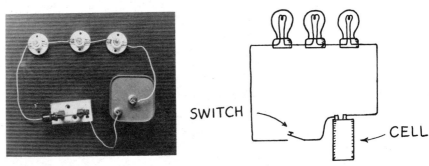

Fig. 21.13 A simple series circuit. A 6-volt cell provides 2 volts across each lamp.

SWITCH

CELL

The circuit shown in Figure 21.13 illustrates the following important characteristics of series connections:

1. Electric current has but a single pathway through the circuit; therefore, the current passing through the resistances in the electrical devices is the same everywhere in the circuit.

2. This current is resisted by the resistance in the first device, the resistance in the second, and in the third in turn, so that the total resistance to current flow in the circuit is the sum of the individual resistances along the circuit path.

3. The current flowing in the circuit is numerically equal to the voltage supplied by the source divided by the total resistance of the circuit. This follows from Ohm's law.

4. The total voltage impressed across a series circuit divides among the individual electrical devices in the circuit so that the sum of the "voltage drops" across the resistance of each individual device equals the total voltage supplied by the source. This follows from the fact that the amount of energy used to move each unit of charge through the entire circuit equals the sum of the energies used to move that unit of charge through each electrical device in turn.

5. The voltage drop across each device is proportional to its resistance. This follows from the fact that more energy is used to move a unit of charge through a large resistance than through a small resistance.

Questions

1. What happens to other lamps if one lamp in a series circuit burns out?*

2. What happens to the light intensity of each lamp in a series circuit when more lamps are added to the circuit?†

*****Answer**   If one of the lamp filaments burns out, the path connecting the terminals of the voltage source will break and no current will flow. All lamps will go out.

†**Answer**   The addition of more lamps in a series circuit results in a greater circuit resistance, which decreases the current in the circuit and therefore in each lamp, causing dimming; energy is divided among more lamps so the voltage drop across each lamp decreases.

It is easy to see the chief disadvantage of the series circuit; all the devices must be operating at the same time, because if any one device is turned off, the circuit will break and all current flow will cease. Most circuits are wired such that it is possible to operate electrical devices independently of one another. In the home, for example, a lamp can be turned on or off without affecting the operation of other lamps or electrical devices. This is because these devices are connected not in series but in parallel to one another.

Parallel Circuits

A simple **parallel circuit** is shown in Figure 21.14. Three lamps are connected to the same two points, A and B, in the electrical circuit. Electrical devices connected to the same two points of an electrical circuit are said to be connected in parallel. Electrons leaving the negative terminal of the battery need travel through only *one* lamp filament before returning to the positive terminal of the battery. In this case there are three separate pathways over which the electrons can travel through the circuit. A break in any one path does not interrupt the flow of charge in the other paths. Each device operates independently of the other devices connected in the circuit.

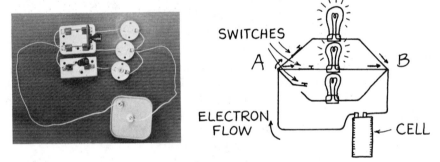

Fig. 21.14 A simple parallel circuit. A 6-volt cell provides 6 volts across each lamp.

This circuit illustrates the following major characteristics of parallel connections:

1. Each device connects the same two points, A and B, of the circuit; therefore, the voltage across each device is the same.

2. The total current in the circuit divides among the parallel branches. Current flows more readily into devices of low resistance, so the amount of current in each branch is inversely proportional to the resistance of the branch.

3. The total current flowing in the circuit equals the sum of the currents flowing in its parallel branches.

4. As the number of parallel branches is increased, the overall resistance of the circuit is decreased. The resistance to current flow is lowered with each added path between any two points of a circuit. It follows that the combined resistance of resistors in parallel to the flow of current between the two points of the circuit that they connect is less than the resistance of any one of the branches.

Questions

1. What happens to other lamps if one of the lamps in a parallel circuit burns out?*

2. What happens to the light intensity of each lamp in a parallel circuit when more lamps are added in parallel to the circuit?†

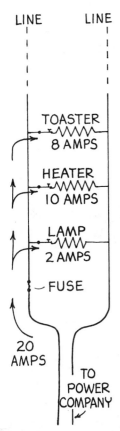

Fig. 21.15 Circuit diagram for appliances connected to a household supply line. The resistances of the appliances are indicated by the sawtooth lines.

Parallel circuits and overloading   Electricity is usually fed into the home via two lead wires called *lines*, which are connected to the wall outlets in each room. About 110–120 volts are impressed on these lines by generators at the power company. This voltage is applied to appliances and other devices connected in parallel by plugs to these lines. As more devices are connected to the lines, more pathways for current flow result in lowering of the combined resistance of the circuit. Thus, a greater amount of current flows in the supply lines. The resistance of the lines themselves is small enough to be neglected. Lines that carry more than a safe amount of current are said to be *overloaded*. The resulting heat may be sufficient to melt the insulation and start a fire.

We can see how overloading occurs by considering the circuit of Figure 21.15. The supply line is connected to an electric toaster that draws 8 amperes, to an electric heater that draws 10 amperes, and to an electric lamp that draws 2 amperes. When only the toaster is operating and drawing 8 amperes, the total line current is 8 amperes. When the heater is also operating, the total line current increases to 18 amperes (8 amperes to the toaster and 10 amperes to the heater). Turning on the lamp increases the line current to 20 amperes, and connecting any more electrical devices draws still more line current. Connecting too many devices into the same line results in dangerous overheating that may cause fires.

*****Answer**   If one lamp burns out, the other lamps will be unaffected. The current in each branch, according to Ohm's law, is equal to voltage/resistance, and since neither voltage nor resistance is affected in the branches, the current in those branches is unaffected. The total current in the overall circuit, however, is lessened by an amount equal to the current drawn by the lamp in question before it burned out. But the current in any other single branch is unchanged.

†**Answer**   The light intensity for each lamp is unchanged as other lamps are introduced (or removed). Only the total resistance and total current in the total circuit change. As lamps are introduced, more paths are available between the battery terminals, which effectively decreases total circuit resistance. This decreased resistance is accompanied by an increased current flow, the increase of which supplies the lamps as they are introduced. Although changes of resistance and current occur for the circuit as a whole, no changes occur in any individual branch of the circuit.

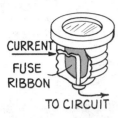

CURRENT
FUSE
RIBBON
TO CIRCUIT

Fig. 21.16 A safety fuse.

Safety fuses To prevent overloading in circuits, fuses are connected in series along the supply line. In this way the entire line current flows through the fuse. A fuse is constructed with a ribbon of wire that has a resistance greater than the supply line but lower than those of the circuit devices and, mainly, a low melting point (Figure 21.16). If the current in the line becomes dangerously large, both line and fuse become hot. The fuse is hottest because of its higher resistance and melts or "blows out" and breaks the circuit. Utility companies have fuses to protect their lines all the way back to the generators.

Fuses are constructed to melt or blow out as soon as the current reaches a certain limit. Thus, a "20-ampere fuse" blows out when the current in the circuit exceeds 20 amperes. Before the blown fuse is replaced, the cause of the overloading should be determined and remedied. Often, insulation separating the wires in a circuit wears away and allows the wires to touch, in effect shortening the path of the circuit. This is called a *short circuit*. The decreased resistance results in an increase in current.

Circuits may also be protected by circuit breakers, which use magnets or bimetallic strips to open the switch.

Summary of Terms

Potential difference The difference in voltage between two points, which can be compared to a difference in water level between two containers. If two containers having different water levels are connected by a pipe, water will flow from the one with the higher level to the one with the lower level until the two levels are equalized. Similarly, if two points with a difference in potential are connected by a conductor, a charge will flow from the one with the greater potential to the one with the smaller potential until the potentials are equalized. Measured in volts.

Electromotive force (emf) Any force that gives rise to an electric current; a battery or generator is a source of emf.

Electric current The flow of electric charge that transports energy from one place to another. Measured in amperes, where 1 ampere is the flow of $6\frac{1}{4} \times 10^{18}$ electrons per second.

Electrical resistance That property of a material to resist the flow of an electric current through it. Measured in ohms.

Ohm's law The statement that the current in a circuit varies in direct proportion to the potential difference of emf and in inverse proportion to resistance:

$$\text{Current} = \frac{\text{voltage}}{\text{resistance}}$$

A potential difference of 1 volt across a resistance of 1 ohm produces a current of 1 ampere.

Electric power The rate of energy transfer, or the rate of doing work; the ratio of energy per time, which electrically can be measured by the product of current and voltage:

$$\text{Power} = \text{current} \times \text{voltage}$$

Measured in watts (or kilowatts), where 1 ampere $\times$ 1 volt = 1 watt.

Direct current (dc) An electric current flowing in one direction only.

Alternating current (ac) Electric current that rapidly reverses in direction; the electric charges vibrate about relatively fixed points, usually at the vibrational rate of 60 hertz.

Series circuit An electric circuit with devices having resistances arranged such that the same electric current flows through each of them.

Parallel circuit An electric circuit having two or more resistances arranged in branches in such a way that any single one completes the circuit independently of all the others.

Review Questions

1. How is the statement "Water seeks its own level" analogous to electricity?

2. What condition is necessary for heat to flow from one end of a metal bar to the other? For the flow of electric charge?

3. What condition is necessary for a sustained flow of electric charge through a conducting medium?

4. Carefully distinguish between an *open circuit* and a *closed circuit*.

5. Distinguish between an *ampere* and a *volt*.

6. What is the effect on current in a circuit of steady resistance when the voltage is doubled? How about if both voltage and resistance are doubled?

7. Distinguish between *ac* and *dc*.

8. Electric current travels at the speed of light in a circuit, yet the individual electrons making up the current travel only a fraction of a centimeter or so per second. Is there a contradiction here? Explain.

9. Where do the electrons that illuminate a common light bulb in an electric circuit originate?

10. Distinguish between a *kilowatt* and a *kilowatt-hour*.

11. Why is it that a single burned-out lamp will disrupt an entire circuit wired in series, but not one wired in parallel?

12. What changes occur in the line current when a lamp burns out in a series circuit? In a parallel circuit?

13. What changes occur in the line current when more lamps or appliances are introduced in a series circuit? In a parallel circuit?

14. How does the line current in a home circuit differ from the current that lights the lamp you are likely reading this by?

15. Why will too many electrical devices operating at one time often blow a fuse?

Home Projects

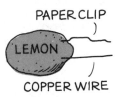

1. Every electric cell is made of two unlike "plates," or strips of metal placed in a conducting solution. A simple $1\frac{1}{2}$-volt cell, equivalent to a flashlight cell, can be made by placing a strip of copper and a strip of zinc in a tumbler of salt water. The voltage of a cell depends on the kinds of materials used and the kind of solution they are placed in, not the size of the plates.

An easy cell to construct is the "citrus cell." Stick a paper clip and a piece of copper wire into a lemon. Hold the ends of the wire close together, but not touching, and place the ends on your tongue. The slight tingle you feel and the metallic taste you experience result from a slight current of electricity pushed by the citrus cell through the wires when your moist tongue closes the circuit.

2. Examine the electric meter in your house. It is probably in the basement or on the outside of your house. You will see that in addition to the clocklike dials in the meter, there is a circular aluminum disk that spins between the poles of magnets when electricity is being used in the house. The more electricity used, the faster the disk turns. The speed of the disk is directly proportional to the number of watts used; for example, it will spin five times as fast for 500 watts as for 100 watts.

You can use the meter to determine how many watts an electrical device uses. First, see that all electrical devices in your home are disconnected (you may leave electric clocks connected, for the 2 watts they use will hardly be noticeable). The disk will be practically stationary. Then connect a 100-watt bulb and note how many seconds it takes for the disk to make five complete revolutions. The black spot painted on the edge of the disk makes this easy. Disconnect the 100-watt bulb and plug in a device of unknown wattage. Again, count the seconds for five revolutions. If it takes the same time, it is a 100-watt device; if it takes twice the time, it's a 50-watt device; half the time, a 200-watt device; and so forth. In this way you can estimate the power consumption of devices with fair accuracy.

Exercises

1. If a conductor is placed in contact with two separated bodies charged to different electrical potential energies, can you say for certain which way charge will flow in the conductor? How about if the bodies are charged to different electric potentials?

2. If an electric current flows from one body to another, what can we say about the electric potentials of the two bodies? (*Hint:* Think of thermal energy versus temperature.)

3. In which of the circuits below does a current exist to light the bulb?

4. Does electric current flow *out of* a battery or *through* a battery? Does it flow *into* a light bulb or *through* a light bulb? Explain.

5. Energy is put *into* electricity by pumping it from a low voltage to a high voltage. How do we get energy *out of* electricity?

6. Sometimes you hear someone say that a particular appliance "uses up" electricity. What is it that the appliance actually "uses up" and what becomes of it?

7. An electron moving in a wire collides again and again with atoms and travels an average distance between collisions that is called the *mean free path*. If the mean free path is less in some metals, what can you say about the resistance of these metals? For a given conductor, what can you do to lengthen the mean free path?

8. Why do wires heat up when they carry electric current?

9. Only a small percentage of the electrical energy fed into a common light bulb is transformed into light. What happens to the rest?

10. Why are thick wires rather than thin wires usually used to carry large currents?

11. Will a lamp with a thick or thin filament draw the most current?

12. Will the current in a light bulb connected to a 220-volt source be greater or less than when connected to a 110-volt source?

13. Which will do less damage—plugging a 110-volt appliance into a 220-volt circuit or plugging a 220-volt appliance into a 110-volt circuit? Explain.

14. Would the resistance of a 100-watt bulb be greater or less than the resistance of a 60-watt bulb? Which bulb has the thicker filament?

15. If a current of one- or two-tenths of an ampere flows into your hand and out the other, you will probably be electrocuted. But if the current flows into your hand and out the elbow above the same hand, you can survive even if the current is large enough to burn your flesh. Explain.

16. Would you expect to find dc or ac in the filament of a light bulb in your home? How about in an automobile?

17. The wattage marked on a light bulb is not an inherent property of the bulb but depends on the voltage to which it is connected, usually 110 or 120 volts. How many amperes flow through a 60-watt bulb connected in a 120-volt circuit?

18. Is a current-carrying wire charged?

19. The damaging effects of electric shock result from the amount of *current* that flows in the body. Why, then, do we see signs that read "Danger—High Voltage" rather than "Danger—High Current"?

20. Comment on the warning sign shown in the sketch.

DANGER !
HIGH RESISTANCE
(1,000,000,000 Ω)

21. Why is the wingspan of birds a consideration in determining the spacing between parallel wires in a power line?

22. Estimate the number of electrons that a power company delivers annually to the homes of a typical city of 50,000 people.

23. If electrons flow very slowly through a circuit, why does it not take a noticeably long time for a lamp to glow when you turn on a distant switch?

24. Why is the speed of electricity so much greater than the speed of sound?

25. If a glowing light bulb is jarred and oxygen leaks inside, it will momentarily brighten considerably before burning out. Putting excess current through a light bulb will also burn it out. What physical change occurs when a light bulb burns out?

26. Rearrange the equation Current = voltage/resistance to express *resistance* by itself. Then solve the following: A certain device in a 120-volt circuit has a current rating of 20 amps. What is the resistance of the device (how many ohms)?

27. Why should you not use a copper penny in place of a safety fuse that has blown out?

28. Using the formula Power = current × voltage, find the current drawn by a 1200-watt hair dryer connected to 120 volts. Then using the method you used in Exercise 26, find the resistance of the hair dryer.

29. As more and more bulbs are connected in series to a flashlight battery, what happens to the brightness of each bulb? How about when more and more bulbs are connected in parallel?

30. Why are household appliances almost never connected in series?

31. If a 60-watt bulb and a 100-watt bulb are connected in series in a circuit, through which bulb will the greater current flow? How about if they are connected in parallel?

32. If a 60-watt bulb and a 100-watt bulb are connected in series in a circuit, through which bulb will there be the greater voltage drop? How about if they are connected in parallel?

33. The useful life of an automobile battery without recharging is given in terms of ampere-hours. A typical 12-volt battery has a rating of 60 ampere-hours, which means that a current of 60 amperes can be drawn for 1 hour, 30 amperes can be drawn for 2 hours, and so forth. Suppose you forget to turn off the headlights in your parked automobile. If each headlight draws 3 amperes, how long will it be before your battery is "dead"?

34. How much does it cost to operate a 100-watt lamp continuously for 1 month if the power utility rate is 8¢ per kilowatt-hour?

35. A 4-watt night light is plugged into a 120-volt circuit and operates continuously for 1 year. Find the following: (*a*) the current it draws, (*b*) the resistance of its filament, (*c*) the energy consumed in a year, and (*d*) the cost of its operation for a year at the utility rate of 8¢ per kilowatt-hour.

22 Magnetism

The term *magnetism* stems from certain stones found by the early Greeks more than 2000 years ago in the region of Magnesia. These *lodestones,* as they were called, had the unusual property of attracting pieces of iron. Magnets were first used for navigation by the Chinese in the twelfth century. In the sixteenth century, William Gilbert, Queen Elizabeth's physician, made artificial magnets by rubbing pieces of iron against lodestone, and suggested that a compass always points north and south because the earth itself has magnetic properties. Later, in 1750, John Michell in England found that magnetic poles obey the inverse-square law, and his results were confirmed by Coulomb. The subjects of magnetism and electricity developed independently of each other until in 1820 a Danish high school physics teacher named Hans Christian Oersted discovered in a classroom demonstration that an electric current affects a magnetic compass.[1] He saw that magnetism was related to electricity. Shortly thereafter, the French physicist Ampere proposed that electric currents are the source of all magnetism. The explanation for this was given in 1905 by Albert Einstein.

Magnetic Forces

We have discussed the forces that electrically charged particles exert on one another, wherein the force between any two charged particles depends on the magnitude of the charge on each and their distance of separation, as specified in Coulomb's law. But Coulomb's law is not precisely true when the charges are moving with respect to each other. The force between electric charges depends also, in a complicated way, on the motion of the charges. We find that in addition to the force we call *electrical,* there is a force due to the motion of the charges that we call the **magnetic force**. Both electrical and magnetic forces are actually different aspects of the same phenomenon of electromagnetism.

Nature of the Magnetic Field

We stated in Chapter 20 that a region of space is altered when an electric charge is introduced. The space contains energy; this energy is contained in the electric field that originates from the electric charge. An electric field surrounds every charge. If the charge is moving, the region of space surrounding the charge is further altered. This alteration due to the motion of a charge is called a **magnetic field**. We say a moving charge is surrounded by both an electric and a magnetic field. Like the electric field, the

---

[1]We can only speculate about how often such relationships become evident when they "aren't supposed to" and are dismissed as "something wrong with the apparatus." Oersted, however, had the insight characteristic of a good physicist to see that nature was revealing another of its secrets.

Fig. 22.1   A horseshoe magnet.

magnetic field is a storehouse of energy. The greater the motion of the charge, the greater the magnitude of the magnetic field surrounding the charge. A magnetic field is produced by the motion of electric charge.

Interestingly enough, since motion is relative, the magnetic field is relative. For example, when a charge moves by you, there is a definite magnetic field associated with the moving charge. But if you move along with the charge, so that there is no motion relative to you, you will find no magnetic field associated with the charge. Magnetism is relativistic.

We are all familiar with the common iron horseshoe magnet (Figure 22.1). The magnetic field between the ends can be extremely strong, as evidenced by its interaction with paper clips, nails, and other iron substances. But if a magnetic field is produced by the motion of electric charges, where is the electrical motion in a solid stationary magnet? Although the magnet as a whole may be stationary, the atoms making up the magnet are in constant motion. More important than the motion of the atoms is the motion of electrons within the atoms. Electrons behave as if they move in an orbital-like motion about the atomic nuclei. This behavior of electrons produces a magnetic field. In addition to this orbital-like motion, electrons spin about their own axes like tops. A spinning electron constitutes a charge in motion and therefore creates a magnetic field. In every atom, at least twice as many magnetic fields exist as electrons, since each electron produces two fields: one for orbital motion and one for spinning. The field due to the spinning of the electron is more predominant.[2]

Every electron is a tiny electromagnet. A pair of electrons spinning in the same direction makes up a stronger electromagnet. A pair of electrons spinning in opposite directions, however, has the opposite effect. The magnetic field of each cancels the other. This is why most substances are not magnets. In most atoms, the various fields cancel each other because the electrons spin in opposite directions. In materials such as iron, nickel, and cobalt, however, the fields do not cancel each other entirely. Each iron atom has four electrons whose spin magnetism is uncanceled. Each iron atom, then, is a tiny magnet. The same is true to a slightly lesser degree for the atoms of nickel and cobalt.

Magnetic Domains

The magnetic field of individual iron atoms is so strong that interaction among adjacent atoms causes large clusters of them to line up with each other. These clusters of aligned atoms are called **magnetic domains**. Each domain is perfectly magnetized, being made up of millions of aligned

---

[2]Most common magnets are made from alloys containing iron, nickel, or cobalt. In these the electron spin contributes virtually all the magnetic properties. In the rare earth metals like gadolinium, the orbital motion is more significant.

atoms. The domains are extremely small, and a crystal of iron contains many of them.

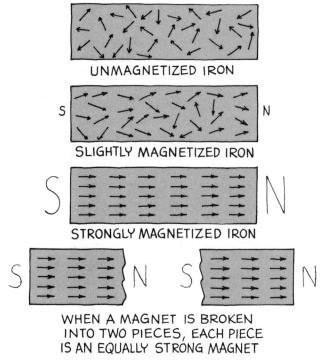

UNMAGNETIZED IRON

SLIGHTLY MAGNETIZED IRON

STRONGLY MAGNETIZED IRON

WHEN A MAGNET IS BROKEN
INTO TWO PIECES, EACH PIECE
IS AN EQUALLY STRONG MAGNET

Fig. 22.2 Pieces of iron in successive stages of magnetism. The arrows represent domains, the head being a north pole and the tail a south pole. Poles of neighboring domains neutralize each other's effects, except at the ends.

Not every piece of iron, however, is a magnet. This is because the domains in the material are not aligned. Consider a common iron nail: the domains in the nail are randomly oriented; they can be induced into alignment, however, by bringing a magnet nearby. (It is interesting to listen with an amplified stethoscope to the clickity-clack of domains undergoing alignment in a piece of iron when a strong magnet is brought nearby.) The domains align themselves much as electrical charges in a piece of paper align themselves in the presence of a charged rod. When we remove the nail from the permanent magnet, ordinary thermal motion causes most or all of the domains of the nail to return to a random arrangement. If the field of the permanent magnet is very strong, however, the nail may retain some permanent magnetism of its own even after the two are separated. Magnets are made in this way by simply placing pieces of iron in strong magnetic fields. It helps to tap the iron to nudge any stubborn domains into alignment. Another way of making a permanent magnet is to stroke a piece of iron with a magnet. The stroking motion aligns the domains in the iron. If a permanent magnet is dropped or heated, some of the domains are jostled out of alignment and the magnet becomes weaker.

Magnetic Poles

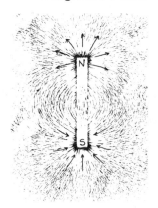

Fig. 22.3 Magnetic field configurations, shown about a bar magnet.

Magnets exhibit both attractive and repulsive forces. In electricity we have positive and negative charges, whereas in magnetism we have north and south poles. The Chinese were perhaps the first to discover that a suspended magnet will line up in a north-south direction. Such a magnet, called a *compass*, has obvious navigational advantages. In suspending a permanent magnet, we find that one end always points northward. This shows that there is something different about the two ends.

The end that points northward is called the *north-seeking pole*, and the end that points southward is the *south-seeking pole*. For the sake of simplicity, these are called the *north* and *south* poles. Figure 22.3 shows the magnetic field about a bar magnet. Iron filings that have been sprinkled on a piece of paper covering the magnet reveal the magnetic field configurations.[3] We say that the magnetic field direction exterior to the magnet is from the north pole to the south pole. The magnetic field lines form closed loops emanating from the north pole and entering the south pole, where they are directed through the magnet and out the north pole again.

If the north pole of one magnet is brought near the north pole of another magnet, repulsion occurs. The same is true of a south pole near a south pole. If opposite poles are brought together, however, attraction occurs.[4] We can say

Like poles repel; unlike poles attract.

Theoretical physicists have speculated for some fifty years about the possible existence of discrete magnetic "charges," called **magnetic monopoles**. These tiny particles would carry either a single north or a single south magnetic pole and would be the counterparts to the positive and negative charges in electricity. Various attempts have been made to find monopoles, but none has proved successful. A magnetic monopole is reported to have been detected on Valentine's Day at Stanford University in 1982. It may have been spurious. All known magnets always have at least one north and one south pole.

Question   How can a magnet attract a piece of iron that is not magnetized?*

---

*__Answer__ Domains in the unmagnetized piece of iron are induced into alignment by the magnetic field of the nearby magnet. See the similarity of this with Figure 20.10. Like the pieces of paper, pieces of iron will jump to a strong magnet when it is brought nearby. But unlike the paper, they are not then repelled. Can you think of the reason why?

[3]Domains align in the individual filings, causing them to act like tiny compasses. The poles of each "compass" are pulled in opposite directions, producing a torque that twists each filing into alignment with the external magnetic field.

[4]The force of interaction between unit magnetic poles is given by $F \sim pp'/d^2$, where p and p' represent magnetic pole strengths, and d represents the distance between them. Note the similarity of this relationship with Coulomb's law.

Earth's Magnetic Field

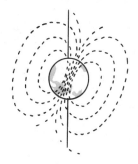

Fig. 22.4 The earth is a magnet.

Fig. 22.5 A compass.

Fig. 22.6 Convection currents in the molten parts of the earth's interior may produce the magnetic field of the earth.

A suspended magnet or compass points northward because the earth itself is a huge magnet. The compass aligns with the magnetic field of the earth. The magnetic poles of the earth, however, do not coincide with the geographical poles, nor are they very close to the geographical poles. The magnetic pole in the northern hemisphere, for example, is located 1750 kilometers from the geographical pole, somewhere in the Hudson Bay region of northern Canada. The other pole is located south of Australia (Figure 22.4).

It is of some interest to note that the south pole of the earth's magnet is near the northern geographical pole, while the north pole of the earth's magnet is near the southern geographical pole. Since unlike poles attract each other, the north poles of compasses are attracted toward the northern part of the world where the magnetic south pole is located. Hence, the north pole of a compass points in a northerly direction.

Since a compass points toward the earth's magnetic pole rather than to its geographical pole, the compass direction generally deviates from the true north. Mariners and others who use compasses must allow for this deviation in determining the true north. The discrepancy between the orientation of a compass and true north is known as the *magnetic declination*.

We do not know exactly why the earth itself is a magnet. The configuration of the earth's magnetic field is like that of a strong bar magnet placed near the center of the earth. But the earth is not a magnetized chunk of iron like a bar magnet. It is simply too hot for individual atoms to hold to a proper orientation. The earth has an outer rocky mantle almost 3000 kilometers deep. Below that is the liquid part of the core, about 2000 kilometers thick, which surrounds the solid center. Most scientists think that the motion of charges in the liquid part of the earth's core creates the magnetic field. Because of the earth's great size, the speed of moving charges looping around within the earth need only be somewhat less than a thousandth of a meter per second. Some scientists think that these currents are caused by the earth's rotation. This idea is supported by the following facts: Jupiter, which has a 10-hour day, has a much stronger magnetic field than the earth. Venus, which rotates more slowly still, has no measurable magnetic field, according to space-probe measurements.

Another likely source of the earth's magnetism may be heat rising from the earth's central core. This heat may be the cause of convection currents of molten material in the liquid part of the earth's core. The motion of ions and electrons in this molten material would produce a magnetic field (Figure 22.6). Probably such convection currents combined with the rotational effects of the earth produce the earth's magnetic field. In geologic history, an unsteady flow of heat might have caused a subsiding and momentary ceasing of convection currents—resulting in a collapse of the earth's magnetic field. Regeneration of heat would have again caused convection currents, but not necessarily in the same direction. This would account for prior magnetic fields reversed from those that exist today.

Whatever the cause, the magnetic field of the earth is not stable but has wandered throughout geologic time. This is known from analysis of the magnetic properties of rock strata. Iron atoms in a molten state tend to align themselves with the magnetic field of the earth. When the atoms solidify, the direction of the earth's magnetic field is recorded by the orientation of domains in the rock. The slight magnetism resulting can be measured with sensitive instruments. As samples of rock are tested from different strata formed throughout geologic time, the magnetic field of the earth for different periods can be charted. This evidence shows that there have been times when the magnetic field of the earth has diminished to zero and then reversed itself. More than twenty reversals have taken place in the past 5 million years; the most recent occurred 700,000 years ago. Prior reversals happened 870,000 and 950,000 years ago. Japanese work on field reversals from studies of deep-sea sediments shows that the field was virtually switched off for 10,000 to 20,000 years just over 1 million years ago. This was the time that modern humans emerged.

We cannot predict when the next reversal will occur because the reversal sequence is not regular. But there is a clue in recent measurements that show a decrease of over 5 percent of the earth's magnetic field strength in the last 100 years. If this change is maintained, we may well have another reversal within 2000 years.

Varying ion winds in the atmosphere cause more rapid but much smaller fluctuations in the earth's magnetic field. Ions in this region are produced by the energetic interactions of solar ultraviolet rays and X rays with atmospheric atoms. The motion of these ions produces a small but important part of the earth's magnetic field. Like the lower layers of air, the ionosphere is churned by winds. The variations in these winds are responsible for nearly all fast fluctuations in the earth's magnetic field.

Magnetic Forces on Moving Charged Particles

A charged particle at rest will not interact with a static magnetic field. But if the charged particle moves in a magnetic field, the field associated with its own motion does interact with the magnetic field in which it moves. A charged particle therefore experiences a force when moving through a magnetic field. The force is greatest when the particle moves in a direction perpendicular to the magnetic field lines. At other angles, the force lessens and becomes zero when the charged particle moves along the magnetic field lines. In any case, the direction of the force is always perpendicular to the magnetic field lines and the velocity of the charged particle (Figure 22.7). So a moving charge experiences a sideways deflection when it crosses through a magnetic field.[5] While it is traveling along the field direction, no deflection occurs.

---

[5]When particles of electric charge q and velocity v move perpendicularly into a magnetic field of strength B, the force F on each particle is simply the product of the three variables: $F = qvB$.

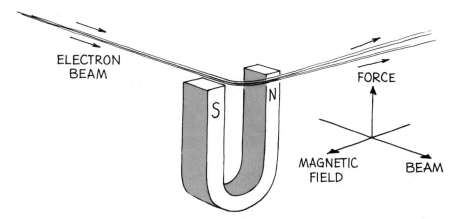

Fig. 22.7 A beam of electrons is deflected by a magnetic field.

This sideways deflection is very different from other interactions like gravitation between two bodies, the electrostatic forces between charges, and the forces between magnetic poles. We see here that the deflecting force does not act in a direction *between* the sources of interaction but instead acts perpendicular to both the magnetic field and the electron beam.

An understanding of the forces on moving charged particles in a magnetic field underlies an understanding of a wide range of phenomena, from electric motors to radiation belts around the earth.

The universe is a shooting gallery of charged particles. They are called cosmic "rays" and are the nuclei of atoms stripped bare of their electrons. Their origin is uncertain; perhaps they are boiled-off stars or are nuclei that did not condense to form stars. In any event, they travel through space at fantastic speeds and make up the cosmic radiation that is hazardous to astronauts. Fortunately for those of us on the earth's surface, most of these charged particles are deflected away by the magnetic field of the earth. Some of them are trapped in the outer reaches of the earth's magnetic field and make up the Van Allen radiation belts (Figure 22.8).

Fig. 22.8 The Van Allen radiation belts, shown here un-distorted by the solar wind.

The Van Allen radiation belts consist of two doughnut-shaped rings, named after James A. Belts, who suggested their existence from data gathered by the U.S. satellite *Explorer I* in 1958.[6] The inner ring is about 3000 kilometers from the earth, and the outer ring, which is a larger and wider doughnut, is about 15,000 kilometers from the earth. Astronauts orbit at safe distances well below these belts of radiation. Most of the charged particles, protons and electrons, trapped in the outer belt probably come from the sun. Storms on the the sun hurl charged particles out in great fountains, many of which pass near the earth and are trapped by its magnetic field. The trapped particles follow corkscrew paths around the magnetic field lines of the earth and bounce between the earth's magnetic poles high above the atmosphere. Disturbances in the earth's field often allow the ions to dip into the atmosphere, causing it to glow like a fluorescent lamp. This is the beautiful aurora borealis (northern lights).

The particles trapped in the inner belt probably originated from the earth's atmosphere. This belt is now masked by electrons that were produced by high-altitude hydrogen bomb explosions in 1962.

In spite of the earth's protective magnetic field, many cosmic rays are incident upon the earth. Cosmic ray bombardment is maximum at the poles, because charged particles incident there do not travel *across* the magnetic field lines, but rather *along* the field lines and therefore are not deflected. Incidence decreases away from the poles and is minimum in equatorial regions. At sea level, incidence is from one to three particles per square centimeter each minute; this number increases rapidly with altitude. So cosmic rays are penetrating your body as you are reading this. (And even when you aren't reading this!)

Magnetism and Evolution

The magnetic changes of the earth may have played an important role in the evolution of life forms. One theory is that when life was passing through its earliest phases, the magnetic field of the earth was strong enough to hold off cosmic and solar radiations that were violent enough to destroy life. The field has gone through periods of zero strength during the many magnetic pole reversals in the recent geologic past. During these periods, cosmic radiation on the earth's surface increased, and the Van Allen belts spilled their cargo of radiation, which increased mutation of the primitive life forms then existing. Sudden bursts of radiation may have been as effective in changing life forms as X rays have been in the famous heredity studies of fruit flies. The coincidences of the dates of increased life changes and the dates of the magnetic pole reversals lend support to this theory.

---

[6]Humor aside, the name is actually James A. Van Allen.

Biomagnetism

Fig. 22.9 The pigeon can sense direction because it has a built-in magnetic "compass" within its skull.

Certain bacteria biologically produce single-domain magnetite grains that they string together to form internal compasses. They then use these compasses to detect the earth's magnetic field. Equipped with a sense of direction, the organisms are able to locate food supplies. Amazingly, these bacteria south of the equator build the same single-domain magnets as their counterparts north of the equator but then align them in the opposite direction to coincide with the oppositely directed magnetic field in the southern hemisphere! Bacteria are not the only living organisms with built-in magnetic compasses: pigeons have recently been found to have multiple-domain magnetite magnets within their skulls that are connected with a large number of nerves to the pigeon brain. Pigeons have a magnetic sense, and not only can they discern longitudinal directions along the earth's magnetic field, but they can detect latitude by the dip of the earth's field as well. Magnetic material has also been found in the abdomens of bees, whose behavior is affected by small magnetic fields. Searches of human beings to date show them to possess no such magnetic sense.

Summary of Terms

Magnetic force   (1) Between magnets, it is the attraction of unlike magnetic poles for each other and the repulsion between like magnetic poles. (2) Between a magnetic field and a moving charge, the moving charge is deflected from its path in the region of a magnetic field; the deflecting force is perpendicular to the motion of the charge and perpendicular to the magnetic field lines. This force is maximum when the charge moves perpendicular to the field lines and is minimum (zero) when moving parallel to the field lines.

Magnetic field   The region of "altered space" that will interact with the magnetic properties of a magnet. It is located mainly between the opposite poles of a magnet or in the energetic space about an electric charge in motion.

Magnetic domains   Clustered regions of aligned magnetic atoms. When these regions themselves are aligned with each other, the substance containing them is a magnet.

Magnetic monopole   A hypothetical particle having a single north or south magnetic pole, analogous to the positive or negative electric charge.

Review Questions

1. What kind of energy field surrounds a stationary electric charge? A moving electric charge?

2. What is the origin of magnetic forces?

3. What two kinds of motion are exhibited by electrons in an atom?

4. What is a magnetic domain?

5. Why is iron magnetic and wood not?

6. Why will dropping a magnet on a hard floor make it a weaker magnet?

7. Why does a compass point northward? Will the needle point in the same direction when in the southern hemisphere?

8. what is meant by *magnetic declination?*

9. What three phenomena are thought to produce the earth's magnetic field?

10. An electron always experiences a force in an electric field, but not necessarily in a magnetic field. Explain.

11. In what direction relative to a magnetic field does a charged particle travel in order to experience maximum deflecting force? Minimum deflecting force?

12. What are the Van Allen radiation belts?

13. What are cosmic rays?

14. What influence does the earth's magnetic field have on cosmic rays?

15. Why does cosmic ray intensity on the earth's surface increase during magnetic pole reversals?

Home Projects

1. Find the direction and dip of the earth's magnetic field lines in your locality. Magnetize a steel knitting needle or straight piece of steel wire by stroking it a couple of dozen times with a strong magnet. Run the needle through a cork and float it in a plastic or wooden container of water. The needle will point to the magnetic  pole. Then remove the needle and cork and press an unmagnetized common pin into each side of the cork. Rest the pins on the rims of a pair of drinking glasses so that the needle points to the magnetic pole. It should dip in line with the earth's magnetic field.

2. An iron bar can be easily magnetized by aligning it with the magnetic field lines of the earth and striking it lightly a few times with a hammer. The hammering jostles the domains so they can better fall into alignment with the earth's field. The bar can be demagnetized by striking it when it is in an east-west direction.

3. Bring a magnetic compass near the tops of iron or steel objects in your home (radiators, refrigerators, stoves, lamps, etc.). You will find that the north pole of the compass needle points to the tops of these objects, and the south pole of the compass needle points to the bottoms. This shows that the objects are magnets, having a south pole on top and a north pole on the bottom. What is the explanation for this? You will find that even cans of food that have been in a vertical position in the pantry are magnetized. Turn one over and test to see how many days it takes to lose and change its polarity.

Exercises

1. In what sense are all magnets electromagnets?

2. Since every iron atom is a tiny magnet, why aren't all iron materials themselves magnets?

3. Why will heating or dropping a magnet reduce its magnetic strength?

4. The core of the earth is likely composed of iron and nickel, excellent metals for making permanent magnets. Why is it unlikely that the earth's core is a permanent magnet?

5. Why will a magnet attract an ordinary nail or paper clip, but not a wooden pencil?

6. Will either pole of a magnet attract a paper clip? Explain what is happening inside the attracted paper clip. (*Hint:* Consider Figure 20.10.)

7. One way to make a compass is to stick a magnetized needle into a piece of cork and float it in a wooden bucket full of water. The needle will align itself with the magnetic field of the earth. Since the north pole of this compass is attracted northward, will the needle float toward the northward side of the bucket? Defend your answer.

8. What is the net magnetic force on a compass needle? By what mechanism does a compass needle line up with a magnetic field?

9. Since the iron filings that line up with the magnetic field of the bar magnet shown in Figure 22.3 are not themselves little magnets, by what mechanism do they align themselves with the field of the magnet?

10. The north pole of a compass is attracted to the north pole of the earth, yet like poles repel. Can you resolve this apparent dilemma?

11. Your friend says that when a compass is taken across the equator, it turns around and points in the opposite direction. Your other friend says this is not true, that southern-hemisphere types use the south pole of the compass to find direction. You're on; what do you say?

12. A TV picture is produced by a beam of electrons that is moved back and forth as the electrons strike the screen. By what means do you suppose the beam is moved back and forth?

13. Why will a magnet placed in front of a television picture tube distort the picture? (*Note:* Do NOT try this with a color set. If you succeed in magnetizing the metal mask in back of the glass screen, you will have picture distortion even when the magnet is removed!)

14. Magnet A has twice the magnetic field strength of magnet B and at a certain distance pulls on magnet B with a force of 50 newtons. With how much force, then, does magnet B pull on magnet A?

15. Can an electron at rest be set into motion with a magnetic field? An electric field?

16. Magnetic fields can be used to trap plasmas in "magnetic bottles," but the plasma must be moving. Why?

17. A cyclotron is a device for accelerating charged particles in ever-increasing circular orbits to high speeds. The charged particles are subjected to both an electric field and a magnet field. One of these fields increases the speed of the charged particles, and the other field holds them in a circular path. Which field performs which function?

18. A magnetic field can deflect a beam of electrons, but it cannot do work on the electrons to speed them up. Why?

19. Two charged particles are projected into a magnetic field that is perpendicular to their velocities. If the charges are deflected in opposite directions, what does this tell you about them?

20. A beam of high-energy protons emerges from a cyclotron. Do you suppose there is a magnetic field associated with these particles? Why or why not?

21. Inside a laboratory room there is said to be either an electric field or a magnetic field but not both. What experiments might be performed to establish what kind of field is in the room?

22. Why do astronauts keep to altitudes beneath the Van Allen radiation belts when doing space walks?

23. Residents of northern Canada are bombarded by more intense cosmic radiation than are residents of Mexico. Why is this so?

24. In a mass spectrograph, ions are directed into a magnetic field, where they are deflected and strike a detecting screen. If a variety of singly ionized atomic nuclei travel at the same speed through the magnetic field, would you expect all nuclei to be deflected by the same amount? Or would different nuclei be bent different amounts? What would you expect?

25. One way to shield a habitat in outer space from cosmic rays is with an absorbing blanket of some kind, like the atmosphere that protects the earth. Speculate on a second way for shielding that is also similar to earth shielding.

26. If you had two bars of iron—one magnetized and the other not—and no other equipment, how could you tell which bar was the magnet?

23 Electromagnetic Interactions

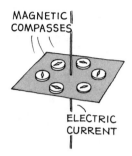

MAGNETIC COMPASSES

ELECTRIC CURRENT

Fig. 23.1 The compasses show the circular shape of the magnetic field surrounding the current-carrying wire.

A moving charge produces a magnetic field. A current of charges, then, produces a magnetic field. The magnetic field surrounding a current-carrying conductor can be demonstrated by arranging an assortment of magnetic compasses around a wire (Figure 23.1) and passing a current through it. The compasses line up with the magnetic field produced by the moving charges and show that the configuration of magnetic field lines forms concentric circles about the wire. When the current reverses direction, the compasses turn completely around, showing that the direction of the magnetic field changes also.

If the wire is bent into a loop, the magnetic field lines that surrounded the wire are bunched up inside the loop (Figure 23.2). If the wire is bent into another loop, overlapping the first, the concentration of magnetic field lines inside the double loop is twice as much as in the single loop. It follows that the magnetic field intensity in this region is increased as the number of loops is increased. The magnetic field intensity is appreciable for a current-carrying coil of wire with many loops. If a piece of iron is placed in such a coil, the magnetic domains in the iron are induced into alignment, which further increases the magnetic field intensity. And we have an electromagnet!

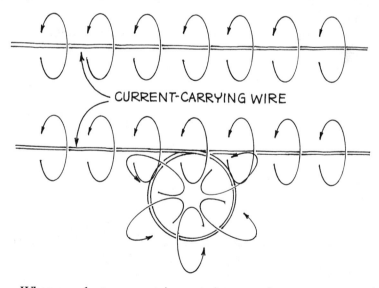

CURRENT-CARRYING WIRE

Fig. 23.2 Magnetic field lines about a current-carrying wire crowd up when the wire is bent into a loop.

When an electromagnet is cooled to very low temperatures by being immersed in a bath of 4 K liquid helium, for example, electric current in the coil runs virtually free of resistance, and we have a powerful supercon-

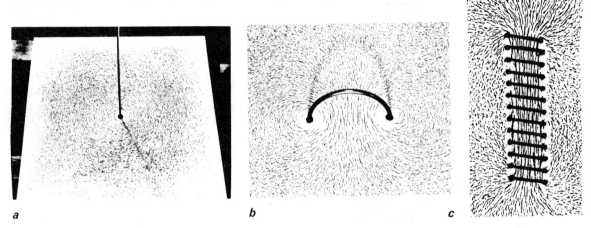

Fig. 23.3   Iron filings sprinkled on paper reveal the magnetic field configurations about (*a*) a current-carrying wire, (*b*) a current-carrying loop, and (*c*) a coil of loops.

ducting magnet. These powerful electromagnets are used for research purposes such as controlling charged particle beams in high-energy accelerators. More impressively, they levitate and propel high-speed trains, now in the developmental stage (Figure 23.4).

Magnetic Force on a Current-Carrying Wire

A charged particle moving through a magnetic field experiences a deflecting force. A current of charged particles, then, moving through a magnetic field experiences a deflecting force. If the moving charged particles are trapped inside a wire when they respond to the deflecting force, the wire will also move (Figure 23.5).

Fig. 23.4   Conventional trains vibrate as they ride on rails at high speeds. This Japanese magnetically levitated train is capable of vibration-free high speeds, even in excess of 200 km/hr.

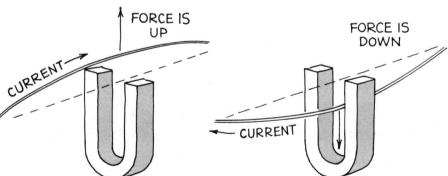

Fig. 23.5   A current-carrying wire experiences a force in a magnetic field. (Compare this figure with Figure 22.7.)

If we reverse the direction of current, the deflecting force acts in the opposite direction. The force is maximum when the current is perpendicular to the magnetic field lines. The direction of force is not along the magnetic field lines or along the direction of current but is perpendicular to both field lines and current. It is a sideways force.

We see that just as a current-carrying wire deflects a magnetic compass, as discovered by Oersted in 1820, a magnet will deflect a current-carrying wire. This discovery created much excitement, for almost immediately people began designing methods to harness this force for useful purposes. Electric motors marked the beginning of a new era. The principle of the electromagnetic motor is shown in bare outline in Figure 23.6. A permanent magnet is used to produce a magnetic field in a region where a rectangular loop of wire is mounted so that it can turn about an axis as shown. When a current flows through the loop, it flows in opposite directions in, say, the upper and lower side of the loop, resulting in opposite forces on the wire. The upper portion of wire is forced to the left, while the lower portion of the wire is forced to the right, causing the loop to turn. To maintain rotation, the current is reversed during each half-revolution by means of stationary contacts on the shaft. In this way, the current in the loop alternates so that the forces in the upper and lower regions do not change directions as the loop rotates. Small dc motors are made this way. Larger motors, dc or ac, are usually made by replacing the permanent magnet with an electromagnet that is energized by the power source. Of course, more than a single loop is used. Many loops of wire are wound about a cylinder, which then rotates when energized with electric current. This cylinder of many windings is called an *armature*.

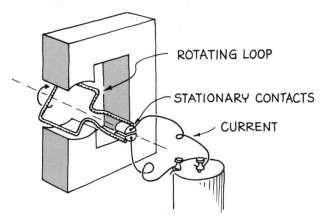

ROTATING LOOP

STATIONARY CONTACTS

CURRENT

Fig. 23.6 A simplified motor.

Electrical meters work in this fashion. Several wire loops are mounted on a core, which constitutes an armature that can rotate in a magnetic field (Figure 23.7). A spring holds it in a "zero" position. When current flows in the armature, it is forced to rotate. If a needle is attached to the core, its deflection (which is proportional to the amount of current) can be readily observed. Such instruments are called **galvanometers**. Voltmeters and ammeters are modified galvanometers.

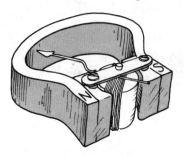

Fig. 23.7 A galvanometer. This instrument may be calibrated to measure current (amperes), in which case it is called an *ammeter,* or to measure electric potential (volts), in which case it is called a *voltmeter.*

Question   What fundamental fact underlies the operation of all electric motors?*

Electromagnetic Induction

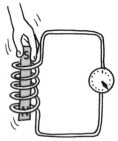

Fig. 23.8 When the magnet is plunged into the coil, voltage is induced, as shown on the meter.

Fig. 23.9 Charges in the wire are set in motion where the magnetic field moves through a stationary wire or the wire moves through a magnetic field.

The discovery that magnetism could be produced with electrical wires was both a theoretical and a technological turning point. The question arose as to whether electricity could be produced from magnetism. When this question was answered, the world was never again the same. That discovery transformed Western civilization. In 1831 Joseph Henry of the United States and Michael Faraday of Scotland independently discovered that when a magnet was plunged into a coil of wire, a voltage was induced.[1] Electric current could be made to flow in a wire by simply moving a magnet in or out of a coil of wire (Figure 23.8). This phenomenon is called **electromagnetic induction**.

Electromagnetic induction in a conductor depends only on the relative motion between the conductor and the magnetic field. Voltage is induced whether the magnetic field of a magnet moves by a stationary conductor or the conductor moves in a stationary magnetic field (Figure 23.9). The results are the same whether either or both move.

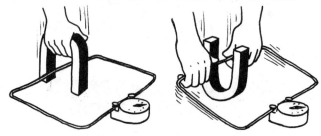

*__Answer__   A current-carrying conductor experiences a force in a magnetic field.

[1]Faraday is reported to have once given a lecture on electromagnetic induction to a group of London citizens, whereupon a member of the audience asked what use it could possibly have. Faraday replied with a line that has become a byline for research scientists: "Of what use is a newborn baby?"

The greater the number of loops of wire moving relative to a magnetic field, the greater the induced voltage and the greater the current in the wire (Figure 23.10). Pushing a magnet into twice as many loops will induce twice as much voltage; ten times as many loops will induce ten times as much voltage; and so on. It may seem as if we get something (energy) for nothing by simply increasing the number of loops in a coil of wire. But we don't: you'll find it is more difficult to push the magnet into a coil with more loops. Think of it this way: each additional current loop is an additional electromagnet to resist the motion of your magnet. You do more work to induce more voltage. The amount of voltage induced also depends on how quickly the magnetic field lines are traversed by the wire. Very slow motion produces hardly any voltage at all. Quick motion induces a greater voltage.

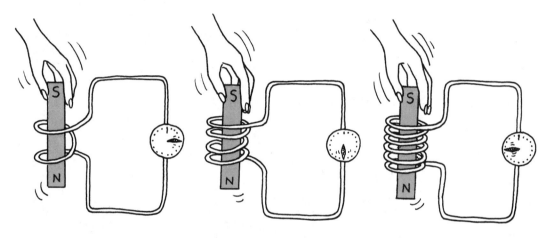

Fig. 23.10 When a magnet is plunged into a coil having twice the number of loops as another, twice as much voltage is induced. If the magnet is plunged into a coil with three times as many loops, then three times as much voltage is induced. (It is more difficult to push the magnet into a coil with more loops because the magnetic field of each current loop resists the motion of your magnet.)

If a magnet is plunged in and out of a coil of wire, the voltage induced in the coil alternates in direction. As the magnetic field strength inside the coil is increased (magnet entering), the induced voltage in the coil is directed one way; when the magnetic field strength diminishes (magnet leaving), the voltage is induced in the opposite direction. The greater the frequency of field change, the greater the induced voltage. The frequency of the changing magnetic field within the loop is equal to the frequency of alternating voltage induced.

Electromagnetic induction can be summarized in the statement

The induced voltage in a coil is proportional to the product of the number of loops and the rate at which the magnetic field changes within those loops.

This statement is called **Faraday's law**.

The amount of current produced by electromagnetic induction depends not only on the induced voltage but also on the resistance of the coil and circuit that it connects.[2] For example, we can plunge a magnet in and out of a closed loop of rubber and in and out of a closed loop of copper. The voltage induced in each is the same, providing each intercepts the same number of magnetic field lines. But the current in each is quite different. The electrons in the rubber feel the same voltage as those in the copper, but their bonding to the fixed atoms prevents a current such as that which so freely ensues in copper.

The discovery that electric currents in a magnetic field are deflected, and the discovery of induced voltages ten years later by Faraday and Henry, both stem from the same single fact: that moving charges experience a force that is perpendicular to both their motion and the magnetic field lines they traverse (Figure 23.11). On the one hand, motion is provided by some source that drives electrons along the wire, while, on the other hand, motion is provided by some mechanical force that pushes electrons, wire and all, across magnetic field lines. In the first case, we have a "motor effect," where the magnetic force on the moving charges is perpendicular to the wire; in the second case, we have a "generator effect," where the magnetic force on the moving electrons is along the wire. In both cases, the magnetic force is perpendicular to both the electron motion and the magnetic field.

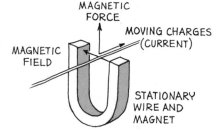 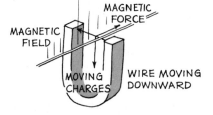

Fig. 23.11 Can you see the relationship between these phenomena?

Motor Effect	Generator Effect
When a current moves to the right, there is a perpendicular upward force on the electrons. Since there is no conducting path upward, the wire is tugged upward along with the electrons.	When a wire with no initial current is moved downward, the charges in the wire experience a deflecting force perpendicular to their motion. There *is* a conducting path in this direction, and the electrons follow it, thereby constituting a current.

[2]Current also depends on the "reactance" of the coil. Reactance is similar to resistance and is important in ac circuits; it depends on the number of loops in the coil and on the frequency of the ac source, among other things. We will not treat this complication here.

Question   If you push a magnet into a coil, as shown in Figure 23.10, you'll feel a resistance to your push. Why is this resistance greater in a coil with more loops?*

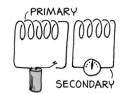

Fig. 23.12 Whenever the primary switch is opened or closed, voltage is induced in the secondary circuit.

So far, we have considered induction with the use of permanent magnets. In the foregoing examples and diagrams, the permanent magnets may be replaced with current-carrying coils of wire—electromagnets. Furthermore, it is not even necessary to have "motion" in the ordinary sense of the word. It is the field, rather than the magnet, that must move. In the case of the electromagnet, we can cause the field to expand and collapse by simply opening and closing an electric switch (Figure 23.12). A coil connected to a battery is placed alongside another coil connected to a galvanometer. It is customary to refer to the coil connected to the source of power as the *primary* (input) and to the other as the *secondary* (output). The instant the switch is closed in the primary and current begins to flow, a current flows in the secondary even though there is no material connection between the two coils. Only a brief surge of current occurs in the secondary, however. When the switch in the primary is opened, a surge of current again registers in the secondary but in the opposite direction.

This is the explanation: A magnetic field builds up around the primary when the current begins to flow through the coil. This means that the magnetic field is growing (that is, *changing*) about the primary. But since the coils are near each other, this changing field extends to the secondary coil, thereby inducing a voltage in the secondary. This induced voltage is only temporary, for when the current and the magnetic field of the primary reach a steady state—that is, when the magnetic field is no longer changing—no further voltage is induced in the secondary. But when the switch is turned off, the current in the primary drops to zero. The magnetic field about the coil collapses, thereby inducing a voltage in the secondary coil, which senses the change. We see that a voltage is induced whenever a magnetic field is *changing* around the conductor, regardless of the reason.

In short, we can say that voltage can be induced in a wire in three different ways: by moving the wire near a magnet, by moving a magnet

*****Answer**   Simply put, more work is required to induce the greater voltage induced by more loops. We can also look at it this way: When the magnetic fields of two magnets (electro or permanent) overlap, the two magnets are either forced together or forced apart. In the case in which one of the fields is induced by motion of the other, the polarity of the fields is always such as to force the magnets apart. This is the resistive force you feel. Inducing more current in more coils simply increases the induced field and hence the resistive force.

near the wire, or by changing a current in a nearby wire. In each case we have the important ingredient: relative motion between the magnetic field and the charges in the wire.[3]

Transformers

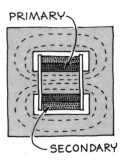

Fig. 23.13 A practical transformer.

If we place an iron core in a coil of current-carrying wire, the magnetic field is intensified by the alignment of magnetic domains in the iron. If we wrap primary and secondary coils one atop the other about an iron core shaped to guide and contain magnetic field lines, the magnetic field the primary produces will be completely intercepted by the secondary. To induce voltage in the secondary, we must change the current and hence the magnetic field in the primary. But instead of opening and closing a switch to produce this change, suppose we use alternating current to power the primary. Then the rate at which the magnetic field changes in the primary (and hence the secondary) equals the frequency of the alternating current. We have a **transformer** (Figure 23.13).

If the primary and secondary coils have an equal number of wire loops, then the input and output alternating voltages will be equal. We get no change. But if the secondary coil has more loops than the primary, the alternating voltage produced across the secondary coil will be *greater* than across the primary; in this case, we say the voltage is "stepped up" in the transformer. If the secondary coil has twice as many loops as the primary, for example, the induced voltage in the secondary will be twice that across the primary. This stepped-up voltage may light a neon sign. If the secondary has fewer loops than the primary, the alternating voltage produced across the secondary will be *lower* than the voltage across the primary; we say the voltage is "stepped down." This stepped-down voltage may safely move a toy electric train. If the secondary has half as many loops as the primary, for example, then only half as much voltage is induced in the secondary. So electric energy can be fed into the primary at a given alternating voltage and taken from the secondary at a greater or lower alternating voltage, depending on the relative number of loops in the primary and secondary coil windings.

The relationship between primary and secondary voltages with respect to the relative coil windings is given by

$$\frac{\text{Primary voltage}}{\text{Number of primary loops}} = \frac{\text{secondary voltage}}{\text{number of secondary loops}}$$

It might seem that we are getting something for nothing with a transformer that steps up the voltage, for we can supply a small voltage to the primary and take a large voltage off the secondary. But this is not without

---

[3]Relative motion must be such that "new" magnetic field lines are intercepted by the wire. Back-and-forth motion of the wire along its length, for example, or motion along and parallel to magnetic field lines will induce no voltage. Maximum voltage is induced when motion is such that a maximum number of field lines are "cut" by the wire (as in Figure 23.9).

recompensation, which is increased current drawn by the primary. The transformer actually transforms energy from one coil to the other. The rate at which energy is transferred is called *power*. The power used in the secondary is supplied by the primary. The primary gives no more than the secondary uses (conservation of energy). If we neglect slight power losses (heating of the core), we can say

$$\text{Power into primary} = \text{power out of secondary}$$

Electrical power is equal to the product of voltage and current, so we can say

$$(\text{Voltage} \times \text{current})_{\text{primary}} = (\text{voltage} \times \text{current})_{\text{secondary}}$$

The ease with which voltages can be stepped up or down with transformers is the principal reason that most electric power is ac rather than dc.

Questions

1. If 100 volts are put across a 100-loop transformer primary, what will be the voltage output if the secondary has 200 loops?*

2. Assuming the answer to the first question is 200 volts, and the secondary is connected to a floodlamp with a resistance of 50 ohms, what current flows in the secondary circuit?†

3. What is the power in the secondary coil?‡

4. What is the power in the primary coil?§

5. What is the current drawn by the primary coil?‖

Self-Induction

Current-carrying loops in a coil interact not only with loops of other coils but also with loops of the same coil. Each loop in a coil interacts with the magnetic field around the current in other loops in the same coil. This is called **self-induction**. A self-induced voltage is produced. This voltage is always in a direction opposing the changing voltage that produces it and is commonly called the "back emf."[4] We won't treat self-induction and back emfs here, except to acknowledge a common and dangerous effect. A coil

---

*__Answer__ From 100 volts/100 primary loops = (?) volts/200 secondary loops, we see that the secondary puts out 200 volts.

†__Answer__ From Ohm's law, 200 volts/50 ohms = 4 amps.

‡__Answer__ Power = 200 volts × 4 amps = 800 watts.

§__Answer__ The power in the primary is the same: 800 watts (conservation of energy).

‖__Answer__ 800 watts = 100 volts × (?) amps; so we see that the primary draws 8 amps. (Note that the voltage is stepped up from primary to secondary and that the current is correspondingly stepped down.)

[4]This opposing direction of induction is called *Lenz's law* and is a consequence of the conservation of energy.

with a large number of turns has a large self-inductance. Suppose such a coil is used as an electromagnet and is powered with a dc source, perhaps a small battery. Current in the coil is then accompanied by a strong magnetic field. When we disconnect the battery by opening a switch, we had better be prepared for a surprise. When the switch is opened, the current in the circuit falls rapidly to zero and the magnetic field in the coil undergoes a sudden decrease (Figure 23.14). What happens when a magnetic field suddenly changes in a coil—even if it is the same coil that produced it? The answer is that a voltage is induced. The rapidly collapsing magnetic field with its store of energy may induce an enormous voltage, large enough to develop an arc across the switch—or you, if you are opening the switch! For this reason, electromagnets are connected to a circuit that absorbs excess charge and prevents the current from dropping too suddenly. This reduces the self-induced voltage. This is also, by the way, why you should not disconnect appliances by pulling out the plug instead of using the switch. The circuitry in the switch may provide a nonsudden change in current.

Fig. 23.14 When the switch is opened, the magnetic field of the coil collapses. This sudden change in the field can induce a huge voltage.

Power Production

Turbogenerator Power

Fifty years after Faraday and Henry discovered electromagnetic induction, Nikola Tesla and George Westinghouse put their findings to practical use and showed the world that electricity could be generated reliably and in sufficient quantities to light entire cities. Tesla's generators, like those of today, had armatures consisting of bundles of copper wires located within strong magnetic fields that were made to spin by means of a turbine, which in turn was spun by the energy of falling water or steam. The rotating loops of wire in the armature cut through the magnetic field of the surrounding electromagnets, thereby inducing alternating voltage and current.

We can look at this process from an atomic point of view. When the wires in the spinning armature cut through the magnetic field, oppositely directed electromagnetic forces act on the negative and positive charges.

Electrons respond to this force by momentarily swarming relatively freely in one direction throughout the crystalline copper lattice; while the copper atoms, which are actually positive ions, are forced in the opposite direction. But the ions are anchored in the lattice so their migration is nil. Only the electrons move, sloshing back and forth in alternating fashion with each rotation of the armature. The energy of this electronic sloshing is tapped at the electrode terminals of the generator.

The construction of a generator is in principle identical to that of a motor. Only the roles of input and output are reversed. In a motor, electrical energy is the input and mechanical energy the output; in a generator, mechanical energy is the input and electrical energy the output. Both devices simply transform energy from one form to another.

MHD Power

An MHD (magnetohydrodynamic) generator does away with a turbine and spinning armature altogether. Instead of making charges move in a magnetic field via a rotating armature, a plasma of electrons and positive ions expands through a nozzle and moves at supersonic speed through a magnetic field. Like the armature in a turbogenerator, the motion of charges through a magnetic field gives rise to an electromotive force and flow of current in accordance with Faraday's law of induction. Whereas in a conventional generator, "brushes" carry the current to the external load circuit, in the MHD generator the same function is performed by "electrodes" (Figure 23.15). Unlike the turbogenerator, the MHD generator can operate at any temperature to which the plasma can be heated, either by combustion or nuclear processes. The higher temperature results in a high

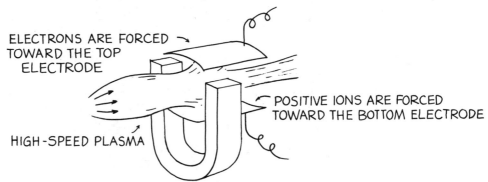

ELECTRONS ARE FORCED TOWARD THE TOP ELECTRODE

POSITIVE IONS ARE FORCED TOWARD THE BOTTOM ELECTRODE

HIGH-SPEED PLASMA

Fig. 23.15 A simplified MHD generator. Motion of the charged particles in the high-speed plasma through the magnetic field induces an electromotive force between the top and bottom electrodes. The oppositely directed forces on positive and negative particles in the plasma beam effectively result in an induced direct electron current that flows from the bottom electrode, through the plasma, to the top electrode, then through the electrical circuit (not shown), and back to the bottom electrode. There are no moving parts; only the plasma moves. In practice, superconducting electromagnets are used.

thermodynamic efficiency, which means more power for the same amount of fuel and less waste heat. Efficiency is further boosted when the "waste" heat is used to turn water into steam and run a conventional steam-turbine generator.

This substitution of a flowing plasma for rotating copper coils in a generator has become operational only recently because the technology to produce plasma of high-enough temperatures is new. Current plants use a high-temperature plasma formed by combustion of fossil fuels in air or oxygen.

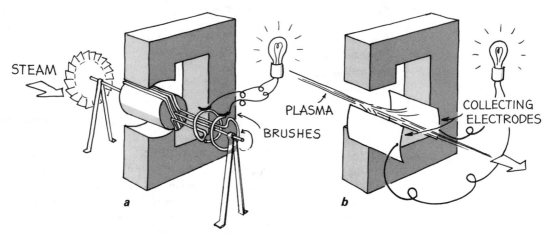

Fig. 23.16   A comparison of the conventional turbine-driven generator and the MHD generator. In both generators, moving electrons are deflected by a magnetic field. (*a*) In the turbogenerator, the electrons are in the copper wire of the spinning armature. (*b*) In the MHD generator, the electrons are in the plasma.

Power Transmission

Almost all electric energy sold today is in the form of alternating current because of the ease with which it can be transformed from one voltage to another. Power is transmitted great distances at high voltages and correspondingly low currents, a process that otherwise would result in large energy losses owing to the heating of the wires. Power may be carried from the power plants to the cities at about 120,000 volts or more, stepped down to about 2200 volts in the city, and finally stepped down to 220 or 110 volts for safe home use.

Energy, then, is transformed from one system of conducting wires to another by electromagnetic induction. It is but a short step further to find that these same principles account for sending energy from a radio-transmitter antenna to a radio receiver many kilometers away, and just a tiny step further to the transformation of energy of vibrating electrons in the sun to life energy on earth. The effects of electromagnetic induction are far-reaching. As far as is conceivable!

Fig. 23.17 Power transmission.

Field Induction

We have discussed electromagnetic induction thus far as the production of voltages and currents. Actually, the more fundamental fields underlie both voltages and currents. The modern view of electromagnetic induction holds that electric and magnetic *fields* are induced, which in turn give rise to the voltages we have considered. Induction takes place whether or not a conducting wire or any material medium is present. In this general sense, Faraday's law states

An electric field is induced in any region of space in which a magnetic field is changing with time. The magnitude of the induced electric field is proportional to the rate at which the magnetic field changes. The direction of the induced electric field is at right angles to the changing magnetic field.

There is a second effect, which is the counterpart to Faraday's law. This effect, known as **Maxwell's counterpart to Faraday's law**, was advanced by James Clerk Maxwell in about 1860:

A magnetic field is induced in any region of space in which an electric field is changing with time. The magnitude of the induced magnetic field is proportional to the rate at which the electric field changes. The direction of the induced magnetic field is at right angles to the changing electric field.

These statements are two of the most important statements in physics. They underlie our understanding of electromagnetic radiation.

In Perspective*

The ancient Greeks discovered that when a piece of amber (a natural plastic-like mineral) was rubbed, it picked up little pieces of papyrus. They found strange rocks on the island of Magnesia that attracted iron. Probably because the air in Greece was relatively humid, they never noticed and studied the static electrical charge effects common in dry climates. Further development of electrical and magnetic phenomena did not take place

---

*Adapted from R. P. Feynman, R. B. Leighton, and M. Sands, *The Feynman Lectures on Physics,* © 1965, Vol. II, California Institute of Technology, Chap. 1, pp. 1-10 and 1-11. Used by permission from Addison-Wesley Publishing Company, Inc., Reading, Massachusetts, U.S.A. Richard P. Feynman is a Nobel laureate in physics and professor of physics at CalTech.

until 400 years ago. The human world shrank, as more was learned about electricity and magnetism. It became possible to signal by telegraph over long distances, then to talk to another person many kilometers away through wires, then not only to talk but also to send pictures over many kilometers with no connections in between.

Energy, so vital to civilization, could be transmitted over hundreds of kilometers. The energy of elevated rivers was diverted into pipes that fed giant "waterwheels" connected to assemblages of twisted and interwoven copper wires that rotated about specially designed chunks of iron—revolving monsters called generators. Out of this, energy was pumped through copper rods as thick as your wrist and sent to huge coils wrapped around transformer cores, boosting it to high voltages for efficient long-distance transmission to cities. Then the transmission lines split into branches . . . then to more transformers . . . then more branching and spreading, until finally the energy of the river was spread throughout whole cities—turning motors, making heat, making light, working gadgetry. There was the miracle of hot lights from cold water hundreds of kilometers away—a miracle made possible by specially designed bits of copper and iron that turned because people had discovered the laws of electromagnetism.

These laws were discovered at about the time the American Civil War was being fought. From a long view of human history, there can be little doubt that events such as the American Civil War will pale into provincial insignificance in comparison with the more significant event of the nineteenth century: the discovery of the electromagnetic laws.

Summary of Terms

Electromagnetic induction The induction of voltage when a magnetic field changes with time. If the magnetic field within a closed loop changes in any way, a voltage is induced in the loop:

Voltage induced $\sim$ $-$ no. of loops $\times$ $\dfrac{\text{mag. field change}}{\text{time}}$

This is a statement of Faraday's law. The induction of voltage is actually the result of a more fundamental phenomenon: the induction of an electric *field*. We will define Faraday's law for this more general case below.

Transformer A device for transforming electrical power from one coil of wire to another by means of electromagnetic induction.

Faraday's law An electric field is induced in any region of space in which a magnetic field is changing with time. The magnitude of the induced electric field is proportional to the rate at which the magnetic field changes. The direction of the induced field is at right angles to the changing magnetic field.

Maxwell's counterpart to Faraday's law A magnetic field is induced in any region of space in which an electric field is changing with time. The magnitude of the induced magnetic field is proportional to the rate at which the electric field changes. The direction of the induced magnetic field is at right angles to the changing electric field.

Review Questions

1. How do the directions of magnetic field lines about a current-carrying wire differ from the directions of electric field lines about a charge?

2. Why is the magnetic field strength greater inside a current-carrying loop of wire than about a straight section of wire?

3. Why does a piece of iron in a current-carrying loop yield an even greater magnetic field strength?

4. In what way is a galvanometer similar to an electric motor?

5. How are the motor effect and generator effect similar?

6. Why is *change* a key word in electromagnetic induction?

7. In what three ways can a voltage be induced in a wire?

8. What are the principal differences between an MHD generator and a conventional generator?

9. What is the principal advantage of ac over dc?

10. What exactly does a transformer "transform"?

11. Why does a transformer require alternating current?

12. Can an efficient transformer step up energy? Explain.

13. What name is given to the rate at which energy is transferred?

14. What is induced by the rapid alternation of a magnetic field?

15. What is induced by the rapid alternation of an electric field?

Exercises

1. Why does an iron core increase the magnetic induction of a coil of wire?

2. Why are the armature and field windings of an electric motor usually wound on an iron core?

3. Why is a generator armature harder to rotate when it is connected to and supplying electric current to a circuit?

4. Will a cyclist coast farther if the lamp connected to his generator is turned off? Explain.

5. Two separate but similar coils of wire are mounted a few meters apart, as in Figure 23.12. The first coil is connected to a battery and has a direct current flowing through it. The second coil is connected to a galvanometer. What do the galvanometer readings show when the current in the first coil is increasing? Decreasing? Remaining steady?

6. Why will more voltage be induced with the apparatus shown in Figure 23.12 if an iron core is inserted in the coils?

7. If your metal car moves over a wide closed loop of wire embedded in a road surface, will the magnetic field of the earth within the loop be altered? Will this produce a current pulse? Can you think of a practical application for this?

8. What is the difference between an electric motor and an electric generator?

9. How could the motor in Figure 23.6 be used as a generator of electricity?

10. How is the current induced in a wire loop affected by the frequency of the changing magnetic field?

11. Does the voltage output increase when a generator is made to spin faster? Explain.

12. A motor acts as both a motor and a generator at the same time. Is the same true of a generator? Explain.

13. Can a current-carrying loop be oriented in a uniform magnetic field in such a way that it does not tend to rotate? Explain.

14. Why would long-distance transmission of electric power done at low voltages incur heavy losses?

15. Why is it important that the core of a transformer pass through both coils?

16. Does the current in the secondary windings of a transformer double or halve when the induced voltage in the secondary is doubled? Explain.

17. How does the current in the secondary compare to the current in the primary when the secondary voltage is doubled?

18. In what sense can a transformer be thought of as an electrical "lever"?

19. Why can a hum usually be heard when a transformer is operating?

20. What is wrong with this scheme? To generate electricity without fuel, arrange a motor to run a generator that will produce electricity that is stepped up with transformers so that the generator can run the motor and simultaneously furnish electricity for other uses.

21. If a bar magnet is thrown into a loop of wire, will it slow down? Why or why not?

22. A model electric train requires 6 volts to operate. If the primary coil of its transformer has 240 windings, how many windings should the secondary have if the primary is connected to a 120-volt household circuit?

23. Neon signs require about 12,000 volts for their operation. What should be the ratio of the number of loops in the secondary to the number of loops in the primary for a neon-sign transformer that operates off 120-volt lines?

24. An induction coil in an automobile is actually a form of transformer that boosts 12 volts to about 24,000 volts. What is the ratio of secondary windings to primary windings? (The 12 volts in a car is dc. In order to make the induction coil operate, the dc must be interrupted frequently. This is controlled by a switch in the distributor.)

25. Would electromagnetic waves exist if changing magnetic fields could produce electric fields, but changing electric fields could not in turn produce magnetic fields? Explain.

26. If you place a metal ring in a region where a magnetic field is rapidly alternating, the ring may become hot to your touch. Why?

24 Electromagnetic Radiation

We commonly associate the word *radiation* with radioactivity, reactions involving the atomic nucleus and characteristic of nuclear power plants and nuclear bombs. That kind of radiation we'll cover in later chapters. Here we are concerned with **electromagnetic radiation**, the transfer of energy by the rapid oscillations of electromagnetic fields in space. These oscillations travel in the form of waves, called *electromagnetic waves,* which are familiar to us as radio and television waves, microwaves, sunlight, and medical X rays.

Even though we are familiar with how waves are propagated on the surface of water, or sound in air, it is difficult to imagine waves propagating through a vacuum. That's why early investigators of light and radio waves explained wave motion by envisaging a material medium they called the *ether* to fill the voids of space as water fills a pond or molecules make up the air. Yet if we accept that the force of gravity acts through a vacuum, or that a magnet affects a compass needle even if it is inside an evacuated chamber, we can more easily conceptualize the idea of oscillating electromagnetic fields emanating from vibrating electric charges and propagating through empty space. Electromagnetic waves—whether they are radio waves, light waves, or high-energy X rays—are the propagation of energy via induced electric and magnetic fields oscillating through space with the same rhythm of oscillation as the vibrating charges that set them into being.

Electromagnetic Wave Velocity

The existence of electromagnetic waves was first deduced by James Clerk Maxwell in about 1860 in conjunction with his theoretical work in linking electricity and magnetism. His highly mathematical ideas about the nature of electromagnetic waves are among the most abstract in physics. Nevertheless, it is possible to understand his ideas conceptually.

Consider an isolated electric charge vibrating to and fro at a certain frequency. The moving charge constitutes an electric current that changes in direction with the frequency of the vibrating charge. Surrounding this electric current, then, is a magnetic field, which also changes direction with each oscillation of the vibrating charge. The law of induction tells us that a changing magnetic field will induce an electric field. These fields mutually induce each other, the changing magnetic field inducing a changing electric field, which in turn induces a changing magnetic field, and so on. Hence, there are induced magnetic and electric fields about the vibrating charge. These fields are not localized, but emanate outward from the vibrating charge. The intensity of the induced fields very much depends on this speed of emanation. Let's see why this is so.

Consider first the initial magnetic field induced by the moving charge. This changing magnetic field induces a changing electric field, which in turn induces a magnetic field. The magnitude of this further induced magnetic field depends not only on the vibrational rate of the electric field but also on the *motion* of the electric field, or the speed at which the induced field emanates from the vibrating charge. The higher the speed, the greater the magnetic field it induces. At low speeds, electromagnetic regeneration would be short-lived because a slow-moving electric field would induce a weak magnetic field, which would induce in turn a weaker electric field. The induced fields would become successively weaker, causing the mutual induction to die out. But what of the energy in such a case? The fields contain energy acquired from the vibrating charge; if the fields disappeared with no means of transferring energy to some other form, energy would be destroyed. We see that low-speed emanation of electric and magnetic fields is incompatible with the law of energy conservation. At emanation speeds too high, on the other hand, the fields would be induced to ever-increasing magnitudes, with a crescendo of ever-increasing energies—again clearly a no-no with respect to energy conservation. At some critical speed, however, mutual induction would continue indefinitely, with neither a loss nor a gain in energy.

From his equations of electromagnetic induction, Maxwell calculated the value of this critical speed—300,000 kilometers per second. But this is the velocity of light! Maxwell at once realized that he had discovered the solution to one of the greatest mysteries of the universe—the nature of light. For if the electric charge is set into oscillation within the incredible frequency range of 4.3×10^{14} to 7×10^{14} vibrations per second, the resulting electromagnetic wave will activate the "electrical antennae" in the retina of the eye. Light is simply electromagnetic radiation in this range of frequencies! The lower frequency appears red, and the higher frequency appears violet.

Fig. 24.1 Shake a charged object to and fro and you produce electromagnetic waves.

Question   The singular velocity of electromagnetic waves is a remarkable consequence of what central principle in physics?*

Electromagnetic Spectrum   All electromagnetic waves travel at the same speed in a vacuum. They differ from one another in their frequency and wavelength. The frequency of the wave as it vibrates through space is identical to the frequency of the oscillating electric charge generating it. The differences in wavelength result from the differences in frequency. For example, since the speed of

---

***Answer**   The underlying principle that dictates the speed of light is the *conservation of energy.*

the wave is 300,000 kilometers per second, an electric charge oscillating once per second (1 hertz) will produce a wave with a wavelength 300,000 kilometers long. This is because only one wavelength is generated in 1 second. If the frequency of oscillation were 10 hertz, then 10 wavelengths would be formed in 1 second and the corresponding wavelength would be 30,000 kilometers long. A frequency of 10,000 hertz would produce wavelengths 30 kilometers long. So the higher the frequency of the vibrating charge, the shorter the wavelength of radiation.[1]

The wavelength of standard AM radio waves ranges from 190 to 560 meters, and FM and television waves from 2.7 to 3.7 meters. The wavelength of visible light waves ranges from 35 to 70 millionths of a centimeter. The shortest waves speculated are cosmic rays of wavelength 10^{-31} centimeter, about one-billionth of a billionth the size of a proton. (Cosmic rays consist of electromagnetic radiation as well as nuclear particles.)

Electromagnetic waves in principle can have any frequency from zero to infinity. The classification of electromagnetic waves according to frequency is called the **electromagnetic spectrum** (Figure 24.2). Radiation has been detected with a frequency as low as 0.01 hertz. Electromagnetic waves with frequencies on the order of several thousand hertz (kilocycles per second) are classified as radio waves. The VHF (very high frequency) television band of waves starts at about 50 million hertz (megacycles per second). Still higher frequencies are called microwaves, followed by infrared waves, often called "heat waves." Further still is visible light, which makes up less than a millionth of 1 percent of the measured electromagnetic spectrum. Beyond light, the higher frequencies extend into the ultraviolet, X-ray, and gamma-ray regions. There is no sharp distinction between these regions, which actually overlap each other. The spectrum is simply broken up into these arbitrary regions for classification.

Fig. 24.2 The electromagnetic spectrum (not to scale) is a continuous range of radiation extending from radio waves to gamma rays. The descriptive names of the sections are merely an historical classification. In all sections the waves are the same in nature, differing only in frequency and wavelength; all have the same speed.

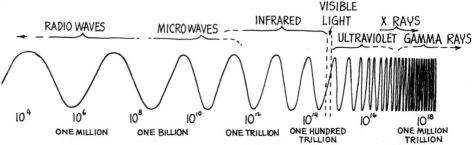

FREQUENCY IN HERTZ

Production of Electromagnetic Waves

Of considerable interest is the way in which the oscillating frequency of the electric charges that produce these waves is controlled. Consider the oversimplified antenna in Figure 24.3. The rotating device in the center

---

[1]The relationship is $c = f\lambda$, where c is the wave velocity (constant), f is the frequency, and λ is the wavelength.

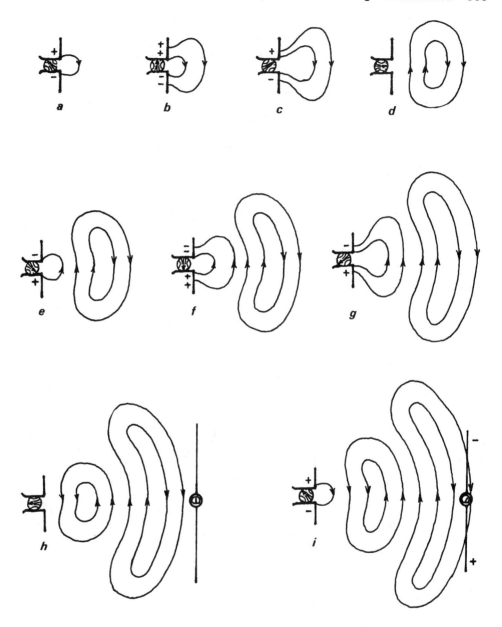

Fig. 24.3 The rotating device alternately charges the upper and lower parts of the antenna positively and negatively. Successive views, (a) through (i), indicate how acceleration of the charges up and down the antenna transmits electromagnetic waves. Only sample electric field lines of the waves are shown; the magnetic field lines are perpendicular to the electric field lines and extend into and out of the page. When the waves are incident on a receiving antenna, (h) and (i), electric charges in it vibrate in rhythm with the field variations.

alternately charges the upper and lower halves of the antenna positively and negatively. Electric charges accelerate up and down in the antenna and send out electromagnetic waves that have a frequency equal to the rotational frequency of the rotor. If there is another wire antenna at some distance from the transmitting antenna, the electric charges in that wire will vibrate to and fro in response to the variation in the electric and magnetic fields impinging on the wire. These pulsations of current can be easily detected by a very sensitive galvanometer.

The frequencies of these pulsations for commercial broadcasting range from 550 to 1500 kilohertz. No mechanical devices can rotate at such high frequencies, so in place of rotors small crystals are used. Small crystals vibrate naturally at these high frequencies—the natural frequency depending on the elastic properties of the crystal, its size, and its shape—just as the natural frequency of a bell depends on much the same variables. Broadcasting stations use small crystals of quartz or other materials that are ground to the proper size and shape for controlling the frequency with which electric charge is driven up and down in the transmitting antenna.

Each radio station has its own crystal, a quartz wafer about the size of a 5-cent coin, which vibrates hundreds of thousands of times each second, ensuring a constant frequency of radiation. The sound signal to be communicated is accomplished by superimposing the much-lower-frequency sound waves on the higher-frequency electromagnetic radio waves, or **carrier waves** (Figure 24.4). When the *amplitude* of the carrier wave is modulated (varied), we called it AM, or **amplitude modulation**; when the *frequency* of the carrier wave is modulated, we called it FM, or **frequency modulation**. Amplitude modulation is like changing the intensity of a constant-color light bulb. Frequency modulation is like changing the color of a constant-intensity light bulb. In radio or TV broadcasting, the resonant frequency of the receiver is adjusted to match and amplify the

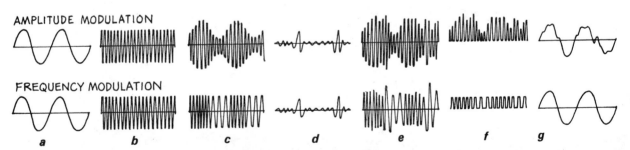

AMPLITUDE MODULATION

FREQUENCY MODULATION

a   b   c   d   e   f   g

Fig. 24.4 AM and FM radio signals. (*a*) Sound waves enter microphone. (*b*) Radio-frequency carrier wave produced by transmitter without sound signal. (*c*) Carrier wave modulated by signal. (*d*) Static interference. (*e*) Carrier wave and signal affected by static. (*f*) Radio receiver cuts out negative half of carrier wave. (*g*) Signal remaining is rough for AM because of static but is smooth for FM because the tips of the wave form are clipped without loss to the signal.

carrier wave and to feed the imposed modulated wave to the loudspeaker or TV picture tube. In this way the output message is a duplicate of that put into the transmitter.

Smaller crystals vibrate at higher frequencies, just as smaller bells emit higher-pitched tones. Smaller crystals are used to transmit higher-frequency radio and television waves. Radar waves require crystals as small as 0.02 centimeter. Beyond this point, crystals become too small to manage.

The crystal size necessary to generate frequencies corresponding to infrared or heat waves is approximately that of a molecule. This suggests that heat waves are generated by vibrating atoms and molecules carrying electric charges. Obtaining frequencies corresponding to visible light requires particles still smaller than molecules and atoms. These particles turn out to be the electrons within the atom. High-frequency vibrating electrons emit light. Higher electron frequencies produce ultraviolet radiation.

At still higher frequencies, electrons emit X rays. X rays are produced not only by the acceleration of vibrating electrons but also by the rapid deceleration of high-speed electrons as they hit a target. They undergo a tremendous acceleration by being stopped or deflected very suddenly. These positive and negative accelerations are accompanied by radiation of extremely high frequency.

Beyond the X-ray region of the spectrum lies the gamma-ray region. These rays accompany radioactive disintegration of the atomic nucleus. Their production indicates that within the nucleus there are electric fields of enormous strengths wherein nuclear charges are accelerated at incredibly high rates. Many of these high-frequency gamma rays originate in the cosmos and are called *cosmic rays*. There seems to be no upper limit to their frequencies.

The origins of all these electromagnetic radiations are alike in that they are produced by accelerating electric charges. All have the same speed in a vacuum. They differ only in their frequencies of vibration and in their corresponding wavelengths. As the frequency increases, the wavelength decreases such that the product of frequency and wavelength is a constant: the velocity of light.

An interesting property of radiation is penetrability. Radio waves go through air but not through metals. Lower-frequency radio waves bounce off the ionic layers high in the atmosphere, while higher-frequency waves pass through and out into space (Figure 24.5). At night when the ionic layers are out of the sunlight and more settled, radio reception is improved, as distant signals are reflected from the layers and are picked up by local radio receivers. Higher-frequency radiation carrying television signals penetrates the ionic layers, however, and transmission must be accomplished on a "line of sight" basis. Infrared radiation goes through dry air but not through water vapor. If passed through a sample of a person's

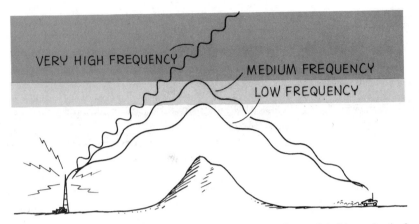

Fig. 24.5 The stratified layers extending from about 50 km upward in the ionosphere transmit very high-frequency electromagnetic waves (FM and TV) but reflect lower-frequency waves (standard AM).

breath, transmission is altered if the person has been drinking alcohol—a fact that enables a very precise, quick, and easy means to determine a person's blood alcohol level. Most things, like people, absorb radiation in the infrared region. That is why we call them "heat waves": because they heat us up with the energy that we absorb from them. Visible light goes through glass, while ultraviolet does not. X rays go through any kind of matter but with some absorption. Any given material, depending on its structure, absorbs, reflects, or transmits a given radiation selectively.

Questions

1. What is the source of all electromagnetic waves?*
2. What determines the frequency of electromagnetic waves?†

Electromagnetic Waves Are Everywhere

Accelerating electric charges, then, radiate energy in waves of electric and magnetic fields. Atoms in every substance are continually being bounced and jostled about; as a result, the electrons of these atoms vibrate and radiate vibrations of electromagnetic energy. Every physical object in the universe—the stars, the planets, rocks, trees, your body—radiates electromagnetic waves at a frequency proportional to the rate of molecular and atomic vibratory motion, which is to say in proportion to the temperature. If the temperature is high enough, part of the radiation is visible as light. At lower temperatures, the bulk of the radiation falls in the infrared portion of the spectrum, with some amounts in the microwave-radio region. Though invisible, these radiations can be detected and measured by various devices.

---

*Answer　Oscillating electric charges.

†Answer　The frequency of the oscillating electric charges determines and is the same as the frequency of the electromagnetic waves so produced.

Fig. 24.6 A thermographic scanner.

Photographic film made with an emulsion sensitive to the infrared region of the spectrum shows different temperatures as different colors. High-altitude photos of the earth do more than detect forest fires. Subterranean lava channels not visible on the surface of the earth, underground geothermal springs, and anything having a higher or lower temperature than the surrounding environment can be detected by infrared photography. We are all literally glowing with our own invisible light. A "heat picture" of the infrared radiations we emit can reveal regions slightly warmer or cooler than normal that may hide a malignancy. Devices called thermographic scanners can produce an on-the-spot TV-like view of the infrared emission of various bodies (Figure 24.6). Different rocks and minerals radiate infrared waves that differ slightly. This fact enables geologists to chart the mineral makeup of the earth from airplanes. Spacecraft can analyze the composition of soils on various parts of the moon and planets. Diseased vegetation radiates a slightly different frequency of radiation than healthy vegetation, which shows up clearly on infrared film. Farmers value this knowledge.

a

b

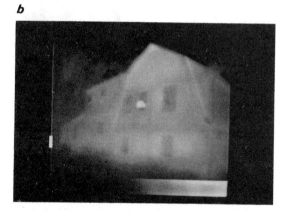

Fig. 24.7 Invisible heat waves from (*a*) are detected by the thermographic scanner and appear as in (*b*).

Cooler substances radiate lower-frequency electromagnetic waves. Icebergs, for example, radiate microwaves. Although icebergs radiate a lower-frequency radiation than water, they emit slightly more radiation than water. Ice-patrol fleets take advantage of this fact, and microwave-detection devices have become highly efficient iceberg spotters along the shipping lanes of the North Atlantic. Microwaves are also used to map heavy rain-bearing clouds in weather studies as well as soil-moisture and water-distribution patterns in hydrological and agricultural surveys.

We tend to think of space as being empty. But this is only because we cannot see the montages of electromagnetic waves that permeate every part of our surroundings. We see some of these waves, of course, and we call them light. These waves constitute only a microportion of the electromagnetic spectrum. For example, we are unconscious of radio waves

Fig. 24.8 An infrared photograph. Diseased portions of the trees will radiate a slightly different frequency and show up as a different color on such a photograph.

that engulf us every moment. Free electrons in every piece of metal on the earth's surface continually dance to the rhythms of these waves. They jiggle in unison with the electrons being driven up and down along radio- and television-transmitting antennae. A radio or television receiver is simply a device that sorts and amplifies these tiny currents. There is radiation everywhere. Our first impression of the universe is one of matter and void—when actually the universe is a dense sea of radiation in which occasional concentrates are suspended.

Summary of Terms

Electromagnetic radiation   The transfer of energy by the rapid oscillations of electromagnetic fields, which travel in the form of waves, called *electromagnetic waves*.

Electromagnetic spectrum   The range of frequencies over which electromagnetic radiation can be propagated. The lowest frequencies are associated with radio waves, then microwaves, infrared waves, light, ultraviolet radiation, X rays, and gamma rays in progression.

Carrier wave   The wave, usually of radio frequency, whose characteristics are modified in the process of modulation.

Modulation   The process of impressing one wave system upon another of higher frequency.

Amplitude modulation (AM)   A type of modulation in which the amplitude of the carrier wave is varied above and below its normal value by an amount proportional to the amplitude of the impressed signal.

Frequency modulation (FM)   A type of modulation in which the frequency of the carrier wave is varied above and below its normal frequency by an amount that is proportional to the amplitude of the impressed signal. In this case the amplitude of the modulated carrier wave remains constant.

Review Questions

1. What is light?

2. What bearing does the conservation of energy have on the speed of light?

3. What is the source of all electromagnetic waves?

4. What determines the frequency of an electromagnetic wave?

5. What is the relationship among wave speed, frequency, and wavelength?

6. What is the wavelength of an electromagnetic wave having a frequency of 1 Hz? Of 100,000 Hz?

7. What does it mean to say a wave is "modulated"?

8. Distinguish between *AM* and *FM radio waves.*

9. Which standard radio wave has the higher frequency—AM or FM? Which has the shorter wavelength?

10. What is the principal difference between radio waves and light waves? Between light waves and X rays?

11. Why is radio reception of distant stations better at night than during the day?

12. How is the frequency of radio waves controlled at broadcasting stations?

Exercises

1. Can electromagnetic waves travel through a vacuum? What evidence can you cite to support your answer?

2. How do sound waves and radio waves differ? How are they alike?

3. Are the wavelengths of radio waves longer or shorter than those detectable by your eyes?

4. Why are infrared waves called heat waves?

5. Use the ideas of forced vibrations, resonance, and electromagnetic induction to explain how microwave ovens are able to cook food.

6. About how many more times greater is the frequency of a 100-MHz-carrier radio wave than a medium-range sound wave?

7. What is the frequency of an electromagnetic wave that has a wavelength of 300,000 kilometers?

8. If you charge a comb by rubbing it through your hair, and then shake it up and down, are you producing electromagnetic waves? With what frequency would you have to shake the comb to produce visible light?

9. A rattlesnake perceives infrared radiation. Can a rattlesnake, then, "see" in the dark? Discuss.

10. If all objects radiate energy, why can't we see objects in a darkened room?

11. Why does the frequency of electromagnetic waves that emanate from a hot source depend on the temperature of the source?

12. The speed of any wave is given by the formula Speed = frequency × wavelength. Revise this formula to give wavelength in terms of speed and frequency. Your new expression will be useful for the next six exercises.

13. Electromagnetic radiation has been detected with a frequency as low as 0.01 Hz. What is the wavelength of such a wave?

14. What is the wavelength of the carrier wave received at FM station 100 MHz on your radio dial?

15. Radiation emitted in outer space by hydrogen atoms have a wavelength of 21 centimeters. What is the frequency of this radiation?

16. FM radio is broadcast in a frequency range from 88 MHz to 108 MHz. What range of wavelengths is spanned by FM radio?

17. An antenna for optimum reception of a 27-MHz CB radio broadcast is designed to be $\frac{5}{8}$ wavelength long. How long is this in centimeters?

18. A microwave oven operates at a frequency of 900 MHz. If the microwave energy penetrates to a depth of $\frac{1}{5}$ the wavelength, how thick can a roast be and still be adequately cooked?

19. A tiny slice of the overall electromagnetic spectrum that ranges from 4×10^{14} to 7×10^{14} Hz is the part we call light. It turns out that this is also the range of maximum-intensity radiation from the sun. Why do you suppose this is so?

20. An ordinary light bulb gives off visible white light. If you were moving toward it at nearly the speed of light, it would appear to be emitting X rays; while if you were moving away from it at the same high speed, it would appear to be emitting radio waves. In both cases you could not see the bulb with your eyes. Why?

PART 6 Light

25 Reflection and Refraction

Less than one millionth of 1 percent of the measured electromagnetic spectrum is the part that seems to fascinate us most—visible light. We begin our study of light in this chapter by considering how light behaves. In the next chapter we will discuss how it appears—color. Then we will learn about its wave nature in Chapter 27, and its quantum nature in Chapters 28 and 29.

Most of the things we see around us do not emit their own light. They are visible because they reflect part of the light that falls upon them from a primary source such as the sun or a lamp or from a secondary source such as the illuminated sky. When light falls on the surface of a material, it is either re-emitted without change in frequency or is absorbed in the material and turned into heat.[1] Usually both of these processes occur in various degrees. When the re-emitted light is returned into the medium from which it was incident, we call the process *reflection*. When the re-emitted light bends from its original course and proceeds in straight lines into a transparent material, we say it is *refracted*.

Reflection

When this page is illuminated with sunlight or lamplight, a general electron vibration is set up in the normal energy states of its atoms; whole electron clouds vibrate in response to the oscillating electric fields of the illuminating light. Even under bright sunlight, the amplitudes of these vibrations are less than 1 percent of the radius of the atomic nucleus. It is these tiny electron vibrations that re-emit the light by which we see the page. When the page is illuminated by white light, it appears white, which reveals the fact that the electrons are set into vibrations at all the visible frequencies. Very little absorption occurs. The ink on the page is a different story. Except for a bit of reflection, it absorbs all the visible frequencies and therefore appears black.

So on the surfaces of all the objects around us, the electron clouds of the atoms undergo slight vibrations under the influence of illuminating light. These tiny vibrations over a variety of frequency ranges emit the various colors of light by which we see these objects. Simply put, we say we see these objects by the light they reflect.

Fig. 25.1 Light interacts with atoms as sound interacts with tuning forks.

[1] Another less common fate is absorption followed by re-emission at lower frequencies—fluorescence—as will be discussed in Chapter 28.

Principle of Least Time*

We all know that light ordinarily travels in straight lines. In going from one place to another, light will take the most efficient path and travel in a straight line. This is true if there is nothing to obstruct the passage of light between the places under consideration. If light is reflected from a mirror, the kink in the otherwise straight-line path is described by a simple formula. If light is refracted, like going from air into water, still another formula describes the alteration of light from the straight-line path. Before thinking of light with these formulas, we will first consider an *idea* that underlies all the formulas that describe light paths. This idea was discovered by the French scientist Pierre Fermat in about 1650, and it is called the *principle of least time,* or *Fermat's principle*. His idea is this: Out of all possible paths that light might take to get from one point to another, it takes the path that requires the *shortest time.*

*This material and many of the examples of least time are adapted from R. P. Feynman, R. B. Leighton, and M. Sands, *The Feynman Lectures on Physics,* Vol. I, Chap. 26 (Reading, Mass.: Addison-Wesley, 1965).

Law of Reflection

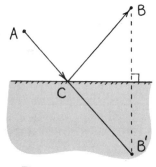

Fig. 25.2

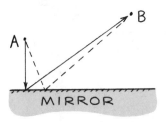

Fig. 25.3

We can understand **reflection** by way of the principle of least time. Consider the following situation. In Figure 25.2 we see two points, A and B, and an ordinary plane mirror beneath. What is the way to get from A to B in the shortest time? The answer is simple enough—go straight from A to B! But if we add the condition that the light must strike the mirror in going from A to B in the shortest time, the answer is not so easy. One way would be to go as quickly as possible to the mirror and then to B, as shown by the solid lines in Figure 25.3. This gives us a short path to the mirror but a very long path from the mirror to B. If we instead consider a point on the mirror a little to the right, we slightly increase the first distance, but we greatly decrease the second distance, and so the total path length shown by the dashed lines, and therefore the travel time, is less. How can we find the exact point on the mirror for which the time is shortest? We can find it very nicely by a geometrical trick.

We construct on the opposite side of the mirror an artificial point, B′, which is the same distance "through" and below the mirror as the point B is above the mirror (Figure 25.4). The shortest distance between A and this artificial point B′ is simple enough to determine: it's a straight line. Now this straight line intersects the mirror at a point C, the precise point of reflection for the shortest path and hence the path of least time for the passage of light from A to B. Inspection will show that the distance from C to B equals the distance from C to B′. We see that the length of the path from A to B′ through C is equal to the length of the path from A to B bouncing off point C along the way.

Further inspection and a little geometrical reasoning will show that the angle of incident light from A to C is equal to the angle of reflection from C to B. This is the **law of reflection**, and it holds for all angles (Figure 25.5). It is stated as

The angle of incidence equals the angle of reflection.

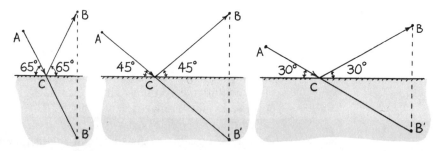

Fig. 25.5 Reflection.

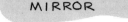

Fig. 25.4

The law of reflection is illustrated with arrows representing light rays in Figure 25.6. Instead of measuring the angles of incident and reflected rays from the reflecting surface, it is customary to measure them from a line perpendicular to the plane of the reflecting surface. This imaginary line is called the *normal*. The incident ray, the normal, and the reflected ray all lie in the same plane.

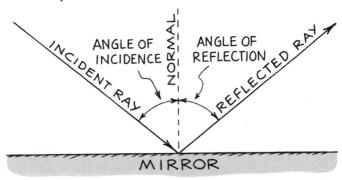

Fig. 25.6   Reflection.

Exercise   The constructions of artificial points B′ in Figures 25.4 and 25.5 show how light encounters point C in reflecting from A to B. By similar construction, show that light that originates at B and reflects to A also encounters the same point C.*

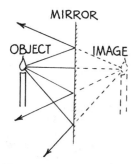

Fig. 25.7 A virtual image is formed behind the mirror and is located at the position where the extended reflected rays (broken lines) converge.

Plane Mirrors

Suppose a candle flame is placed in front of a plane mirror. Rays of light are sent from the flame in all directions. Figure 25.7 shows only four of the infinite number of rays leaving one of the infinite number of points on the candle. When these rays encounter the mirror, they are reflected at angles equal to their angles of incidence. The rays diverge from the flame and on reflection diverge from the mirror. These divergent rays *appear* to emanate from behind the mirror, from a point located where the rays seem to diverge (broken lines). An observer sees an image of the candle at this point. The light rays do not actually come from this point, so the image is called a **virtual image**. The image is as far behind the mirror as the object is in front of the mirror, and image and object have the same size. When you view yourself in a mirror, for example, the size of your image is the same as the size your twin would appear if located as far behind the mirror as you are in front.

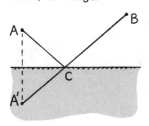

*Answer   Similarly, construct an artificial point A′ as far below the mirror as A is above; then draw a straight line from B to A′ to find C, as shown at the left. Both constructions superimposed, at right, show that C is common to both. We see that light will follow the same path if it goes in the opposite direction. Whenever you can see somebody else in a mirror, be assured that they can also see you.

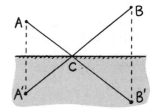

It is interesting to note that the eye-brain system cannot ordinarily tell the difference between an object and its reflected virtual image. So the illusion that an object exists behind a mirror is merely due to the fact that the light that enters the eye is entering in exactly the same manner, physically, as it would have entered if there really were an object there.

Question   If you wish to take a picture of your image while standing 5 meters in front of a plane mirror, for what distance should you set your camera to provide sharpest focus?*

Only part of the light that is incident upon a surface is reflected. On a surface of clear glass, for example, only about 4 percent is reflected from each surface; while on a clean and polished aluminum or silver surface, about 90 percent of incident light is reflected.

Diffuse Reflection

When light is incident on a rough surface, it is reflected in many directions. This is called *diffuse reflection* (Figure 25.8). If the surface is so smooth that the distances between successive elevations on the surface are less than about one-eighth the wavelength of the light, there is very little diffuse reflection, and the surface is said to be polished. A surface therefore may be polished for radiation of long wavelength but not polished for light of short wavelength. The wire-mesh "dish" shown in Figure 25.9 is very rough compared to light waves and is hardly mirrorlike. But for long-wavelength radio waves it is "polished" and is an excellent reflector.

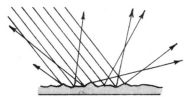

Fig. 25.8 Diffuse reflection. Although the reflection of each single ray obeys the law of reflection, the many different angles that light rays encounter in striking a rough surface cause reflection in many directions.

Fig. 25.9 The open-mesh parabolic dish is a diffuse reflector for short-wavelength light but polished for long-wavelength radio waves.

*****Answer**   10 m; the situation is equivalent to your standing 5 m in front of an open window and viewing your twin standing 5 m in back of the window.

Light reflecting from this page is diffuse. The page may be smooth to a long radio wave, but to a fine light wave it is rough. Rays of light incident on this page encounter millions of tiny flat surfaces facing in all directions. The incident light therefore is reflected in all directions. This is a desirable circumstance. It enables us to see objects from any direction or position. Most of our environment is seen by diffuse reflection.

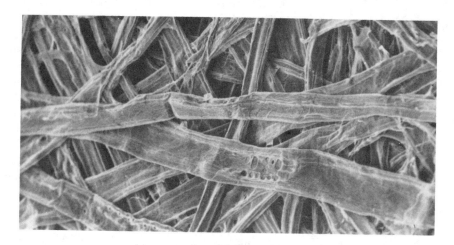

Fig. 25.10 The surface of ordinary paper as seen with a scanning electron microscope.

Refraction

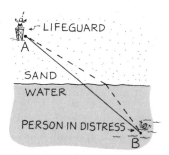

Fig. 25.11 Refraction.

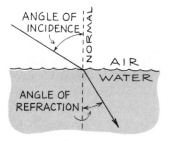

Fig. 25.12 Refraction.

Light travels at different speeds in different materials. It travels at 300,000 kilometers per second in a vacuum, at slightly less speed in air, and at about three-fourths that speed in water. In a diamond, light travels at about 40 percent its speed in a vacuum. Light bends when it passes obliquely from one medium to another. This is **refraction**. It is a common observation that a ray of light bends and takes a longer path when it encounters glass or water at a grazing angle. But the longer path taken is nonetheless the path requiring the least time. A straight-line path would take a longer time. We can illustrate this with the following situation.

Imagine that you are a lifeguard at a beach and you spot a person in distress in the water. We show the relative position of you, the shoreline, and the person in distress in Figure 25.11. You are at point A, and the person is at point B. You can run faster than you can swim. Should you travel in a straight line to get to B? A little thought will show that a straight-line path would not be the best choice, because if you instead spent a little bit more time traveling farther on land, you would save a lot more time in swimming a lesser distance in the water. The path of shortest time is shown by the dashed-line path, which clearly is not the path of shortest distance. The amount of bending at the shoreline of course depends on how much faster you can run than you can swim. The situation is similar for a ray of light incident upon a body of water, as shown in Figure 25.12. The angle of incidence is larger than the angle of refraction by an amount that depends on the relative speeds of light in air and in water.

Question Suppose our lifeguard in the preceding example were a seal instead of a human being. How would its path of least time from A to B differ?*

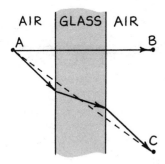

AIR | GLASS | AIR

Fig. 25.13 Refraction through glass.

Consider the pane of thick window glass in Figure 25.13. When light goes from point A through the glass to point B, it will go in a straight-line path. In this case, light encounters the glass perpendicularly, and we see that the shortest distance through both air and glass corresponds to the shortest time. But what about light that goes from point A to point C? Will it travel in the straight-line path shown by the dotted line? The answer is no, for if it did so it would be spending more time inside the glass, where light travels slower than in air. The light will instead take a less-inclined path through the glass. The time saved by taking the resulting shorter path through the glass more than compensates for the added time required to travel the slightly longer path through the air. The overall path is the path of least time. The result is a parallel displacement of the light beam, because the angles in and out are the same. You'll notice this displacement when you look at an angle through a thick pane of glass. The more acute the angle, the more pronounced the displacement.

Another example of interest is the prism, where opposite faces of the glass are not parallel (Figure 25.14). Light that goes from point A to point B will not follow the straight-line path shown by the dashed line, because too much time would be spent in the glass. Instead, the light will follow the path shown by the solid line—a path that is a bit farther through the air—and pass through a thinner section of the glass to make its trip to point B. By this reasoning, one might think that the light should take a path closer to the upper vertex of the prism and seek the minimum thickness of glass. But if it did, the extra distance through the air would result in an overall longer time of travel. The angles of refraction provide the single path of least time.

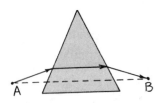

Fig. 25.14 A prism.

Interestingly enough, a properly curved prism will provide many paths of equal time from a point A on one side to a point B on the opposite side (Figure 25.15). The curve compensates for the extra distances light travels to points higher on the surface by decreasing correctly the thickness of the glass. For appropriate positions of A and B and for the appropriate curve on the surfaces of this modified prism, all light paths are of exactly equal

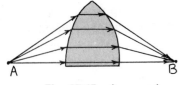

Fig. 25.15 A curved prism.

*****Answer** The seal can swim faster than it can run and its path would bend as shown; likewise with light emerging from the bottom of a piece of glass into air.

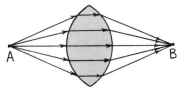

Fig. 25.16 A converging lens.

time. In this case, all the light from A that is incident on the glass surface is *focused* on point B. We see that this shape is simply the upper half of a converging lens (treated in more detail later in this chapter).

Whenever we watch a sunset, we see the sun for several minutes after it has sunk below the horizon. The earth's atmosphere is thin at the top and dense at the bottom. Since light travels faster in thin air than it does in dense air, light from the sun can get to us more quickly if, instead of just going in a straight line, it avoids as much as possible the denser air by taking a higher and longer path to penetrate the atmosphere at a steeper tilt (Figure 25.17). Since the density of the atmosphere changes gradually, the light path bends gradually to produce a curved path. Interestingly enough, this path of least time provides us with a slightly longer period of daylight each day. Furthermore, when the sun (or moon) is near the horizon, the rays from the lower edge are bent more than the rays from the upper edge. This produces a shortening of the vertical diameter, causing the sun to appear elliptical (Figure 25.18).

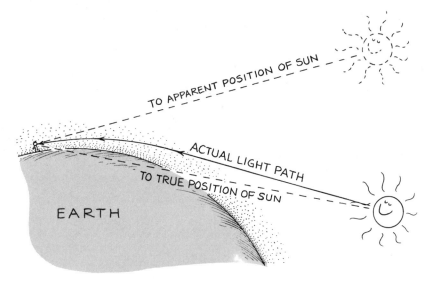

Fig. 25.17 Because of atmospheric refraction, the sun appears higher in the sky when near the horizon.

Fig. 25.18 The sun is distorted by differential refraction.

We are all familiar with the mirage we see while driving on a hot road. The sky appears to be reflected from water on the distant road, but when we get there, the road is dry. Why is this so? The air is very hot just above the road surface and cooler above. Light travels faster through the less dense and thinner hot air than in the cool region. So instead of light coming to us from the sky in straight lines, it also has least-time paths by

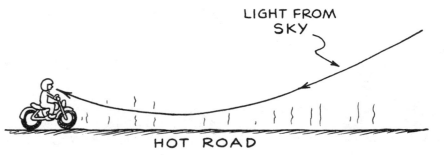

Fig. 25.19 A mirage is the result of refraction through the air.

which it curves down into the hotter region near the road for a while before reaching our eyes (Figure 25.19). A mirage is not, as many people mistakenly believe, a "trick of the mind." A mirage is formed by real light and can be photographed, as shown in Figure 25.20.

Fig. 25.20 The apparent wetness of the road in the photo at the left is not reflection of the sky by water but refraction of sky light through the air. The road is seen to be dry when viewed from a slightly higher elevation, as in the photo at the right.

When we look at an object over a hot stove or over a hot pavement, we see a wavy, shimmering effect. This is due to the various least-time paths of light as it passes through varying temperatures and therefore varying densities of air. The twinkling of stars results from similar phenomena in the sky, where light passes through unstable layers in the atmosphere.

Question   If the speed of light were the same in the various densities of air, would there still be mirages, slightly longer daytimes, and a squashed sun at sunset?*

---

*Answer   No.

In the foregoing examples, how does light seemingly *know* what conditions exist and what compensations a least-time path requires? When approaching window glass at an angle, how does light know to travel a bit farther in air to save time in taking a less-inclined angle and therefore shorter path through the pane of glass? In approaching a prism or a lens, how does light know to travel a greater distance in air to reach a thinner portion of the glass? How does light from the sun know to travel above the atmosphere an extra distance before taking a shortcut through the air to save time? How does sky light above know that it can reach us in minimum time if it dips toward a hot road before tilting upward to our eyes? The principle of least time appears to be noncausal. It seems that light has a mind of its own; that it can "smell" all the possible paths, calculate the times for each, and choose the one that requires the least time. Is this the case? Well, it turns out that light can sense the path ahead of it but only a distance on the order of its wavelength. We will now see that the path that light takes in refracting through transparent media can be understood causally by investigating the interactions of light waves and photons with the molecules that compose the media. To understand this, we need to know why light slows down in transparent materials such as glass and water.

Velocity of Light in a Transparent Medium

The speed of light is a constant in nature; that is, its speed in a vacuum is a constant 300,000 kilometers per second.[2] We shall call the speed of light c. Light has less speed in a transparent medium. In water, for example, light travels at 75 percent of its speed in a vacuum, or $0.75c$; in glass, about $0.67c$, depending on the type of glass; in a diamond, only $0.41c$. When light emerges from these media, it again travels at its original speed, c. From what we have learned about energy, this seems like strange behavior. For example, if a bullet is fired through a board, the bullet slows down in passing through the board and emerges at a speed less than its incident speed. It loses some of its kinetic energy while interacting with the fibers and splinters of wood that make up the board material. We would certainly be surprised if, after slowing down in the board, the bullet emerged with a speed equal to its original speed! Yet this is apparently what happens in the case of light. A beam of light is incident upon glass at speed c, slows down in the glass, and emerges at speed c!

How light passes through a transparent medium involves complex models beyond the scope of this chapter. Theoretical physicists devise conceptual models to understand nature, particularly at the submicroscopic level. A model must be consistent with and explain observations; it is useful if it predicts what may happen. We can partially understand how light slows down in transparent materials by the following simplified

---

[2]The presently accepted value is 299,792 km/s, which we round off here and in the rest of this book to 300,000 km/s. This corresponds to 186,000 miles per second.

model. Imagine a beam of light incident upon the surface of a pane of glass. The glass is composed mostly of silicon atoms with electron clouds that are easily forced to vibrate over the entire range of visible frequencies. When incident light is absorbed, the electron clouds are set into vibration with a frequency equal to that of the absorbed light. The vibrating electrons in turn emit light of their own at the same frequency. This light is indistinguishable from the incident light except for two factors: (1) it is less intense because some of the incident light from the beam was absorbed and turned into heat in the process, and (2) most importantly, it has not traveled as far as the undisturbed incident light would have traveled in the same time. Why? Because the process of absorption/re-emission is not instantaneous; some time is required for the process and as a result the *average* speed of light through the glass is less than c. Does this contradict the idea that the speed of light is a constant in nature?[3] No, because the *instantaneous* speed of light between atoms is still c. Whether in the vast regions of free interstellar space or in the tiny free spaces between atoms in a dense medium, the instantaneous speed of light is c. When we speak about the velocity of light in a transparent medium, we speak of the average speed, that which takes into account the lag associated with the process of absorption and re-emission.

So unlike the bullet passing through the board, light does not "burrow" through glass. If our model for light is one of corpuscles or tiny light particles called *photons*, we see that the photons emerging from a glass surface are not the same photons that entered the glass. The view you see when looking out a window is composed of photons ejected from the inner surface of the glass!

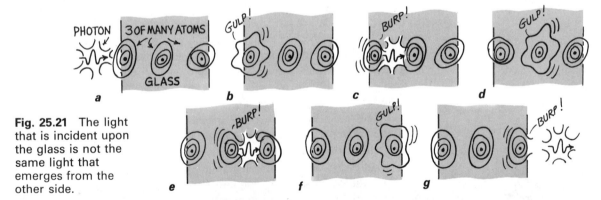

Fig. 25.21 The light that is incident upon the glass is not the same light that emerges from the other side.

---

[3] Just how much the average speed of light differs from its speed in a vacuum is given by the index of refraction, n, of the material; n is the ratio of the speed of light in a vacuum to the speed of light in the material:

$$n = \frac{\text{speed of light in vacuum}}{\text{speed of light in material}}$$

For example, the speed of light in a diamond is $(\frac{1}{2.4})c$; and so $n = 2.4$. For a vacuum, $n = 1$.

Question   When you look at a star, light that left the star thousands or millions of years ago impinges on the retina at the back of your eye. Where do the *photons* that interact with your retina originate?*

Cause of Refraction

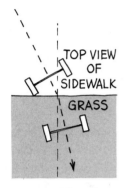

Fig. 25.22 The direction of the rolling wheels changes when one part slows down before the other part.

Fig. 25.23 The direction of the light waves changes when one part of the wave slows down before the other part.

When light bends as it passes obliquely from one medium to another, we call the process *refraction*. The cause of refraction is the changing of light speed in going from one transparent medium to another. We can understand this by considering the action of a pair of toy cart wheels connected to an axle as the wheels roll from a smooth sidewalk onto a grass lawn. If the wheels meet the grass at some angle (Figure 25.22), they will be deflected from their straight-line course. The direction of the rolling wheels is shown by the dashed line. Note that on meeting the lawn, where the wheels roll slower owing to interaction with the grass, the left wheel slows down first. This is because it meets the grass while the right wheel is still on the smooth sidewalk. The faster-moving right wheel tends to pivot about the slower-moving left wheel. It travels farther during the same time the left wheel travels a lesser distance in the grass. This action bends the direction of the rolling wheels toward the "normal," the lightly dashed line perpendicular to the grass-sidewalk border.

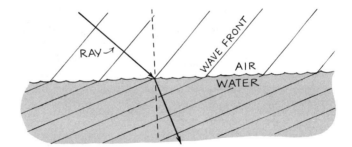

A light wave bends in a similar way (Figure 25.23). We see that the direction of light is shown by the solid arrow (the light ray) and that light waves are at right angles to the ray. (If the light source were close, the waves would be circular; but if we assume the sun is the source, the wave crests practically form straight lines.) In any case, we see that the waves are everywhere perpendicular to the light rays. If the wave is incident on a transparent surface at an angle, the left portion of the wave slows down before the part of the wave still in the air. The ray or beam of light remains

---

***Answer**   The photons originate in your eye! A photon absorption/re-emission process cascades all the way from the star to your retina, the last photon being ejected from an atom in the vitreous humor inside your eye. Although beams of light originate outside your eye, the photons at the ends of the beams that stimulate your sense of vision come from within your eye. How about that!

NORMAL

MIRROR

Fig. 25.24 A beam of light bends toward the normal when entering a denser medium and away from the normal when leaving.

Fig. 25.25

Fig. 25.26 Because of refraction, a submerged object appears to be nearer to the surface than it actually is.

perpendicular to the wave front and bends just as the direction of the wheels bends. When the speed of light decreases in going from one medium to another, the light bends toward the normal. When the speed of light increases in traveling from one medium to another, the light ray bends away from the normal.[4] This is shown in Figure 25.24.

The refraction of light is responsible for many illusions; one of them is the apparent bending of a stick partly immersed in water. Another illusion is shown in Figure 25.25. The full glass mug appears to hold more root beer than it actually does. Light from the root beer is refracted through the sides of the thick glass, making the glass appear thinner than it is. The eye, accustomed to perceiving light traveling along straight lines, perceives the root beer to be at the outer edge of the glass, along the broken lines.

When we stand on a bank and view a fish in the water, the fish appears to be nearer the surface than it actually is. It will also seem closer (Figure 25.26). Because of refraction, submerged objects appear to be magnified. If we look straight down into water, an object submerged 4 meters beneath the surface will appear to be 3 meters deep.

We see that we can interpret the bending of light at the water in at least two ways: We can say that the light that reflects from the fish to reach the observer's eye deliberately takes a shorter path upward toward the water surface and a correspondingly longer path through the air in such a way that minimum time results. We can as well say that the waves of light that happen to be directed upward at an angle toward the water surface are bent off-kilter as they speed up when reaching the air; the resulting path turns out to be, equivalently, a least-time path. Whichever view we choose, the results are the same.

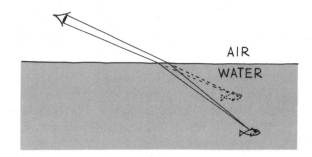

AIR

WATER

---

[4]The quantitative law of refraction, called *Snell's law,* was first worked out in 1621 by W. Snell, a Dutch astronomer and mathematician: $n \sin \theta = n' \sin \theta'$, where n and n' are the indices of refraction of the media on either side of the surface, and θ and θ' are the respective angles of incidence and refraction. If three of these values are known, the fourth can be calculated from this relationship.

Question   If the speed of light were the same in all media, would refraction still occur when light passes from one medium into another?*

Dispersion in a Prism

We discussed how the average speed of light is less than c in a transparent medium; how much less depends on the nature of the medium and on the frequency of the light. The speed of light in a transparent medium is frequency-dependent. Frequencies of light that are closer to the natural frequency of the electron oscillators in the atoms and molecules making up the transparent medium interact more often in the absorption/re-emission sequence and therefore travel slower. Since the natural or resonant frequency of most transparent materials is in the ultraviolet part of the spectrum, the higher frequencies of visible light travel slower than the lower frequencies. Violet travels about 1 percent slower in ordinary glass than does red light. The colors between red and violet travel at their own respective speeds.

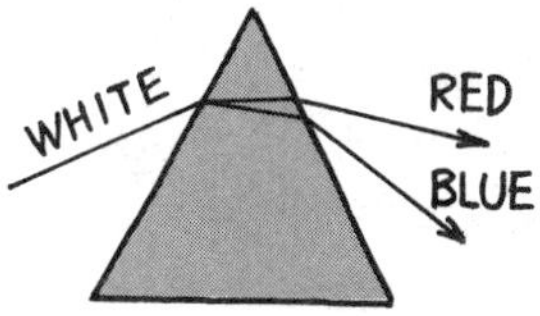

Different frequencies of light travel at different speeds in transparent materials; because they travel at different speeds, they refract differently and bend by different amounts. When light is bent twice, as in a prism, the separation of the different colors of light is quite noticeable. This separation of light into colors arranged according to their frequency is called **dispersion** (Figure 25.27).

Fig. 25.27 Dispersion through a prism makes the components of white light visible.

Rainbows

A most spectacular illustration of dispersion is the rainbow. The conditions for seeing a rainbow are that the sun be shining in one part of the sky and that rain be falling in the opposite part of the sky. When we turn our backs toward the sun, we see the spectrum of colors in a bow. Seen from an airplane, the bow may form a complete circle. The colors are dispersed from the sunlight by thousands of tiny drops that act like prisms.

Fig. 25.28 The rainbow is seen in a part of the sky opposite the sun and is centered on the "antisun."

*Answer   No.

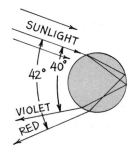

Fig. 25.29 Dispersion of sunlight by a single raindrop.

To understand how light is dispersed by the spherical raindrops, consider an individual raindrop as shown in Figure 25.29. Follow the ray of sunlight as it enters the drop near its top surface. Some of the light here is reflected (not shown), and the remainder is refracted into the water. At this first refraction, the light is dispersed into its spectrum colors—violet being deviated the most and red the least. Reaching the opposite side of the drop, each color is partly refracted out into the air (not shown) and partly reflected back into the water. Arriving at the lower surface of the drop, each color is again reflected (not shown) and refracted into the air. This second refraction is similar to that of a prism, where refraction at the second surface increases the dispersion already produced at the first surface.

Although each drop disperses a full spectrum of colors, an observer is in a position to see only a single color from any one drop (Figure 25.30). If violet light from a single drop is incident upon the eye of an observer, red light from the same drop is incident elsewhere toward her feet. To see red light, she must look to a drop higher in the sky. She will see the color red when the angle between a beam of sunlight and the dispersed light is 42°. The color violet is seen when the angle between the sunbeams and dispersed light is 40°.

Fig. 25.30 Sunlight is incident on two sample raindrops as shown and emerges from them as dispersed light. The observer sees the red light from the upper drop and the violet light from the lower drop. Millions of drops produce the whole spectrum.

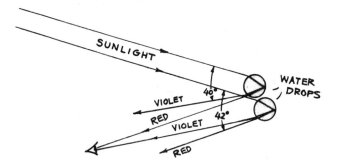

Why does the light dispersed by the raindrops form a bow? The answer to this involves a little geometrical reasoning. For simplicity, we will consider only the dispersion of red light. An observer can see red by looking upward so that there is a 42° angle between the sunbeams and the dispersed light that reaches her eyes. But she can also see red by looking sideways and in other directions at the same angle. Figure 25.31 shows that all the drops that disperse light at the same angle form a bow. Drops elsewhere may form somebody else's rainbow, but not the bow that our observer presently sees. Do you see why a rainbow is bow-shaped?

Often a larger, secondary bow with colors reversed can be seen at greater angles about the primary bow. We won't treat this secondary bow except to

Fig. 25.31 Only raindrops along the dashed line disperse red light to the observer at a 42° angle; hence, the light forms a bow.

Fig. 25.32 Under the rainbow, the sky is bright where thousands of overlapping rainbows combine into a whitish glow.

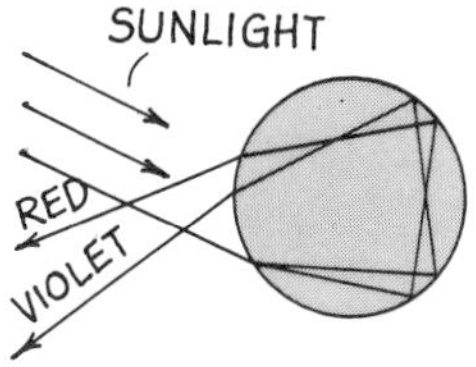

SUNLIGHT

RED

VIOLET

Fig. 25.33 Double reflection in a drop produces a secondary bow.

say that it is formed by similar circumstances and is a result of double reflection within the raindrops (Figure 25.33). Because of this extra reflection (and extra refraction loss), the secondary bow is much dimmer.

Question   If light traveled at the same speed in raindrops as it does in air, would we still have rainbows?*

Lenses   A very practical case of refraction occurs in lenses. We can understand a lens by analyzing equal-time paths as we did earlier, or we can assume that it consists of a set of several matched prisms and blocks of glass arranged in the order shown in Figure 25.34. The prisms and blocks refract incoming parallel light rays so they converge (or diverge) to a focal point. The arrangement shown in Figure 25.34*a* converges the light, and we call such a lens a **converging lens**. Note that it is thicker in the middle.

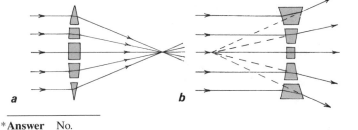

Fig. 25.34 A lens may be thought of as a set of prisms.

a

b

_____

Answer   No.

The arrangement in *b* is different. The middle is thinner than the edges, and it diverges the light; such a lens is called a **diverging lens**. Note that the prisms diverge the incident rays in a way that makes them appear to come from a single point in front of the lens. In both lenses the greatest deviation of rays occurs at the outermost prisms, for they have the greatest angle between the two refracting surfaces. No deviation occurs exactly in the middle, for in that region the glass faces are parallel to each other. Real lenses are not made of prisms, of course, as is indicated in Figure 25.34, but of a solid piece of glass with surfaces ground usually to a circular curve. In Figure 25.35 we see how smooth lenses refract waves.

Fig. 25.35 Wave fronts travel slower in glass than in air. In (*a*), the waves are retarded more through the center of the lens, and convergence results. In (*b*), the waves are retarded more at the edges, and divergence results.

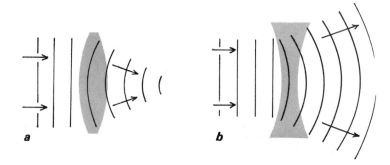

The refraction of light rays through a lens enables the lens to produce an image of a distant object, which may be magnified. Magnification occurs when an image is observed through a wider angle with the use of a lens than without the lens. With unaided vision, an object far away is seen through a relatively small angle of view, while the same object when closer is seen through a larger angle of view, enabling the perception of more detail. A magnifying lens simply increases the angle of view, thereby making objects appear to be closer or larger than they really are.

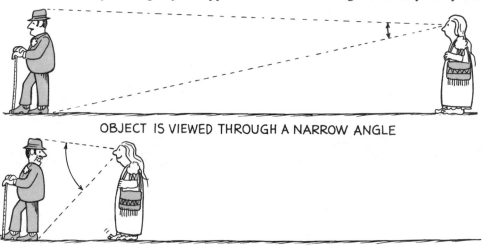

OBJECT IS VIEWED THROUGH A NARROW ANGLE

Fig. 25.36

OBJECT IS VIEWED THROUGH A WIDE ANGLE

No lens provides a perfect image. The distortions in an image are called **aberrations**. By combining lenses in certain ways, aberrations can be minimized. For this reason, most optical instruments use compound lenses, each made up of several simple lenses, instead of single lenses.

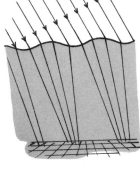

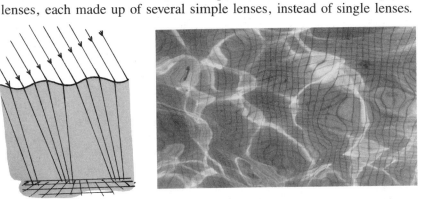

Fig. 25.37 The moving patterns of bright and dark lines result from the uneven surface of water, which behaves as a blanket of undulating lenses. A waterbug looking upward at the sun from the bottom of the pool would see the sun shimmering in intensity. Because of similar irregularities in the atmosphere, we see the stars twinkle.

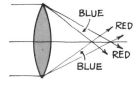

Fig. 25.38 Spherical aberration.

Spherical aberration results from light passing through the edges of a lens and focusing at a slightly different place than light passing through the center of the lens (Figure 25.38). This can be remedied by covering the edges of a lens, as with a diaphragm in a camera. Spherical aberration is corrected in good optical instruments by a combination of lenses.

Chromatic aberration is the result of the different speeds of various colors and hence the different refractions they undergo. In a simple lens, red light and blue light do not come to focus in the same place. *Achromatic lenses,* which combine simple lenses of different kinds of glass, correct this defect. Mirrors are preferable to lenses in telescopes because there is no chromatic aberration with reflection. (Why?)

The pupil of the eye regulates the amount of light reaching the retina by changing its size. Vision is sharpest when the pupil is smallest because light then passes through only the center of the eye's lens, where spherical and chromatic aberrations are less than toward the periphery.[5]

Astigmatism of the eye is a defect that results when the cornea is curved more in one direction than another, somewhat like the side of a barrel. Because of this defect, the eye does not form sharp images. The remedy is eyeglasses with cylindrical lenses, which have more curvature in one direction than in another.

Fig. 25.39 Chromatic aberration.

Question   If light traveled at the same speed in different media, would glass lenses still alter the directions of light rays?*

Answer   No.

[5]If you ever misplace your eyeglasses or find it difficult to read small print, like in a telephone book, hold a pinhole (in a piece of paper or whatever) in front of your eye close to the page. You'll see the print clearly—a pinhole magnifier. Try it and see!

Total Internal Reflection

Some Saturday night when you're taking your bath, fill the tub extra deep and bring a waterproof flashlight into the tub with you. Put the bathroom light out. Shine the submerged light straight up and then slowly tip it. Note how the intensity of the emerging beam diminishes and how more light is reflected from the water surface to the bottom of the tub. At a certain angle, called the *critical angle,* you'll notice that the beam no longer emerges into the air above the surface. It instead grazes the surface. For water, the critical angle is 48° from the normal to the surface. When the flashlight is tipped beyond the critical angle, you'll notice that the beam is *totally internally reflected.* The light obeys the law of reflection: the angle of incidence is equal to the angle of reflection. The only light emerging from the water surface is that which is diffusely reflected from the bottom of the bathtub. This procedure is shown in Figure 25.40. The proportion of light refracted and light internally reflected is indicated by the relative lengths of the arrows.

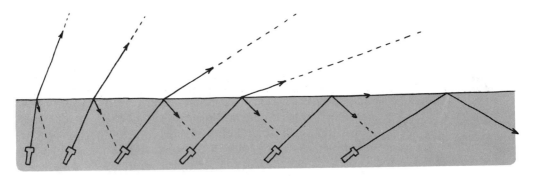

Fig. 25.40 Light emitted in the water is partly refracted and partly reflected at the surface. The length of the arrows indicates the proportions refracted and the beam is totally internally reflected.

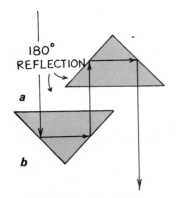

Fig. 25.41 Total internal reflection in a prism.

The critical angle for glass is about 43°, depending on the type of glass. This means that within glass, light that is incident at angles greater than 43° will be totally internally reflected. No light will escape beyond this angle; instead, all of it will be reflected back into the glass. Whereas a silvered or aluminumized mirror absorbs about 10 percent of the light on reflection, the glass prism shown in Figure 25.40 is more efficient. Light is incident on the slanted face at 45° and is totally reflected. Moreover, all the light incident on the slanted face is reflected and is not marred by any dirt or dust on the outside surface. This is the principal reason why prisms are used instead of mirrors in many optical instruments.

A pair of prisms each reflecting light through 180° is shown in Figure 25.41. Binoculars use pairs of prisms to lengthen the light path between lenses because they eliminate the need for long barrels. This makes it

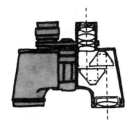

Fig. 25.42 Prism binoculars.

possible for a compact pair of binoculars to be as effective as a longer telescope (Figure 25.42). Another advantage of prisms is that whereas the image of a straight telescope is upside down, reflection by the prisms in binoculars reinverts the image, so we see things right side up.

Another interesting use of total internal reflection involves optical fibers, or light pipes (Figure 25.43). An optical fiber "pipes" light from one place to another by a series of total internal reflections, much as a bullet ricochets down a steel pipe. Optical fibers are used in decorative table lamps and to illuminate instrument displays on automobile dashboards from a single bulb. Dentists use them with flashlights to get light where they want it. Bundles of thin glass fibers are used to see what is going on in inaccessible places, such as the interior of a motor or a patient's stomach. Light shines down some of the fibers to illuminate the scene and is reflected back along others.

Fig. 25.43 The light is "piped" from below by a succession of internal reflections until it emerges at the top ends.

Optical fibers are important in communications because they offer a practical alternative to copper wires and cables. In metropolitan areas, thin glass fibers are more and more replacing thick, bulky, and expensive copper cables to carry thousands of simultaneous telephone messages between the major switching centers. In some aircraft, control signals are fed from the pilot to the control surfaces by means of optical fibers. Signals are carried in the modulations of laser light. Unlike electricity, light is indifferent to temperature and fluctuations in surrounding magnetic fields, so the signal is clearer. Also, it cannot be tapped by eavesdroppers.

Summary of Terms

Fermat's principle of least time   Light will take the path that requires the least time when it goes from one place to another.

Reflection   The return of light rays from a surface in such a way that the angle at which a given ray is returned is equal to the angle at which it strikes the surface. When the reflecting surface is irregular, light is returned in irregular directions; this is *diffuse reflection*.

Refraction   The bending of an oblique ray of light when it passes from one transparent medium to another. This is caused by a difference in the speed of light in the transparent media. When the change in medium is abrupt (say, from air to water), the bending is abrupt; when the change in medium is gradual (say, from cool air to warm air), the bending is gradual, accounting for mirages.

Photon   A light corpuscle, or the basic packet of electromagnetic radiation. Just as matter is composed of atoms, light is composed of photons (quanta).

Converging lens   A lens that is thicker in the middle than at the edges and refracts parallel rays passing through it to a focus.

Diverging lens   A lens that is thinner in the middle than at the edges, causing parallel rays passing through it to diverge.

Aberration   A limitation on perfect image formations inherent to some degree in all optical systems.

Total internal reflection   The total reflection of light traveling in a medium when it is incident on the surface of a less dense medium at an angle greater than the critical angle.

Suggested Reading

Epstein, L. *Thinking Physics*. San Francisco: Insight Press, 1983.

Greenler, R. *Rainbows, Halos, and Glories*. New York: Cambridge University Press, 1980.

Review Questions

1. What is Fermat's principle of least time?

2. What is diffuse reflection and how does it depend on the wavelength of incident radiation?

3. Does a beam of light travel at different speeds in different transparent materials? Do photons travel at different speeds in different materials? Why are your answers different?

4. As light travels from one medium to another, does its velocity change? Its frequency? Its wavelength?

5. When a beam of light emerges from water to air at an angle, does the beam bend toward or away from the normal? What causes this bending?

6. What conditions produce a mirage?

7. Why do stars "twinkle"?

8. Does atmospheric refraction tend to lengthen or shorten the day?

9. Why do blue light waves bend more than red light waves when passing obliquely from air to water?

10. Would dispersion occur if the speed of light were not frequency-dependent? Would rainbows exist?

11. What are the necessary conditions for a rainbow?

12. What causes a secondary rainbow? Why is it dimmer than the primary bow?

13. Why does light converge when passing through a converging lens? Why does light diverge when passing through a diverging lens?

14. What is meant by *critical angle*?

15. Light travels in straight lines, but appears to bend in a light pipe. Explain.

Home Projects

1. Make a pinhole camera, as illustrated below. Cut out one end of a small cardboard box, and cover the end with tissue or onionskin paper. Make a clean-cut pinhole at the other end. (If the cardboard is thick, make it through a piece of tinfoil placed over an opening in the cardboard.) Aim the camera at a bright object in a darkened room, and you will see an upside-down image on the tissue paper. If in a dark room you replace the tissue paper with unexposed photographic film, cover the back so it is light-tight, and cover the pinhole with a removable flap, you are ready to take a picture. Exposure times differ depending principally on the kind of film and amount of light. Try different exposure times, starting with about 3 seconds. Also try boxes of various lengths. You'll find everything in focus in your photographs, but the pictures will not have clear-cut, sharp outlines. The principal difference between your pinhole camera and a commercial one is the glass

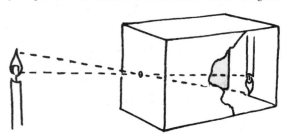

lens, which is larger than the pinhole and therefore admits more light in less time. It is because a lens camera is so fast that the pictures it takes are called "snapshots."

2. If you don't have a prism, you can produce a spectrum by placing a tray of water in bright sunlight. Lean a pocket mirror against the inside edge of the pan and adjust it until a spectrum appears on the wall or ceiling.

3. Stand a pair of mirrors on edge with the faces parallel to each other. Place an object such as a coin between the mirrors and look at the reflections in each mirror. Neat?

4. Set up two pocket mirrors at right angles and place a coin between them. You'll see four coins. Change the angle of the mirrors and see how many images of the coin you can see. With the mirrors at right angles, look at your face. Then wink. What do you see? You now see yourself as others see you. Hold a printed page up to the double mirrors and contrast its appearance with the reflection of a single mirror.

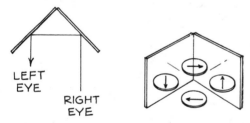

LEFT
EYE

RIGHT
EYE

5. Determine the magnification power of a lens by focusing on the lines of a ruled piece of paper. Count the spaces between the lines that fit into one magnified space, and you have the magnification power of the lens. You can do the same with binoculars and a distant brick wall. Hold the binoculars so that only one eye looks at the bricks through the eyepiece while the other eye looks directly at the bricks. The number of bricks seen with the unaided eye that will fit into one magnified brick gives the magnification of the instrument.

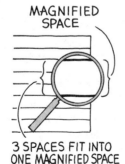

MAGNIFIED
SPACE

3 SPACES FIT INTO
ONE MAGNIFIED SPACE

6. Look at the reflections of overhead lights from the two surfaces of eyeglasses, and you will see two fascinatingly different images. Why are they different?

Exercises

1. Fermat's principle is of least time rather than of least distance. Would least distance apply as well for reflection? For refraction? Why are your answers different?

2. A solar eclipse occurs when the moon is between the earth and the sun. Make a diagram to show why some areas of the earth see a total eclipse, while other areas see a partial eclipse, and most of the earth sees no eclipse.

3. A lunar eclipse occurs when the earth is between the sun and the moon. Make a diagram to show why all areas on the nighttime side of the earth see the same eclipse.

4. Lunar eclipses are always eclipses of a full moon. That is, the moon is always at its fullest just before and after the earth's shadow passes over it. Would it be possible to have a lunar eclipse when the moon is in its crescent or half-moon phase? Explain.

5. Does the reflection of a scene in calm water look the same as the scene itself only upside down? (*Hint:* Can you ever see the reflection of the top of a stone that extends above water?)

6. Cowboy Joe wishes to shoot his assailant by ricocheting a bullet off a mirrored metal plate. To do so, should he simply aim at the mirrored image of his assailant? Explain.

7. Show with a simple diagram that when a mirror is rotated through a certain angle when a fixed beam is incident on it, the reflected beam is rotated through an angle twice as large. (This doubling of displacement makes irregularities in ordinary window glass more evident.)

8. Her eye at point P looks into the mirror. Which of the numbered cards can she see reflected in the mirror?

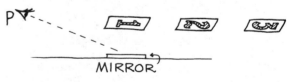

9. What must be the minimum length of a plane mirror in order for you to see a full view of yourself?

10. What effect does your distance from the plane mirror have in the above answer? (Try it and see!)

11. Suppose that you walk 1 meter per second toward a mirror. How fast do you and your image approach each other? (The answer is *not* 1 meter per second.)

12. What is wrong with the cartoon of the man looking at himself in the mirror? (Have a friend face a mirror as shown, and you'll see.)

13. Why is the lettering on the front of some vehicles "backward"?

ƎƆИ∀⅃UᗺMA

14. Does the frequency of light change on reflection? (Cite an everyday example to support your answer.)

15. A person in a dark room looking through a window can clearly see a person outside in the daylight, whereas the person outside cannot see the person inside. Explain.

16. Why is it difficult to see the roadway in front of you when driving on a rainy night?

17. Why does reflected light from the sun or moon appear as a column in the body of water as shown? How would it appear if the water surface were perfectly smooth?

18. When you look at yourself in the mirror and wave your right hand, your beautiful image waves the left hand. Then why don't the feet of your image wiggle when you shake your head?

19. In about 1675 the Danish astronomer Olaus Roemer found that light from eclipses of Jupiter's moon took an extra 1000 seconds to travel 300,000,000 kilometers across the diameter of the earth's orbit around the sun. Show how this finding provided the first reasonably accurate measurement for the speed of light.

20. Pretend a man can walk only at a certain pace—no faster, no slower. If you time his uninterrupted walk across a room of known length, you can calculate his walking speed. If, however, he stops momentarily along the way to greet others in the room, the extra time spent in his brief interactions yields an average speed across the room that is less than his actual speed at any instant he is moving. In what way is this analogous to light traversing a transparent medium? In what way is it not?

21. A pair of toy cart wheels are rolled obliquely from a smooth surface onto two plots of grass, a rectangular plot as shown in *a* and a triangular plot as shown in *b*. The ground is on a slight incline so that after slowing down in the grass, the wheels will speed up again when emerging on the smooth surface. Finish each sketch by showing some positions of the wheels inside the plots and on the other sides, thereby indicating the direction of travel.

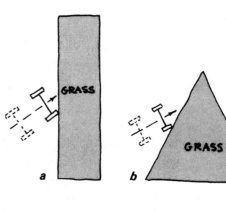

22. If light of all frequencies traveled at the same speed in glass, how would white light appear after passing through a prism?

23. A beam of light bends as shown in *a*, while the edges of the immersed square bend as shown in *b*. Do these pictures contradict each other? Explain.

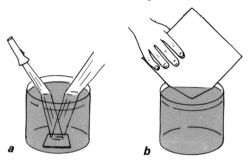

a   *b*

24. If you were spearing a fish with a laser, would you aim above, below, or directly at the observed fish to make a direct hit? Defend your answer.

25. If you were to send a beam of laser light to a space station above the atmosphere and just above the horizon, would you aim the laser above, below, or at the visible space station? Defend your answer.

26. Does a dolphin underwater see a ball held above the water surface as farther away or closer than it really is? Explain.

27. Two observers standing apart from one another do not see the "same" rainbow. Explain.

28. If you walk while viewing a rainbow, the rainbow moves with you. Why?

29. A rainbow viewed from an airplane may form a complete circle. Will the shadow of the airplane appear at the center of the circle? Explain.

30. In the elliptically setting sun of Figure 25.18, is the vertical diameter shortened or is the horizontal diameter lengthened? Defend your answer.

31. How is a rainbow similar to the halo sometimes seen around the moon on a night when ice crystals are in the upper atmosphere?

32. What is responsible for the rainbow-colored fringe commonly seen at the edges of a spot of white light from the beam of a lantern or slide projector?

33. What accounts for the large shadows cast by the ends of the thin legs of the water strider? What accounts for the ring of bright light around the shadows?

34. When a fish looks upward at an angle of 45°, does it see the sky or the reflection of the bottom beneath? Defend your answer.

35. No glass is perfectly transparent. Mainly because of reflections, about 92 percent of light energy is transmitted by an average sheet of clear, dust-free windowpane. The 8-percent loss is not noticeable through a single sheet, but through several sheets it is apparent. How much light is transmitted by two sheets, each of which transmits 92 percent?

26 Color

Roses are red and violets are blue; colors intrigue artists and physicists too. To the physicist, the colors of objects are not in the substances of the objects themselves or even in the light they emit or reflect. Color is a physiological experience and is in the eye of the beholder. So when we say that light from a rose is red, in a stricter sense, we mean that it *appears* red. Many organisms, including people with defective color vision, will not see the rose as red at all.

The colors we see depend on the frequency of the light we see. Different frequencies of light are perceived as different colors; the lowest frequency is red, the highest violet, and between them ranges the infinite number of hues that make up the color spectrum of the rainbow. Except for light sources such as lamps, lasers, and gas discharge tubes (which we will treat in Chapter 28), most of the objects around us reflect rather than emit light. They reflect only part of the light that is incident upon them, the part that gives them their color.

Selective Reflection

Fig. 26.1

Fig. 26.2 Atoms and molecules behave as optical tuning forks.

A rose, for example, doesn't emit light; it reflects light (Figure 26.1). If we pass sunlight through a prism and then place a deep-red rose in various parts of the spectrum, the rose will appear brown or black in all parts of the spectrum except in the red. In the red part of the spectrum, the petals will appear red, but the green stem and leaves will appear black. This shows that the red rose has the ability to reflect red light, but it cannot reflect other kinds of light; the green leaves have the ability to reflect green light and likewise cannot reflect other kinds of light. When the rose is held in white light, the petals appear red and the leaves green, because the petals reflect the red part of the white light and the leaves reflect the green part of the white light. To understand why objects reflect specific colors of light, we must turn our attention to the atom.

Recall from our study of sound how the vibrations from a distant source could include a tuning fork of matched vibrating frequency into oscillation—into **resonance**. A tuning fork can also be made to oscillate, although at somewhat lesser amplitudes, even when the frequencies are not matched—when the driving frequency is not too far below or above the resonant frequency. Then the tuning fork is undergoing **forced vibration**. The same is true of atoms and molecules. We can think of atoms as three-dimensional tuning forks with electrons that behave as tiny oscillators that can vibrate as if attached by invisible springs (Figure 26.2). They can be forced into vibration by the vibrating energies of electromagnetic waves. Like acoustical tuning forks, once vibrating, they send out their own energy waves in all directions.

Fig. 26.3 This square reflects all the colors illuminating it. In sunlight it is white. When illuminated with red light, it is red.

This square absorbs all the colors illuminating it. In sunlight it is warmer than the white square.

Atoms (and molecules) have their own characteristic frequencies; electrons of one kind of atom can be set into oscillation over a certain narrow range of frequencies, while the electrons of other atoms oscillate over a different range. An electron set into vibration will radiate light with a frequency equal to its oscillation frequency. If an object reradiates (reflects) all the light incident upon it, its color will be that of the illuminating light. Usually it will absorb some and reflect the rest. This is *selective reflection*. If an object absorbs all the light incident upon it, it reflects none and has no color; it is black.

Fig. 26.4 Fur around a panda's eyes, ears, and legs absorbs all the radiant energy in incident sunlight and therefore is black. Fur on other parts of the body reflects radiant energy and therefore is white.

An object can only reflect frequencies that are present in the illuminating light. The appearance of a colored object therefore depends on the kind of light used. A candle flame gives off light that is deficient in blue; it gives off a yellowish light. An incandescent lamp gives off light that is richer toward the lower frequencies, enhancing the reds; while a fluorescent lamp is richer in the higher frequencies, enhancing the blues. With this kind of illumination, it is difficult to tell the "true" color of objects. Colors appear different in daylight than when illuminated with either of these lamps (Figure 26.5).

Fig. 26.5 Color depends on the light source.

Selective Transmission

The color of a transparent object depends on the color of the light it transmits. A red piece of glass appears red because it absorbs all the colors that compose white light, except red, which it *transmits*. Similarly, a blue piece of glass appears blue because it transmits only blue and absorbs all the other colors composing white light. From an atomic view, electrons in the pigment molecules are set into oscillation by the frequencies of blue light, and they reradiate this energy from molecule to molecule in the glass. The other frequencies are absorbed by the molecules as a whole and are not reradiated. The kinetic energy of the molecules is increased by the absorbed frequencies, and the glass is warmed (Figure 26.6). An ordinary windowpane is colorless; it transmits all the colors of white light equally well. Atomically, the electron oscillators vibrate at all the visible frequencies.

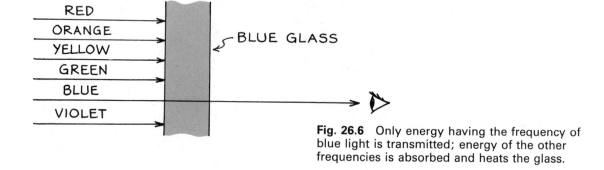

Fig. 26.6 Only energy having the frequency of blue light is transmitted; energy of the other frequencies is absorbed and heats the glass.

Questions
1. When a rose is illuminated with red light, why do the leaves become warmer than the petals?*

2. When the rose is illuminated with green light, which become warmer—the petals or the leaves?†

3. If you hold a match, a candle flame, or any small source of white light in front of a piece of red glass, you'll see two reflections from the glass: one from the front surface and one from the back surface. What color reflections will you see?‡

Color Mixing
The white light from the sun is actually a composite of all the visible frequencies; this is easily demonstrated by passing sunlight through a prism and observing the rainbow-colored spectrum. The distribution of solar frequencies is uneven, being most intense in the yellow-green part of the spectrum. It is interesting to note that our eyes have evolved to have maximum sensitivity in this range, which is shown in the radiation curve (Figure 26.7a). Most whites produced from reflected sunlight share this frequency distribution.

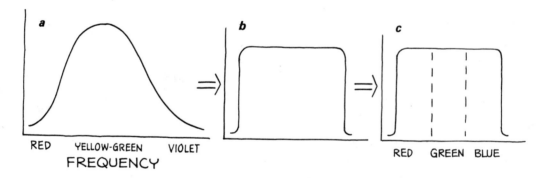

Fig. 26.7 (a) Radiation curve of sunlight (b) simplified and (c) divided into three regions.

So all the visible frequencies mixed together produce white. Interestingly enough, white also results from the combination of only red, green, and blue light. We can understand this by greatly simplifying the solar radiation curve to that shown in Figure 26.7b and dividing it into

*Answer   The leaves absorb rather than reflect the energy of red light; hence, the leaves become warmer.

†Answer   The petals absorb the energy of green light and become warmer, whereas the leaves reflect the green light.

‡Answer   Reflection from the top surface is white, because the light doesn't go far enough into the colored glass for absorption of nonred light. The back surface has only red to reach it, so red is what you see reflected.

three regions as in Figure 26.7c. The lowest third of the spectral distribution appears *red,* the middle third appears *green,* and the high-frequency third appears *blue* (Figure 26.8).

Fig. 26.8 The different regions of the simplified radiation curve for sunlight appear as the colors red, green, and blue. These are called the *additive primary colors.*

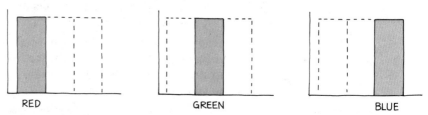

If we project these three colors on a screen, overlapping each other, their colors will add to produce white. When the light beams reflect off a white screen, the light seen is an additive mixture because the lights are added together before the observer sees them. If two of the three colors are added, then another color will be produced (Figure 26.9).

Fig. 26.9 Red + green = yellow. Red + blue = magenta (purple). Green + blue = cyan (turquoise). Yellow, magenta, and cyan are called the *subtractive primary colors.*

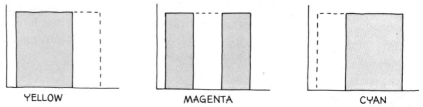

By adding various amounts of red, green, and blue, we can, in fact, generate any color in the spectrum (Figure 26.10). For this reason, red, green, and blue are called the **additive primary colors.** A close examination of the picture on most color television tubes will reveal that the picture is composed of an assemblage of tiny spots, each less than a millimeter across. When the screen is lit, some of the spots are red, some green, some blue; the mixtures of these primary colors at a distance provide a complete range of colors, plus white.[1]

Every artist knows that if you mix red, green, and blue paint, the result will not be white, but a muddy, dark brown. Red and green paint certainly do not combine to form yellow as is the rule for additive mixtures. The mixing of paints is an entirely different process from the mixing of colored lights. Paint is composed of pigments—finely divided solid particles that produce their characteristic colors by the processes of selective absorption or selective transmission of frequencies. When red and green pigments are combined, every color in the white light that illuminates them is absorbed, and the mixture is black. (In practice, absorption is less

---

[1]It's interesting to note that the ''black'' you see on the darkest scenes on a black-and-white TV tube is simply the color of the tube face itself, which is more a light gray than black. Because our eyes are sensitive to the contrast with the illuminated parts of the screen, we see this gray as black.

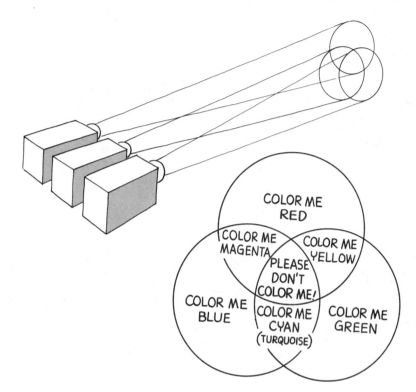

Fig. 26.10 Color addition by the mixing of colored lights. When three projectors shine red, blue, and green light on a white screen, the overlapping parts produce different colors. The *addition* of the three primary colors produces white light. See this in full color on the inside back cover.

than 100 percent, so the mixture appears brown rather than black.) We can understand this by placing our special distribution "blocks" for red and green one atop the other as in Figure 26.11. Let white light be incident from above. The red pigment will absorb the green and blue parts of the white light, so these sections are shaded to represent absorption. The red pigment passes red, but this is absorbed by the green pigment (which absorbs red and blue and passes green). So incident white light is absorbed everywhere in the spectrum; no color is reflected.

Fig. 26.11 When white light is incident on red pigment, represented by the top three blocks, the green and blue components are absorbed. The green pigment, represented by the bottom three blocks, absorbs the red and blue components. A mixture of red and green pigments absorbs all the incident white light, and none is reflected.

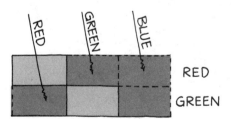

Consider, on the other hand, the mixing of cyan and yellow in Figure 26.12. This combination absorbs red and blue but reflects green. Cyan and yellow paints mixed together produce green. Similarly, the mixing of

Fig. 26.12 When white light is incident on cyan pigment, the red component is absorbed. Yellow pigment absorbs the blue component. When cyan and yellow pigments are mixed, green is not absorbed and is the only component of white light reflected.

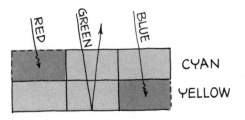

magenta and yellow paints produces red (Figure 26.13). By mixing magenta, cyan, and yellow pigments, we can produce the additive primaries—red, green, and blue. For this reason, we can summarize the results of color mixing by noting that any of the three additive primaries can be

Fig. 26.13 Magenta pigment absorbs green; yellow pigment absorbs blue. Only red is not absorbed by the mixture of magenta and yellow pigments. The mixture therefore appears red.

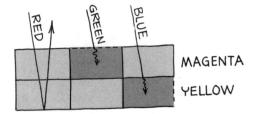

produced by a combination of any two of cyan, magenta, or yellow pigments. The mixture of absorbing pigments results in a *subtraction* of colors; the observer sees the light left over after absorption has taken place. For this reason, the latter three are called the **subtractive primary colors**.

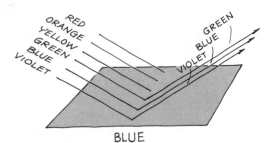

Fig. 26.14 Color subtraction by the mixing of colored pigments. Blue pigment reflects not only blue, but also colors to either side of blue, namely, green and violet. It absorbs red, orange, and yellow.

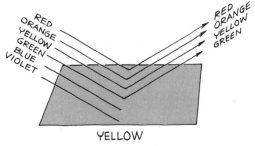

Yellow pigment reflects not only yellow, but also red, orange, and green. It absorbs blue and violet.

When blue and yellow pigments are mixed, the only common color reflected is green. The other colors have been *subtracted* from the incident white light.

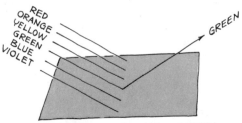

BLUE MIXED WITH YELLOW → GREEN

In painting, the primaries are often said to be red, yellow, and blue. Here one is loosely speaking of magenta, yellow, and cyan. By painting with them, one can produce any color in the spectrum.

Exercise   Complete the block diagram and predict the color that results from the mixing of cyan and magenta paints. Label and shade the appropriate areas, like Figures 26.12 and 26.13.

Color printing is an interesting application of color mixing. Three photographs (color separations) are taken of the illustration to be printed: one through a magenta filter, one through a yellow filter, and one through a cyan filter. This produces three negatives, each with a different pattern of exposed areas, which correspond to the filter used and the color distribution in the original illustration. Light is shone through these negatives onto metal plates specially treated to hold printer's ink only in areas that have been exposed to light. The ink deposits are regulated on different parts of the plate by tiny dots. Examine the colored pictures in a magazine with a magnifying glass and see how the overlapping dots of three colors give the appearance of many colors. Or look at a billboard up close.

Rules for Color Mixing

All the rules of color addition and subtraction can be deduced from the overlapping circles in Figure 26.10, and on the inside of the back cover of the book. The additive primaries are the red, green, and blue circles, and the subtractive primaries are located where the primary circles overlap. We see that a subtractive primary is the combination of two additive primaries.

Notice that when opposite subtractive primary colors combine—magenta with green, for example—they produce white. From the figure we can also see that yellow plus blue produces white[2]—cyan plus red does also. Any two colors that add together to produce white are called **complementary colors**.

Questions   1. From Figure 26.10, find the complements of cyan, of yellow, and of red.*

2. Red + blue = ____.†

3. White − red = ____.‡

4. White − blue = ____.§

---

*__Answer__ Red, blue, cyan.   †__Answer__ Magenta.   ‡__Answer__ Cyan.   §__Answer__ Yellow.

[2]This follows logically: since yellow is red + green, so blue + red + green = white.

Why the Sky Is Blue

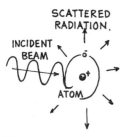

Fig. 26.15 A beam of light falls on an atom and causes the electrons in the atom to vibrate. The vibrating electrons, in turn, radiate in various directions.

If a beam of a particular frequency sound is directed to a tuning fork of similar frequency, the tuning fork will be set into vibration and effectively redirect the beam in multiple directions. The tuning fork *scatters* the sound—similarly with the scattering of light.

Atoms, molecules, and minute particles behave like tiny optical tuning forks and selectively scatter light waves of appropriate frequencies. The tinier the particle, the higher its natural frequency, and the higher the frequency of light it will scatter. This is similar to small bells that ring with higher notes than large bells. The nitrogen and oxygen molecules that predominate in the atmosphere are like tiny bells that "ring" with high frequencies when energized by sunlight. Most ultraviolet sunlight is absorbed by a protective layer of ozone gas in the upper atmosphere. The remaining ultraviolet light that passes through is scattered by atmospheric nitrogen and oxygen molecules. Of the visible frequencies, violet is scattered the most, followed by blue, green, yellow, orange, and red, in that order. Red is scattered only a tenth as much as violet. Although violet light is scattered more than blue, our eyes are not very sensitive to violet light. That's why we see a blue sky!

When the air is filled with innumerable particles of dust and particles larger than nitrogen and oxygen molecules, the lower frequencies of light are scattered more, and the sky seems less blue and takes on a more whitish appearance. But after a heavy rainstorm when the particles have been washed away, the sky becomes a deeper blue.

Fig. 26.16 When the air is clear, the scattering of high-frequency light gives us a blue sky. When the air is full of particles larger than molecules, lower frequencies are also scattered. These add to give white.

Why Sunsets Are Red

The lower frequencies of light are scattered the least by nitrogen and oxygen; therefore, red, orange, and yellow light is transmitted more readily than blue. Red, which is scattered the least, traverses more atmosphere relatively unhindered than any other color. Thus, in passing through a thick atmosphere, the higher frequencies are scattered while the lower frequencies are transmitted. This is why the white sun appears reddish at sunset. At noon it appears yellowish, as the path through the atmosphere is at a minimum, and a smaller amount of blue is scattered from the white light. As the day progresses, and the sun is lower in the sky (Figure 26.17), the path through the atmosphere is longer, and more blue is scattered from the sunlight. The sun becomes progressively redder, going from yellow to orange and finally to deep red-orange at sunset.

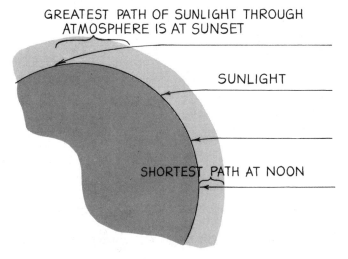

GREATEST PATH OF SUNLIGHT THROUGH ATMOSPHERE IS AT SUNSET

SUNLIGHT

SHORTEST PATH AT NOON

Fig. 26.17 More light is scattered from a sunbeam at sunset than at noontime because of the atmospheric path difference.

The colors of the sunset are consistent with our rules for color mixing. When blue is subtracted from white light, the complementary color that is left is yellow; when higher-frequency violet is subtracted, the resulting complementary color is orange; and for medium-frequency green, magenta is left. The combinations of resulting colors vary with atmospheric conditions, which change from day to day and give us a variation of sunsets. The next time you view a beautiful sunset, think about all those ultra-tiny optical tuning forks vibrating away; you'll appreciate the sunset even more!

Question

If the atmosphere were appreciably thicker than it presently is, how would the apparent color of the sun differ?*

---

*__Answer__   At noon the sun would range from a deeper yellow to a red, depending on the thickness of the atmosphere. Sunsets would correspondingly be much redder.

Color Vision

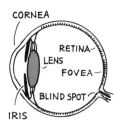

Fig. 26.18 The human eye.

Color is not in the world around you; color is in your head. The world around you is filled with a montage of vibrations—electromagnetic waves that stimulate the sensation of color when the vibrations interact with the cone-shaped receiving antennae in the retina of your eye. We will therefore direct our attention to the most remarkable optical instrument known—your own eye. A diagram of the eye is shown in Figure 26.18.

Light enters the eye through the transparent cover called the *cornea* that does about 70 percent of the necessary bending before the light passes through the pupil (which is an aperture in the iris) and then passes through the lens, which is used only to provide the extra bending power needed to focus images of nearby objects on the layer at the back of the eye. This layer—the *retina*—is more sensitive to light than any artificial detector made. Different parts of the retina receive light from different parts of the visual field outside. The retina is not uniform. There is a spot in the center of our field of view at which we have the greatest acuity of vision; it is called the *fovea*, or region of most distinct vision. Much greater detail can be seen here than at the side parts of the eye. There is also a spot in the retina where the nerves carrying all the information run out; this is the *blind spot.* You can demonstrate that you have a blind spot in each eye if you hold this book at arm's length, close your left eye, and look at Figure 26.19 with your right eye only. You can see both the circle and the X at this distance. If you now move the book slowly toward your face, with your right eye fixed upon the circle, you'll reach a position about 20 to 25 centimeters from your eye where the X disappears. To establish the blind spot in your left eye, close your right eye and similarly look at the X with your left eye so that the circle disappears. With both eyes open, you'll find no position where either the X or the circle disappears because one eye "fills in" the part to which the other eye is blind. Aren't you glad you have two eyes?

Fig. 26.19 The blind-spot experiment.

Fig. 26.20 Rods and cones as they are seen with a scanning electron microscope.

The retina is composed of tiny antennae that resonate to the incoming light. There are basically two kinds of antennae, the *rods* and the *cones* (Figure 26.20). As the names imply, some of the antennae are rod-shaped and some cone-shaped. The rods predominate closer to the periphery of the retina, while the cones are denser toward the fovea. The cones are very dense in the fovea itself, and since they are packed so tightly, they are much finer or narrower than elsewhere in the retina. Color vision is possible because of the cones. Hence we see color most acutely by focusing an image on the fovea. There are no rods in the fovea. Primates and a species of ground squirrel are the only mammals that have cones and experience color vision. The retinas of other mammals consist only of rods, which are sensitive only to lightness or darkness, like a black-and-white photograph or motion picture.

Very few people are totally color blind, but about 10 percent of the population has difficulty of some kind in seeing colors. These difficulties have to do with cones that are either missing or malfunctioning. Most forms of partial color blindness are inherited sex-linked genetic anomalies: very few women are color blind, for, unlike with men, both of a woman's parents must carry the defective gene.

The simplest way to describe color blindness is by the number of colors that a person must mix together in order to match all the hues of the rainbow. People with normal vision need only the three additive primary colors to reproduce all the colors of the spectrum. People who are partially color blind need only two hues to reproduce all the colors they can see. The majority of partially color blind people can match all the hues they can see with combinations of only blue and yellow; they cannot distinguish between reds and greens, which take on a grayish appearance. (How many automobile accidents could have been avoided if traffic engineers had taken color blindness into account when stoplights were first designed?) A few people are yellow-blue blind, and all the colors they can see can be reproduced by combinations of red and green. The very few individuals who are totally color blind need only one color (any color at all—they all seem the same) to reproduce all they can see. These few see no color at all—their world looks like a black-and-white movie.

Color blindness is not usually an all-or-nothing thing. There are many people who actually can see all the colors and who need three hues to match the spectrum but who need abnormal amounts of one of the hues. A person who is green-weak sees the spring grass as being a grayish green, but at least is able to perceive green as a distinct color. A person who is red-green *blind*, however, cannot distinguish green at all, no matter how intense the source is.

In the human eye, the number of cones decreases as we move away from the fovea. It's interesting that the color of an object disappears if viewed on the periphery of vision. This can be tested by having a friend enter your periphery of vision with some brightly colored objects. You will find that you can see the objects before you can see what color they are.

Another interesting fact is that the periphery of the retina is very sensitive to motion. Although our vision is poor from the corner of our eye, we are sensitive to anything moving there. We are "wired" to look for something jiggling to the side of our visual field, a feature that must have been important in our evolutionary development. So have your friend shake those brightly colored objects when she brings them into the periphery of your vision. If you can just barely see the objects when they shake, but not when they're held still, then you won't be able to tell what color they are (Figure 26.21). Try it and see!

Fig. 26.21 On the periphery of your vision you can see an object only if it is moving; you cannot see its color at all.

Another distinguishing feature of the rods and cones is the intensity of light to which they respond. The cones require more energy before they will "fire" an impulse through the nervous system. If the intensity of light is very low, the things we see have no color. We see low intensities with our rods. Dark-adapted vision is almost entirely due to the rods, while vision in bright light is due to the cones. Stars, for example, look white to us. Yet most stars are actually brightly colored. A time exposure of the stars with a camera reveals reds and oranges for the "cooler" stars and blues and violets for the "hotter" stars. The starlight is too weak, however, to fire the color-perceiving cones in the retina. So we see the stars with our rods, and perceive them as white or, at best, as only faintly colored. Females have a slightly lower threshold of firing for the cones, however, and can see a bit more color than males. So if she says she sees colored stars and he says she doesn't, she is probably right!

The sensitivity of the rods decreases in bright light. We find that the rods see better toward the blue end of the color spectrum than the cones. The cones can see a deep red where the rods see no light at all. Red light may as well be black as far as the rods can tell. Thus, if you have two colored objects, say, blue and red, the blue will appear much brighter than the red in dim light, though the red might be much brighter than the blue in bright light. The effect is quite interesting. Try this: In a dark room, find a magazine or something that has colors, and, before you know for sure what the colors are, judge the lighter and darker areas. Then carry the magazine into the light. You should see a remarkable shift between the brightest and dimmest colors.[3]

The rods and cones in the retina are not connected directly to the optic nerve, but, interestingly enough, are connected to many other cells that are connected to each other. Many of these cells are interconnected, while only a few carry information to the optic nerve. Through these interconnections a certain amount of information is combined from several visual receptors and "digested" in the retina. In this way the light signal is "thought about" before it goes to the optic nerve and then to the main body of the brain. So some brain functioning occurs in the eye itself. The eye does some of our "thinking."

This thinking is betrayed by the iris, the colored part of the eye that expands and contracts and regulates the size of the pupil for admitting more or less light as the intensity of light changes. It so happens that the relative size of this enlargement or contraction is also related to our emotions. If we see, smell, taste, or hear something that is pleasing to us, our pupils automatically increase in size. If we see, smell, taste, or hear something repugnant to us, our pupils automatically contract. Many a card player has betrayed the value of a hand by the size of his pupils![4]

SHE LOVES YOU...

SHE LOVES YOU NOT?

Fig. 26.22

---

[3]This phenomenon is called the *Purkinje effect.*

[4]The study of the size of the pupil as a function of attitudes is called *pupilometrics.*

The human eye can perceive brightness that ranges from about 500 million to 1. The difference in brightness between the sun and moon, for example, is about 1 million to 1. But, because of an effect called *lateral inhibition,* we don't perceive gradual differences in brightness. Lateral inhibition prevents the brightest places in the visual field from outshining the rest. Whenever a receptor cell on your retina sends a brightness signal to your brain, it also signals neighboring cells to dim their responses. In this way, you even out your visual field, and you can discern detail in very bright areas and in dark areas as well. Lateral inhibition exaggerates the difference in brightness at the edges of places in your visual field. Edges, by definition, separate one thing from another. So we accentuate differences rather than similarities. The gray square on the left in Figure 26.23 appears dimmer than the gray square on the right when the edge that separates them is in our view. But cover the edge with your pencil or your finger and they look equally bright. That's because both squares *are* equally bright; each square is shaded lighter to darker, moving from left to right. Your eye concentrates on the boundary where the dark edge of the left square joins the light edge of the right square, and your eye-brain system assumes that the rest of the square is the same.

Is the way the eye picks out edges and makes assumptions about what lies beyond similar to the way we sometimes make judgments about other

Fig. 26.23 Both squares are equally bright. Cover the boundary between them with your pencil and see.

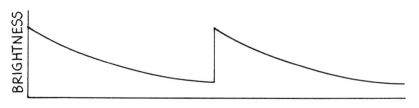

Fig. 26.24 Graph of brightness levels for the squares in Figure 26.23.

cultures and other people? Don't we in the same way tend to exaggerate the differences on the surface while ignoring the similarities and subtle differences within?

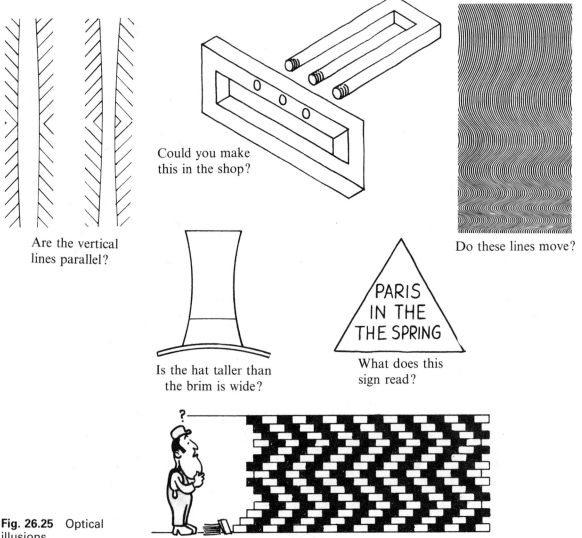

Are the vertical lines parallel?

Could you make this in the shop?

Do these lines move?

Is the hat taller than the brim is wide?

PARIS
IN THE
THE SPRING

What does this sign read?

Fig. 26.25 Optical illusions.

Are the tiles really crooked?

Summary of Terms

Additive primary colors The three colors—red, blue, and green—that when added in certain proportions will produce any color in the spectrum.

Subtractive primary colors The three colors of absorbing pigments—magenta, yellow, and cyan—that when mixed in certain proportions will reflect any color in the spectrum.

Complementary colors Any two colors that, when added, produce white light.

Review Questions

1. What is the difference between light that appears red and light that appears blue?

2. What is the difference between an object that appears red and one that appears blue?

3. Why will the temperature of a rose increase more when illuminated with green light than with red light?

4. Why do the colors of various objects appear different when illuminated by various light sources?

5. Which will undergo the greater increase in temperature when illuminated with white light—clear glass or colored glass? Explain.

6. What are the three additive primary colors? What are the three subtractive primary colors? Why are these not the same?

7. Close inspection of a color television tube reveals that the range of colors seen is composed of red, blue, and green dots. Close inspection of a magazine photograph in full color reveals magenta, yellow, and greenish-blue dots. Why are these different?

8. In what sense is scattering similar to the forced vibration or resonance of tuning forks?

9. Why is the sky blue at midday and reddish at sunup and sundown?

10. What is a principal difference between the retina of a human eye and that of a dog?

11. Why are fewer women than men color blind?

12. Most stars are colored. Why do they appear to be white?

Home Projects

1. Stare at a piece of colored paper for a minute or so. The cones in your retina receptive to the color of the paper become fatigued, so that when you look at a white area, you see an afterimage of the complementary color. This is because the fatigued cones send a weaker signal to the brain. All the colors produce white, but all the colors minus one produce the complementary to the missing color. Try it and see!

2. Which eye do you use more? To test which you favor, hold a finger up at arm's length. With both eyes open, look past it at a distant object. Now close your right eye. If your finger appears to jump to the right, then you use your right eye more.

3. Cut a disk a few centimeters or so in diameter from a piece of cardboard; punch two holes a bit off-center, big enough for a piece of string loop as shown in the sketch.

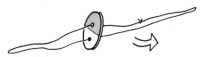

Twirl the disk as shown, so the string winds up like a rubber band on a model airplane. Then tighten the string by pulling outward and the disk will spin. If half the disk is colored yellow and the other half blue, when it is spun the color will be mixed and appear nearly white (how close to white depends on the hues of the colors). Try this for other complementary colors.

4. Fashion a cardboard tube with closed ends of metal foil in which you punch holes with a pencil, one about 3 or so millimeters and the other twice as big. Look at the colors of things against the black background of the tube. You'll see colors very much different than they appear against ordinary backgrounds.

5. Simulate your own sunset: add a few drops of milk to a glass of water and look through it to a light bulb. The bulb appears to be red or pale orange, while light scattered to the side appears blue. Try it and see.

6. Monitor the change of size of your own eye pupil. Here's how. Punch a small hole in a piece of paper with a small-size paper clip. Hold the paper a few centimeters from your eye and look at overhead lights or the sky through the hole. Don't focus on the pinhole, but relax and "stare" at the hole and put it so close that you wouldn't be able to focus on it anyway (remove your glasses if you wear any). You see a nice circle. But that circle is not the hole. It's your pupil! To demonstrate this, cover the other eye with your free hand. Both pupils are locked so that if one changes, the other does as well. Covering the other eye and diminishing light to it make that pupil expand. You'll note they both expand as you see the circle of light expand while you look through the pinhole. Neat?

7. Compare the size of the moon on the horizon with its size higher in the sky. One way to do this is to hold various objects at arm's length that will just barely block out the moon. Experiment until you find something just right, perhaps a thick pencil or pen. You'll find the object will be less than a centimeter, depending on the length of your arms. Is the moon really bigger when near the horizon?

Exercises

1. Why will the leaves of a rose be heated more than the petals when illuminated with red light? What does this have to do with people in the hot desert wearing white clothes?

2. In a dress shop having only fluorescent lighting, a customer insists on taking dresses into the daylight at the doorway. Is she being reasonable? Explain.

3. If the sunlight were green instead of white, what color garment would be most advisable on an uncomfortably hot day? On a very cold day?

4. Why do we not list black and white as colors?

5. Why are the interiors of optical instruments black?

6. What color would red cloth appear if it were illuminated by sunlight? By light from a neon sign? By cyan light?

7. Why does a white piece of paper appear white in white light, red in red light, blue in blue light, and so on for every color?

8. A spotlight is coated so that it won't transmit yellow light from its white-hot filament. What color is the emerging beam of light?

9. How could you use the spotlights at a play to make the yellow clothes of the performers suddenly change to black?

10. Suppose two flashlight beams are shone on a white screen, one beam through a pane of blue glass and the other through a pane of yellow glass. What color appears on the screen where the two beams overlap? Suppose, instead, that the two panes of glass are placed in the beam of a single flashlight. What then?

11. Does a color television work by color addition or by color subtraction? Which dots must be struck by electrons to create the color yellow? Magenta? White?

12. By reference to Figure 26.10, complete the following equations:

Yellow light + blue light = _____ light.
Green light + _____ light = white light.
Magenta + yellow + cyan = _____ light.

13. Check Figure 26.10 to see if the following three statements are accurate. Then fill in the last statement. (All colors are of light.)

Red + green + blue = white.
Red + green = yellow = white − blue.
Red + blue = magenta = white − green.
Green + blue = cyan = white − _____ .

14. Why can't we see stars in the daytime?

15. Why is the sky a darker blue at high altitudes? (*Hint:* What color is the "sky" on the moon?)

16. Can stars be seen from the moon in the "daytime" when the sun is shining?

17. Will people in space habitats see the sun against a background of stars, or will they see stars only when they are in the shadow of their habitat? Explain.

18. If the sky were composed of atoms that scattered orange light instead of blue, what color would sunsets be?

19. Tiny particles, like tiny bells, scatter high-frequency waves. Large particles, like large bells, scatter low frequencies. Intermediate-size particles and bells scatter intermediate frequencies. Can you see why clouds appear white? Explain.

20. Very big particles, like droplets of water, absorb more radiation than they scatter. Explain why rain clouds appear dark.

21. If the atmosphere of the earth were about fifty times thicker, would ordinary snowfall still seem white or would it be some other color? What color?

22. The atmosphere of Jupiter is more than 1000 kilometers thick. From the surface of this planet, would you expect to see a white sun?

23. Why do distant dark-colored mountains appear blue? Why do distant snow-covered mountains appear yellowish?

24. The refraction of all the sunsets around the world causes the redness of the moon during a lunar eclipse. Explain how this is so.

25. Light from a location on which you concentrate your attention is incident on your fovea, which contains only cones. If you wish to observe a weak source of light, like a faint star, should you look *directly at* the source? Explain.

26. Why do objects illuminated by moonlight lack color?

27. Why do we not see color at the periphery of our vision?

28. From your experimentation with Figure 26.19, is your blind spot located noseward from your fovea or to the outside of it?

29. When she holds you and looks at you with her contracting eye pupils and says, "I love you," what inference can you draw?

30. Can we infer that a person with large pupils is generally happier than a person with small pupils? Why not?

27 Light Waves

Throw a rock in a quiet pool, and it produces waves along the surface of the water. Strike a tuning fork, and waves of sound emanate in all directions. Light a match, and waves of light similarly expand in all directions—at the enormously high speed of 300,000 kilometers per second. In this chapter we will study the wave nature of light. In the next chapter we will see that light has particle properties as well, but for now we will investigate some of the wave properties of light: diffraction, interference, and polarization.

Huygens' Principle

A wave theory of light was proposed by the English physicist Robert Hooke in 1665 and improved 20 years later by the Dutch scientist and mathematician Christian Huygens. Huygens stated that light waves spreading out from a point source could be regarded as the superposition of tiny secondary wavelets. This idea holds for water waves as well. In Figure 27.1 we see parts of circularly expanding waves. We direct our attention to the leading edges of these wave segments, which we call *wave fronts*. Huygens went on to state that every point on any wave front may be regarded as a new point source of secondary waves (Figure 27.2). This idea is called **Huygens' principle**.

Fig. 27.1 Wave fronts of water waves.

Fig. 27.2 Huygens' principle. Every point on a wave behaves as if it were a new source of waves. Secondary wavelets starting at b,b,b,b form a new wavefront d-d; secondary wavelets starting at d,d,d,d form still another new wave front DCEF. (From Christian Huygens, *Treatise on Light*.)

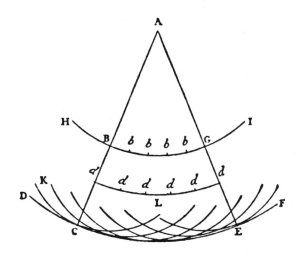

Consider the spherical wave front in Figure 27.3. We can see that if all points along the wave front AA′ are sources of new wavelets, a short time later the new overlapping wavelets will form a new surface, BB′, which can be regarded as the *envelope* of all the wavelets. In the figure we show

435

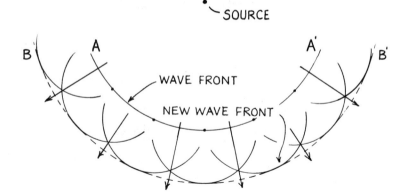

Fig. 27.3 Huygens' principle applied to a spherical wave front.

only a few of the infinite number of wavelets from a few secondary point sources that combine to produce the smooth envelope BB'. As the wave spreads, a segment appears less curved. Very far from the original source, the waves nearly form a plane, as do waves from the sun, for example.

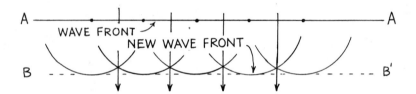

Fig. 27.4 Huygens' principle applied to a plane wave front.

A Huygens wavelet construction for plane wave fronts is shown in Figure 27.4. We see the laws of reflection and refraction illustrated via Huygens' principle in Figure 27.5.

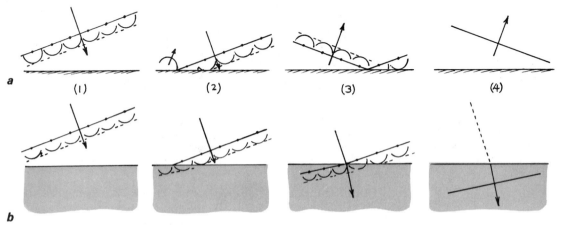

Fig. 27.5 Huygens' principle applied to (a) reflection and (b) refraction.

Plane waves can be generated in water by successively dipping a horizontally held straightedge such as a meter stick into the surface. The pictures in Figure 27.6 are top views of a ripple tank, where plane waves are produced in water by an oscillating straightedge. The straightedge is not shown, but the plane waves produced are shown incident upon openings of various sizes. In *a*, where the opening is wide, we see the plane waves continue through the opening without change—except at the corners where the waves are bent into the shadow region as predicted by Huygens' principle. As the width of the opening is narrowed as in *b*, less and less of the incident wave is transmitted, and the spreading of waves into the shadow region becomes more pronounced. When the opening is small compared to the wavelength of the incident wave as in *c*, Huygens' idea that every part of a wave front can be regarded as a source of new wavelets becomes quite apparent. As the waves are incident upon the narrow opening, the water sloshing up and down in the opening is easily seen to act as a "point" source of the new waves that fan out on the other side of the barrier. We say the waves are *diffracted* as they spread into the shadow region.

Diffraction

In the last chapter we saw that light can be bent from its ordinary straight-line path by reflection and refraction, and now we see another way light bends. Any bending of light by means other than reflection and refraction is called **diffraction**. The pictures in Figure 27.6 show the diffraction of plane water waves through various openings. When the size of the opening is large compared to the wavelength, the spreading effect is small, and as the opening becomes smaller, the spreading of waves is more appreciable. The same occurs for all kinds of waves, including light waves.

When light passes through an opening that is large compared to the wavelength of light, it casts a shadow such as that shown in Figure 27.7. We see a rather sharp boundary between the light and dark area of the

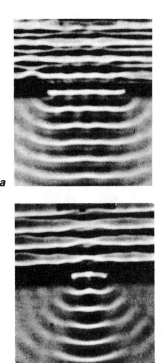

a

b

c

Fig. 27.6 Straight waves passing through openings of various sizes. The smaller the opening, the greater the bending of the waves at the edges.

Fig. 27.7 Diffraction is subtle.

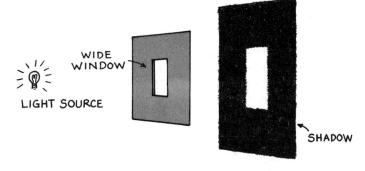

Fig. 27.8 Diffraction is pronounced.

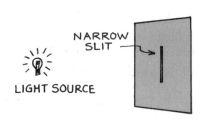

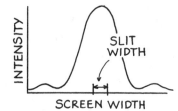

Fig. 27.9 Graphical interpretation of diffracted light through a single thin slit.

Fig. 27.10 Laser light that bends around a tea strainer leaves diffraction patterns in its wake. The fuzzy edges of most shadows are diffraction patterns too fine to be ordinarily seen.

shadow. But pass light through a thin razor slit in a piece of opaque cardboard, and we see that the light diffracts (Figure 27.8). The sharp boundary between the light and dark area disappears, and the light spreads out like a fan to produce a bright area that fades into darkness without sharp edges. The light is diffracted.

A graph of the intensity distribution for diffracted light through a single thin slit appears in Figure 27.9. Because of diffraction, there is a gradual increase in light intensity rather than an abrupt change from dark to light. A photodetector sweeping across the screen would sense a gradual increase from no light to maximum light. (Actually, there are slight fringes of intensity to either side of the main pattern; we will see shortly that these are evidence of interference that is more pronounced with a double slit or multiple slits.)

Diffraction is not confined by narrow slits or to openings in general but can be seen for all shadows. On close examination, even the sharpest shadow is blurred slightly at the edge (Figure 27.10).

The amount of diffraction depends on the wavelength of the wave compared to the size of the obstruction. Radio waves are very long, ranging from 180 to 6000 meters for the standard AM broadcast band. As a result, radio waves readily bend around objects that might otherwise obstruct them. The radio waves of the FM band range from 2.7 to 3.7 meters and don't bend as much as AM waves. This is one of the reasons that FM reception is often poor in localities where AM comes in loud and clear. Diffraction therefore aids radio reception.

A negative consequence of diffraction is the limitation it imposes on microscopy. The shadows of small objects become less and less well defined as the size of the object approaches the wavelength of light illuminating it. If the object is smaller than a single wavelength, no structure—or anything—can be seen. No amount of magnification or perfection of microscope design can defeat this fundamental diffraction limit. To circumvent this problem, microscopists illuminate tiny objects with electron beams rather than light. They use electron microscopes, which take advantage of the fact that all matter has wave properties: a beam of electrons has a wavelength smaller than visible light. (We will look at this in the next chapter.) So the diffraction limit of an electron microscope is much less than that of an optical microscope that is restricted to visible light. We use electric and magnetic fields, rather than optical lenses, to focus and magnify images.

Fig. 27.11 (*a*) The shadow of a screw in laser light shows fringes produced by destructive interference of the diffracted light. (*b*) A longer exposure of the screw shows fringes within the shadow produced by constructive interference.

a   *b*

Question   Why does a microscopist use blue light to illuminate the objects viewed?*

Interference   Spectacular illustrations of diffraction are shown in Figure 27.11. C. K. Manka of Sam Houston State University made these by placing pieces of photographic film in the shadow of a screw illuminated with spatially filtered light. The fringes appear in both figures. These fringes are produced by **interference**, which we discussed in Chapter 17. Constructive and destructive interference is reviewed in Figure 27.12. We see that the superposition of a pair of identical waves in phase with each other produces a wave of the same frequency but with twice the amplitude. If the

a + = REINFORCEMENT   *b* + = CANCELATION   *c* + = PARTIAL CANCELATION

*Answer   Less diffraction results from the short waves of blue light compared to other, longer waves.

Fig. 27.13 Interference of water waves.

pair of waves are exactly one-half wavelength out of phase, their superposition results in complete cancelation. If they are out of phase by other amounts, partial cancelation occurs.

The interference of water waves is a common sight, as shown on the cover of this book, and in closer detail in Figure 27.13. In some places crests overlap crests, while in other places crests overlap troughs of other waves.

Fig. 27.14 Interference of two sets of identical waves. Arrows mark regions where waves have canceled each other.

Under more carefully controlled conditions, interesting patterns are produced when two sources of waves are placed side by side (Figure 27.14). Drops of water are allowed to fall at a controlled frequency into ripple tanks while their patterns are photographed from above. Note that

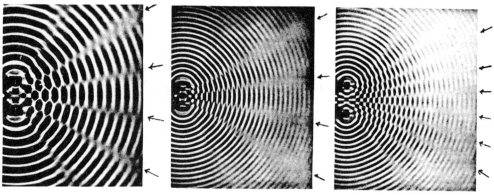

areas of constructive and destructive interference are incident on the right-side edges of the ripple tanks and that the number of these regions and the distances between them depend on the distance between the wave sources and on the wavelength (or frequency) of the waves.

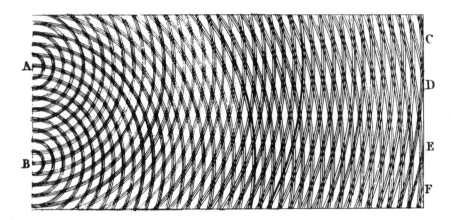

Fig. 27.15 The Thomas Young drawing of a two-source interference pattern.

Fig. 27.16 Interference fringes.

In 1801 the wave nature of light was convincingly demonstrated when the British physicist and physician Thomas Young performed his now-famous interference experiment. Young found that light directed through two closely spaced pinholes recombined to produce fringes of brightness and darkness on a screen behind. The bright fringes of light resulted from light waves from the two holes arriving crest to crest, while the dark areas resulted from light waves arriving trough to crest. Figure 27.15 shows Young's drawing of the pattern of superimposed waves from the two sources. His experiment is now done with two closely spaced slits instead of pinholes, so the fringe patterns are straight lines (Figure 27.16).

We see in Figure 27.17 how the series of bright and dark lines results from the different path lengths from the slits to the screen. For the central bright fringe, the paths from each slit are the same length, and the waves arrive in phase and reinforce each other. The dark fringes on either side of the central fringe result from one path being longer (or shorter) by one-half wavelength, where the waves arrive a half-wavelength, or 180°, out of

Fig. 27.17 Bright fringes occur when waves from both slits arrive in phase; dark areas result from the superposition of waves that are out of phase.

phase. The other sets of dark fringes occur where the paths differ by odd multiples of a half-wavelength: $\frac{3}{2}$, $\frac{5}{2}$, and so on. We see this illustrated in Figure 27.18.

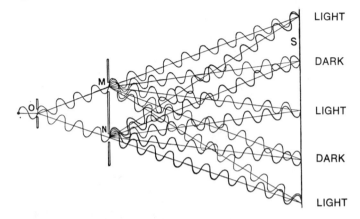

LIGHT

DARK

LIGHT

DARK

LIGHT

Fig. 27.18 Light from O passes through slits M and N and produces an interference pattern on the screen S.

Questions

1. If the double slits were illuminated with monochromatic red light, would the fringes be more widely or more closely spaced, as compared to illumination with monochromatic blue light?*

2. Why is it important that monochromatic light is used?†

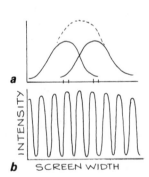

Fig. 27.19 The light that diffracts through each of the double slits does not form a super-position of intensities as suggested in (a). The intensity pattern, because of interference, is as shown in (b).

In performing this double-slit experiment, suppose we cover one of the slits so that light passes through only a single slit. Then light will fan out and illuminate the screen to form a simple diffraction pattern as discussed earlier (Figure 27.9). If we covered the other slit and allowed light to pass only through the slit just uncovered, we would get the same illumination on the screen, only displaced somewhat because of the difference in slit location. If we didn't know better, we might expect that with both slits open, the pattern would simply be a superposition of the single-slit diffraction patterns, as suggested in Figure 27.19a. But this doesn't happen. Instead, the pattern formed is one of alternating light and dark bands, as shown in b. We have an interference pattern. Interference of light waves does not, by the way, create or destroy light energy; it merely redistributes it.

****Answer**   More widely spaced. Can you see in Figure 27.18 that a slightly longer and therefore a slightly more displaced path from the entrance slit to the screen would be required for the longer waves of red light?

†**Answer**   If light of various wavelengths were diffracted by the slits, dark fringes for one wavelength would be filled in with bright fringes for another, resulting in no distinct fringe pattern. If you haven't seen this, be sure to ask your instructor to demonstrate it.

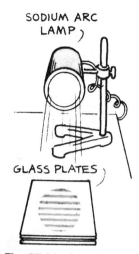

SODIUM ARC LAMP

GLASS PLATES

Fig. 27.20 Interference fringes.

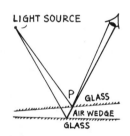

LIGHT SOURCE

P / GLASS
AIR WEDGE
GLASS

Fig. 27.21 Reflection from the upper and lower surfaces of a "thin film of air."

Another way interference fringes can be produced is by the reflection of light from the top and bottom surfaces of a thin film. A simple demonstration can be set up with a monochromatic light source and a couple of pieces of glass. A sodium-vapor lamp provides a good source of monochromatic light. The two pieces of glass are placed one atop the other, as shown in Figure 27.20. A very thin piece of paper is placed between the plates at one edge. This leaves a very thin, wedge-shaped film of air between the plates. If the eye is in a position to see the reflected image of the lamp, the image will not be a continuous one but will be made up of dark and bright bands.

The cause of these bands is the interference between the waves reflected from the glass on the top and bottom surfaces of the air wedge, as shown in the exaggerated view in Figure 27.21. The light reflecting from point P comes to the eye by two different paths. In one of these paths the light is reflected from the top of the air wedge; in the other path it is reflected from the lower side. If the eye is focused on point P, both rays reach the same place on the retina of the eye. But these rays have traveled different distances and may meet in phase or out of phase depending on the thickness of the air wedge—that is, on how much farther one ray has traveled than the other. When we look over the whole surface of the glass, we see alternate dark and bright regions—the dark portions where the air thickness is just right to produce destructive interference and the bright portions where the air wedge is just the proper amount thinner or thicker to result in the reinforcement of light. So the dark and bright bands are caused by the interference of light waves reflected from the two sides of the thin film.[1]

If the surfaces of the glass plates used are perfectly flat, the bands are uniform. But if the surfaces are not perfectly flat, the bands are distorted. The interference of light provides an extremely sensitive method for testing the flatness of surfaces. Surfaces that produce uniform fringes are said to be *optically flat*—that is, flat compared with the wavelength of visible light (Figure 27.22).

When a plano-convex lens is placed on an optically flat plate of glass and illuminated from above with monochromatic light, a series of light and dark rings is produced. This pattern is known as *Newton's rings*

---

[1]*Phase shifts* at some reflecting surfaces also contribute to interference. For simplicity and brevity, our concern with this topic will be limited to this footnote: in short, when light in a medium is reflected at the surface of a second medium in which the speed of transmitted light is less (when there is a greater index of refraction), there is a 180° phase shift, but no phase shift occurs when the second medium is one that transmits light at a higher speed (and there is a lower index of refraction). In our air-wedge example, no phase shift occurs for reflection at the upper glass-air surface, and a 180° shift does occur at the lower air-glass surface. So at the apex of the air wedge where the thickness approaches zero, the phase shift produces cancelation, and the wedge is dark. Likewise with a soap film so thin that its thickness is appreciably smaller than the wavelength of light.

Fig. 27.22 An optical flat used for testing the flatness of surfaces.

(Figure 27.23). These light and dark rings are the same kinds of fringes observed with plane surfaces. This is a useful technique for polishing precision lenses.

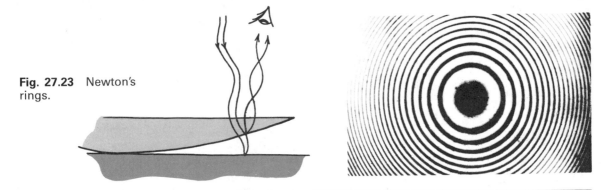

Fig. 27.23 Newton's rings.

Question How would the spacings between Newton's rings differ when illuminated by red and by blue light?*

Interference Colors by Reflection from Thin Films

We have all noticed the beautiful spectrum of colors reflected from a soap bubble or from gasoline on a wet street. We have seen the colors reflected from some types of bird feathers that seem to change in hue as the bird moves. These colors are produced by the *interference* of light waves. This phenomenon is often called *iridescence* and is observed in thin films.

---

*__Answer__ The rings would be more widely spaced for longer-wavelength red light than for the shorter waves of blue light. Do you see the geometrical reason for this?

A soap bubble appears iridescent in white light when the thickness of the soap film is about the same as the wavelength of light. Light waves reflected from the outer and inner surfaces of the film travel different distances. When illuminated by white light, the film may be just the right thickness at one place to cause the destructive interference of, say, yellow light. When yellow light is subtracted from white light, the mixture left will appear as the complementary color of yellow, which is blue. At another place where the film is thinner, a different color may be canceled by interference, and the light seen will be its complementary color. The same thing happens to gasoline on a wet street (Figure 27.24). Light reflects from both the upper gasoline surface and the lower gasoline-water surface. If the incident beam is monochromatic blue, as the figure suggests, then the gasoline surface appears dark to the eye. If the incident beam is white, then the gasoline surface appears yellow to the eye. This is because the blue is subtracted from the white, leaving the complementary color, yellow. The different colors, then, correspond to different thicknesses of the thin film, providing a vivid "contour map" of microscopic differences in surface "elevations."

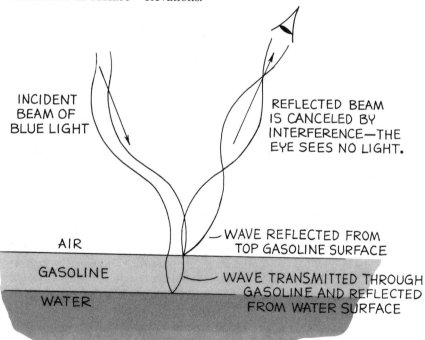

Fig. 27.24 Light reflected from the top of the thin film of gasoline is canceled by light of the same wavelength reflected from the water.

INCIDENT BEAM OF BLUE LIGHT

REFLECTED BEAM IS CANCELED BY INTERFERENCE—THE EYE SEES NO LIGHT.

AIR

GASOLINE

WATER

WAVE REFLECTED FROM TOP GASOLINE SURFACE

WAVE TRANSMITTED THROUGH GASOLINE AND REFLECTED FROM WATER SURFACE

Over a wider field of view, different colors can be seen even if the thickness of the gasoline film is uniform. This has to do with the "apparent thickness" of the film: light reaching the eye from different parts of the surface is reflected at different angles and traverses different thicknesses.

If the light is incident at a grazing angle, for example, the ray transmitted to the gasoline's lower surface travels a longer distance. Longer waves are canceled in this case, and different colors appear.

Dishes washed in soapy water and poorly rinsed have a thin film of soap on them. Hold such a dish up to a light source so that interference colors can be seen. Then turn the dish to a new position, keeping your eye on the same part of the dish, and the color will change. Light reflecting from the bottom surface of the transparent soap film is canceling light reflecting from the top surface of the soap film. Different wavelengths of light are canceled for different angles.

The wavelengths of light and other regions of the electromagnetic spectrum are measured with interference techniques. The principle of interference provides a means of measuring extremely small distances with great accuracy. Instruments called *interferometers*, which use the principle of interference, are the most accurate instruments known for measuring small distances.

Question What color will appear to be reflected from a soap bubble in sunlight when its thickness is such that green light is canceled?*

Polarization

Interference and diffraction provide the best evidence that light is wavelike in nature. Wave motion in general can be either longitudinal or transverse. Sound in air travels in longitudinal waves, where the vibratory motion is *along* the direction of wave propagation. When we shake a taut rope, the vibratory motion that propagates along the rope is perpendicular, or *transverse*, to the rope. Both longitudinal and transverse waves exhibit interference and diffraction effects. Are light waves, then, longitudinal or transverse? The fact that light waves can be *polarized* demonstrates that they are transverse.

If we shake a taut rope up and down as in Figure 27.25, a transverse wave travels along the rope in a plane. We say that such a wave is *plane-polarized*.[2] If the rope is shaken up and down vertically, the wave is vertically plane-polarized; that is, the waves traveling along the rope are confined to a vertical plane. If we shake the rope from side to side, we produce a horizontally plane-polarized wave.

A vibrating electron emits an electromagnetic wave that is also plane-polarized. The plane polarization will be along the direction of the vibrat-

Fig. 27.25 A vertically polarized plane wave and a horizontally polarized plane wave.

*__Answer__ The composite of all the visible wavelengths except green results in the complementary color, magenta. See Figure 26.10.

[2]Light may also be circularly polarized and elliptically polarized, which are also transverse polarizations, but we will not consider these cases.

ing charge. A vertically accelerating electron, then, emits light that is vertically polarized, while a horizontally accelerating electron emits light that is horizontally polarized (Figure 27.26).

Fig. 27.26 (a) A vertically polarized plane wave from a charge vibrating vertically. (b) A horizontally polarized plane wave from a charge vibrating horizontally.

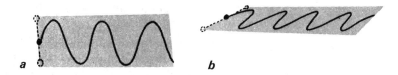

A common light source such as an incandescent lamp, a fluorescent lamp, a candle flame, or an arc lamp emits light that is nonpolarized. This is because there is no preferred vibrational direction for the accelerating charges producing the light. The number of planes of vibration might be as numerous as the accelerating charges producing them. A few planes are represented in Figure 27.27a. We can represent all these planes by radial lines (Figure 27.27b) or, more simply, by vectors in two mutually perpendicular directions (Figure 27.27c), as if we had resolved all the vectors of Figure 27.27b into horizontal and vertical components.[3] This simpler schematic representation is a useful shorthand. In this case, the diagram represents nonpolarized light. Polarized light would be represented by a single vector.

Fig. 27.27 Graphical representations of planes of plane-polarized waves.

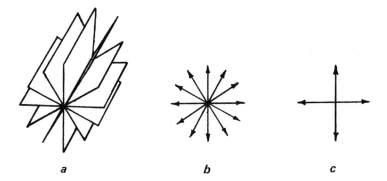

All transparent crystals of a noncubic natural shape have the property of transmitting light through two mutually perpendicular planes. Certain crystals[4] not only produce two internal beams polarized at right angles to each other but also strongly absorb one beam while transmitting the other (Figure 27.28). Tourmaline is one such common crystal, but unfortunately the transmitted light is colored. Herapathite, however, does the job without discoloration. Microscopic crystals of herapathite are embedded between

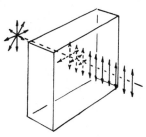

Fig. 27.28 One component of the incident nonpolarized light is absorbed, resulting in emerging polarized light.

---

[3]Refer to Appendix III, "Vectors."
[4]Called *dichroic.*

cellulose sheets in uniform alignment and are used in making Polaroid filters. Some Polaroid sheets consist of certain aligned molecules rather than tiny crystals.[5]

If you look at nonpolarized light through a Polaroid filter, you can rotate the filter in any direction, and the light will appear unchanged. But if this light is polarized, then as you rotate the filter, you can progressively cut off more and more of the light until it is blocked out. An ideal Polaroid will transmit 50 percent of incident nonpolarized light. That 50 percent is, of course, polarized. When two Polaroids are arranged so that their polarization axes are aligned, light will be transmitted through both (Figure 27.29). If their axes are at right angles to each other, no light will penetrate the pair. Actually, some of the shorter wavelengths do get through, but not to any significant degree. When Polaroids are used in pairs like this, the first one is called the *polarizer,* and the second one the *analyzer.*

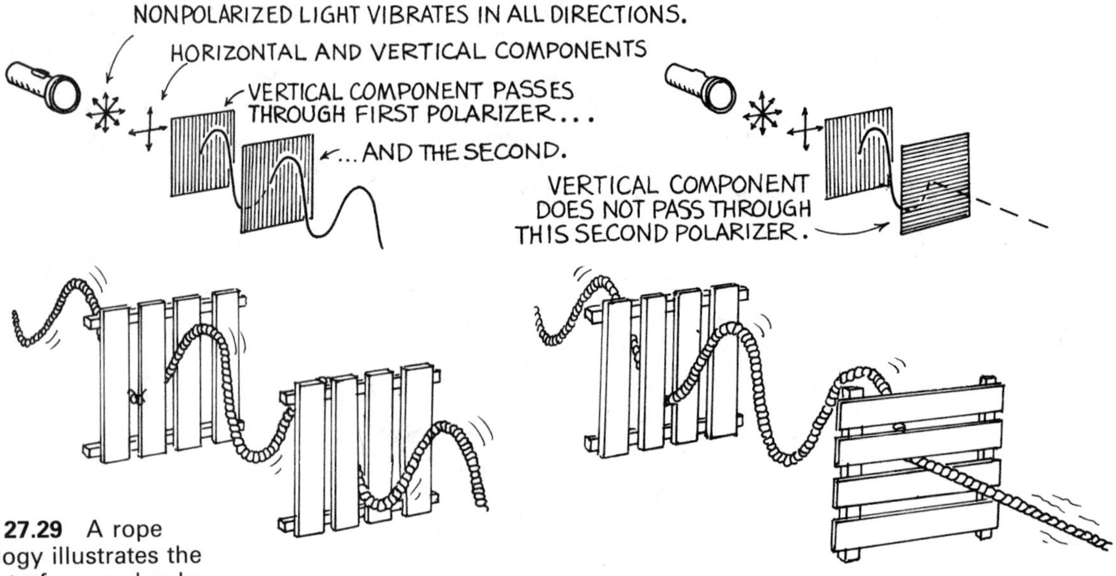

NONPOLARIZED LIGHT VIBRATES IN ALL DIRECTIONS.

HORIZONTAL AND VERTICAL COMPONENTS

VERTICAL COMPONENT PASSES THROUGH FIRST POLARIZER...

...AND THE SECOND.

VERTICAL COMPONENT DOES NOT PASS THROUGH THIS SECOND POLARIZER.

Fig. 27.29 A rope analogy illustrates the effect of crossed polarizers.

Much of the light reflected from nonmetallic surfaces is polarized. The glare from glass or water is a good example. Except for normal incidence, the reflected ray contains more vibrations parallel to the reflecting surface, while the transmitted beam contains more vibrations at right angles to these (Figure 27.31). Skipping flat rocks off the surface of a pond is analogous. When the rocks are incident parallel to the surface, they easily reflect; but if they are incident with their faces at right angles to the surface, they ''refract'' into the water. The glare from reflecting surfaces

[5]The molecules are polymeric iodine in a sheet of polyvinyl alcohol or polyvinylene.

can be appreciably diminished with the use of Polaroid sunglasses. The polarization axes of the lenses are vertical, as most glare reflects from horizontal surfaces. Properly aligned Polaroid eyeglasses enable us to see the projection of stereoscopic movies or slides on a flat screen in three dimensions.

Fig. 27.30 Polaroid sunglasses block out horizontally vibrating light. When the lenses overlap at right angles, no light gets through.

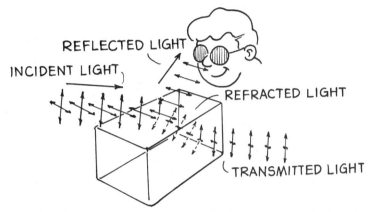

Fig. 27.31 Most glare from nonmetallic surfaces is polarized. Here we see that the components of incident light that are parallel to the surface are reflected, while the components that are perpendicular to the surface are refracted into the medium. Since most of the glare we encounter is from horizontal surfaces, the polarization axes of Polaroid sunglasses are vertical.

a   b   c

Fig. 27.32 Light is transmitted when the axes of the Polaroids are aligned (a), but absorbed when Dotty Jean rotates one so that the axes are at right angles to each other (b). When she inserts a third Polaroid at an angle between the crossed Polaroids, light is again transmitted (c). Why? (For the answer after you have given this some thought, see Appendix III, "Vectors.")

Three-Dimensional Viewing

Vision in three dimensions is based primarily on the fact that both eyes give their impressions simultaneously, each eye viewing the scene from a slightly different angle. To convince yourself that each eye sees a different view, hold an upright finger at arm's length and see how it appears to switch back and forth in front of the background as your eyes are alternately closed. Figure 27.33 illustrates two such views of a well. The

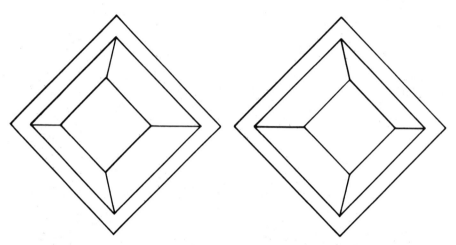

Fig. 27.33   If you look at these figures with your eyes focused for distant viewing, each figure will appear double. If you change your focus, the two inside images can be superimposed and will then appear in relief like a well sinking into the page. You can more easily achieve this effect by holding a card or piece of paper between the eyes and between the two figures so that the left eye can see only the left figure and the right eye only the right figure. (If you instead *cross* your eyes to superimpose the figures, why does the image appear like a solid standing *up* from the page?)

familiar hand-held stereoscopic viewer (Figure 27.35) simulates the effect of depth. In this device, there are two photographic transparencies (or slides) taken from slightly different positions. When they are viewed at the same time, the arrangement is such that the left eye sees the scene as

Fig. 27.34 Look at the pair of statements with your eyes adjusted as for Figure 27.33. With your eyes focused for distant viewing, the second and fourth lines appear to be farther away; cross-eyed, they appear closer.

```
I believe most simply in the
nobility of this great effort
  to understand nature,
and what we can of ourselves,
  that is science.
        J. Robert Oppenheimer
```

```
I believe most simply in the
  nobility of this great effort
to understand nature,
  and what we can of ourselves,
that is science.
          J. Robert Oppenheimer
```

Fig. 27.35 A stereoscopic viewer.

photographed from the left, and the right eye sees it from the right. As a result, the objects in the scene sink into relief in correct perspective, giving so-called depth to the picture. The viewer is constructed so that each eye sees only the proper view. There is no chance for one eye to see both views. If you remove the slides from the hand viewer and project each view on a screen by slide projector (so that the views are superimposed), a blurry picture results. This is because each eye sees both views simultaneously. This is where Polaroid filters come in. If you place the Polaroids in front of the projectors so that one is horizontal and the other vertical, and you view the polarized image with polarized glasses of the same orientation, each eye will see the proper view as with the stereoscopic viewer (Figure 27.36). You then will see an image in three dimensions.

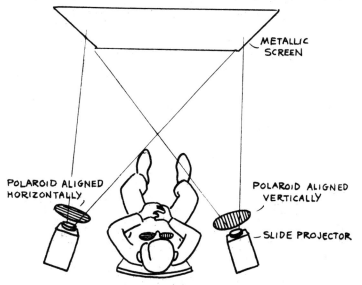

Fig. 27.36 A three-dimensional slide show using Polaroids. The left eye sees only light from the left projector, and the right eye sees only light from the right projector.

Colors by Transmission Through Polarizing Materials

Colors as vivid as those produced by interference from reflecting surfaces can also be produced under somewhat comparable conditions by transmission through certain transparent crystals. It so happens that all transparent crystals of a noncubic natural shape have the property of transmitting light in two mutually perpendicular planes and, most importantly, at two slightly different speeds. When light emerges from these crystals, the vibrations are confined to these planes. They are polarized.

When materials that are themselves composed of noncubic transparent crystals, like sugar, are placed between crossed Polaroids and illuminated with white light, colored light emerges. Various thicknesses of mica sheets between crossed Polaroids produce a wide assortment of colors. More spectacular is cellophane, which is composed of chain molecules that behave optically like microscopic noncubic crystals. When the cellophane is crumpled, a great variety of vivid colors is produced.

The explanation for this is not simple and requires a fairly good understanding of vectors. So unless you're into vectors, you may want to skip or skim the following paragraph and Figures 27.37 and 27.38.

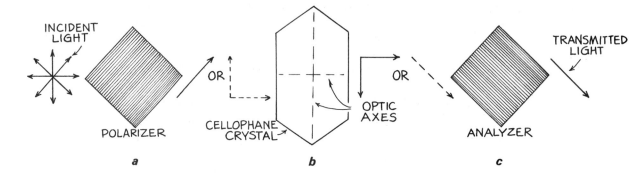

Fig. 27.37 The cellophane should be sandwiched between the crossed Polaroids but is spread out for convenience. Follow the transformations of the light vector from incidence (*a*) to transmission (*c*). Figure 27.38 shows how the vertical component changes direction in the cellophane.

Consider a single wavelength of light that travels through cellophane that is sandwiched between two crossed Polaroids. We show this in a spread-out fashion in Figure 27.37. Illuminating light is incident upon the first Polaroid, the polarizer, which directs polarized light into the cellophane before finally emerging from the second Polaroid, the analyzer. We assume that the optic axes of the cellophane are at 45° with respect to either of the Polaroid axes. Incoming light passes through the polarizer and is polarized at 45° with respect to the horizontal and vertical axes of the cellophane. The 45° vector has a horizontal and vertical component, as shown by the broken lines, which will be transmitted along the axes of the cellophane. These components will travel through the cellophane at different speeds. As a result, one wave component gets ahead of the other. In the figure, we suppose that the thickness of the cellophane is such that the vertical component gains one-half wavelength over the horizontal component. The emerging vertical component in this case points down instead of up. The result of the vectors is effectively rotated through 90° from the vector passed by the polarizer. It is then in a direction corresponding to the axis of the analyzer and passes through at maximum transmission. Only for this particular wavelength and this particular cellophane thickness will a 90° rotation of the incident vector occur. A wave diagram illustrating this case is shown in Figure 27.38. If the analyzer is rotated to a different position, some other rotated wavelength will correspond to the alignment with the axis of the analyzer. As the analyzer is turned, then, different colors are transmitted by the system.

The variety of colors that emerge from crumpled cellophane sandwiched between crossed Polaroids is spectacular—especially when projected on a large screen and one of the Polaroids is rotated in rhythm with your favorite music!

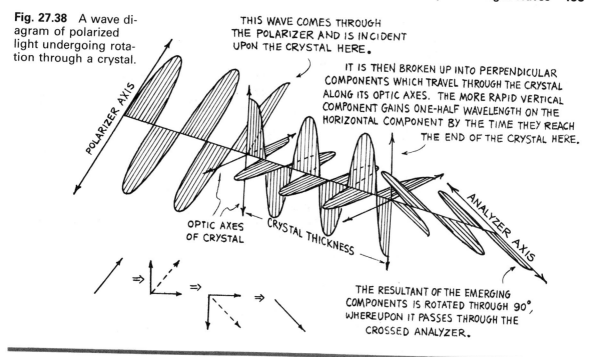

Fig. 27.38 A wave diagram of polarized light undergoing rotation through a crystal.

THIS WAVE COMES THROUGH THE POLARIZER AND IS INCIDENT UPON THE CRYSTAL HERE.

IT IS THEN BROKEN UP INTO PERPENDICULAR COMPONENTS WHICH TRAVEL THROUGH THE CRYSTAL ALONG ITS OPTIC AXES. THE MORE RAPID VERTICAL COMPONENT GAINS ONE-HALF WAVELENGTH ON THE HORIZONTAL COMPONENT BY THE TIME THEY REACH THE END OF THE CRYSTAL HERE.

POLARIZER AXIS

ANALYZER AXIS

OPTIC AXES OF CRYSTAL

CRYSTAL THICKNESS

THE RESULTANT OF THE EMERGING COMPONENTS IS ROTATED THROUGH 90°, WHEREUPON IT PASSES THROUGH THE CROSSED ANALYZER.

Question   If all the crystals in a sample were perfectly aligned, and the plane of incident polarized light were coincident with one of the transmission axes of the crystals, would colored light emerge through the second Polaroid?*

Holography   Perhaps the most exciting illustration of interference is the *hologram,* a two-dimensional photographic plate illuminated with laser light that allows you to see a faithful reproduction of a scene in three dimensions. The hologram was invented and named by Dennis Gabor in 1947, ten years before lasers were invented. *Holo* in Greek means "whole," and *gram* in Greek means "message" or "information." A hologram contains the whole message or entire picture. When illuminated with laser light, the image is so realistic that you can actually look around the corners of objects in the image and see the sides.

In ordinary photography, a lens is used to form an image of an object on photographic film. Light reflected from each point on the object is directed by the lens only to a corresponding point on the film. And all the light that reaches the film comes from the object being photographed. In

---

*__Answer__   No, because the incident beam would not split into two components and travel at different speeds through both crystal axes to emerge in a shifted phase and rotated for color separation.

the case of holography, however, no image-forming lens is used. Instead, each point of the object being "photographed" reflects light to the *entire* photographic plate, so *every* part of the plate is exposed with light reflected from every part of the object. Most importantly, the light used to make a hologram must be of a single frequency and all parts exactly in phase: it must be *coherent*. Only a laser can produce such light. Holograms are made with laser light.

A conventional photograph is a recording of an image, but a hologram is a recording of the interference pattern resulting from the combination of two sets of wave fronts. One set of wave fronts is from light reflected by the object, and the other set is from the *reference beam,* which is diverted from the illuminating beam and sent directly to the photographic plate (Figure 27.39). The developed photograph has no recognizable image. The hologram is simply a hodgepodge of whirly lines, tiny fringed areas of varying density—dark where wave fronts from the object and reference beam arrived in phase, and light where the wave fronts were out of phase. The hologram is a photographic pattern of microscopic interference fringes.

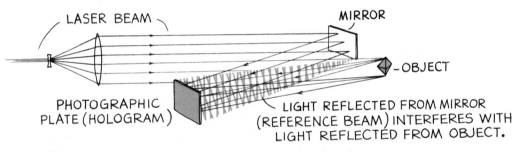

LASER BEAM

MIRROR

PHOTOGRAPHIC
PLATE (HOLOGRAM)

LIGHT REFLECTED FROM MIRROR
(REFERENCE BEAM) INTERFERES WITH
LIGHT REFLECTED FROM OBJECT.

OBJECT

Fig. 27.39 A simplified arrangement for making a hologram. The laser light that exposes the photographic plate consists of two parts: the reference beam reflected from the mirror, and light reflected from the object. The wave fronts of these two parts interfere to produce microscopic fringes on the photographic plate. The exposed and developed plate is then a hologram.

When a hologram is placed in a beam of coherent light, the light is diffracted through the microscopic fringes to produce wave fronts identical in form to the original wave fronts reflected by the object. When viewed with the eye or any other optical instrument, the diffracted wave fronts produce the same effect as the original reflected wave fronts. You look through the hologram and see a full, realistic three-dimensional image as though you were viewing the original object through a window. Parallax is evident when you move your head and see down the sides of the object or when you lower your head and look underneath the object. Holographic pictures are extremely realistic.

Interestingly enough, if the hologram is made on film, you can cut it in half and still see the entire image. And you can cut one of the pieces in half again, and again. This is because every part of the hologram has received and recorded light from the entire object. Because vast amounts

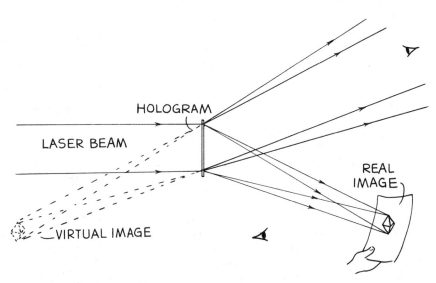

Fig. 27.40 When laser light is transmitted through the hologram, the divergence of diffracted light produces a three-dimensional image that can be seen when looking through the hologram, like looking through a window. This is called a *virtual image*, for it appears to be only in back of the hologram, like your virtual image in a mirror. By refocusing your eyes, you can see all parts of the virtual image, near and far, in sharp focus. Converging diffracted light produces a real image in front of the hologram that can be projected on a screen. Since the image is three-dimensional, you cannot see the entire image in sharp focus for any single position on a flat screen.

of information are recorded in a tiny area, the film used for holograms must have a resolving power much greater than ordinary fine-grain photographic film. The optical storage of information via holograms is finding wide application in computers.

Even more interesting is holographic magnification. If holograms are made using a short-wavelength light and viewed with a longer-wavelength light, the resulting image is magnified in the same proportion as the wavelengths. Holograms made with X rays would be magnified thousands of times when viewed with visible light and appropriate geometrical viewing arrangements. Because holograms require no lenses, the possibilities of an X-ray microscope are particularly attractive.

And as for television, today's two-dimensional screens may appear as quaint to your children as your grandparents' early radio receivers appear to you now.

Light is interesting—especially when it is diffracted through the interference fringes of a hologram.

Summary of Terms

Huygens' principle   The theory wherein light waves spreading out from a point source can be regarded as the superposition of tiny secondary wavelets.

Diffraction   The bending of light around an obstacle or through a narrow slit in such a way that fringes of light and dark or colored bands are produced.

Interference   The superposition of waves producing regions of reinforcement and regions of cancelation. *Constructive interference* refers to regions of reinforcement; *destructive interference* refers to regions of cancelation. The interference of selected wavelengths of light produces colors, knows as *interference colors*.

Polarization   The alignment of the electric vectors comprising electromagnetic radiation. Such waves of aligned vibrations are said to be polarized.

Holography   The process in which three-dimensional optical images are produced by laser light.

Suggested Reading

Falk, D. S., D. R. Brill, and D. Stork, *Seeing the Light: Optics in Nature*. New York: Harper & Row, 1985.

Review Questions

1. What is Huygens' principle?

2. What are some of the assets and liabilities of diffraction?

3. By what means are the light and dark bands in diffraction patterns produced?

4. What did Thomas Young demonstrate in his famous experiment?

5. What is the cause of Newton's rings?

6. What causes the spectrum of colors seen in gasoline splotches on a wet street? Why are these not seen on a dry day?

7. Why do coated lenses appear bluish by reflected light?

8. What becomes of the energy of waves that undergo destructive interference?

9. What phenomenon distinguishes longitudinal and transverse waves?

10. Why will light pass through a pair of Polaroids when the axes are aligned but not when the axes are at right angles to each other?

11. Distinguish between *polarized light* and *nonpolarized light*. Is sound ever polarized?

12. What does the phenomenon of diffraction have to do with a hologram?

Home Projects

1. With a razor blade, cut a slit in a card and look at a light source through it. You can vary the size of the opening by bending the card slightly. See the interference fringes? Try it with two closely spaced slits.

2. Next time you're in the bathtub, froth up the soapsuds and notice the colors of highlights from the illuminating light overhead on each tiny bubble. Notice that different bubbles reflect different colors, due to the different thicknesses of soap film. If a friend is bathing with you, compare the different colors that you each see reflected from the same bubbles. You'll see that they're different—for what you see depends on your point of view!

3. Do this one at your kitchen sink. Dip a dark-colored coffee cup (dark makes the best background for viewing interference colors) in dishwashing detergent and then hold it sideways and look at the reflected light from the soap film that covers its mouth. Swirling colors appear as the soap runs down to form a wedge that grows thicker at the bottom with time. The top becomes thinner, so thin that it appears black. This tells us that its thickness is less than one-fourth the thickness of the shortest waves of visible light. Whatever its wavelength, light reflecting from the inner surface reverses phase, rejoins light reflecting from the outer surface, and cancels. The film soon becomes so thin it pops.

4. When you're wearing Polaroid sunglasses, view the glare from nonmetallic surfaces such as a road or body of water, tip your head from side to side, and see how the glare intensity changes as you vary the number of electric vector components aligned with the polarization axis of the glasses. Also notice the polarization of different parts of the sky.

5. Place a bottle of corn syrup between two sheets of Polaroid. Place a white light source behind the syrup. Then look through the Polaroids and syrup and view spectacular colors as you rotate one of the Polaroids.

6. See spectacular interference colors with a polarized-light microscope. Any microscope, including an inexpensive toy microscope, can be converted into a polarized-light microscope by fitting a piece of Polaroid inside the eyepiece and taping another onto the stage of the microscope. Mix drops of naphthalene and benzene on a slide and watch the growth of crystals. Rotate the eyepiece and change the colors. Or simply observe the interference colors of pieces of cellophane.

7. Make some slides for a slide projector by sticking some crumpled cellophane onto pieces of slide-sized Polaroid. (Also try strips of cellophane tape overlapped at different angles and experiment with different brands of transparent tape.) Project them onto a large screen or white wall and rotate a second slightly larger piece of Polaroid in front of the projector lens in rhythm with your favorite music. You'll have your own light show.

Exercises

1. Diffraction for sound waves is much more evident in our everyday environment than for light waves. Why is this so?

2. How are interference fringes of light analogous to the sound of beats?

3. Why do radio waves diffract around buildings, while light waves do not?

4. Can you think of a reason why TV channels of lower numbers give better pictures in regions of poor TV reception?

5. Two loudspeakers a meter or so apart emit pure tones of the same frequency and loudness. When a listener walks past in a path parallel to the line that joins the loudspeakers, the sound is heard to alternate from loud to soft. What is going on?

6. In the preceding exercise, suggest a path along which the listener could walk so as not to hear alternate loud and soft sounds.

7. Light illuminates two closely spaced thin slits and produces an interference pattern on a screen behind. How will the distance between the fringes of the pattern differ for red light as compared to blue light?

8. What happens to the distance between interference fringes if the two slits are moved farther apart?

9. Why is Young's experiment more effective with slits than the pinholes he first used?

10. For complete cancelation of light reflected out of phase from the two surfaces of a thin film, what else besides frequency and wavelength must be the same for both parts of the recombined wave?

11. A pattern of fringes is produced when monochromatic light passes through a pair of thin slits. Would such a pattern be produced by three parallel thin slits? By thousands of such slits? If the slit width and distance between slits were the same for all cases, how might the patterns of fringes differ? (Do a bit of research and find out more about diffraction gratings.)

12. The colored wings of many butterflies are due to pigmentation, but in others, such as the Morpho butterfly, the colors do not result from any pigmentation. When the wing is viewed from different angles, the colors change. How are these colors produced?

13. Why do the iridescent colors seen in some seashells (such as abalone shells) change as the shells are viewed from different positions?

14. When dishes are not properly rinsed after washing, different colors are reflected from their surfaces. Explain.

15. Why are interference colors more apparent for thin films than for thick films?

16. Will the light from two very close stars produce an interference pattern? Explain.

17. We explain the vivid colors seen when double reflection occurs by interference. Does interference also underlie the explanation for the colors produced by transmission with the aid of Polaroids? Explain.

18. Why do Polaroid sunglasses reduce glare, whereas unpolarized sunglasses simply cut down on the total amount of light reaching the eyes?

19. How can you determine the polarization axis for a single sheet of Polaroid?

20. Most of the glare from nonmetallic surfaces is polarized, the axis of polarization being parallel to that of the reflecting surface. Would you expect the polarization axis of Polaroid sunglasses to be horizontal or vertical? Why?

21. How can a single sheet of Polaroid film be used to show that the sky is partially polarized?

22. Why will an ideal Polaroid filter transmit 50 percent of incident nonpolarized light?

23. What percentage of light would be transmitted by two ideal Polaroids sandwiched with their polarization axes aligned? With their axes at right angles to each other?

24. Why did practical holography have to await the advent of the laser?

25. How is magnification accomplished with holograms?

26. If you are viewing a hologram and you close one eye, will you still perceive depth? Explain.

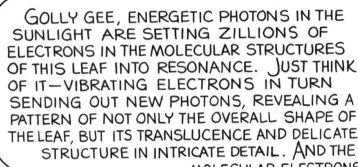

28 Light Emission

If energy is pumped into a metallic antenna in such a way as to cause free electrons to vibrate to and fro a few thousand times per second, a radio wave is emitted. If free electrons could be made to vibrate to and fro on the order of a million billion times per second, a visible light wave would be emitted. But light is not produced from metallic antennae, nor is it exclusively produced by atomic antennae via subtle oscillations of electron shells, as is the case for reflection. The details of light emission from atoms involve the transitions of electrons from higher to lower energy states within the atom. This emission process can be understood in terms of the familiar planetary model of the atom that we discussed in Chapter 9. Just as each element is characterized by the number of electrons that occupy the shells surrounding the atomic nucleus, so also each element possesses its own characteristic pattern of electron shells, or energy states. These states are found only at certain radii and energies. We say that these noncontinuous energy states are *discrete*. We call these discrete states *quantum states,* and we'll return to them in detail in the next two chapters. For now we'll concern ourselves only with their role in light emission.

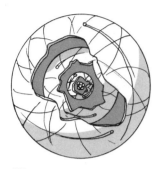

Fig. 28.1 Electrons occupy discrete shells about the nucleus of an atom.

Excitation

An electron farther from the nucleus has a greater electric potential energy with respect to the nucleus than an electron nearer to the nucleus. We say that the farthermost electron is at a higher *energy level*. In a sense, this is similar to the energy of a spring door or a pile driver. The wider the door is pulled open, the greater its spring potential energy; the higher the ram of a pile driver is raised, the greater its gravitational potential energy.

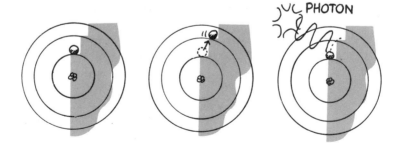

Fig. 28.2 The different orbits in an atom are like steps. When an electron is raised to a higher orbit, the atom is excited. When the electron falls to its original step, it releases energy in the form of light.

When an electron is in any way raised to a higher energy level, the atom is said to be *excited*. The electron's higher level is only momentary, for like the pushed-open spring door, it soon returns to its stable state. The

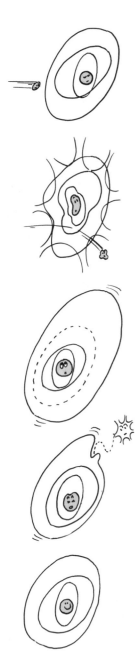

atom loses its temporarily acquired energy when the electron returns to a lower level and releases this energy as radiation. The atom has undergone the process of **excitation** and **de-excitation**.

As each element has its own number of electrons, each element also has its own characteristic set of energy levels. Electrons dropping from higher to lower energy levels in an excited atom emit with each jump a throbbing pulse of electromagnetic radiation called a **photon**, the frequency of which is related to the energy transition of the jump. We think of this photon as a localized corpuscle of pure energy, which is ejected from the atom. The frequency of the photon is directly proportional to its energy. In shorthand notation,

$$E \sim f$$

A photon in a beam of red light, for example, carries an amount of energy that corresponds to its frequency. A photon of twice the frequency has twice as much energy and is found in the ultraviolet part of the spectrum. If many atoms in a material are excited, many frequencies are emitted that correspond to the many different levels excited. These frequencies combine to give a characteristic color of light from each excited element.

The light emitted in the glass tubes in an advertising sign is a familiar example of excitation. The different colors in the sign correspond to the excitation of different gases, although it is common to refer to any of these as "neon." Only the red light is that of neon. At the ends of the glass tube that contains the neon gas are electrodes. Electrons are boiled off these electrodes and are jostled back and forth at high speeds by a high ac voltage. Millions of high-speed electrons vibrate back and forth inside the glass tube and smash into millions of target atoms; each smash boosts orbital electrons into higher energy levels by an amount of energy equal to the decrease in kinetic energy of the bombarding electron, and this energy is radiated as the characteristic red light of neon when the electrons fall back to their stable orbits. The process occurs and recurs many times, as neon atoms are continually undergoing a cycle of excitation and de-excitation. The overall result of this process is the transformation of electrical energy into radiant energy.

The colors of various flames are due to excitation. Different atoms in the flame emit colors characteristic of their energy-level spacings. Common table salt placed in a flame, for example, produces the characteristic yellow of sodium. Every element, excited in a flame or otherwise, emits its own characteristic color—a composite of the many colors that make up its characteristic spectrum.

Street lamps provide another example. Most city streets used to be illuminated by incandescent lamps but are now illuminated with the light emitted by gases such as mercury vapor. Not only is the light brighter, it is less expensive. Whereas most of the energy in an incandescent lamp is converted to heat, most of the energy put into a mercury-vapor lamp is converted to light. The light from these lamps is rich in blues and violets

Fig. 28.3 Excitation and de-excitation.

and therefore is a different "white" from the light from an incandescent lamp. See if your instructor has a spare prism you can borrow or, better yet, a diffraction grating.[1] Look through the prism or grating at the light from the street lamp and see the discreteness of the colors, showing the discreteness of the atomic levels. The light from these gas-discharge lamps, however, cannot be understood from the points of view of classical mechanics and classical optics, which would have the electrons going in all possible energy states and emitting all possible colors, not just the few discrete ones you observe. Also note that the colors from different mercury-vapor lamps are identical, showing that the atoms of mercury are identical. That would not be the case for classical planetary systems.

The excitation/de-excitation process can be accurately described only by quantum mechanics. An attempt to view the process in terms of classical physics runs into contradictions. Classically, an accelerating electric charge emits electromagnetic radiation. Interestingly enough, an electron does accelerate in a transition from a higher to a lower energy level, for just as the innermost planets of the solar system have greater orbital speeds than those in the outermost orbits, the electrons in the innermost orbits of the atom have greater speeds. An electron gains speed in dropping to lower energy levels. Fine—the accelerating electron radiates a photon! But not so fine—the electron is continually undergoing acceleration (centripetal acceleration) in any orbit, whether or not it changes energy levels. According to classical physics, it should continually radiate energy. But it doesn't. All attempts to explain the emission of light by an excited atom in terms of a classical model have been unsuccessful. We have no simple model for conceptualizing this process. We shall simply say that light is emitted when electrons in an atom make a transition from a higher to a lower energy level and that the energy and frequency of the emitted photon are described by the relationship $E \sim f$.

Question Suppose a friend suggests that for a first-rate operation, the gaseous neon atoms in a neon tube should be periodically replaced with fresh atoms because the energy of the atoms tends to be used up with continued excitation, producing dimmer and dimmer light. What do you say to this?*

***Answer** The neon atoms don't give out any energy that is not imparted to them by the electric current in the tube and therefore don't get "used up." Any single atom may be excited and re-excited without limit. If the light is, in fact, becoming dimmer and dimmer, it is probably because a leak exists. Otherwise there is no advantage whatsoever in changing the gas in the tube, for a "fresh" atom is indistinguishable from a "used" one. Both are ageless and older than the solar system.

[1]Diffraction gratings consisting of thousands of thin slits are also used to separate colors. A phonograph record acts as a diffraction grating when it separates a grazing beam of white light into its component colors.

Emission Spectra

Every element has its own characteristic pattern of electron energy levels and therefore emits its own characteristic pattern of light frequencies when excited. This pattern can be seen when light is passed through a prism or, better, when it is first passed through a thin slit and then focused through a prism onto a viewing screen behind. Such an arrangement of slit, focusing lenses, and prism (or diffraction grating) is called a **spectroscope**, one of the most useful instruments of modern science (Figure 28.4).

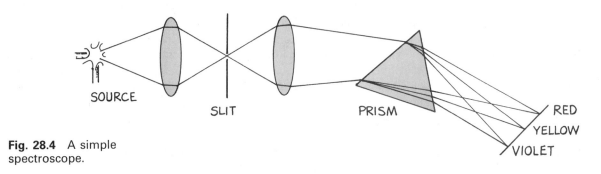

Fig. 28.4 A simple spectroscope.

Each component color is focused at a definite position according to its frequency and forms an image of the slit on the screen, photographic film, or appropriate detector. The different-colored images of the slit are called *spectral lines*. A black-and-white picture of some typical spectral patterns labeled by wavelengths is shown in Figure 28.5, and more in their natural colors are shown on the inside of the back cover of this book. It is custom-

Fig. 28.5 Some typical spectral patterns.

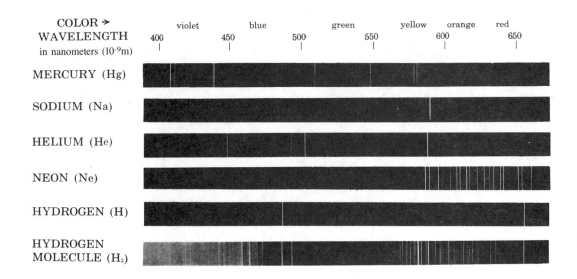

ary to refer to colors in terms of their wavelengths rather than their frequencies. A given frequency corresponds to a definite wavelength.[2]

If the light given off by a sodium-vapor lamp is analyzed in a spectroscope, a single yellow line is produced—a single image of the slit. If we narrow the width of the slit, we find that this line is really composed of two very close lines. These lines correspond to the two predominant frequencies of light emitted by excited sodium atoms. The rest of the spectrum is dark. (Actually, there are many other lines, often too dim to be seen with the naked eye.)

The same happens with all glowing vapors. The light from a mercury-vapor lamp shows a pair of bright yellow lines close together (but in different positions from those of sodium), a very intense green line, and several blue and violet lines. A neon tube produces a more complicated pattern of lines. We find that the light emitted by each element in the vapor state produces its own characteristic pattern of lines. These lines correspond to the electron transitions between atomic energy levels and are as characteristic of each element as are fingerprints of people. The spectroscope therefore is widely used in chemical analysis.

The next time you see evidence of atomic excitation, perhaps the green flame produced when a piece of copper is placed in a fire, squint your eyes and see if you can imagine electrons jumping from one energy level to another in a pattern characteristic of the atom being excited—a pattern that gives off a color unique to that atom. Because that's what's happening.

Question   Spectral patterns are not shapeless smears of light but, instead, consist of fine and distinct straight lines. Why is this so?*

Incandescence   Light that is produced as a result of high temperature is called **incandescent**, from a Latin word that means "to grow hot." Most incandescent light we are familiar with is white, like that from a common incandescent lamp. This seems to be quite different from the bright red light emitted from a cool neon gas tube. We know that a composite of all visible

---

*__Answer__   The spectral lines are simply images of the slit, which is itself a thin, straight opening through which light is admitted before being diffracted by the prism (or diffraction grating). When the slit is adjusted for its narrowest opening, closely spaced lines can be resolved (distinguished from one another). A wider slit admits more light, which permits easier detection of dimmer radiation but at the expense of resolution of closely spaced lines.

[2]Recall from Chapter 17 that $v = f\lambda$, where v is the wave velocity, f is the wave frequency, and λ (lambda) is the wavelength. For light, v is the constant c, so we see from $c = f\lambda$ the relationship between frequency and wavelength; namely, $f = \frac{c}{\lambda}$ and $\lambda = \frac{c}{f}$.

frequencies appears white, so does this mean that an infinite number of energy levels characterizes the tungsten atoms making up the filament of the incandescent lamp? The answer is no; if the filament were vaporized and then excited, the tungsten gas would emit a finite number of frequencies and produce an overall bluish color. Light emitted by atoms far apart from each other in the gaseous state is quite different from the light emitted by the same atoms closely packed in the solid state. This is analogous to the differences in sound from an isolated ringing bell and a box crammed with ringing bells (Figure 28.6). In a gas the atoms are far apart. Electrons undergo transitions between energy levels within the atom quite unaffected by the presence of neighboring atoms. But when the atoms are closely packed, as in a solid, electrons of the outer orbits make transitions not only within the energy levels of their "parent atoms" but also between the levels of neighboring atoms. These energy-level transitions are no longer well defined but are altered by interactions between neighboring atoms, resulting in an infinite variety of transitions—hence the infinite number of radiation frequencies.

Fig. 28.6 The sound of an isolated bell rings with a clear and distinct frequency, whereas the sound emanating from a box of bells crowded together is discordant. Likewise with the difference between the light emitted from atoms in the gaseous state and that from atoms in the solid state.

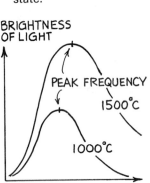

Fig. 28.7 Radiation curves for an incandescent solid.

As we might expect, the interactions of electrons between neighboring atoms in a solid are proportional to temperature. The greater the molecular kinetic energy, the greater the interactions and resulting radiation. A plot of radiated energy over a wide range of frequencies for two different temperatures is shown in the pair of radiation curves in Figure 28.7. As the solid is heated further, wider energy-level transitions take place and higher frequencies of radiation are emitted. The most predominant frequency of emitted radiation, the *peak frequency,* is directly proportional to the absolute temperature of the emitter:

$$\overline{f} \sim T$$

We use the bar above the *f* to indicate *peak* frequency, for many frequencies of radiation are emitted from incandescent sources. If the temperature of an object (in kelvins) is doubled, the peak frequency of emitted radiation is doubled. The electromagnetic waves of violet light have twice the frequency of red light waves. A violet-hot star therefore has twice the temperature of a red-hot star.[3] The temperature of incandescent bodies, whether they be stars or blast-furnace interiors, can be determined by measuring the peak frequency (or color) of radiation they emit.

Question   From the radiation curves shown in Figure 28.7, which emits the higher average frequency of radiation—the 1000° source or the 1500° source? Which emits more radiation?*

Absorption Spectra

When we view white light from an incandescent source with a spectroscope, we see a continuous rainbow-colored spectrum. If a gas is placed between the source and the spectroscope, however, careful inspection will show that the spectrum is not quite continuous. There are dark lines distributed throughout it; these dark lines against a rainbow-colored background are like emission lines in reverse. These are *absorption lines*.

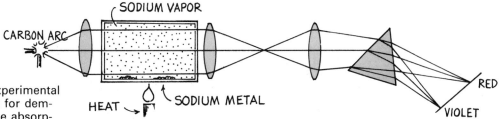

Fig. 28.8 Experimental arrangement for demonstrating the absorption spectra of a gas.

Atoms absorb light as well as emit light. An atom will most strongly absorb light having the frequencies to which it is tuned—the same frequencies it emits. When a beam of white light passes through a gas, the atoms of the gas absorb selected frequencies from the beam. This absorbed light is reradiated but in *all* directions instead of only in the

---

*Answer   The 1500° radiating source emits the higher average frequencies, as noted by the extension of the curve to the right. The 1500° source is the brighter and also emits more radiation, as noted by its greater vertical displacement.

[3]If you study this topic further, you will find that the time rate at which an object radiates energy is proportional to the fourth power of its kelvin temperature. So a doubling of temperature corresponds to a doubling of the frequency of radiation but a sixteen-fold increase in the rate of radiation. So it would emit sixteen times as many photons per second and therefore be sixteen times brighter.

direction of the incident beam. When the light remaining in the beam spreads out into a spectrum, the frequencies that were absorbed show up as dark lines in the otherwise continuous spectrum. The positions of these dark lines correspond exactly to the positions of lines in an emission spectrum of the same gas (Figure 28.9).

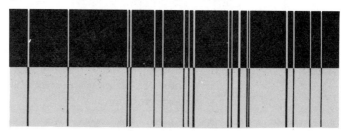

Fig. 28.9 Emission and absorption spectra.

Although the sun is a source of incandescent light, the spectrum it produces, upon close examination, is not continuous. There are many absorption lines, called *Fraunhofer lines* in honor of the Bavarian optician, J. D. Lines, who first observed and mapped them accurately. Similar lines are found in the spectra produced by the stars. These lines indicate that the sun and stars are each surrounded by an atmosphere of cooler gases that absorb some of the frequencies of light coming from the main body. Analysis of these lines reveals the chemical composition of the atmospheres of such sources. We find from these analyses that the stellar elements are the same elements that exist on earth. An interesting sidelight is that when spectroscopic measurements were made of the sun, spectral lines different from those measured in the laboratories were found. These lines identified a new element, which was named *helium,* after Helios, the sun. Helium was discovered in the sun before it was discovered on earth. How about that!

We can determine the speed of stars by studying the spectra they emit. Just as a moving sound source produces a Doppler shift in its pitch (Chapter 17), a moving light source produces a Doppler shift in its light frequency. The frequency (not the speed!) of light emitted by an approaching source increases, while the frequency of light from a receding source decreases. The corresponding spectral lines are displaced toward the violet end of the spectrum for approaching sources and toward the red end of the spectrum for receding sources. Since the universe is expanding, the galaxies show a red shift in their spectra.

We shall see in Chapter 31 how the spectra of elements enable us to determine atomic structure.

Fluorescence   So we see that thermal agitation or bombarding particles such as high-speed electrons are not the only means of imparting excitation energy to an atom. An atom may be excited by absorbing a photon of light. From

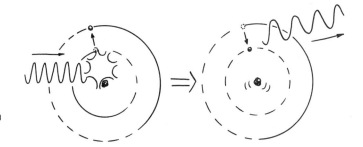

Fig. 28.10 In fluorescence some of the absorbed energy of an ultraviolet photon goes into excitation and the remainder goes into heat. The photon then emitted is less energetic and therefore of a lower frequency than the ultraviolet photon.

the relationship $E \sim f$, we see that high-frequency light, such as ultraviolet, which lies beyond the visible spectrum, delivers much more energy per photon than lower-frequency light. Many substances undergo excitation when illuminated with ultraviolet light.

Some materials that are excited by ultraviolet light emit visible light upon de-excitation. The action of these materials is called **fluorescence**. In these materials a photon of ultraviolet light collides with an atom of the material and gives up its energy in two parts. Part of the energy produces heat, increasing the kinetic energy of the entire atom. The other part of the energy goes into excitation, boosting an electron to a higher energy state. Upon de-excitation, this part of the energy is released as a photon of light. Since this photon has only part of the energy of the ultraviolet photon, its frequency is lower and falls into the visible part of the spectrum.

In some other materials, an electron boosted to a higher energy level by a photon of ultraviolet light returns to its stable orbit in several steps. Since the photon energy released at each step is less than the total energy originally in the ultraviolet photon, lower-frequency photons are emitted. Hence, ultraviolet light shining on such a material may cause it to glow an overall red, yellow, or whatever color is characteristic of the material. Fluorescent dyes are used in paints and fabrics to make them glow when bombarded with ultraviolet photons in sunlight. These are the so-called DayGlo colors, which are spectacular when illuminated with an ultraviolet lamp.

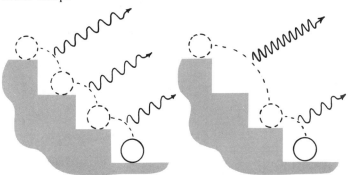

Fig. 28.11 An excited atom may de-excite in several combinations of jumps.

Question   Why would it be impossible for a fluorescent material to emit ultraviolet light when illuminated by infrared light?*

Detergents that make the claim of cleaning your clothes "whiter than white" use the principle of fluorescence. Such detergents contain a fluorescent dye that converts the ultraviolet light in sunlight into blue visible light, so clothes dyed in this way appear to reflect more blue light than they otherwise would. This makes the clothes appear whiter.

The next time you visit a natural science museum, go to the geology section and take in the exhibit of minerals illuminated with ultraviolet light. You'll notice that different minerals radiate different colors. This is to be expected because different minerals are composed of different elements, which in turn have different sets of electron energy levels. Seeing the radiating minerals is a beautiful visual experience, which is even more fascinating when integrated with your knowledge of nature's sub-microscopic happenings. High-energy ultraviolet photons impinge on the surface of the minerals, causing the excitation of atoms in the mineral structure; and the radiation of light frequencies corresponds exactly to the tiny energy-level spacings—and every excited atom emits its characteristic frequency, with no two different minerals giving off exactly the same color light. Beauty is in both the eye and the mind of the beholder.

Fluorescent Lamps

Light emitted by a fluorescent lamp is produced by primary and secondary excitation processes. The primary process is excitation of gas by electron bombardment, which produces ultraviolet light, and the secondary process is excitation of *phosphors* on the inner surface of the glass tube by the ultraviolet photons—fluorescence.

The common fluorescent lamp consists of a cylindrical glass tube with electrodes at each end (Figure 28.12). Like the neon sign tube, electrons are boiled from the electrodes and forced to vibrate to and fro at high speeds within the tube by the ac voltage. The tube is filled with very low-pressure mercury vapor, which is excited by the impact of the high-speed electrons. As the energy levels in mercury are relatively far apart, the resulting emission of light is of high frequency, mainly in the ultraviolet

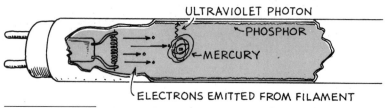

Fig. 28.12 A fluorescent tube.

ULTRAVIOLET PHOTON
PHOSPHOR
MERCURY
ELECTRONS EMITTED FROM FILAMENT

*__Answer__ Photon energy output would be greater than photon energy input, which would violate the conservation of energy.

region. This is the primary excitation process. The secondary process occurs when the ultraviolet light impinges upon the phosphors on the inner surface of the tube. The phosphors are excited by the absorption of the ultraviolet photons and give off a multitude of lower frequencies that combine to produce white light. Different phosphors can be used to produce different colors of light.

Phosphorescence

When excited, certain crystals as well as some large organic molecules remain in a state of excitement for a prolonged period of time. Their electrons are boosted into higher orbits and become "stuck." As a result, there is a time delay between the process of excitation and de-excitation. Materials that exhibit this peculiar property are said to be **phosphorescent**. The element phosphorus is a good example. Such materials are used in luminous clock dials and in other objects that are made to glow in the dark. Atoms in these objects are excited by incident visible light. Rather than de-exciting immediately as fluorescent materials do, many of the atoms remain in a state of excitement, sometimes for as long as several hours, although most undergo de-excitation rather quickly. If the source of excitation is removed—for example, if the lights are put out—the phosphorescent object will glow for some time while millions of atoms spontaneously undergo gradual de-excitation.

Lasers

The phenomena of excitation, fluorescence, and phosphorescence underlie the operation of a most intriguing instrument, the **laser** (**l**ight **a**mplification by **s**timulated **e**mission of **r**adiation).[4] Stimulated emission is a technique only recently made possible, although Einstein predicted it in 1917. To understand how a laser operates, we must first discuss coherent light.

Light emitted by a common lamp is incoherent; that is, photons of many frequencies and many phases of vibration are emitted. The light is as incoherent as the footsteps on an auditorium floor when a mob of people are chaotically rushing about. Incoherent light is chaotic. A beam of incoherent light spreads out after a short distance, becoming wider and wider and less intense with increased distance.

Fig. 28.13 Incoherent white light contains waves of many frequencies and wavelengths that are out of phase with one another.

Even if the beam is filtered so that it is a single frequency (monochromatic), it is still incoherent, for the waves are out of phase with one another. The slightest differences in their directions result in a spreading with increased distance.

---

[4]A word constructed from the initials of a phrase is called an *acronym*.

A beam of photons having the same frequency, phase, and direction—that is, a beam of photons that travel exactly alike—is said to be coherent. Only a beam of coherent light will not noticeably spread and diffuse.

Fig. 28.15 Coherent light: all the waves are identical and in phase.

A laser is an instrument that produces a beam of coherent light. One model, the earliest, consists of a ruby crystal rod a few centimeters long that has been "doped" with impurity atoms of chromium. The ruby rod is surrounded by a photoflash tube that sends high-intensity green light into the ruby (Figure 28.16). Excitation of the chromium atoms occurs as the outer electrons are boosted to higher energy levels. The electrons fall back to an intermediate level and then, more slowly, to their lower level. At this last stage they emit a photon of red light; the ruby fluoresces.

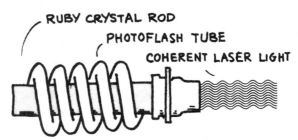

Fig. 28.16 A ruby crystal laser.

Photons of red light within the crystal trigger the de-excitation of neighboring atoms that are still in the excited state. The photons stimulated into radiation will be identical in frequency, phase, and direction to the incident photons. These photons then may further stimulate the radiation of other excited atoms, thereby producing a beam of coherent light. Most of this light initially escapes through the sides of the crystal in random directions. Light traveling along the crystal axis, however, is reflected from mirrors that have been coated to selectively reflect the desired wavelength of light. One mirror is totally reflecting, while the other is partially reflecting. The reflected waves reinforce each other after each round-trip reflection between the mirrors, thereby setting up a resonance condition in which the light builds up to an appreciable intensity. The light that escapes in pulses through the more transparent-mirrored end makes up the laser beam.

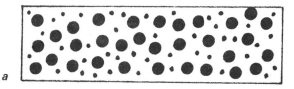

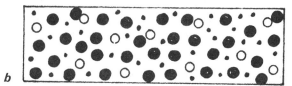

a

b

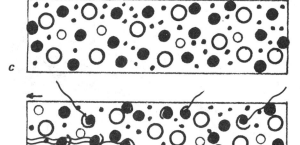

c

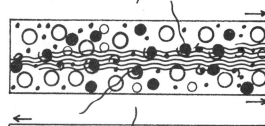

d

e

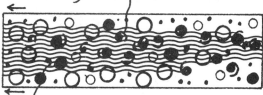

f

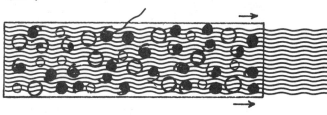

g

Fig. 28.17 Laser action in a helium-neon laser.

(*a*) The laser consists of a narrow Pyrex tube that contains a low-pressure gas mixture consisting of 85% helium (small black dots) and 15% neon (large black dots).

(*b*) When a high-voltage current zaps through the tube, it excites both helium and neon atoms to their usual states, which immediately undergo de-excitation, except for one state in the helium that is characterized by a prolonged delay before de-excitation. This is called a *metastable state*. Since this state is relatively stable, a sizable population of excited helium atoms (fine open circles) is built up. These wander about in the tube and act as an energy source for neon, which has an otherwise hard-to-come-by metastable state very close to the energy of the excited helium.

(*c*) When excited helium atoms collide with neon atoms in the ground state, the helium gives up its energy to the neon, which is boosted to its metastable state (bold open circles). The process continues, and the population of excited neon atoms soon outnumbers neon in the ground state. This inverted population in effect is waiting to radiate its energy.

(*d*) Some neon atoms eventually de-excite and radiate red photons in the tube. When this radiation passes other excited neon atoms, the latter are stimulated into emitting photons exactly in phase with the radiation that stimulated the emission. Photons pass out of the tube in irregular directions, giving it the familiar neon glow.

(*e*) Photons parallel to the tube reflect from specially coated parallel mirrors at the ends of the tube. The reflected photons stimulate the emission of other neon atoms, thereby producing a photon avalanche of the same frequency, phase, and direction.

(*f*) The photons flash to and fro between the mirrors, becoming amplified with each pass.

(*g*) Some "leak" out one of the mirrors, which are only partially reflecting. These make up the laser beam.

Shortly after the development of the ruby laser, physicists made a gas laser that produced a continuous beam. The first such laser used a mixture of helium and neon gases. When a high voltage is applied to electrodes at the ends of a tube of the mixed gases, the electrical discharge through the helium gas excites the atoms of the neon gas into a prolonged excited state so that the lasing process can take place. In addition to crystal lasers and gas lasers, other new types have taken their places in the laser family: glass lasers, chemical and liquid lasers, and semiconductor lasers. Present models produce beams ranging from infrared through ultraviolet. Some models can be tuned to various frequency ranges. Most exciting is the prospect of an X-ray laser beam.

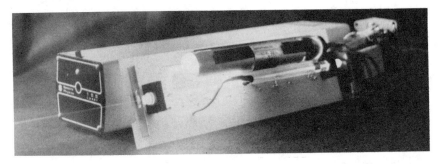

Fig. 28.18   A helium-neon laser.

The laser is not a source of *energy*. It is simply a converter of energy that takes advantage of the process of stimulating emission to concentrate a certain fraction of its energy (commonly 1 percent) into radiation of a single frequency moving in a single direction. Like all devices, a laser can put out no more energy than is put in.

The list of practical applications of the laser is growing rapidly. In its simplest application, a laser beam is an unbendable chalk line and an unstretchable measuring stick, which provides accuracy for the machinist, speed for construction workers, and economy for the surveyor. Lasers were used to line up the dredging equipment for the underwater tunnel in San Francisco Bay and inside the tunnel to ensure its straightness. Laser beams have been bounced off the moon and returned in order to measure the exact distance between the earth and the moon and to provide information on continental drift.

Some lasers are so intense (and concentrated) that eye surgeons use them to "weld" detached retinas back into place without making an incision. The light is simply brought to focus in the region where the welding is to take place.

Whereas radio wavelengths span hundreds of meters and television waves span many centimeters, laser wavelengths are measured in millionths of a centimeter. As a result, an enormous number of messages bunched into a very narrow band of frequencies can be carried on these ultrashort waves. Communications can be carried in a laser beam directed

Fig. 28.19 Six of the twenty Shiva laser amplifier chains of the Shiva laser system that delivers more than 30 trillion watts of optical power in less than 1 billionth of a second to a tiny fusion target the size of a grain of sand. Shiva, the world's most powerful laser, is one in a series of high-power laser systems built by the Lawrence Livermore Laboratory.

through space, through the atmosphere, or through optical fibers (light pipes) that can be bent like cables.

Lasers are used with computers. A laser gun "burns" tiny, microscopic holes each of which represents a bit of information into the surface of a computer tape. A second laser later "reads" them. In this way a trillion bits of data can be stored on a single standard-length reel of computer tape. A telephone directory for all the world's inhabitants would contain as much information.

The laser is at work in many supermarket checkout counters, where code-reading machines scan the universal product code (UPC) symbol printed on packages (Figure 28.20). Laser light is reflected from the bars and spaces and converted to an electric signal as the symbol is scanned. The signal rises to a high value when reflected from a bright space and falls to a low value when reflected from a dark bar. The high and low values are converted to 1's and 0's, respectively, and the information is in the binary code ready for computer processing.

Lasers measure and detect pollutants in exhaust gases. Different gases absorb light at characteristic wavelengths and leave their "fingerprints" on a reflected beam of laser light. The specific wavelength and amount of light absorbed are analyzed by a computer, which produces an immediate tabulation of the pollutants.

The beam of a laser is used as a nonwearing "optical needle" for video and phonograph records, as a knife to rapidly and accurately cut cloth in garment factories, in a much-improved technique for fingerprint detection, and for the production of holograms. The list goes on and on.

Fig. 28.20 A product's unique identification is coded in the UPC symbol. The price and name of the product are held in computer memory.

A most important application of the laser is concentrating huge amounts of energy on hydrogen pellets and bringing them up to thermonuclear temperatures. The laser may someday soon allow us to control thermonuclear fusion.

Just after its invention in 1958 the laser was touted as being a solution looking for a problem. It didn't have to look very far or wait very long. It has ushered in a whole new technology the promise of which we have only begun to tap. The future for laser applications seems unlimited.

Summary of Terms

Excitation   The process of boosting one or more electrons in an atom or a molecule from a lower to a higher energy level. An atom in an excited state will usually decay (de-excite) rapidly into a lower state by the emission of radiation. The energy of the radiation is proportional to its frequency: $E \sim f$.

Emission spectrum   A continuous or partial spectrum of wavelengths resulting from the characteristic dispersion of light from a luminous source.

Spectroscope   An optical instrument that separates light into its constituent frequencies in the form of spectral lines.

Incandescence   The state of glowing while at a high temperature, caused by electrons in vibrating atoms and molecules that are shaken in and out of their stable energy levels, emitting radiation in the process. The peak frequency of radiation is proportional to the absolute temperature of a heated substance:

$$\bar{f} \sim T$$

Absorption spectrum   A continuous spectrum, like that of white light, interrupted by dark lines or bands that result from the absorption of certain frequencies of light by a substance through which the radiation passes.

Fluorescence   The property of absorbing radiation of one frequency and re-emitting radiation of lower frequency. Part of the absorbed radiation goes into heat, and the other part into excitation; hence, the emitted radiation has a lower energy, and therefore lower frequency, than the absorbed radiation.

Phosphorescence   A type of light emission that is the same as fluorescence except for a delay between excitation and de-excitation, which provides an afterglow. The delay is caused by atoms being excited to energy levels that do not decay rapidly. The afterglow may last from fractions of a second to hours, or even days, depending on the type of material, temperature, and other factors.

Laser (**l**ight **a**mplification by **s**timulated **e**mission of **r**adiation)   An optical instrument that produces a beam of coherent monochromatic light.

Review Questions

1. What is meant by stating that an atom is "excited"?

2. Does an atom possess more energy or less energy when it is excited? What becomes of this energy?

3. Why do elements emit their own characteristic colors when undergoing de-excitation?

4. What does the relation $E \sim f$ have to do with excitation?

5. How do emission and absorption spectra differ?

6. What are Fraunhofer lines?

7. By how much does the peak frequency of radiation increase when the absolute temperature of an incandescent solid is tripled?

8. A fluorescent material absorbs photons of high energy and emits photons of lower energy. What becomes of the remaining energy?

9. What is the principal difference between fluorescent materials and phosphorescent materials?

10. What is monochromatic light?

11. How does coherent light differ from common light?

12. How does laser light differ from regular light?

Home Project

Borrow a diffraction grating from your physics instructor. The common kind looks like a photographic slide, and light passing through it or reflecting from it is diffracted into its component colors by thousands of finely ruled lines. In the evening or at night, look through the grating at the light from a sodium-vapor street lamp. If it's a low-pressure lamp, you'll see the nice yellow spectral "line" that dominates sodium light (actually, it's two closely spaced "lines"). If the street lamp is round, you'll see circles instead of lines; look through a slit cut in cardboard or whatever, and you'll see lines. What happens with the now-common high-pressure sodium lamps is more interesting. Because of the collisions of excited atoms, you'll see a smeared-out spectrum that is nearly continuous, almost like that of an incandescent lamp. Right at the yellow location where you'd expect to see the sodium line is a dark area. This is the sodium absorption band. It is due to the cooler sodium, in a lower-pressure part of the lamp, between the high-pressure emission region and you, the observer. You should view this a block or so away so that the "line," or circle, is small enough to allow the resolution to be maintained. Try this. It is very easy to see!

Exercises

1. Have you ever watched a fire and noticed that the burning of different materials often produces flames of different colors? Why is this so?

2. Ultraviolet light causes sunburns, whereas visible light does not. Why is this so?

3. If we double the frequency of light, we double its energy. If we instead double the wavelength of light, what happens to its energy?

4. Why doesn't a neon sign finally "run out" of excited atoms and produce dimmer and dimmer light?

5. If light were passed through a round hole instead of a thin slit in a spectroscope, how would the spectral "lines" appear? Why would a hole be disadvantageous compared to a slit?

6. If we investigate the light from a common neon tube and from a helium neon laser with a prism or diffraction grating, what striking difference do we see?

7. What is the evidence for the claim that iron exists in the cool gas surrounding the sun?

8. How might the Fraunhofer lines in the spectrum of sunlight that are due to absorption in the sun's atmosphere be distinguished from absorption by gases in the earth's atmosphere?

9. Does atomic excitation occur in solids as well as in gases? How does the radiation from an incandescent solid differ from the radiation emitted by an excited gas?

10. A lamp filament is made of tungsten. Why do we get a continuous spectrum rather than a tungsten line spectrum when light from an incandescent lamp is viewed with a spectroscope?

11. Why are fluorescent colors so bright?

12. Why do different fluorescent minerals emit different colors when illuminated with ultraviolet light?

13. Your friend reasons that if ultraviolet light can activate the process called *fluorescence*, infrared light ought to also. Your friend looks to you for approval or disapproval of this idea. What is your position?

14. The forerunner to the laser involved microwaves rather than visible light. What does *maser* mean?

15. A laser cannot put out more energy than is put into it. A laser can, however, produce *pulses* of light with more power output than the power input required to run the laser. Explain.

16. We know that a lamp filament at 2500 K emits white light. Does the lamp filament emit radiation at room temperature?

17. Since every object has some temperature, every object radiates energy. Why, then, can't we see objects in the dark?

18. How do the relative temperatures compare for reddish, bluish, and whitish stars?

19. Most of the radiation emitted by a red-hot star is in the infrared. Most of the radiation emitted by a violet-hot star is in the ultraviolet. Why is it there are no green-hot stars? (*Hint:* If there were a green-hot star, what would this tell you about the relative rates of individual molecular vibrations in the star? What would its radiation curve look like? Is this likely?)

20. Figure *a* below shows a radiation curve of an incandescent solid and its spectral pattern as produced with a spectroscope. Figure *b* shows the "radiation curve" of an excited gas and its emission spectral pattern. Figure *c* shows the curve produced when a cool gas is between an incandescent source and the viewer; the corresponding spectral pattern is left as an exercise for you to construct. Figure *d* shows the spectral pattern of an incandescent source as seen through a piece of green glass; you are to sketch in the corresponding radiation curve.

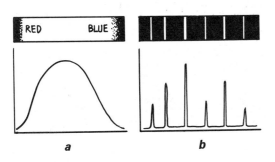

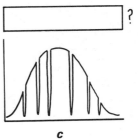

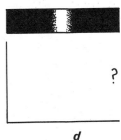

a　　　b　　　c　　　d

29 Light Quanta

The classical physics that we have so far studied deals with two categories of phenomena: particles and waves. According to our everyday experience, "particles" are tiny objects like bullets. They have mass and obey Newton's laws in that they travel in straight lines unless a force acts upon them. Likewise, according to our everyday experience, "waves," like waves in the ocean, are phenomena that extend in space. When a wave travels through an opening or around a barrier, different parts of the wave interfere and diffraction results. Therefore, particles and waves are easy to distinguish from each other. In fact, they have properties that are mutually exclusive. Nonetheless, the question of how to classify light has been a mystery for centuries.

One of the earliest recorded theories about the nature of light is that of Plato, who thought that light consisted of streamers emitted by the eye. This view was also held by Euclid. The Pythagoreans, somewhat later, believed that light emanated from luminous bodies to the eye in the form of very fine particles, while Empedocles taught that light is composed of high-speed waves of some sort. In 1704 Isaac Newton described light as a stream of particles or corpuscles. He maintained this view despite his later experiment with light reflecting from glass plates, where he noticed fringes of brightness and darkness (Newton's rings) and recognized that particles of light have some sort of wave nature. Christian Huygens, a contemporary of Newton, strongly advocated a wave theory of light.

With all this history as background, Thomas Young in 1801 performed "the double-slit experiment," which seemed to prove, finally, that light was a wave phenomenon. This view was reinforced in 1862 by Maxwell's announcement that light is energy carried in electric and magnetic fields of electromagnetic waves. Furthermore, the wave view of light was confirmed experimentally 25 years later by Hertz. In 1905, however, Albert Einstein published a Nobel Prize–winning paper that *confuted* the wave theory of light by demonstrating that light in its interactions with matter is not confined in continuous waves as Maxwell envisioned but in tiny packets of energy called *light quanta*, or **photons**. This strange behavior of light was just the initial glimpse of the quantum rules that govern the world of the atom. In this chapter we shall enter the world of the very small and look into some of the strange and exciting aspects of quantum reality.

Birth of the Quantum Theory

At the turn of the century, technology reached a level that enabled scientists to design experiments that were sensitive to the behavior of very small particles. With the discovery of the electron in 1897 and the inves-

Quantization and Planck's Constant

tigation of radioactivity about the same time, experimenters began to probe the atomic structure of matter. In 1900 the German theoretical physicist Max Planck announced his theoretical discovery that the energies of vibrating electrons in incandescent light sources were restricted to distinct values. As a result, radiation was emitted in discrete bunches of energy, which he called **quanta** (*quanta* is the plural form of *quantum*, just as *momenta* is the plural form of *momentum*). Planck's discovery began a revolution of ideas that completely changed the way we think about the physical world. We will see that the rules we apply to the everyday macroworld, the Newtonian laws that work so well for large objects like baseballs and planets, simply don't apply to events in the microworld of the atom. Whereas in the macroworld the study of motion is called *mechanics,* in the microworld the study of the motion of quanta is called **quantum mechanics**. More broadly, the body of laws developed from 1900 to the late 1920s that describe all quantum phenomena of the microworld have become known as *quantum physics.*

Quantization, the idea that physical quantities come in discrete bunches, is certainly not a new idea to physics. Matter is quantized; the mass of a brick of gold, for example, is equal to some whole-number multiple of the mass of a single gold atom. Electricity is quantized, as all electric charge is some whole-number multiple of the charge of a single electron.

Quantum physics tells us that other quantities are also quantized, quantities such as energy and angular momentum. This is quite a difference from the everyday physics we have already learned. In Newton's physics a system can have any value of energy. In quantum physics, by contrast, only certain values are possible. There are some energy values that are simply impossible for a system to possess. Energy in a light beam is quantized. The energy comes in packets, or quanta, and only a whole number of quanta can exist. The quanta of light, or electromagnetic radiation in general, are the photons.

Recall from the last chapter that the frequency of a photon f is proportional to its energy: $E \sim f$. When the energy of a photon is divided by its frequency, the single number that results is the proportionality constant, called **Planck's constant**, h.[1] We shall see that Planck's constant is a fundamental constant of nature that serves to set a lower limit on the smallness of things. It ranks with the velocity of light as a basic constant of nature and appears again and again in quantum physics. We can insert this constant in the above proportion and express it as an exact equation:

$$E = hf$$

This equation gives the smallest amount of energy that can be converted to light with frequency f. The radiation of light is not emitted continuously but is emitted as a stream of photons, each with an energy hf.

---

[1]Planck's constant, h, has the numerical value 6.6×10^{-34} joule-second.

Question   How much total energy is in a monochromatic beam composed of *n* photons of frequency *f*?*

The new physics tells us that the physical world is a rough, grainy place of discrete packets rather than the smooth, continuous place with which we are familiar. How can the "commonsense" world described by classical physics result from quantum physics? It is because the quantum graininess is on a very small scale compared to the size of things in the familiar world. This is reflected in the fact that Planck's constant is small in terms of familiar units. For example, the smooth blends of black, white, and gray in the photographs in this book are not smooth at all when viewed through a magnifying glass. You'll see that a photograph consists of many tiny dots. The image is quantized, something you do not notice from a distance. Using special tools, quantum physics provides a description of the grainy world of atoms and energy that blurs to make the world we know.

Question   What does the term *quantum* mean?†

Physicists were reluctant to adopt Planck's revolutionary quantum notion. Before it could be taken seriously the quantum idea would have to be verified by some means independent of light emission from an incandescent source. The verification was supplied less than five years later by Einstein, who extended Planck's ideas to explain the photoelectric effect (this was the Nobel Prize–winning paper we mentioned earlier).

Photoelectric Effect   In the latter part of the nineteenth century, several investigators had noticed that light was capable of ejecting electrons from various metal surfaces. This is the **photoelectric effect**, now used in electric eyes, in the photographer's light meter, and in the sound track of motion pictures.

An arrangement for observing the photoelectric effect is shown in Figure 29.1. Light shining on the negatively charged photosensitive metal surface liberates electrons, which are attracted to the positive plate to produce a measurable current. If we instead charge this plate negatively so that it repels electrons, the current can be stopped. We can then calculate the energies of the ejected electrons from the easily measured potential difference between the plates.

---

*Answer   The energy in a beam of light containing *n* quanta is $E = nhf$.

†**Answer**   A *quantum* is the smallest elemental unit of a quantity. Radiant energy, for example, is composed of many quanta, each of which is called a *photon*. So the more photons in a beam of light, the more energy in that beam.

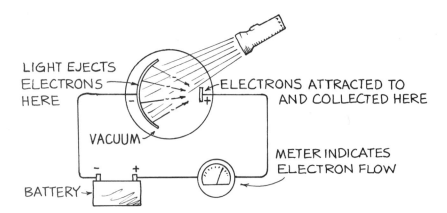

Fig. 29.1 An apparatus used for measuring the photoelectric effect.

The photoelectric effect was not particularly surprising to early investigators. The ejection of electrons could be accounted for by classical mechanics, which pictures the incident light waves building an electron's vibration up to greater and greater amplitudes until it finally breaks loose from the metal surface, just as water drops boil off the surface of hot water. For a weak source it should take considerable time to heat the surface sufficiently to boil off electrons. It was instead found that electrons are ejected immediately—but not as many as with a strong source. Careful examination of the behavior of the photoelectric effect led, however, to observations that were quite contrary to the classical wave picture:

1. The time lag between turning on the light and the ejection of the first electrons was independent of the brightness and frequency of the light.

2. The effect was easy to observe with violet or ultraviolet light but not with red light.

3. The rate at which electrons were ejected was proportional to the brightness of the light.

4. The maximum energy of the ejected electrons was independent of the brightness of the light, but there were indications that the energy did depend on the frequency of the light.

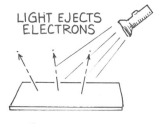

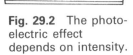

Fig. 29.2 The photoelectric effect depends on intensity.

The lack of any appreciable time lag was especially difficult to understand in terms of the wave picture, according to which an electron in dim light should, after some delay, build up enough vibrational energy for ejection, while an electron in bright light should be ejected immediately. However, this didn't happen. It was not unusual to observe an electron being ejected immediately even in the dimmest of lights. The observation that the brightness of light in no way affected the energies of ejected electrons was perplexing. The stronger electric fields of brighter light did not cause electrons to be ejected at greater speeds. More electrons were ejected in brighter light, but not at greater speeds. A weak beam of ultraviolet light, on the other hand, produced a smaller number of ejected electrons but at much higher kinetic energies. This was most puzzling.

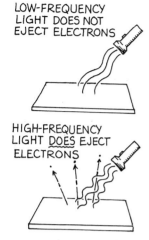

LOW-FREQUENCY LIGHT DOES NOT EJECT ELECTRONS

HIGH-FREQUENCY LIGHT DOES EJECT ELECTRONS

Fig. 29.3 The photo-electric effect depends on frequency.

Einstein produced the answer in 1905, the same year he explained Brownian motion and set forth his theory of special relativity. His clue was Planck's quantum theory of radiation. Planck had assumed that the emission of light in quanta was due to restrictions on the vibrating electrons that produced the light, which he considered to be discrete pulses of the electromagnetic waves described by Maxwell. Einstein, on the other hand, attributed the quantum properties to light itself and viewed radiation as a hail of particles. To emphasize this particle aspect, we speak of *photons* (by analogy with electrons, protons, and neutrons) whenever we are thinking of the particle nature of light. One photon is completely absorbed by each electron ejected from the metal. The absorption is an all-or-nothing process and is immediate; so there is no delay as "wave energies" build up.

A wave has a broad front, and its energy is distributed along this front rather than concentrated so as to eject a single electron from a metal surface. This is as improbable as an ocean wave hitting a beach and knocking a single seashell far inland—and with an energy equal to the energy of the whole wave. Instead of light being regarded as a continuous train of waves, the photoelectric effect suggests that we conceive it as a succession of corpuscles, or photons. The number of photons in a light beam has to do with the brightness of the beam, while the energy of each photon is related to the frequency of light.

Experimental verification of Einstein's explanation of the photoelectric effect was made 11 years later by the American physicist Robert Millikan. Every aspect of Einstein's interpretation was confirmed.

The photoelectric effect proves conclusively that light has particle properties. We cannot conceive of the photoelectric effect on the basis of waves. On the other hand, we have seen that the phenomenon of interference demonstrates convincingly that light has wave properties. We cannot conceive of interference in terms of particles. In classical physics, this appears to be and is contradictory. From the point of view of quantum physics, light has properties resembling both. It is "just like a wave" or "just like a particle," depending on the particular experiment. Perhaps we should think of light not as a particle or a wave but as a "wavicle."

Questions

1. Will brighter light eject more electrons from a photosensitive surface than dimmer light of the same frequency?*

2. Will high-frequency light eject a greater number of electrons than low-frequency light?†

---

*__Answer__ Yes. The number of ejected electrons depends on the number of incident photons.

†__Answer__ Not necessarily. The energy (not the number) of ejected electrons depends on the frequency of the illuminating photons. A bright source of blue light, for example, may eject more electrons than a dim violet source, but not more energetic electrons.

Wave-Particle Duality

The wave and particle nature of light is evident in the formation of optical images. We understand the photographic image produced by a camera in terms of light waves, which spread from each point of the object, refract as they pass through the lens system, and converge to focus on the photographic film. The path of light from the object through the lens system and to the focal plane can be calculated using methods developed from the wave theory of light.

But now consider carefully the way in which the photographic image is formed. The photographic film consists of an emulsion that contains grains of silver halide crystal, each grain containing about 10^{10} silver atoms. Each photon that is absorbed gives up its energy hf to a single grain in the emulsion and is stored as potential energy. This energy activates surrounding crystals in the entire grain and is used in development to complete the photochemical process. Many photons activating many grains produce the usual photographic exposure. If a photograph is taken with exceedingly feeble light, we find that the image is built up by individual photons arriving independently and seemingly random in their distribution. We see this strikingly illustrated in Figure 29.4, which shows how an exposure progresses photon by photon.

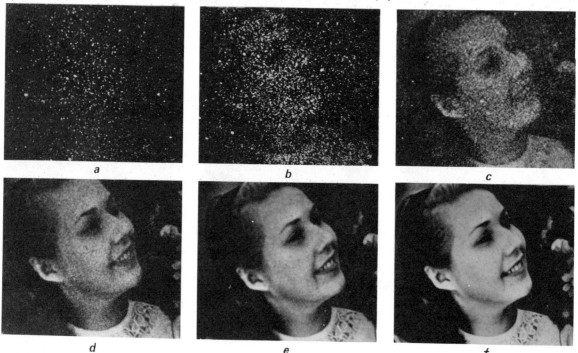

Fig. 29.4 Stages of exposure revealing the photon-by-photon production of a photograph. The approximate numbers of photons at each stage were (*a*) 3×10^3, (*b*) 1.2×10^4, (*c*) 9.3×10^4, (*d*) 7.6×10^5, (*e*) 3.6×10^6, and (*f*) 2.8×10^7.

Double-Slit Experiment

Let's return to Thomas Young's double-slit experiment, which we discussed in terms of waves in Chapter 27. Recall that when we pass monochromatic light through a pair of closely spaced thin slits, we produce an interference pattern (Figure 29.5). Now let's consider the experi-

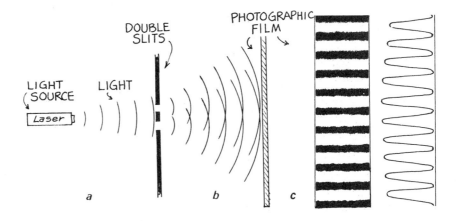

Fig. 29.5 (*a*) Arrangement for double-slit experiment. (*b*) Photograph of interference pattern. (*c*) Graphical representation of pattern.

ment in terms of photons. Suppose we dim our light source so that in effect only one photon at a time is incident upon the barrier with the thin slits. If the film is exposed to the light for a very short time, the film becomes exposed as simulated in Figure 29.6*a*. Each spot represents the place where the film has been exposed to a photon. If the light is allowed to expose the film for a longer time, a pattern of fringes begins to emerge as in Figure 29.6*b* and *c*. This is quite amazing. Spots on the film are seen to progress photon by photon to form the same interference pattern characterized by waves!

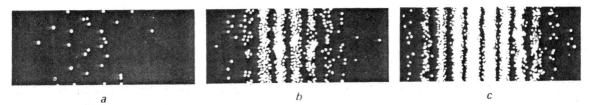

Fig. 29.6 Stages of two-slit interference pattern. The pattern of individually exposed grains progresses from (*a*) 28 photons to (*b*) 1,000 photons to (*c*) 10,000 photons. As more photons hit the screen, a pattern of interference fringes appears.

If we cover one slit so that photons that hit the photographic film can only pass through a single slit, the tiny spots on the film accumulate to form a single-slit diffraction pattern (Figure 29.7). We find that photons hit the film at places they would not hit if both slits were open! If we view

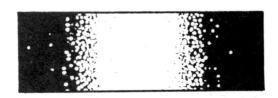

Fig. 29.7 Single-slit diffraction pattern.

this classically, we are perplexed and may ask how photons passing through the single slit "know" that the other slit is covered and therefore fan out to produce the wide single-slit diffraction pattern. Or, with both slits open, how do photons traveling through one slit "know" that the other slit is open and avoid certain regions, proceeding only to areas that will ultimately fill to form the fringed double-slit interference pattern?[2] We have no satisfying answer to these questions that presuppose that light obeys the rules of classical physics. What we do know is that light behaves as a stream of photons when it interacts with the photographic film or other detectors and behaves as a wave in traveling from a source to the place where it is detected. Photons strike the film at places where we would expect to see constructive interference of waves. The fact that light exhibits both wave and particle behavior is one of the interesting surprises that physicists have discovered in this century. Even more surprising has been the discovery that objects with mass also exhibit a dual wave-particle behavior.

Particles as Waves: Electron Diffraction

If light can have particle properties, why can't particles have wave properties? This question was posed by the French physicist Louis de Broglie while still a graduate student in 1924. His answer constituted his doctoral thesis in physics, and later won him the Nobel Prize in physics. According to de Broglie, every particle of matter is somehow endowed with a wave to guide it as it travels. Under the proper conditions, then, every particle will produce an interference or diffraction pattern. All bodies—electrons, protons, atoms, mice, you, planets, suns—have a wavelength that is related to their momentum by

$$\text{Wavelength} = \frac{h}{\text{momentum}}$$

where h is Planck's constant. A body of large mass and ordinary speed has such a small wavelength that interference and diffraction are negligible;

[2]From a Newtonian point of view, this wave-particle duality is indeed mysterious. This leads some people to believe that quanta have some sort of consciousness, with each photon or electron having "a mind of its own." The mystery, however, is like beauty. It is in the mind of the beholder rather than in nature itself. We conjure models to understand nature, and when inconsistencies arise, we sharpen or change our models. The wave-particle duality of light doesn't fit the model devised by Newton. An alternate model is that quanta have minds of their own. Another model is quantum physics. In this book we subscribe to the latter.

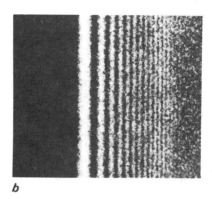

Fig. 29.8 (a) Fringes produced by the diffraction of light and (b) of an electron beam.

a   *b*

rifle bullets fly straight and do not pepper their targets far and wide with detectable interference patches.[3]

Experimental verification for de Broglie's wave nature of the electron was provided in 1926 by the American physicists Clinton J. Davisson and Lester H. Germer, who scattered low-energy electrons off the surface of a metal crystal. The scattered electrons formed diffraction patterns characteristic of the patterns produced by diffracted light. A comparison of light and electron diffraction is shown in Figure 29.8.

Fig. 29.9 An electron microscope makes practical use of the wave nature of electrons. The wavelength of electron beams is typically thousands of times shorter than the wavelength of visible light, so the electron microscope is able to distinguish detail not possible with optical microscopes.

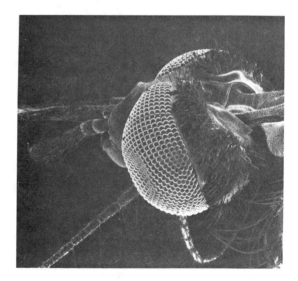

Fig. 29.10 Detail of a female mosquito head as seen with a scanning electron microscope at a "low" magnification of 200 times.

---

[3] A bullet of mass 0.02 kg traveling at 330 m/s, for example, has a de Broglie wavelength of $\frac{6.6 \times 10^{-34} \text{ J} \cdot \text{s}}{(0.02 \text{ kg})(330 \text{ m/s})} = 10^{-34}$ m, an incredibly small size a million million million millionth the size of a hydrogen atom. An electron traveling at 2 percent the speed of light, on the other hand, has a wavelength 10^{-10} m, which is equal to the diameter of the hydrogen atom. Diffraction effects for electrons are measurable, whereas diffraction effects for bullets are not.

Beams of electrons directed through double slits exhibit interference patterns just as photons do. The double-slit experiment discussed in the last section can be performed with electrons as well as photons. For electrons the apparatus is more complex, but the procedure is essentially the same. The intensity of the source can be reduced so as to direct electrons one at a time through a double-slit arrangement to produce the same remarkable results as with photons. Like photons, electrons arrive at the screen as particles, but the pattern of arrival is wavelike. The angular deflection of electrons to form the interference pattern agrees perfectly with calculations using de Broglie's equation for the wavelength of an electron. This wave-particle duality is not restricted to photons and electrons. In Figure 29.11 we see the results of a similar procedure that uses a standard electron microscope and an electrostatic biprism. The electron beam of very low current density is directed at the prism, and a pattern of fringes produced by individual electrons builds up step by step and is displayed on a TV monitor. The image is gradually filled by electrons to produce the interference pattern customarily associated with waves. Neutrons, protons, whole atoms, and to an immeasurable degree even high-speed rifle bullets exhibit a duality of particle and wave behavior.

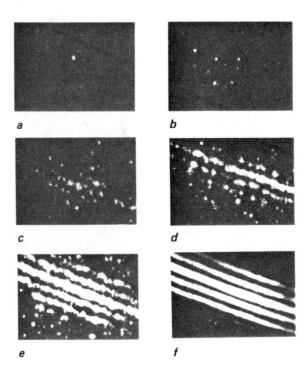

a   b

c   d

e   f

Fig. 29.11 Electron interference patterns filmed from a TV monitor, showing the diffraction of a very low-intensity electron-microscope beam through an electrostatic biprism.

Questions

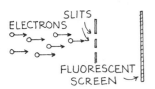

ELECTRONS
SLITS
FLUORESCENT
SCREEN

1. If electrons behaved only like particles, what pattern would you expect on the screen after the electron passed through the double slits?*

2. We don't notice the de Broglie wavelength for a pitched baseball. Is this because the wavelength is very large or very small?†

3. If an electron and a proton have the same de Broglie wavelength, which particle has the greater speed?‡

Uncertainty Principle

The wave-particle duality of quanta has led to interesting discussions about the limits of our ability to accurately measure the properties of small objects. The discussion centers around the idea that the measuring device in some way affects the quantity being measured.

We know, for example, that if we place a relatively cool thermometer in a cup of hot coffee, the temperature of the coffee is altered as it gives heat to the thermometer. The measuring device alters the quantity being measured. But we can correct for these errors in measurement if we know the initial temperature of the thermometer, the masses and specific heats involved, and so forth. Such corrections fall well within the domain of classical physics. In addition, however, there is an intrinsic limit in our ability to make accurate measurements—a limit stemming from the wave-particle duality of matter. Associated with every measurement is a quantum uncertainty: we find that any measurement that in any way probes a system necessarily disturbs the system by at least one quantum of action, *h*—Planck's constant. So any measurement that involves interaction between the measurer and that being measured is subject to this minimum inaccuracy.

If we look at a cup of hot coffee across the room and judge that it is hot by observing the steam rising from it, this measurement involves no probing. This act of "measuring" neither adds nor subtracts energy from the coffee. Placing a thermometer in it is a different story. We physically interact with the coffee and thereby subject it to alteration. The quantum contribution to this alteration, however, is completely dwarfed by classical uncertainties and is negligible. Quantum uncertainties are significant only in the atomic and subatomic realm.

---

*__Answer__ If electrons behaved only like particles, they would form two bands, as indicated in *a*. Because of their wave nature, they actually produce the pattern shown in *b*.

a   *b*

†__Answer__ We don't notice the wavelength of a pitched baseball because it is extremely small, on the order of 10^{-20} times smaller than the atomic nucleus.

‡__Answer__ The same wavelength means that the two particles have the same momentum. This means that the less massive electron must travel faster than the heavier proton.

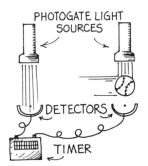

PHOTOGATE LIGHT
SOURCES

DETECTORS

TIMER

Fig. 29.12 The ball's speed is measured by dividing the distance between the photogates by the time difference in crossing light paths. Photons hitting the ball alter its motion much less than the motion of an oil supertanker is altered when a few fleas bump into it.

Suppose we wish to measure the speed of a pitched baseball by passing it through a pair of photogates that are a known distance apart. The ball is timed as it interrupts beams of light in the gates. The accuracy of the ball's measured speed has to do with uncertainties in distance between the gates and the timing mechanisms. Interactions between the macroscopic ball and the photons it encounters are insignificant. But not so in the case of measuring submicroscopic things like electrons. Even a single photon bouncing off an electron appreciably alters the motion of the electron—and in an unpredictable way. If we wished to observe an electron and determine its whereabouts with light, the wavelength of the light would have to be very short. We fall into a dilemma. A short wavelength corresponds to a large quantum of energy, which greatly alters the electron's state of motion. If, on the other hand, we use a long wavelength that corresponds to a smaller quantum of energy, the change we induce to the electron's state of motion will be smaller, but the determination of its position by the coarser wave will be less accurate. The act of observing something as tiny as an electron produces a considerable uncertainty in either its position or its motion. Although this uncertainty is completely negligible for measurements of position and motion regarding everyday (macroscopic) objects, it is a predominant fact of life in the atomic domain.

The uncertainty of measurement in the atomic domain was first stated mathematically by Werner Heisenberg of Germany and is called the **uncertainty principle**. It is a fundamental principle in quantum mechanics. Heisenberg found that when the uncertainties in the measurement of momentum and position for a particle are multiplied together, the product must be equal to or greater than Planck's constant, h, divided by 2π, which is represented as $\hbar$ (called *h-bar*). We can state the uncertainty principle in a simple formula:

$$\Delta P \, \Delta X \geq \hbar$$

The Δ symbol here means "uncertainty of": ΔP is the uncertainty of momentum (the symbol for momentum is conventionally P), and ΔX is the uncertainty of position. The product of these two uncertainties can be equal to or greater than ($\geq$) the size of $\hbar$. For minimum uncertainties, the product will equal $\hbar$; the product of larger uncertainties will be greater than $\hbar$. But in no case can the product of the uncertainties be less than $\hbar$. The significance of the uncertainty principle is that even in the best of conditions, the lower limit of uncertainty is $\hbar$. This means that if we wish to know the momentum of an electron with great accuracy, the corresponding uncertainty in position will be large. Or if we wish to know the position with great accuracy, the corresponding uncertainty in momentum will be large. The sharper one of these quantities, the less sharp the other. Only in the classical limit where $\hbar$ goes to zero could the uncertainties of both position and momentum be arbitrarily small. Planck's constant is greater than zero, and we cannot in principle simultaneously know both quantities with absolute certainty.

The uncertainty principle operates similarly with *energy* and *time*. We cannot measure a particle's energy with complete precision in an infinitely short span of time. The uncertainty in our knowledge of energy, ΔE, and the duration taken to measure the energy, Δt, are related by the expression[4]

$$\Delta E \; \Delta t \geqslant \hbar$$

The greatest accuracy we can ever hope to attain is that case in which the product of the energy and time uncertainties equals $\hbar$. The more accurately we determine the energy of a photon, an electron, or a particle of whatever kind, the more uncertain we will be of the time it has that energy.

The uncertainty principle is relevant only to quantum phenomena. The inaccuracies in measuring the position and momentum of a baseball due to the interactions of observation, for example, are completely negligible. But the inaccuracies in measuring the position and momentum of an electron are far from negligible. The inaccuracies in measuring the energy and time of fall for the ram of a pile driver are negligible, but the inaccuracies in measuring simultaneously the energy release and time of de-excitation of an atom are not negligible. This is because the uncertainties in the measurements of these subatomic quantities are comparable to the magnitudes of the quantities themselves.[5]

There is a danger in applying the uncertainty principle to areas outside of quantum mechanics. Some people conclude from statements of the interaction between the observer and the observed that the universe does not "exist out there" independently of all acts of observation and that reality is created by the observer. Others interpret the uncertainty principle as nature's shield of forbidden secrets. Some critics of science use the uncertainty principle as evidence that science itself is uncertain. The real-

---

[4]We can see that this is consistent with the uncertainty in momentum and position. Recall that Δ momentum = force $\times$ Δ time and that Δ energy = force $\times$ Δ distance. Then

$$\hbar = \Delta \text{ momentum } \Delta \text{ distance}$$
$$= (\text{force } \times \Delta \text{ time}) \; \Delta \text{ distance}$$
$$= (\text{force } \times \Delta \text{ distance}) \; \Delta \text{ time}$$
$$= \Delta \text{ energy } \Delta \text{ time}$$

[5]The uncertainties in measurements of momentum, position, energy, or time that are related to the uncertainty principle for a pitched baseball are only 1 part in about 10 million billion billion billion (10^{-34}). Quantum effects are negligible even for the swiftest bacterium, where the uncertainties are about 1 part in a billion (10^{-9}). Quantum effects become evident for atoms, where the uncertainties are about 1 part in a hundred (10^{-2}). For electrons moving in an atom, quantum uncertainties dominate, and we are in the full-scale realm of the quantum world.

ity of the universe whether observed or not, nature's secrets, and the uncertainties of science have very little to do with Heisenberg's uncertainty principle. The profundity of the uncertainty principle has to do with the unavoidable interaction between nature at the atomic level and the means by which we probe it.

Questions   **1.** Is Heisenberg's uncertainty principle applicable to the practical case of using a thermometer to measure the temperature of a glass of water?*

2. A Geiger counter measures radioactive decay by registering the electrical pulses produced in a gas tube when radiation particles pass through it. The radiation particles emanate from a radioactive source, say, radium. Does the act of measuring the decay rate of radium alter the radium or its decay rate?†

3. Can the quantum principle that we cannot observe something without changing it be reasonably extrapolated to support the claim that you can make a stranger turn around and look at you by staring intently at his back?‡

---

*Answer   No. Although we likely subject the temperature of water to a change by the act of probing it with a thermometer, especially one appreciably colder or hotter than the water, the uncertainties that relate to the precision of the thermometer are quite within the domain of classical physics. The role of uncertainties at the subatomic level is inapplicable here.

†Answer   Not at all, because the interaction involved is between the Geiger counter and the radiation, not between the Geiger counter and the radium. It is the behavior of the radiation particles that is altered by measurement, not the radium from which they emanate. See how this ties into the next question.

‡Answer   No. Here we must be careful in defining what we mean by *observing*. If our observation involves probing (giving or extracting energy), we indeed change to some degree that which we observe. For example, if we shine a light source onto the person's back, our observation consists of probing, which, however slight, physically alters the configuration of atoms on his back. If he senses this, he may turn around. But simply staring intently at his back is observing in the passive sense. The light you receive or block by blinking, for example, has already left his back. So whether you stare, squint, or close your eyes completely, you in no physical way alter the atomic configuration on his back. Shining a light or otherwise probing something is not the same thing as passively looking at something. A failure to make the simple distinction between *probing* and *passive observation* is at the root of much nonsense that is said to be supported by quantum physics. Better support for the above claim would be positive results from a simple and practical test rather than the assertion that it rides on the hard-earned reputation of quantum theory.

Complementarity

The realm of quantum physics is seemingly confusing. Light waves that interfere and diffract deliver their energy in particle packages of quanta. Electrons that move through space in straight lines and experience collisions like particles distribute themselves spatially in interference patterns as if they were waves. In this confusion there is an underlying order. The behavior of light and electrons is confusing in the same way! Light and electrons both exhibit wave and particle characteristics.

The Danish physicist Niels Bohr, one of the founders of quantum physics, formulated an explicit expression of the wholeness inherent in this dualism. He called his expression of this wholeness the concept of **complementarity**. According to this concept, quantum phenomena exhibit complementary properties—appearing either as particles or as waves—depending on the type of experiment in which they manifest. Experiments designed to examine individual exchanges of energy and momentum bring out particlelike properties, while experiments designed to examine spatial distribution of energy bring out wavelike properties. The wavelike properties of light and particlelike properties of light complement (mutually exclude) one another, but *both* are necessary for the understanding of "light."

The idea that opposites are components of a wholeness is not new. Ancient Eastern cultures incorporated it as an integral part of their world view. This is demonstrated in the yin-yang diagram of T'ai Chi Tu (Figure 29.13). One side of the circle is called *yin,* and the other side is called *yang.* Where there is yin, there is yang. Where there is low, there is also high. Where there is night, there is also day. Where there is birth, there is also death. The yin-yang diagram symbolized the principle of complementarity for Niels Bohr. In later life he wrote many essays on the implications of complementarity in many aspects of life. In 1947 when he was knighted for his contributions to physics, he chose for his coat of arms the yin-yang symbol.

Fig. 29.13 Opposites are seen to complement one another and are symbolized in the yin-yang diagram prevalent in Eastern cultures.

Summary of Terms

Quantum theory The theory that energy is radiated in definite units called *quanta,* or *photons.* Just as matter is composed of atoms, radiant energy is composed of quanta. The theory further states that all material particles have wave properties.

Planck's constant A fundamental constant, h, which relates the energy and the frequency of light quanta:

$$h = 6.6 \times 10^{-34} \text{ joule-second}$$

Photoelectric effect The emission of electrons from a metal surface when light is shined on it.

Uncertainty principle The principle formulated by Heisenberg which states that the ultimate accuracy of measurement is given by the magnitude of Planck's constant, h. Further, it is not possible to measure exactly both the position and the momentum of a particle at the same time, nor the energy and time associated with a particle simultaneously.

Complementarity The principle enunciated by Niels Bohr which states that the wave and particle models of either matter or radiation complement each other and when combined provide a fuller description.

Suggested Reading

Cole, K. C. *Sympathetic Vibrations: Reflections of Physics as a Way of Life*. New York: Morrow, 1984. A delightful unraveling of the theories of Bohr, Einstein, and other developers of quantum physics, with emphasis on the human side of physics.

March, R. H. *Physics for Poets*, 2nd ed. New York: McGraw-Hill, 1978. This textbook presents an interesting account of contemporary physics and its historical roots.

Zukav, Gary. *The Dancing Wu Li Masters: An Overview of the New Physics*. New York: Morrow, 1979. Although mystical in tone, some very readable physics by a non-physicist.

Review Questions

1. What is a quantum of light called?

2. How does the idea of a quantum differ from the idea of a wave?

3. What does it mean to say that a certain quantity is "quantized"?

4. In the last chapter we learned the formula $E \sim f$. In this chapter we learned the formula $E = hf$. Explain the difference between these two formulas. What is h?

5. In the formula $E = hf$, does f stand for wave frequency, as defined in Chapter 17?

6. Which has the smaller quanta—red light or blue light? Radio waves or X rays?

7. Why are photons of violet light more successful than photons of red light in dislodging electrons from a metal surface?

8. Why won't a very bright beam of red light impart more energy to an electron than a feeble beam of violet light?

9. Compare Planck's view of quantization with that of Einstein.

10. What evidence can you cite for the wave nature of light? For the particle nature of light?

11. What evidence can you cite for the wave nature of particles?

12. What is the uncertainty principle?

13. Is the uncertainty principle relevant to the everyday macro world? Why or why not?

14. What is the principle of complementarity?

15. Is light a wave or a particle? Explain briefly.

Exercises

1. Is the following statement true or false? "Light is the only thing we ever really see!"

2. Distinguish between *classical physics* and *quantum physics*.

3. How can a photon's energy be given by the formula $E = hf$ when f in the formula is a wave frequency?

4. Which photon has the greatest energy—an infrared, a visible, or an ultraviolet?

5. We speak of photons of red light and photons of green light. Can we speak of photons of white light? Why or why not?

6. A beam of red light and a beam of blue light have exactly the same energy. Which beam contains the greater number of photons?

7. As a solid is heated and begins to glow, why does it first appear red?

8. Silver bromide (AgBr) is a light-sensitive substance used in some types of photographic film. To cause exposure of the film, it must be illuminated with light having sufficient energy to dissociate the molecules. Why do you suppose this film may be handled without exposure in a darkroom illuminated with red light? How about blue light? How about very bright red light as compared to very dim blue light?

9. Suntanning produces cell damage in the skin. Why is ultraviolet radiation capable of producing this damage, while visible radiation is not?

10. In the photoelectric effect, does brightness or frequency determine the number of electrons ejected per second?

11. If you shine an ultraviolet light on the metal ball of a negatively charged electroscope (shown in Exercise 6 in Chapter 20), it will discharge. But if the electroscope is positively charged, it won't discharge. Can you venture an explanation?

12. Ultraviolet light falls on certain dyes, and visible light is emitted. Why does this not happen when infrared light falls on these dyes?

13. Estimate the relative current readings in the ammeter in Figure 29.1 for illumination of the photosensitive plate by light of various colors. For various intensities.

14. Explain briefly how the photoelectric effect is used in the operation of at least two of the following: an electric eye, a photographer's light meter, the sound track of a motion picture.

15. Does the photoelectric effect *prove* that light is corpuscular? Do interference experiments *prove* that light is composed of waves? (Is there a distinction between what something *is* and how it *behaves?*)

16. Does Einstein's explanation of the photoelectric effect invalidate Young's explanation of the double-slit experiment? Explain.

17. The camera that took the photograph of the woman's face (Figure 29.4) used ordinary lenses that are well known to refract waves. Yet the progression of the image is evidence of photons. How can this be? What is your explanation?

18. How can the wavelength of an electron be given by the de Broglie formula in which a term for *momentum* is included?

19. One electron travels twice as fast as another. Which has the longer wavelength?

20. Does the de Broglie wavelength of a proton become longer or shorter as velocity increases?

21. If a cannonball and a BB have the same speed, which has the longer wavelength?

22. We don't notice the wavelength of moving matter in our ordinary experience. Is this because the wavelength is extraordinarily large or extraordinarily small?

23. What principal advantage does an electron microscope have over an optical microscope?

24. Would a beam of protons in a "proton microscope" exhibit greater or less diffraction than electrons of the same speed in an electron microscope? Defend your answer.

25. If Planck's constant, *h,* were equal to zero rather than a tiny number, how would the uncertainty principle differ?

26. Comment on the idea that the theory one believes determines the meaning of one's observations and not vice versa.

27. There is at least one electron on the tip of your nose. If somebody looks at it, will its motion be altered? How about if it is looked at with one eye closed? With two eyes, but crossed? Does the uncertainty principle apply here?

28. Do we alter that which we attempt to measure in a public opinion survey? Does the uncertainty principle apply here?

29. If a system has been determined by specific causes that are well understood, does it follow that the future of the that system can be predicted? (Is there a distinction between a determined system and a predictable system?)

30. To measure the exact age of Old Methuselah, the oldest living tree in the world, a Nevada professor of dendrology, aided by an employee of the U.S. Bureau of Land Management, in 1965 cut the tree down and counted its rings. Is this an example of the uncertainty principle in its extreme or an example of arrogant and criminal stupidity?

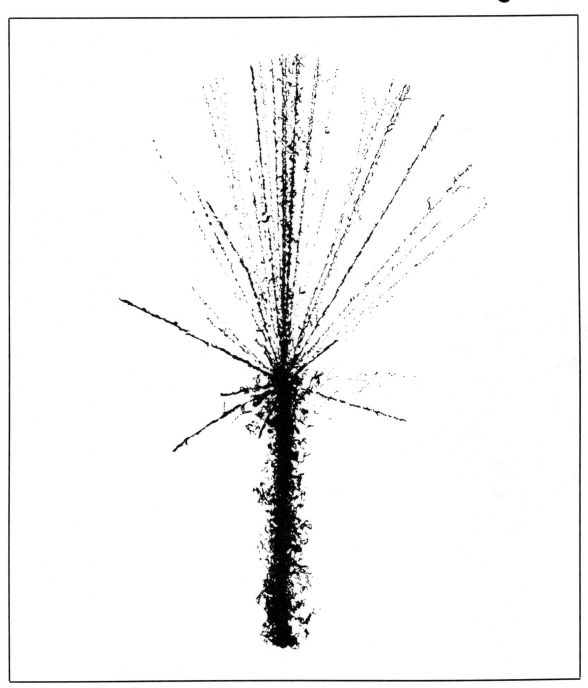

30 The Atom and the Quantum

We discussed the atom as a building block of matter in Chapter 9 and as an emitter of light in the preceding chapters. We know that the atom is composed of a central nucleus surrounded by a complex structure of electrons; the study of this atomic structure is called *atomic physics*. In this chapter we will outline some of the developments that led to our present understanding of atomic physics and trace the start of these developments from classical to quantum physics. Then, in the two following chapters, we will learn about the structure of the atomic nucleus, *nuclear physics*. This knowledge and its implications are presently having a profound impact on human society.

We begin our study of atomic and nuclear physics with a brief account of some of the events at the beginning of the century that led to our present understanding of the atom.

Discovery of the Atomic Nucleus

A few years after Einstein announced the photoelectric effect, the British physicist Ernest Rutherford performed his now-famous gold-foil experiment. This experiment showed that the atom was mostly empty space with most of its mass packed into the central region called the *nucleus*. Rutherford discovered this by directing a beam of positively charged particles (alpha particles) from a radioactive source through a very thin gold leaf and measuring the angle at which the particles were deflected from their straight-line path as they emerged. This was accomplished by noting spots of light on a zinc sulphide screen around the gold leaf (Figure 30.1). As would be expected, most particles continued in a more or less straight-line path through the thin foil. But, surprisingly, some particles were widely deflected. Some were even scattered back along their incident paths. It was like firing bullets at a piece of tissue paper and finding some bullets

Fig. 30.1 The occasional large-angle scattering of alpha particles from the gold atoms led Rutherford to the discovery of the small, very massive nuclei at their centers.

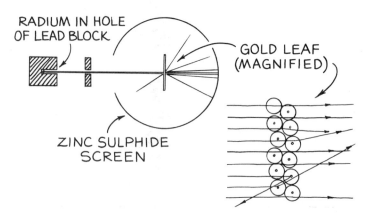

RADIUM IN HOLE OF LEAD BLOCK

GOLD LEAF (MAGNIFIED)

ZINC SULPHIDE SCREEN

bouncing backward. Rutherford reasoned that the particles that were un-deflected traveled through regions of the gold foil that were empty space, while the small number of particles that were deflected were repelled from the center of gold atoms that must also be positively charged. Rutherford had discovered the atomic nucleus.

Atomic Spectra: Clues to Atomic Structure

During the period of Rutherford's experiments, chemists were using the spectroscope (discussed in Chapter 28) for chemical analysis, while physicists were busy trying to find order in the confusing arrays of spectral lines. It had long been noted that the lightest element, hydrogen, had a far more orderly spectrum than the other elements. An important sequence of lines in the hydrogen spectrum runs from a single line in the red region of the spectrum to one in the green, and then to one in the blue, then to several lines in the violet, followed by many in the ultraviolet. Spacing between successive lines becomes less and less from the first in the red to the last in the ultraviolet, until the lines become so close they seem to merge. A Swiss schoolteacher, J. J. Balmer, first expressed the positions of these lines in a single mathematical formula. Balmer could give no reason why his formula worked so successfully. Using his formula, it was possible to locate lines that had not yet been measured. This success spurred the discovery of other formulas that fixed the line positions for other elements.

Fig. 30.2 A portion of the hydrogen spectrum.

Another regularity in atomic spectra was found by J. Rydberg. He noticed that the sum of the frequencies of two lines in the spectrum of hydrogen sometimes equals the frequency of a third line. This relationship was later advanced as a general principle by W. Ritz and is called the **Ritz combination principle**. It states that the spectral lines of any element include frequencies that are either the sum or the difference of the frequencies of two other lines. Like Balmer, Ritz was unable to offer an explanation for this regularity. These regularities were the clues Danish physicist Niels Bohr used to understand the structure of the atom itself.

Bohr Model of the Atom

In 1913 Bohr applied the quantum theory of Planck and Einstein to the nuclear atom of Rutherford and formulated the well-known planetary model of the atom. Bohr reasoned that Planck's quantized electron energy states (discussed in the previous chapter) would correspond to electron orbits of different radii. He was able to calculate the energy states in the hydrogen atom and show that they corresponded to quantized circular

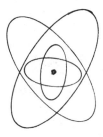

Fig. 30.3 The Bohr model of the atom. Although this model is very oversimplified, it is still useful in understanding light emission.

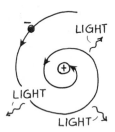

Fig. 30.4 From a classical point of view, an electron accelerating around its orbit should emit radiation. This loss of energy should be accompanied by a decrease in orbital speed and a spiraling of the electron into the nucleus. But this does not happen.

orbits. From the Balmer formula he was able to calculate the angular momentum of an electron in each orbit. Bohr asserted that the angular momentum is a whole-number multiple of $\frac{h}{2\pi}$—where h is Planck's constant, which links the energy and frequency of light.

According to Bohr, electrons orbiting in the farthermost orbits would be at a higher energy state than those in the orbits closer to the nucleus. Bohr reasoned that light is emitted when electrons make a transition from a higher to a lower orbit and that the frequency of emitted radiation is given by the quantum relationship $E = hf$, where E is the difference in energy of the atom when the electron is in the different orbits.

Bohr's planetary model of the atom faced a major difficulty. Accelerating electrons radiate energy; an electron accelerating around a nucleus should radiate energy continuously. This radiating away of energy should cause the electron to lose orbital speed and spiral into the nucleus (Figure 30.4). Bohr boldly broke with classical physics by stating that the electron doesn't radiate light while it accelerates around the nucleus in a single orbit, but that radiation of light takes place only when the electron jumps orbits from a higher energy level to a lower energy level. The farthermost orbits are at higher energy levels than the orbits closer to the nucleus. The energy of the emitted photon is equal to the *difference* in energy between the two energy levels, which obeys the quantum relationship $E = hf$. So the quantization of light energies neatly corresponds to the quantization of electron energies.

Bohr's views, as outlandish as they seemed at the time, accounted for the theretofore-unexplained regularities found in atomic spectra. Bohr's explanation of the Ritz combination principle is shown in Figure 30.5. If an electron is raised to the third energy level, it can return to its initial level by a single jump from the third to the first level—or by a double jump, first to the second level and then to the first level. These two return paths will produce three spectral lines. Note that the energy jump along path A plus B is equal to the energy jump C. Since frequency is proportional to energy, the frequencies of light emitted along path A and path B when added equal the frequency of light emitted when the transition is along path C. Now we see why the sum of two frequencies in the spectrum is equal to a third frequency in the spectrum.

Bohr was able to account for X rays and show that they were emitted when electrons jumped from outermost to innermost orbits. He predicted X-ray frequencies that were later experimentally confirmed. Bohr was also able to calculate the "ionization energy" of a hydrogen atom, the energy needed to knock the electron out of the atom completely. This also was verified by experiment.

Using measured X-ray and visible frequencies, it was possible to map energy levels of all the atomic elements. Bohr's model had electrons orbiting in neat circles (or ellipses) arranged in groups or shells. This model of the atom accounted for the general chemical properties of the elements and predicted properties of a missing element, which led to its discovery (hafnium).

Questions   **1.** What is the maximum number of paths for de-excitation available to a hydrogen atom excited to the $n = 3$ level in going to the ground state?*

2. Two predominant spectral lines in the hydrogen spectrum, an infrared one and a red one, have frequencies 2.7×10^{14} Hz and 4.6×10^{14} Hz respectively. Can you predict a higher-frequency line in the hydrogen spectrum?†

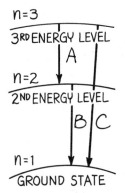

Fig. 30.5 Three of the many energy levels in an atom. The energy jumps A + B equal the energy jump C. The frequencies of light emitted from jump A + jump B therefore equal the frequency of light emitted from jump C.

Relative Sizes of Atoms

The diameters of the electron orbits in the Bohr model of the atom are determined by the amount of electrical charge in the nucleus. For example, the single positively charged proton in the hydrogen atom holds one negatively charged electron in an orbit at a particular radius. If we double the positive charge in the nucleus, the orbiting electron will be pulled into a tighter orbit with half its former radius since the electrical attraction is doubled. This doesn't quite happen, however, because the double charge in the nucleus normally holds a second orbital electron, which diminishes the effect of the positive nucleus. This added electron makes the atom electrically neutral. The atom is no longer hydrogen, but is helium. The two orbital electrons assume an orbit characteristic of helium. An additional proton in the nucleus pulls the electrons into an even closer orbit and, furthermore, holds a third electron in a second orbit. This is the lithium atom, atomic number 3. We can continue with this process, increasing the positive charge of the nucleus and adding successively more electrons and more orbits all the way up to atomic numbers above 100, to the synthetic radioactive elements.[1]

We find that as the nuclear charge increases and additional electrons are added in outer orbits, the inner orbits shrink in size because of the

Answer   Three, as shown in Figure 30.5.

†*Answer*   The addition of frequencies $2.7 \times 10^{14} + 4.6 \times 10^{14} = 7.3 \times 10^{14}$ Hz, which so happens to be the frequency of a violet line in the hydrogen spectrum. Can you see that if the infrared line is produced by a transition similar to path A in Figure 30.5, and the red line by path B, then the violet line corresponds to path C?

[1]Each orbit will hold only so many electrons. A rule of quantum mechanics states that an orbit is filled when it contains a number of electrons given by $2n^2$; where n is 1 for the first orbit, 2 for the second orbit, 3 for the third orbit, and so on. For $n = 1$, there are 2 electrons; for $n = 2$, there are $2(2^2)$ or 8 electrons; for $n = 3$, there are a maximum of $2(3^2)$ or 18 electrons, etc. The number n is called the *principal quantum number*.

stronger nuclear attraction. This means that the heavier elements are not much larger in diameter than the lighter elements. The diameter of the uranium atom, for example, is only about three hydrogen diameters even though it is 238 times more massive. The schematic diagrams in Figure 30.6 are drawn approximately to the same scale.

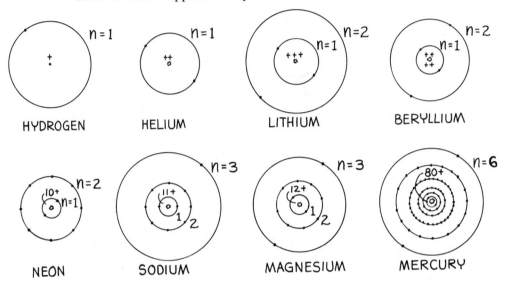

Fig. 30.6 Orbital models for some light and heavy atoms drawn to approximate scale. Note that the heavier atoms are not appreciably larger than the lighter atoms. (Adapted from *Descriptive College Physics* by Harvey E. White. © 1971 Litton Educational Publishing Inc. Reprinted by permission of D. Van Nostrand Co.)

We find that each element has an arrangement of electron orbits unique to that element. For example, the radii of the orbits for the sodium atom are the same for sodium atoms, but different from the radii of the orbits for other kinds of atoms. When we consider the 92 naturally occurring elements, we find that there are 92 distinct patterns or orbits, a different pattern for each element.

Bohr solved the mystery of atomic spectra while providing an extremely useful model of the atom. He was quick to point out that his model was to be interpreted as a crude beginning, and the picture of electrons whirling like planets about the sun was not to be taken literally (to which popularizers of science paid no heed). His discrete orbits were conceptual representations of an atom whose later description involved quantum mechanics. Still, his planetary model of the atom with electrons occupying discrete energy levels underlies the more complex models of the atom today.

Question   What fundamental force dictates the size of an atom?*

Answer   The electrical force.

Explanation of Quantized Energy Levels: Electron Waves

So we see that photons are emitted when electrons make a transition from a higher to a lower energy level, and the frequency of the photon is equal to the energy-level difference divided by Planck's constant, h. If an electron jumps through a large energy-level difference, the emitted photon has a high frequency—perhaps ultraviolet. If an electron jumps through a lesser energy difference, the emitted photon is lower in frequency— perhaps a brief burst of red light. Each element has its own characteristic energy levels; thus, transitions of electrons between these levels result in each element emitting its own characteristic colors. Each of the elements emits its own pattern of spectral lines.

The idea that electrons may occupy only certain levels was very perplexing to early investigators and to Bohr himself. It was perplexing because the electron was considered to be a particle, a tiny BB whirling around the nucleus like a planet whirling around the sun. Just as a satellite can orbit at any distance from the sun, it would seem that an electron should be able to orbit around the nucleus at any radial distance, depending, of course, like the satellite, on its speed. But it doesn't. It can't. Why the electron occupies only discrete levels is understood by considering the electron to be not a particle, but a *wave*.

Louis de Broglie introduced the concept of **matter waves** in 1924. De Broglie went on to show that the values of angular momentum in Bohr's orbits were a natural consequence of standing electron waves. A Bohr orbit exists where an electron wave closes in on itself in phase. In this way it reinforces itself constructively in each cycle, just as the wave on a music string is constructively reinforced by its successive reflections. In this view the electron is not thought of as a particle located at some point in the atom but as though its mass and charge were spread out into a standing wave surrounding the atomic nucleus, the wavelength of which must fit evenly into the circumferences of the orbits (Figure 30.7). The circumference of the innermost orbit is equal to one wavelength of the electron wave. The second orbit has a circumference of two electron wavelengths,

Fig. 30.7 (a) Orbital electrons form standing waves only when the circumference of the orbit is equal to an integral number of wavelengths. In (b) the wave does not close in on itself in phase and therefore undergoes destructive interference.

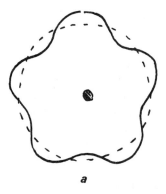

a

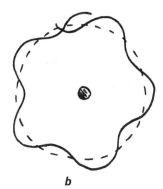

b

the third three, and so forth (Figure 30.8). This is similar to a "chain necklace" made of paper clips. No matter what size necklace is made, its circumference is equal to some multiple of the length of a single paper clip.[2] Since the circumferences of electron orbits are discrete, it follows that the radii of these orbits, and hence the energy levels, are also discrete.

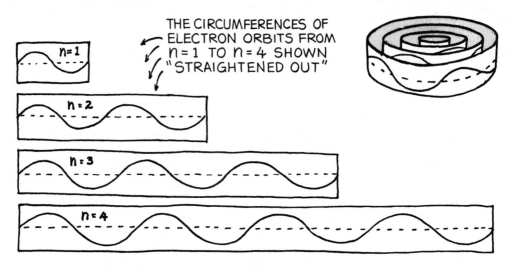

THE CIRCUMFERENCES OF
ELECTRON ORBITS FROM
$n = 1$ TO $n = 4$ SHOWN
"STRAIGHTENED OUT"

$n = 1$

$n = 2$

$n = 3$

$n = 4$

Fig. 30.8 The electron orbits in an atom have discrete radii because the circumferences of the orbits are integral multiples of the electron wavelengths. The wavelengths differ for different elements (and also for different orbits within the elements), resulting in discrete orbital radii, or energy levels, which are characteristic of each element. The figure is greatly oversimplified, as the standing waves make up spherical and ellipsoidal shells rather than flat, circular ones.

This view explains why electrons don't spiral closer and closer to the nucleus, resulting in atoms shrinking in on themselves. If each electron orbit is described by a standing wave, the circumference of the smallest orbit can be no smaller than one wavelength—no fraction of a wavelength is possible in a circular (or elliptical) standing wave.

Quantum Mechanics

The mid-1920s saw many changes in physics. Not only did light have particle properties, but particles were found to have wave properties. Starting with de Broglie's matter waves, the Austrian-German physicist Erwin Schrödinger formulated a wave equation that plays the same role in quantum mechanics that Newton's equation (force = mass × acceleration)

---

[2]Electron wavelengths are successively larger for orbits of increasing radii; so to make our analogy accurate, we would have to use not only *more* paper clips to make increasingly longer necklaces but *larger* paper clips for each longer necklace as well.

plays in classical physics.[3] Schrödinger's equation provides us with a purely mathematical rather than a visual model of the atom, which is therefore quite beyond the scope of this book. So our discussion of it will be brief.[4]

In **Schrödinger's wave equation**, the thing that "waves" is the non-material *matter wave amplitude*—a mathematical entity called a *wave function*, represented by the symbol ψ (the Greek letter psi). The wave function can be calculated from Schrödinger's equation and represents all the possibilities that can occur for a system. For example, the location of the electron in a hydrogen atom may be anywhere from the center of the nucleus to a radial distance of infinity. Its possible position and its probable position at a particular time are not the same. A physicist can calculate its probable position by multiplying the wave function by itself ($|\psi|^2$). This produces a second mathematical entity called a *probability function,* which tells us the probabilities at a given time(s) for each of the possibilities represented by ψ.

Experimentally, there is a finite probability of finding an electron at some particular location at some instant. The value of this probability must lie between the limits 0 and 1. For example, if the probability is 0.4 for finding the electron at the Bohr radius, this would signify a 40 percent chance of finding it there.

So the Schrödinger equation cannot tell a physicist where an electron in an atom *is* at any moment, but only where it is *likely to be.* An electron does not orbit the atomic nucleus at a fixed radius, as the early Bohr model suggested. If the electron's position in the first Bohr energy state for hydrogen is repeatedly measured and the results plotted in the form of dots, the resulting pattern will resemble a sort of electron cloud (Figure 30.9). An individual electron may at various times be detected anywhere in this probability cloud; it even has an extremely small but finite probability of momentarily existing inside the nucleus. It is detected most of its time, however, at an average distance from the nucleus, which fits the orbital radius described by Niels Bohr.

Most—but not all—physicists view quantum mechanics as fundamental to nature. Interestingly enough, Albert Einstein, one of the founders of quantum physics, never accepted it as fundamental; he considered the probabilistic nature of quantum phenomena as the outcome of a deeper physics. He stated, "Quantum mechanics is certainly imposing. But an

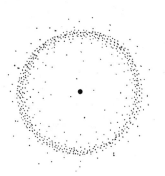

Fig. 30.9 Probability distribution of an electron cloud.

---

[3]Schrödinger's wave equation is

$$\left(-\frac{\hbar^2}{2m} \nabla^2 + V \right) \psi = i\hbar \frac{\partial \psi}{\partial t}$$

Believe it or not, this equation does not lend itself to a visual interpretation.

[4]Our short treatment of this complex subject is hardly conducive to any real understanding of quantum mechanics. At best it serves as a brief overview and possible introduction to further study. The reading suggested at the end of the chapter may be quite useful.

inner voice tells me it is not yet the real thing. The theory says a lot, but does not really bring us closer to the secret of 'the Old One'."

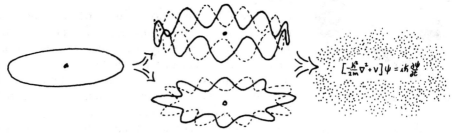

Fig. 30.10 From the Bohr model of the atom to the modified model with de Broglie waves to the mathematical model of Schrödinger.

Correspondence Principle

If a new theory is valid, it must account for the verified results of the old theory. This is the **correspondence principle**, first enunciated by Bohr. New theory and old must correspond; that is, they must overlap and agree in the region where the results of the old theory have been fully verified.[5]

When the techniques of quantum mechanics are applied to macroscopic systems rather than atomic systems, the results are essentially identical to those of classical mechanics. For a large system such as the motion of planets where classical physics is successful, the Schrödinger equation leads to results that differ only by infinitesimal amounts from classical theory. The two domains blend when the de Broglie wavelength is small compared to the dimensions of the system. Classical physics is less cumbersome and more useful when a system's dimensions are large compared with atomic dimensions. But at the atomic level, quantum physics reigns.

So we see that the classical physics of Newton is smoothly linked with quantum physics as it extends into the domain of the very small. We will consider the very large in later chapters and see that Newtonian physics links with relativity theory in the domain of the very massive and/or the very fast.

---

[5]This is a general rule not only for good science, but for all good theory—even in areas as far removed from science as government, religion, and ethics.

Summary of Terms

Ritz combination principle   The spectral lines of the elements have frequencies that are either the sums or the differences of the frequencies of two other lines.

de Broglie matter waves   All particles of matter have associated wave properties. The wavelength of a particle wave is related to its momentum and Planck's constant, h, by the relationship

$$\text{Wavelength} = \frac{h}{\text{momentum}}$$

Quantum mechanics   The branch of quantum physics that deals with finding the probability amplitudes of matter waves, organized principally by Werner Heisenberg (1925) and Erwin Schrödinger (1926).

Schrödinger wave equation   The fundamental equation of quantum mechanics which interprets the wave nature of material particles in terms of probability wave amplitudes. It is as basic to quantum mechanics as Newton's laws of motion are to classical mechanics.

Correspondence principle   The rule that a new theory is valid provided that, when it overlaps with the old, it agrees with the verified results of the old theory.

Suggested Reading

Cline, Barbara L. *The Questioners: Physicists and the Quantum Theory*. New York: Crowell, 1973. A fascinating account of the development of quantum physics with emphasis on the participating physicists.

Pagels, H. R. *The Cosmic Code: Quantum Physics as the Language of Nature*. New York: Simon & Schuster, 1982. An elegant and highly recommended book for the general reader.

Wolf, F. A. *Taking the Quantum Leap: The New Physics for Nonscientists*. San Francisco: Harper & Row, 1981. Another very readable and enjoyable book.

Review Questions

1. Distinguish between *atomic physics* and *nuclear physics*.

2. Are the atomic models of Rutherford and Bohr compatible? Explain.

3. What is the Ritz combination principle?

4. How does the Bohr model of the atom account for the Ritz combination principle?

5. Did Bohr state that atoms are like tiny solar systems? Are they?

6. How can we explain why electrons don't spiral into the attracting nucleus?

7. Why are heavy atoms not appreciably larger than the hydrogen atom?

8. What evidence reveals the wave nature of electrons?

9. Distinguish between a *wave function* and a *probability function*.

10. What is it that "waves" in the Schrödinger wave equation?

11. Why is the correspondence principle a test for the validity of an idea?

12. Would it be correct to say that an electron is a wave or a particle? Or would it be correct to say that an electron, whatever it is, *behaves* as if it were a wave or a particle?

Exercises

1. How does Rutherford's model of the atom account for the back-scattering of alpha particles directed at the gold leaf?

2. Why are spectral lines often referred to as "atomic fingerprints"?

3. Uranium is 238 times more massive than hydrogen. Why, then, isn't the diameter of the uranium atom 238 times that of the hydrogen atom?

4. If all atoms of the same kind were not of the same size, would crystal structure be affected? How?

5. How does the wave model of electrons orbiting the nucleus account for discrete energy values rather than arbitrary energy values?

6. How can a hydrogen atom, which has only one electron, have so many spectral lines?

7. A hypothetical atom possesses four distinct energy levels. Assuming that all transitions between levels are possible, how many spectral lines will this atom exhibit? Which transitions correspond to the highest-energy light emitted? To the lowest-energy light?

8. An electron de-excites from the fourth quantum level to the third and then directly to the ground state. Two photons are emitted. How do their combined energies compare to the energy of the single photon that would be emitted by de-excitation from the fourth level directly to the ground state?

9. A helium atom is more massive than a hydrogen atom. But why is it smaller than a hydrogen atom?

10. Why does no stable electron orbit exist in an atom for a circumference of $2\frac{1}{2}$ de Broglie wavelengths?

11. Distinguish between the *wavelength* and the *amplitude* of a matter wave.

12. If Planck's constant, *h,* were larger, would atoms be larger also? Defend your answer.

13. If the world of the atom is so uncertain, and subject to the laws of probabilities, how can we accurately measure such things as light intensity, electric current, and temperature?

14. Why do we say that light has wave properties? Why do we say that light has particle properties?

15. Why do we say that electrons have particle properties? Why do we say that electrons have wave properties? Is there a contradiction here? Explain.

16. Comment on the statement, "If an electron is not a particle, then it must be a wave." (Do you hear statements like this often?)

17. What principle supports the fact that quantum physics reduces to classical physics for large numbers of quanta?

18. Is the correspondence principle relevant to the everyday macro world?

19. Explain how Newton's first and second laws of motion satisfy the correspondence principle.

20. Richard Feynman in his book *The Character of Physical Law* states: "A philosopher once said, 'It is necessary for the very existence of science that the same conditions always produce the same results.' Well, they don't!" Who was speaking of classical physics, and who was speaking of quantum physics?

31 The Atomic Nucleus and Radioactivity

In the last chapter we were concerned with the clouds of electrons that make up the atom—atomic physics. In this chapter we will burrow beneath the electrons and go deeper into the atom—to the atomic nucleus. The study of the nucleus is connected to the chance discovery of radioactivity in 1896, which in turn is connected to the discovery of X rays two months earlier. We will begin our study of nuclear physics by considering X rays.

X Rays and Radioactivity

Before the turn of the century the German physicist Wilhelm Roentgen discovered a "new kind of ray" produced by an electron beam striking a glass surface. He named these *X rays,* that is, rays of an unknown nature. Roentgen further discovered that a beam of electrons directed against a metal plate produced X rays that could pass through solid materials. Today we know that X rays are high-frequency electromagnetic waves, some of which are emitted by the excitation of the innermost orbital electrons of atoms. Whereas the electron current in a fluorescent lamp excites the outer electrons of atoms and produces visible photons, a more energetic beam of electrons excites the innermost electrons and produces higher-frequency photons of X radiation.

X-ray photons have high energy and can pass through many layers of atoms before being absorbed or scattered. X rays do this when they pass through you to produce pictures of the inside of your body (Figure 31.1).

Fig. 31.1 X rays emitted by excited metallic atoms in the electrode pass through the hand and expose the film.

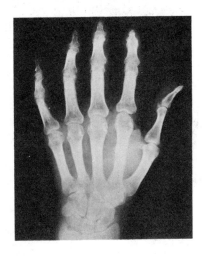

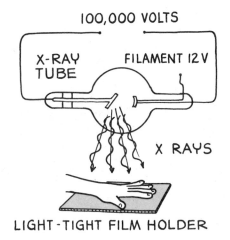

100,000 VOLTS

X-RAY TUBE

FILAMENT 12 V

X RAYS

LIGHT-TIGHT FILM HOLDER

Two months after Roentgen discovered X rays, the French physicist Antoine Henri Becquerel was trying to find out whether any elements spontaneously emitted X rays. To do this, he wrapped a photographic plate in black paper to keep out the light and then put pieces of various elements against the wrapped plate. He thought that if these materials emitted X rays, the rays would go through the paper and blacken the plate. He found that although most elements produced no effect, uranium did give out rays. It was soon discovered that similar rays are emitted by other elements, such as thorium, actinium, and by two new elements discovered by Pierre and Marie Curie, polonium and radium. The emission of these rays was evidence of much more drastic changes in the atom than atomic excitation. These rays were not the result of changes in the electron energy states of the atom, but of changes occurring within the central atomic core—the nucleus. These rays were the result of a spontaneous chipping apart of the atomic nucleus—*radioactivity*.

Alpha, Beta, and Gamma Rays

The radioactive elements emit three distinct types of rays, called by the first three letters of the Greek alphabet, α, β and γ—*alpha, beta,* and *gamma,* respectively. **Alpha rays** have a positive electrical charge, **beta rays** have a negative charge, and **gamma rays** have no charge at all. The three rays are separated by putting a magnetic field across their paths (Figure 31.2). Further investigation has shown that an alpha ray is a stream of helium nuclei, and a beta ray is a stream of electrons. Hence, we often call these **alpha particles** and **beta particles**. A gamma ray is electromagnetic radiation (a stream of photons) whose frequency is higher than that of X rays. Gamma photons provide information about nuclear structure much as visible and X-ray photons give information about atomic electron structure. Gamma photon energies indicate the differences between nuclear energy states as optical photon energies indicate differences in electron energy states in an atom.

Fig. 31.2 The bending of a narrow beam of rays in a magnetic field. The beam comes from a radioactive source placed at the bottom of a hole drilled in a lead block.

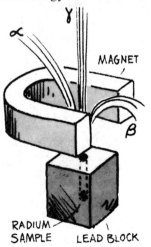

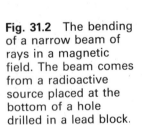

The Nucleus

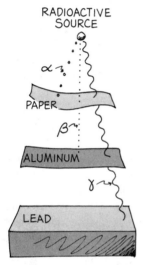

RADIOACTIVE
SOURCE

PAPER

ALUMINUM

LEAD

Fig. 31.3 Alpha particles are the least penetrating and can be stopped by a few sheets of paper. Beta particles will readily pass through paper, but not through a sheet of aluminum. Gamma rays penetrate several centimeters into solid lead.

Study of these rays, together with much other research, indicates that the atom consists of a tiny nucleus surrounded by a number of orbiting electrons as described in earlier chapters. The atomic nucleus occupies only a few quadrimillionths the volume of the atom. Thus, the atom is mostly empty space. The nucleus is composed of **nucleons**, which when electrically charged are protons and when electrically neutral are neutrons. The neutron's mass is slightly greater than the proton's. Nucleons have nearly 2000 times the mass of electrons, so the mass of an atom is practically equal to the mass of its nucleus. The positive charge of the proton is the same in magnitude as the negative charge of the electron. When an electron is ejected from the nucleus (beta emission), a neutron becomes a proton.

Nuclear radii range from about 10^{-13} centimeter for hydrogen to about six times as large for uranium. The shapes of nuclei are generally spherical but deviate from that shape sometimes in the "football" way and sometimes in the "doorknob" way. The protons and neutrons tend to cluster into alpha particles, making a generally lumpy surface. A "skin" of neutrons is thought to make up the outer portion of the heavier nuclei. Just as there are energy levels for the orbital electrons of an atom, there are energy levels within the nucleus. Whereas electrons making transitions to lower orbits emit photons of light, similar changes of energy states within the nucleus result in the emission of gamma-ray photons. The emission of alpha particles can be understood from the viewpoint of quantum mechanics. Just as orbital electrons form a probability cloud about the nucleus, inside the radioactive nucleus there is a similar probability cloud for alpha particles (and other particles as well). Probability is extremely high that the alpha particle will be inside the nucleus, but if the probability wave extends outside, there is a finite chance that the alpha particle will be outside. Once outside, it is hurled violently away by electrical repulsion.

In addition to alpha, beta, and gamma rays, more than 200 various other particles have been detected coming from the nucleus when clobbered by energetic particles. We do not think of these so-called elementary particles as being buried within the nucleus and then popping out, just as we do not think of a spark as being buried in a match before it is struck. They come into being when the nucleus is disrupted. Regularities in the masses of these particles as well as the particular characteristics of their creation can be boiled down to the simple motions and interactions of the members of a fairly compact family of subnuclear particles called **quarks**. It is generally believed that protons and neutrons are each composed of three quarks. Like magnetic poles, none has been isolated and experimentally observed. We believe today that quarks are the elementary particles of which atomic nuclei are made. Evidence for quarks is circumstantial but quite overwhelming. This is a frontier of knowledge and the area of much current excitement and research.

Isotopes

The nucleus of a hydrogen atom consists of a single proton. Helium has two protons, lithium has three, and so forth. In neutral atoms, there are as many protons in the nucleus as there are electrons outside the nucleus. The number of protons in the nucleus is the same as the atomic number. Every succeeding element in the list of elements has one more proton than the preceding element.

The number of neutrons in the nucleus does not follow as simple a rule. In a given element, the number of neutrons may vary somewhat. For example, every atom of chlorine has 17 protons and hence 17 orbital electrons that determine what chemical compounds it can make. But the number of accompanying neutrons varies. Atoms that have like numbers of protons but unlike numbers of neutrons are called **isotopes**. The two chief types of chlorine isotopes have 35 and 37 times the mass of a single nucleon. We denote these by $_{17}^{35}Cl$ and $_{17}^{37}Cl$, where the numbers refer to the **atomic number** and the **atomic mass number** (approximately but not exactly equal to the *atomic weight*).

The mass number corresponds to the total number of nucleons in the nucleus. Since for both isotopes 17 of these are protons, the lighter isotope has $35 - 17 = 18$ neutrons, and the heavier isotope has $37 - 17 = 20$ neutrons. In nature the isotopes of chlorine are found mixed, with three times as many $_{17}^{35}Cl$ atoms as there are $_{17}^{37}Cl$ atoms, so the average atomic mass of naturally occurring chlorine is about 35.5. Chlorine is only one of many elements found in nature that consist of two or more stable isotopes. Even hydrogen, the lightest element, has three known isotopes (Figure 31.4). The double-weight hydrogen isotope $_1^2H$ is called *deuterium*. "Heavy water" is the name usually given to H_2O in which the H's are deuterium atoms. The triple-weight hydrogen isotope $_1^3H$, which is not stable but lives long enough to be a known constituent of atmospheric water, is called *tritium*. There are three isotopes of uranium that naturally occur in the earth's crust, the most common being $_{92}^{238}U$. In briefer notation we can drop the atomic number and more simply say U-238. (We will see in the next chapter that it is the isotope U-235 that undergoes nuclear fission.) Of the 83 elements present on earth in significant amounts, 20 are of a single stable configuration. The others have from 2 to 10 stable isotopes. Taking radioactive isotopes into account, over 1300 distinct isotopes are known.

Fig. 31.4 Three isotopes of hydrogen. Each nucleus has a single proton, which holds a single orbital electron, which, in turn, determines the chemical properties of the atom. The different number of neutrons changes the mass of the atom but not its chemical properties.

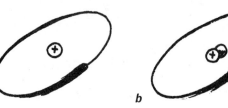

a   b   c

Why Atoms Are Radioactive

The positively charged and closely spaced photons in a nucleus have huge electrical forces of repulsion between them. Why don't they fly apart because of this huge repulsive force? Because there is an even more formidable force within the nucleus—the nuclear force. Both neutrons and protons are bound together by this awesome attractive force. The nuclear force is much more complicated than the electrical force and is only recently becoming understood. The principal part of the nuclear force, the part that holds the nucleus together, is called the *strong interaction*.[1] The strong interaction is an attractive force that acts between protons, neutrons, and particles called *mesons*, all of which are called *hadrons*. The strong interaction between hadrons depends on the distance between them and their velocities and spins. This interaction acts over only a very short distance. It is very strong between nucleons less than 10^{-15} meter apart, but close to zero at greater separations. So the strong nuclear interaction is a short-range force. Electrical interaction, on the other hand, weakens as the inverse square of separation distance and is a relatively long-range force. So as long as protons are close together, as in small nuclei, the nuclear force easily overcomes the electrical force of repulsion. But for distant protons, like those on opposite edges of a large nucleus, the attractive nuclear force may be small in comparison to the repulsive electrical force. Hence, a larger nucleus is not as stable as a smaller nucleus (Figure 31.5).

The presence of the neutrons also plays a large role in nuclear stability. Although protons both attract and repel, neutrons only attract, and therefore may be considered a sort of "nuclear cement." Many of the first 20 or so elements have equal numbers of neutrons and protons.

Even more "nuclear cement" is needed by the heavier elements. In fact, most of the mass of the heavier elements is composed of neutrons. For example, in U-238, which has 92 protons, there are 146 (238 − 92) neutrons. If the uranium nucleus were to have equal numbers of protons and neutrons, 119 protons and 119 neutrons, it would fly apart at once because of the electrical repulsion forces. Relative stability is found only if 27 of these protons are replaced by neutrons. This greater number of neutrons is required to compensate for the repulsion between distant protons. Even so, the U-238 nucleus is still unstable because of the electrical forces.

To put the matter in another way: there is an electrical repulsion between *every pair* of protons in the nucleus, but there is not a substantial nuclear attractive force between every pair (Figure 31.5). Every proton in

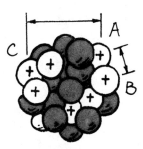

Fig. 31.5 Proton A both attracts (nuclear force) and repels (electrical force) proton B, but mainly repels proton C (because the nuclear attraction is weaker at greater distances). The greater the distance between A and C, the more unstable the nucleus. (The shaded particles are neutrons.)

---

[1]Fundamental to the strong interaction is what is called the *color force* (which has nothing to do with visible color). This color force interacts between quarks and holds them together by the exchange of "gluons." Read more about this in H. R. Pagels, *The Cosmic Code: Quantum Physics as the Language of Nature* (New York: Simon & Schuster, 1982).

the uranium nucleus exerts a repulsion on each of the other 91 protons—those near and those far. However, each proton (and neutron) exerts an appreciable nuclear attraction only on those nucleons that happen to be near it.

We find that all nuclei having more than 82 protons are unstable. In this unstable environment, alpha and beta emissions take place. A nuclear force distinct from the strong force is responsible for beta emission. This weaker nuclear force is called the *weak interaction* and principally affects lighter particles like electrons and still lighter particles called *neutrinos*. These belong to the class of particles called *leptons*. The mechanism for the emission of leptons goes beyond electrical repulsion and involves a quantum-mechanical explanation beyond the scope of this book.

Half-Life

The radioactive decay rate of an element is measured in terms of a characteristic time, the **half-life**. The half-life of a radioactive material is the time needed for half of the active atoms of any given quantity to decay. Radium-226, for example, has a half-life of 1620 years. This means that half of any given specimen of radium-226 will be converted into some other element by the end of 1620 years. In the next 1620 years, half of the remaining radium will decay, leaving only one-fourth the original number of radium atoms. The other three-fourths are converted, by a succession of disintegrations, to lead. After 20 half-lives, an initial quantity of radioactive atoms will be diminished 1 millionfold. The isotopes of some ele-

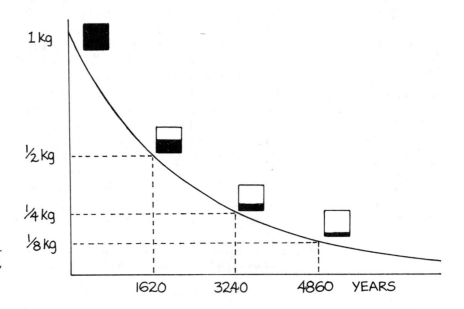

Fig. 31.6 Every 1620 years the amount of radium decreases by half, but some always remains.

ments have a half-life of less than a millionth of a second, while uranium-238, for example, has a half-life of $4\frac{1}{2}$ billion years. The isotopes of each radioactive element have their own characteristic half-lives.

The rates of radioactive decay are remarkably constant and are not affected by any external conditions, however drastic. Wide temperature and pressure extremes, strong electric and magnetic fields, and even violent chemical reactions have no detectable effect on the rate of decay of a given element. Any of these stresses, although severe by ordinary standards, is far too mild to affect the nucleus in the deep interior of the atom.

How do we measure half-lives? Observing a specimen and waiting until half of it is decayed won't work if the half-life is long. The half-life of a radioactive substance is related to its rate of disintegration. In general, the shorter the half-life of an element, the faster is the rate of disintegration and the more active is the element. The half-life is thus often computed from the rate of disintegration, which can be measured in the laboratory.

Half-lives can also be estimated with the use of quantum mechanics. The kinetic energy of an emerging alpha particle tells us its wavelength, which in turn gives its probability of leaking from the nucleus. This information enables the prediction of the half-life of the parent nucleus with some success.

Questions

1. If a sample of radioactive isotopes has a half-life of 1 day, how much of the original sample will be left at the end of the second day?*

2. Which will give a higher reading on a Geiger counter—radioactive material that has a short half-life or radioactive material with a long half-life?†

Radiation Detectors

Ordinary thermal motions of atoms jostling against one another in a gas or liquid are not energetic enough to dislodge electrons, and the atoms remain neutral. But when an energetic particle such as an alpha or a beta particle shoots through matter, electrons one after another are knocked from the target atoms, leaving a trail of freed electrons and positively charged ions. This ionization process is responsible for the harmful effects of high-energy radiation in living cells. It also makes it relatively easy to trace the paths of high-energy particles. We will briefly discuss five radiation detection devices.

*****Answer**   One-quarter.

†**Answer**   The material with the shorter half-life is more active and will give a higher reading on a radiation detector.

1. A Geiger counter consists of a central wire in a hollow metal cylinder filled with gas (Figure 31.7). An electrical voltage is applied across the cylinder and wire so that the wire is more positive than the cylinder. If radiation enters the tube and ionizes an atom in the gas, the freed electron is attracted to the positively charged central wire. This electron collides with other atoms and knocks out more electrons, resulting in a cascade of electrons moving toward the wire. This makes a short pulse of electric current, which activates a counting device connected to the tube.

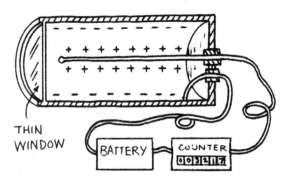

Fig. 31.7 A cutaway view of a Geiger tube.

2. A cloud chamber shows a visible path of ionizing radiation in the form of fog trails. It consists of a cylindrical glass chamber closed at the upper end by a glass window and at the lower end by a movable piston. Water vapor or alcohol vapor in the chamber can be saturated by adjusting the piston.

The radioactive sample is placed inside the chamber as shown in Figure 31.8, or it can be placed outside the thin glass window. When radiation passes through the chamber, ions are produced along the paths. If the saturated air in the chamber is then suddenly cooled, tiny droplets of moisture condense about these ions and form vapor trails showing the paths of the radiation. These are the atomic versions of the ice-crystal trails left in the sky by jet planes.

The cooling of the chamber is accomplished by moving the piston downward suddenly, which results in a sudden expansion of the air in the chamber. Even simpler is the continuous cloud chamber that has a steady supersaturated vapor because it sits on a slab of dry ice so there is a temperature gradient from near room temperature at the top to very low at the bottom. In either version the fog tracks that form are illuminated with a lamp and may be seen or photographed through the glass top. The chamber may be placed in a strong electric or magnetic field, which will bend the paths in a manner that provides information about the charge, mass, and momentum of the radiation particles. Positively and negatively charged particles will bend in opposite directions.

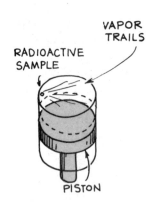

Fig. 31.8 The cloud chamber.

Fig. 31.9 The 72-inch hydrogen bubble chamber at the Lawrence Radiation Laboratory. Liquid hydrogen fills the bottom of the chamber. The rays to be studied enter the chamber through a window.

3. The particle trails seen in a **bubble chamber** are minute bubbles of gas in liquid hydrogen (Figure 31.10). The liquid hydrogen is heated under pressure in a glass chamber to a point just short of boiling. If the pressure

Fig. 31.10 Tracks of elementary particles in a bubble chamber. Two particles have been destroyed where the spirals emanate, and four others created in the collision.

in the chamber is suddenly released at the moment an ion-producing particle enters, a thin trail of bubbles is left along the particle's path. All the liquid erupts to a boil, but in the few thousandths of a second before this happpens, photographs are taken of the particle's brief-lived trail. Like the cloud chamber, the bubble chamber reveals the charge and relative mass of the particles being studied. Bubble chambers have been widely used by researchers in recent decades, but presently there is greater interest in spark chambers.

4. A **spark chamber** is a counting device that consists of an array of closely spaced parallel plates; every other plate is grounded, and plates in between are maintained at a high voltage (about 10 kilovolts). Ions are produced in the gas between the plates as charged particles pass through the chamber. Discharge along the ionic path produces a visible spark between pairs of plates. Many sparks reveal the path of the incident particle. A different design called a *streamer chamber* consists of only two widely spaced plates, between which an electric discharge or "streamer" closely follows the path of the incident charged particle. The principal advantage of spark and streamer chambers over the bubble chamber is that more events can be monitored in a given time.

5. A **scintillation counter** uses the fact that certain substances are easily excited and emit light when charged particles or gamma rays pass through them. Tiny flashes of light or scintillations are converted into electric signals by means of special photo-multiplier tubes. A scintillation counter is much more sensitive to gamma rays than a Geiger counter and, in addition, can measure the energy of charged particles or gamma rays absorbed in the detector.

Natural Transmutation of Elements

When a nucleus emits an alpha or a beta particle, a different element is formed. The changing of one chemical element to another is called **transmutation**. Consider uranium, for example. Uranium has 92 protons. When an alpha particle is ejected, the nucleus is reduced by two protons and two neutrons (an alpha particle is a helium nucleus consisting of two protons and two neutrons). The 90 protons and 144 neutrons left behind are then the nucleus of a new element. This element is *thorium*. We can express this reaction as

$$^{238}_{92}U \rightarrow {}^{234}_{90}Th + {}^{4}_{2}He$$

An arrow is used here to show that the $^{238}_{92}U$ changes into the other elements. When this happens, energy is released, partly in the form of gamma radiation, partly in the kinetic energy of the alpha particle ($^{4}_{2}He$), and partly in the kinetic energy of the thorium atom. In this equation, notice that the mass numbers at the top balance (238 = 234 + 4) and that the atomic numbers at the bottom also balance (92 = 90 + 2).

Thorium-234, the product of this reaction, is also radioactive. When it decays, it emits a beta particle. Recall that a beta particle is an electron—not an orbital electron, but one from the nucleus—from one of the neutrons. In a loose sense, we can think of a neutron as a combined proton and electron; so when the neutron emits an electron, it becomes a proton. A neutron is ordinarily stable when it is locked in the nucleus of an atom, but a free neutron is radioactive and has a half-life of 10.6 minutes. It decays into a proton by beta emission.[2] So in the case of thorium, which has 90 protons, beta emission leaves it with one less neutron and one more proton. The new nucleus then has 91 protons and is no longer thorium, but the element *protactinium*. Although the atomic number has increased by 1 in this process, the mass number (protons + neutrons) remains the same. The nuclear equation is

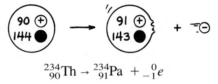

$$^{234}_{90}\text{Th} \rightarrow {}^{234}_{91}\text{Pa} + {}^{0}_{-1}e$$

We write an electron as $_{-1}^{0}e$. The 0 indicates that its mass is insignificant when compared to that of the protons and neutrons that alone contribute to the mass number. The −1 is the charge of the electron. Remember that this electron is a beta particle from the nucleus and not an electron from the surrounding electron cloud.

From the previous two examples, we can see that when an element ejects an alpha particle from its nucleus, the mass number of the resulting atom decreases by 4, and its atomic number *decreases* by 2. The resulting atom belongs to an element two spaces back in the periodic table (see the

---

[2] Beta emission is always accompanied by the emission of a *neutrino*, a neutral particle with about zero mass that travels at about the speed of light. The neutrino (little neutral one) was predicted from theoretical calculations by Wolfgang Pauli in 1930 and detected in 1956. Neutrinos are hard to detect because they interact very weakly with matter. To capture a neutrino is extremely difficult. Whereas a piece of solid lead a few centimeters thick will stop most gamma rays from a radium source, it would take a piece of lead about 8 light-years thick to stop half the neutrinos produced in typical nuclear decays. Thousands of neutrinos are flying through you every second of every day, because the universe is filled with them. Only occasionally, one or two times a year or so, does a neutrino or two interact with the matter of your body.

At this writing, whether or not the neutrino has mass is not known. Present speculation is that if neutrinos do have any mass, they are so numerous that they may comprise about 90 percent of the mass of the universe—enough to halt the present expansion and ultimately close the cycle from the Big Bang to the Big Crunch. Neutrinos may be the "glue" that holds the universe together.

inside front cover). When an element ejects a beta particle (electron) from its nucleus, the mass of the atom is practically unaffected so there is no change in mass number, but its atomic number *increases* by 1. The resulting atom belongs to an element one place forward in the periodic table. Gamma emission results in no change in either the mass number or the atomic number. So we see that radioactive elements can decay backward or forward in the periodic table.[3]

The radioactive decay of $^{238}_{92}$U to $^{206}_{82}$Pb, an isotope of lead, is shown in Figure 31.11. The steps in the decay process are shown in the diagram, where each nucleus that plays a part in the series is shown by a burst. The vertical column containing the burst shows its atomic number, and the horizontal column shows its mass number. Each arrow that slants downward toward the left shows an alpha decay, and each arrow that points to the right shows a beta decay. Notice that some of the nuclei in the series can decay in both ways. This is one of several similar radioactive series that occur in nature.

Artificial Transmutation of Elements

The alchemists of old tried vainly for over two thousand years to cause the transmutation of one element to another. Enormous efforts were expended and elaborate rituals performed in the quest to change lead into gold. They never succeeded. Lead in fact can be changed to gold but not by the chemical means employed by the alchemists. Chemical reactions involve alterations of the outermost shells of the electron clouds of atoms and molecules. To change an element from one kind to another, one must go deep within the electron clouds to the central nucleus, which is immune to the most violent chemical reactions. To change lead to gold, three positive charges must be extracted from the nucleus. Ironically enough, transmutations of atomic nuclei were constantly going on all around the alchemists as they are around us today. Radioactive decay of minerals in rocks has been occurring since their formation. But this was unknown to the alchemists, and even if they had discovered these radiations, they would most likely not have recognized them as evidence of atomic transmutation.

The alchemists did not know about the tiny central nucleus and the surrounding electrons. These whirling electrons provide a sort of atomic skin similar to the apparent disk formed by an electric fan. Only at high speed can a charged particle pass through the electron cloud and enter the atom, like a high-speed bullet passing between the blades of a fan. A slow particle drifting toward an atom is deviated, much as a marble tossed slowly into a spinning fan is knocked back. Had the alchemists used the

[3]Sometimes a nucleus emits a *positron*, which is the "antiparticle" of an electron. In this case, a proton becomes a neutron, and the atomic number is decreased.

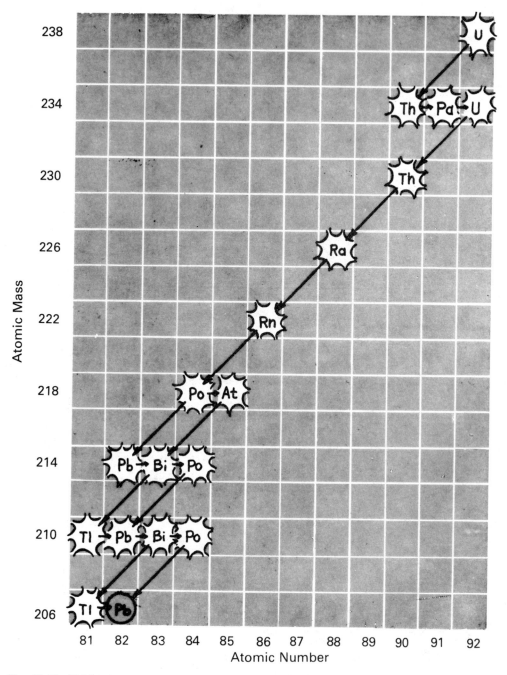

Fig. 13.11 U-238 decays to Pb-206 througn a series of alpha and beta decays.

high-speed particles ejected from radioactive ores as bullets, they would have succeeded in transmuting some of the atoms in a substance. But the atoms so transmuted would most likely have escaped their notice.

Ernest Rutherford, in 1919, was the first of many investigators to succeed in transmuting a chemical element. He bombarded nitrogen nuclei with alpha particles and succeeded in transmuting nitrogen into oxygen:

$$^{14}_{7}\text{N} + ^{4}_{2}\text{He} \rightarrow ^{17}_{8}\text{O} + ^{1}_{1}\text{H}$$

His source of alpha particles was a radioactive piece of ore. From a quarter-of-a-million cloud-chamber tracks photographed on movie film, he showed seven examples of atomic transmutation. Tracks bent by a strong external magnetic field showed that when an alpha particle collided with a nitrogen atom, a proton bounced out and a heavy atom recoiled a short distance. The alpha particle disappeared. The alpha particle was absorbed in the process, transforming nitrogen to oxygen.

Since Rutherford's announcement in 1919, we have made many such nuclear reactions, first with natural bombarding projectiles from radioactive ores and then with still more energetic projectiles, protons and electrons hurled by giant atom-smashing accelerators. Artificial transmutation has been used to produce the hitherto-unknown synthetic elements from atomic number 93 to 109, the first eleven of which are named *neptunium, plutonium, americium, curium, berkelium, californium, einsteinium, fermium, mendelevium, nobelium,* and *lawrencium.* The purported discoverers of elements 104, 105, and 106 have recommended the names *rutherfordium (Rf)* for element 104, and *hahnium (Ha)* for 105. Russian researchers, who also claim the discovery of these elements, have proposed different names. All these artificially made elements have short half-lives. If they ever existed naturally when the earth was formed, they have long since decayed.

Radioactive Isotopes

All elements have isotopes. Some of the isotopes of a given element may be radioactively stable, while most are quite radioactive. Radioactive isotopes of all the elements have been made by the bombardment with neutrons and other particles. As such isotopes have become available and inexpensive, their use in scientific research and industry has expanded tremendously.

A small amount of radioactive isotopes can be mixed with fertilizer before being applied to growing plants. Once the plants are growing, we can easily measure how much fertilizer they have taken up, by using a radiation detector. From such measurements, farmers know the proper

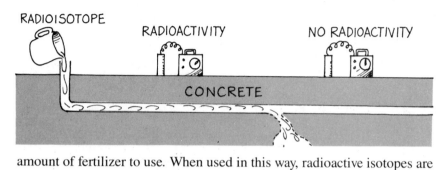

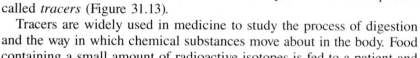

Fig. 31.12 Tracking pipe leaks with radioactive isotopes.

amount of fertilizer to use. When used in this way, radioactive isotopes are called *tracers* (Figure 31.13).

Tracers are widely used in medicine to study the process of digestion and the way in which chemical substances move about in the body. Food containing a small amount of radioactive isotopes is fed to a patient and traced through the body with a radiation detector. The same method can be used to study the circulation of the blood. Large amounts of radioactive iodine taken into the body are used to combat cancer of the thyroid gland. Since iodine tends to collect in the thyroid gland, the radioactive isotopes lodge where they can destroy the malignant cells. Radioactive isotopes are increasingly becoming a powerful tool for the modern biologist and physician.

Fig. 31.13 Radioisotopes are used to check the action of fertilizers in plants and the progress of food in digestion.

Engineers can study how the parts of an automobile engine wear away during use by making the cylinder walls radioactive. While the engine is running, the piston rings rub against the cylinder walls. The tiny particles of radioactive metal that are worn away fall into the lubricating oil where they can be measured with a radiation detector. By repeating this test with different oils, the engineer can determine which oil gives the least wear and longest life to the engine.

Tire manufacturers also employ radioactive isotopes. If a known fraction of the carbon atoms used in an automobile tire is radioactive, the amount of rubber left on the road when the car is braked can be estimated through a count of the radioactive atoms.

There are hundreds more examples of the use of radioactive isotopes. The important thing is that this technique provides a way to detect and count atoms in samples of materials too small to be seen with a microscope.

Carbon Dating

A continuous bombardment of cosmic rays on the earth's atmosphere results in the transmutation of many atoms in the upper atmosphere. Protons and neutrons are scattered throughout the atmosphere. Most of the protons quickly capture stray electrons and become hydrogen atoms in the upper atmosphere, but the neutrons keep going for long distances because they have no charge and therefore do not interact electrically with matter. Sooner or later many of them collide with the nuclei of atoms in the lower atmosphere. If they are captured by the nucleus of a nitrogen atom, the following reaction takes place:

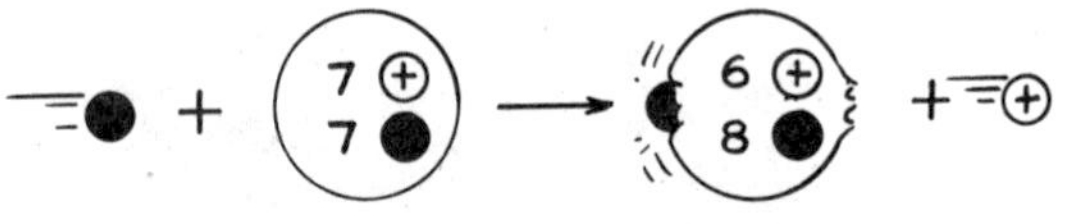

$${}^{14}_{7}\text{N} + {}^{1}_{0}n \rightarrow {}^{14}_{6}\text{C} + {}^{1}_{1}\text{H}$$

The nitrogen nucleus breaks up into an isotope of carbon by emitting a proton ${}^{1}_{1}\text{H}$. Most of the carbon that exists is the stable carbon-12, ${}^{12}_{6}\text{C}$. But because of the cosmic-ray bombardment, less than one-millionth of 1 percent of the carbon in the atmosphere is carbon-14. Both of these isotopes join with oxygen and become carbon dioxide, which is breathed by plants. Hence, all plants have a tiny bit of radioactive carbon-14 in them. All animals eat plants (or eat plant-eating animals), and therefore have a little carbon-14 in them. All living things contain some carbon-14.

Carbon-14 is a beta emitter and decays back into nitrogen by the following reaction:

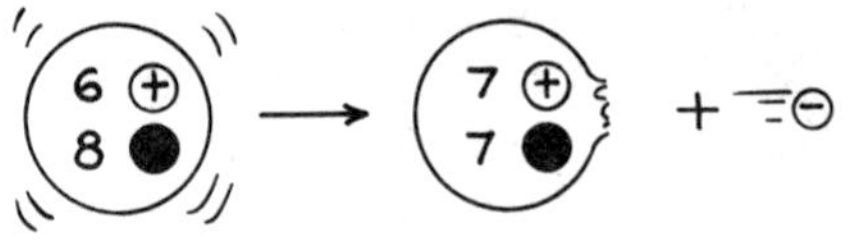

$${}^{14}_{6}\text{C} \rightarrow {}^{14}_{7}\text{N} + {}^{0}_{-1}e$$

So the carbon-14 in living things changes into stable nitrogen. But because living things breathe, this decay is accompanied by a replenishment of carbon-14, and a radioactive equilibrium is reached where there is a fixed ratio of carbon-14 to carbon-12. When a plant or an animal dies, however, replenishment stops. The percentage of carbon-14 steadily decreases—at a known rate. The longer an organism is dead, the less carbon-14 remains.

The half-life of carbon-14 is about 5730 years, which means that half the carbon-14 atoms present in a body decay in that time. Half the remaining carbon-14 atoms decay in the following 5730 years, and so forth. The radioactivity of living things therefore gradually decreases at a steady rate after they die (Figure 31.14).

Fig. 31.14 The radioactive carbon isotopes in the skeleton diminish by one-half every 5730 years.

20,945 B.C.   15,215 B.C.   9485 B.C.   3755 B.C.   1985 A.D.

We can measure the radioactivity of plants and animals today and compare this with the radioactivity of ancient organic matter. If we extract a small, but precise, quantity of carbon from an ancient wooden ax handle, for example, and find it has one-half as much radioactivity as an equal quantity of carbon extracted from a living tree, then the old wood must have come from a tree that was cut down or made from a log that died 5730 years ago. In this way, we can probe as much as 50,000 years into the past to find out such things as the age of ancient civilizations or when the ice ages covered the earth.

Carbon dating would be an extremely simple and accurate dating method if the amount of radioactive carbon in the atmosphere had been constant over the ages. But it hasn't been. Fluctuations in the sun's magnetic field as well as changes in the strength of the earth's magnetic field have affected cosmic-ray intensities in the earth's atmosphere, which in turn produce fluctuations in the production of carbon-14. In addition, changes in the earth's climate affect the amount of carbon dioxide in the atmosphere. The oceans are great reservoirs of carbon dioxide. When the oceans are cold, they release less carbon dioxide into the atmosphere than when they are warm. These complications require complex corrective procedures for the accurate dating of ancient organic objects. Nevertheless, recalibration and refined techniques make carbon dating the most versatile chronometric method we have. Present laser-enrichment techniques using samples containing only a few milligrams of carbon are soon expected to double the current 50,000-year range. Similar results are now possible with a technique that bypasses a radioactive measure altogether. It involves an improved atomic acceleration-mass spectrometer device that effectively makes a carbon-12/carbon-14 count directly. There are also other techniques for assigning dates to relics of the past.

Question   Suppose an archeologist extracts a gram of carbon from an ancient ax handle and finds it one-fourth as radioactive as a gram of carbon extracted from a freshly cut tree branch. How old is the ax handle?*

Uranium Dating

The dating of older, but nonliving, things is accomplished with radioactive minerals, such as uranium. The naturally occurring isotopes U-238 and U-235 decay very slowly and ultimately become isotopes of lead—but not the common lead isotope Pb-208. For example, U-238 decays through several stages to finally become Pb-206; whereas U-235 finally becomes the isotope Pb-207. Some of the lead isotopes 206 and 207 that exist were at one time uranium. The older the uranium-bearing rock, the higher the

---

*Answer   The ax handle is two half-lives of carbon-14, about 11,460 years old.

percentage of these remnant isotopes. We know the half-lives of the uranium isotopes, and we can determine the percentage of each radiogenic lead isotope in a rock sample, so we can calculate the time of decay for uranium in the rock. In this way the ages of rocks have been found to be as much as 3.7 billion years. Samples from the moon, where there has been less obliteration of the early rocks than occurs on earth, have been dated at 4.2 billion years, which is beginning to impinge closely upon the well-established 4.6-billion-year age of the earth and solar system.

Effects of Radiation on Humans

The cells of living tissue are composed of intricately structured molecules in a watery, ion-rich brine. When X radiation or nuclear radiation encounters this highly ordered soup, it produces chaos on the atomic scale. A beta particle, for example, passing through living matter collides with a small percentage of the molecules and leaves a randomly dotted trail of altered or broken molecules along with newly formed, chemically active ions and free radicals. Free radicals are unbonded, electrically neutral, very chemically active atoms or molecular fragments. The ions and free radicals may break even more molecular bonds or they may quickly form new strong bonds, creating molecules that may be useless or harmful to the cell. Gamma radiation produces a similar effect. As a high-energy gamma-ray photon moves through matter, it may rebound from an electron and give it a high kinetic energy. The electron then may career through the tissue, creating havoc in the ways described above. All types of high-energy radiation break or alter the structure of some molecules and create conditions in which other molecules will be formed that may be harmful to life processes.

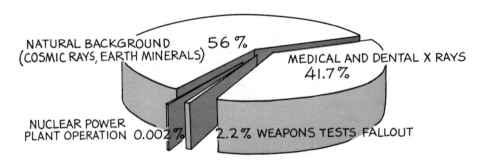

NATURAL BACKGROUND (COSMIC RAYS, EARTH MINERALS) 56%

MEDICAL AND DENTAL X RAYS 41.7%

NUCLEAR POWER PLANT OPERATION 0.002%

2.2% WEAPONS TESTS FALLOUT

Fig. 31.15 Origins of radiation exposure for an average individual in the United States.

The cells are able to repair most kinds of molecular damage if the radiation is not too intense. A cell could survive an otherwise lethal dose of radiation if the dose were small and there were intervals for healing. When radiation is sufficient to kill a cell, the dead cells can be replaced by new ones. Important exceptions to this are most nerve cells, which are irreplaceable. Sometimes a cell will survive with a damaged DNA molecule. Defective genetic information will be transmitted to its daughter cell when it reproduces; a cell mutation will occur. This will sometimes be insignificant. But if significant, it will probably result in cells that do not function as well as the original one. In rare cases it will be an improvement. A genetic change of this type could also be part of the cause of a cancer that will develop in the tissue at a much later time.

The concentration of disorder produced along the trajectory of a particle depends upon its energy, charge, and mass. Gamma-ray photons and very energetic beta particles cause the lowest intensity of damage. A particle of their type penetrates deeply with widely separated interactions, like a very fast BB fired through a hailstorm. Slow, massive, highly charged particles such as low-energy alpha particles are most disruptive. They have collisions that are close together, more like a bull charging through a flock of sleepy sheep. They do not penetrate deeply because their energy is absorbed by many closely spaced collisions. Especially damaging particles of this type are the assorted nuclei (called *heavy primaries*) flung outward by the sun in solar flares. These include all the elements found on earth. Particles that approach the earth a few minutes after leaving the sun are partly captured in the earth's magnetic field. The others are absorbed by collisions in the atmosphere, so practically none reaches the earth. We are partly shielded from these dangerous particles by the very property that makes them a threat—their tendency to have many collisions close together. Astronauts do not have this protection, and they absorb large doses of radiation for the amount of time they spend in space. Every few decades there is an exceptionally powerful solar flare that would almost certainly kill any conventionally protected astronaut far from the earth who is unprotected by its atmosphere and magnetic field.

Generally, radiation dosage is measured in **rads** (short for *radiation*), a unit of absorbed dose of ionizing radiation. The number of rads indicates the amount of radiation energy absorbed per gram of exposed material. However, when concerned with the potential ability of radiation to affect human beings, we measure the radiation in **rems** (**r**oentgen **e**quivalent **m**an). In calculating the dosage in rems, we multiply the number of rads by a factor that allows for the different health effects of different types of radiation. For example, 1 rad of slow alpha particles has the same biological effect as 10 rads of fast electrons. We call both of these dosages 10 rems.

The average person in the United States is exposed to about 0.2 rem a year. This comes from within the body itself, from the ground, buildings, cosmic rays, diagnostic X rays, television, and so on. It varies widely from place to place on the planet, but it is strongest near the poles where the earth's magnetic field does not act as a shield. It is also stronger at higher altitudes where the atmosphere provides less protection.

The lethal dose of radiation is on the order of 500 rems; that is, a

CAUTION

RADIATION AREA

Fig. 31.16 This is the internationally used symbol to indicate an area where radioactive material is being handled or produced.

person has about a 50 percent chance of surviving a dose of this magnitude received over a short period of time. Under radiotherapy—the use of radiation to kill cancer cells—a patient may receive localized doses in excess of 200 rems each day for a period of weeks. A typical diagnostic chest X ray exposes a person to from 5 to 30 millirems, less than one ten-thousandth of the lethal dose. However, even small doses of radiation can produce long-term effects due to mutations within the body's tissues. And, because a small fraction of any X-ray dose reaches the gonads, some mutations occasionally occur that are passed on to the next generation. Medical X rays for diagnosis and therapy have a far larger effect on the human genetic heritage than any other artificial source of radiation.

Taking all causes into account, most of us will receive a lifetime exposure of less than 20 rems, distributed over several decades. This makes us a little more susceptible to cancer and other disorders. But more significant is the fact that all living beings have always absorbed natural radiation and that the radiation received in the reproductive cells has produced genetic changes in all species for generation after generation. Small mutations selected by nature for their contributions to survival over billions of years can gradually come up with some interesting inventions—*us*, for example!

Summary of Terms

Alpha ray   A stream of helium nuclei ejected by certain radioactive nuclei.

Beta ray   A stream of beta particles.

Gamma ray   High-frequency electromagnetic radiation emitted by the nuclei of radioactive atoms.

Alpha particle   The nucleus of a helium atom, which consists of two neutrons and two protons, ejected by certain radioactive elements.

Beta particle   An electron (or positron) emitted during the radioactive decay of a nucleus.

Nucleon   A nuclear proton or neutron; the collective name for either or both.

Isotopes   Atoms whose nuclei have the same number of protons but different numbers of neutrons.

Atomic number   The number associated with an atom, the same as the number of protons in the nucleus or as the number of extranuclear electrons.

Atomic mass number   The number associated with an atom, the same as the number of nucleons in the nucleus.

Half-life   The time required for half the atoms of a radioactive element to decay.

Transmutation   The conversion of an atomic nucleus of one element into an atomic nucleus of another element through a loss or gain in the number of protons.

Review Questions

1. What is the principal difference between a beam of X rays and a beam of light?

2. How do the electric charges of alpha, beta, and gamma rays differ?

3. What changes occur in the nucleus of an atom when an alpha particle is emitted? A beta particle?

4. What are isotopes? Are they normally in our environment?

5. Do electric forces tend to hold the nucleus together or push it apart?

6. Why are larger nuclei more unstable than smaller nuclei?

7. Distinguish between a *hadron* and a *lepton*.

8. What is meant by *radioactive half-life?*

9. When thorium, atomic number 90, decays by emitting an alpha particle, what is the atomic number of the resulting nucleus?

10. When thorium decays by emitting a beta particle, what is the atomic number of the resulting nucleus?

11. How does the atomic mass number change for each of the above two reactions?

12. Why do no appreciable amounts of the transuranic elements (those above uranium in the periodic table) occur in the world?

13. How is carbon-14 produced in the atmosphere? What becomes of it after beta emission?

14. Are you carbon-14 radioactive? When will your bones be only half as radioactive?

15. Why is lead always found in natural uranium deposits?

Home Project

Some watches and clocks have luminous hands that continuously glow. These have traces of radium bromide mixed with zinc sulphide (as opposed to safer clock faces that use light rather than radioactive disintegration as a means of excitation and become progressively dimmer in the dark). If you have a luminous watch or clock available, take it into a completely dark room and after your eyes have become adjusted to the dark, examine the luminous hands with a very strong magnifying glass or the eyepiece of a microscope or telescope. You should be able to see individual tiny flashes, which altogether seem to be a steady source of light to the unaided eye. Each flash occurs when an alpha particle ejected by a radium nucleus strikes a molecule of zinc sulphide.

Exercises

1. X rays are most similar to which of the following: alpha, beta, or gamma rays?

2. Why is a sample of radioactive material always a little warmer than its surroundings?

3. Some people say that all things are possible. Is it at all possible for a hydrogen nucleus to emit an alpha particle?

4. Why are alpha and beta rays deflected in opposite directions in a magnetic field? Why are gamma rays undeflected?

5. The alpha particle has twice the electric charge of the beta particle but deflects *less* than the beta in a magnetic field. Why is this so?

6. How would the paths of alpha, beta, and gamma radiations compare in an electric field?

7. Which type of radiation—alpha, beta, or gamma—results in the greatest change in mass number? Atomic number? Charge?

8. Which type of radiation—alpha, beta, or gamma—results in the least change in mass number? Atomic number? Charge?

9. When an alpha particle leaves the nucleus, would you expect it to speed up? Defend your answer.

10. In bombarding atomic nuclei with proton "bullets," why must the protons be accelerated to high energies to make contact with the target nuclei?

11. Why would you expect alpha particles to penetrate materials less readily than beta particles of the same energy?

12. Within the atomic nucleus, which force tends to hold it together and which force tends to push it apart?

13. What evidence supports the contention that the strong nuclear force is stronger than the electrical force at short internuclear distances?

14. If a sample of radioactive isotopes has a half-life of 1 day, how much of the original sample will be left at the end of the second day? Third day? Fourth day?

15. A sample of a particular radioisotope is placed near a Geiger counter, which is observed to count at a rate of 160 per second. Eight hours later the detector counts at a rate of 10 counts per second. What is the half-life of the material?

16. Radiation from a point source obeys the inverse-square law. If a Geiger counter 1 meter from a small sample reads 200 counts per second, what will be its counting rate 2 meters from the source? 3 meters?

17. Why do the charged particles in bubble chambers move in spiral paths rather than the circular or helical paths they would ideally follow?

18. From Figure 31.11, how many alpha and beta particles are emitted in the radioactive decay of a uranium-238 nucleus to lead-206?

19. When radium—atomic number 88—decays by emitting an alpha particle, what is the atomic number of the resulting nucleus? What is the resulting atomic mass?

20. When the element polonium—atomic number 84, atomic mass 218—emits a beta particle, it transforms into a new element. What are the atomic number and atomic mass of this new element? What are they if polonium emits an alpha particle?

21. State the number of neutrons and protons in each of the following nuclei: ^1_1H, $^{14}_6C$, $^{56}_{26}Fe$, $^{197}_{79}Au$, $^{90}_{38}Sr$, and $^{238}_{92}U$.

22. How is it possible for an element to decay "forward in the periodic table"—that is, to decay to an element of higher atomic number?

23. When radioactive phosphorus, $^{30}_{15}$P, decays, it emits a positron. Will the resulting nucleus be another isotope of phosphorus? Or what?

24. How could a physicist test the following statement? "Strontium-90 is a pure beta source."

25. Elements above uranium in the periodic table do not exist in any appreciable amounts in nature because they have short half-lives. Yet there are several elements below uranium in atomic number with equally short half-lives that *do* exist in appreciable amounts in nature. How can you account for this?

26. What is meant by the term *background radiation?*

27. A friend produces a Geiger counter to check the local background radiation. It ticks. Another friend, who normally fears most that which is understood least, makes an effort to keep away from the region of the Geiger counter and looks to you for advice. What do you say?

28. Why is the carbon-dating technique not accurate for estimating the ages of materials older than 30,000 years?

29. The age of the Dead Sea Scrolls was found by carbon dating. Could this technique have worked if they were carved in stone tablets? Explain.

30. The biological damage to human beings and other living organisms is considerably less for high-energy alpha-particle radiation than it is for lower-energy beta-particle radiation. Why is this so?

32 Nuclear Fission and Fusion

Nuclear Fission

THE GREATER FORCE
IS NUCLEAR

CRITICAL
DEFORMATION

THE GREATER FORCE
IS ELECTRICAL

Fig. 32.1 Nuclear deformation may result in repulsive electrical forces exceeding attractive nuclear forces, in which case fission occurs.

In 1939 two German scientists, Otto Hahn and Fritz Strassmann, made an accidental discovery that was to change the world. While bombarding a sample of uranium with neutrons in the hope of creating new and heavier elements, they were astonished to find chemical evidence for the production of barium, an element about half the size of uranium. They were reluctant to believe their own results. News of this discovery reached Lise Meitner and Otto Frisch, then refugees from Nazism working in Sweden, who proposed that the uranium nucleus, activated by neutron bombardment, had split in half. Meitner named the process *fission,* after the similar process of cell division in biology.

Nuclear fission involves the delicate balance within the nucleus between the attraction of nuclear forces and the repulsion of electrical forces. In nearly all nuclei the nuclear forces dominate. In uranium, however, this domination is tenuous. If the uranium nucleus is stretched into an elongated shape (Figure 32.1), the electrical forces may push it into an even more elongated shape. If the elongation passes a critical point, nuclear forces give way to electrical ones, and the nucleus separates. This is fission.

The absorption of a neutron by a uranium atom is apparently enough to cause such an elongation. The resultant fission process may produce any of several combinations of smaller nuclei. A typical example recorded in cloud chambers is

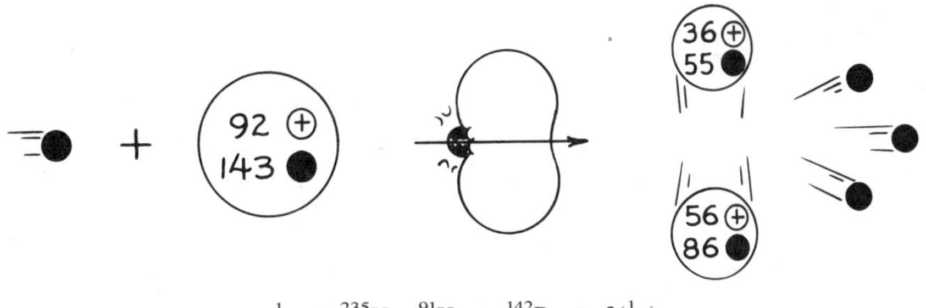

$$_0^1 n + {}_{92}^{235}U \quad {}_{36}^{91}Kr + {}_{56}^{142}Ba + 3({}_0^1 n)$$

This reaction has an energy release of about 200,000,000 electron volts.[1] (The explosion of the TNT molecule releases 30 electron volts). The combined mass of the fission fragments is less than the mass of the uranium nucleus. The tiny amount of missing mass is converted to an awesome amount of energy, according to the relation $E = mc^2$. The

---

[1]The electron volt (eV) is defined as the amount of energy an electron acquires in accelerating through a potential difference of 1 volt.

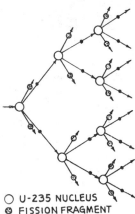

○ U-235 NUCLEUS
⊘ FISSION FRAGMENT
• NEUTRON

Fig. 32.2 A chain reaction.

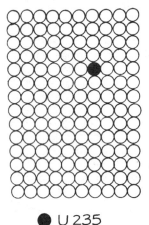

● U 235
○ U 238

Fig. 32.3 Only 1 part in 140 of naturally occurring uranium is U-235.

scientific world was jolted—not only because of the enormous energy release but also because of the extra neutrons liberated in the process. A typical fission reaction releases, on the average, between two and three neutrons. These new neutrons can in turn cause the fissioning of two or three other atoms, releasing more energy and a total of from four to nine more neutrons. If each of these succeeds in splitting just one atom, the next step in the reaction will produce between 8 and 27 neutrons, and so on. Thus, a whole **chain reaction** can proceed at an ever-accelerating rate (Figure 32.2).

Why doesn't a chain reaction occur in naturally occurring uranium deposits? Actually, it has occurred to a small degree millions of years ago when isotopic abundances were different, and in unusually rich concentrations under very unusual circumstances (see *Scientific American*, July 1976). Chain reactions don't ordinarily happen because fission occurs mainly for the rare isotope U-235, which now makes up only 0.7 percent of the uranium in pure uranium metal. The prevalent isotope U-238 absorbs neutrons but does not ordinarily undergo fission, so any chain reaction is quickly snuffed by the neutron-absorbing U-238 nuclei. With rare exceptions, naturally occurring uranium is too ''impure'' to undergo a chain reaction spontaneously.

If a chain reaction began in a tiny piece of pure U-235, the number of neutrons escaping from the surface without being captured might be so numerous that a growing chain reaction would be short-lived. Recall from Chapter 10 that small objects have a disproportionately large surface area compared to their small volumes—the greater the surface area, the greater the number of neutrons escaping. In a large piece of fissionable material, where the surface area is less in proportion to size, the results would be explosive. The minimum size of the material that will sustain a chain reaction is called the *critical size*, the mass of which is called the **critical mass**. If the mass of a fissionable material is greater than the critical mass, an explosion of enormous magnitude may take place.

Consider a large quantity of U-235 in two units, each smaller than the critical size and separated a short distance (Figure 32.4). Because of the relatively large surface area of each unit, neutrons readily escape and a chain reaction cannot develop. But if the pieces are suddenly driven to-

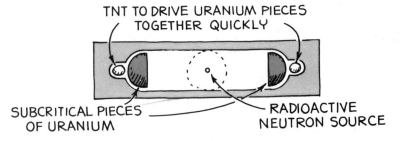

TNT TO DRIVE URANIUM PIECES
TOGETHER QUICKLY

SUBCRITICAL PIECES
OF URANIUM

RADIOACTIVE
NEUTRON SOURCE

Fig. 32.4 Diagram of a simple atomic bomb.

gether, the relative surface area is decreased. If the timing is right and the combined mass is greater than the critical mass, a violent explosion takes place. Such a device is a nuclear bomb.

The uranium used in the Hiroshima blast was U-235 about the size of a baseball. The separation of this much fissionable material from natural uranium was one of the principal and difficult tasks of the secret Manhattan Project during World War II. Project scientists used two methods of isotope separation. One method depended on the fact that lighter U-235 moves slightly faster on the average than U-238 at the same temperature. In gaseous form the faster isotope therefore has a higher rate of diffusion through a thin membrane or small opening, resulting in a slightly enriched U-235 gas on the other side (Figure 32.5). Diffusion through thousands of chambers ultimately produced a sufficiently enriched sample of U-235. The other, and less successful, method consisted of shooting uranium ions into a magnetic field. The smaller-mass U-235 ions were deflected more by the magnetic field than U-238 ions and were collected atom by atom through a slit positioned to catch them. After a couple of years, both methods netted a few kilograms of U-235.

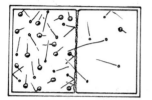

Fig. 32.5 Lighter molecules move faster than heavier ones at the same temperature and diffuse more readily through a thin membrane.

Question   *Why* will gaseous U-235 move slightly faster than U-238 at the same temperature?*

Uranium-isotope separation today is more easily accomplished with a gas centrifuge. Uranium is mixed with fluorine and the uranium-hexafluorine gas is whirled in a drum at tremendously high rim speeds (on the order of 1500 kilometers per hour). Under centrifugal force, the heavier U-238 gravitates to the outside like milk in a dairy separator, and gas rich in the lighter U-235 is extracted from the center. Engineering difficulties, only recently overcome, prevented the use of this method in the Manhattan Project.

Nuclear Reactors   A chain reaction cannot ordinarily take place in *pure* natural uranium, since it is mostly U-238. The neutrons released by fissioning U-235 atoms are fast neutrons, readily captured by U-238, which do not fission. A crucial experimental fact is that *slow* neutrons are far more likely to be captured by U-235 than U-238.[2] If neutrons can be slowed down, there is an increased chance that a neutron released by fission will cause fission in

---

*__Answer__   At the same temperature, both isotopes have the same kinetic energy ($\frac{1}{2}mv^2$). So the U-235 isotope with the lesser mass must have a correspondingly higher velocity.

[2]This is similar to the selective absorption of different frequencies of light. Just as there are characteristic electron energy levels in an atom, there are similar energy levels within the nucleus.

A salute in bronze for Fermi The bronze plaque at Chicago's Stagg Field commemorates Enrico Fermi's historic fission chain reaction. Criticized in Italy for not giving the Fascist salute when presented with the 1938 Nobel Prize, Fermi never returned to Italy, and became an American citizen in 1945.

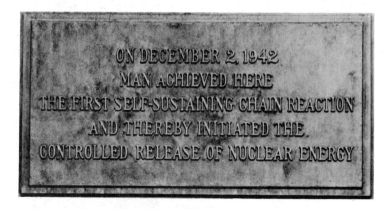

ON DECEMBER 2, 1942
MAN ACHIEVED HERE
THE FIRST SELF-SUSTAINING CHAIN REACTION
AND THEREBY INITIATED THE
CONTROLLED RELEASE OF NUCLEAR ENERGY

Cradle of an age The now-demolished stands at the University of Chicago's Stagg Field as they looked in 1942 when in the racquet courts beneath them, a small group of scientists inaugurated the atomic age. They toasted in solemn silence by sipping wine from paper cups.

another U-235 atom, even amidst the more plentiful and otherwise neutron-absorbing U-238 atoms. This increase may be enough to allow a chain reaction to take place.

The Italian physicist Enrico Fermi reasoned that a chain reaction with ordinary uranium metal might be possible if the uranium were broken up into small lumps and separated by a material that would slow down neutrons. Fermi and his co-workers constructed the first nuclear reactor—or *atomic pile,* as it was called—in the racquet courts underneath the grandstands of the University of Chicago's Stagg Field. They achieved the first self-sustaining controlled release of nuclear energy on December 2, 1942.

Three fates are possible for a neutron in ordinary uranium metal. It may (1) cause fission of a U-235 atom, (2) escape from the metal into nonfissionable surroundings, or (3) be absorbed by U-238 without causing fission. To make the first fate more probable, the uranium was divided into discrete parcels and buried at regular intervals in nearly 400 tons of graphite, a familiar form of carbon. A simple analogy clarifies the function of the graphite: If a golf ball rebounds from a massive wall, it loses hardly any speed; but if it rebounds from a baseball, it loses considerable speed. The case of the neutron is similar. If a neutron rebounds from a

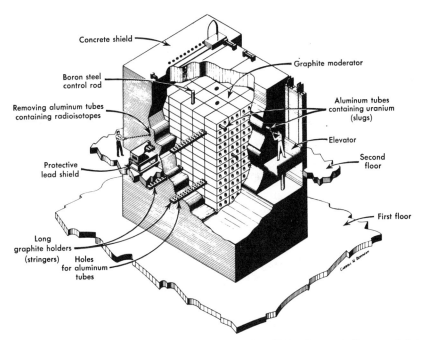

Concrete shield

Graphite moderator

Boron steel
control rod

Aluminum tubes
containing uranium
(slugs)

Removing aluminum tubes
containing radioisotopes

Elevator

Second
floor

Protective
lead shield

First floor

Long
graphite holders
(stringers) Holes
for aluminum
tubes

Fig. 32.6 A detailed view of an atomic reactor. A controlled chain reaction occurs in the uranium slugs embedded in the graphite.

heavy nucleus, it loses hardly any speed; but if it rebounds from a lighter carbon nucleus, it loses considerable speed. The graphite was said to "moderate" the neutrons.[3] The whole apparatus was called a *reactor*.

Now an uncontrolled reactor would not be a reactor, but a technical accident about to happen. A runaway chain reaction is prevented by control rods made of cadmium or boron metal, which easily absorb neutrons. The control rods are inserted into the reactor and absorb a selected fraction of the neutrons so as to maintain a steady rate of energy release. When fully inserted into the reactor, the control rods snuff the chain reaction. If fully removed, the reaction could build to a level capable of melting the reactor. Because "bomb-grade" isotopes are so highly diluted with U-238, an explosion like that of a nuclear bomb is not possible.

Questions

1. What is the function of a *moderator* in a nuclear reactor?*

2. What is the function of *control rods* in a nuclear reactor?†

---

*Answer A moderator slows down neutrons that are normally too fast for absorption by fissionable isotopes.

†**Answer** Control rods absorb neutrons and thereby control the neutron flux and the rate of the chain reaction.

[3]*Heavy water,* which is composed of the heavy hydrogen isotope deuterium, is an even more effector moderator. This is because ordinary hydrogen readily captures neutrons while deuterium doesn't.

The principal use of Fermi's atomic pile was the production of the synthetic element plutonium.[4] Plutonium-239, like U-235, will fission. The atomic bomb exploded over Nagasaki was a plutonium bomb. This element is not found in nature but is manufactured synthetically from U-238. When U-238 captures a neutron, it emits a beta particle and becomes the first transuranic element, neptunium. Neptunium, in turn, emits a beta particle and becomes plutonium (Figure 32.7). The half-life of neptunium is only 2.3 days, while the half-life of plutonium is about 24,000 years. Since plutonium is an element distinct from uranium, it can be separated from uranium in the pile by ordinary chemical methods. Consequently, the pile provides a process for making pure fissionable material far more easily than by separating the U-235 from natural uranium.

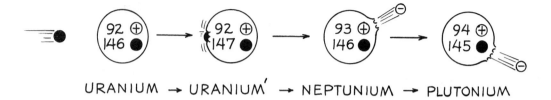

URANIUM → URANIUM' → NEPTUNIUM → PLUTONIUM

Fig. 32.7 After U-238 absorbs a neutron, it emits a beta particle, resulting in a neutron becoming a proton. The atom is no longer uranium, but neptunium. After the neptunium atom, in turn, emits a beta particle, it becomes plutonium.

Although the process is simple in theory, it is very difficult in practice. Not only plutonium is produced in a reactor, but large quantities of radioactive fission products are formed as well. The radioactivity becomes so intense that the materials in the pile begin to fall apart physically. As a result, the pile has to be shut down after only a few grams of plutonium have been produced. The separation of plutonium from uranium must be done by remote control to protect the personnel against radiation.

The principal use of nuclear reactors in the world today is the production of electric power. It is interesting to note that fission reactors are simply nuclear furnaces, which, like fossil-fuel furnaces, do nothing more elegant than boil water to produce steam for a turbine (Figure 32.8). Reactors come in a variety of designs, none of which we will treat here. We will, however, briefly discuss the basic idea of the breeder reactor.

---

[4]Elemental plutonium is chemically toxic in the same sense as are lead and arsenic. It attacks the nervous system and can cause paralysis; death can follow if the dose is sufficiently large. Fortunately, plutonium doesn't remain in its elemental form for long but rapidly oxidizes into three forms: PuO, PuO_2, and Pu_2O_3, all of which are chemically inert, that is, not soluble in water or in biological systems. These plutonium molecules do not attack the nervous system and have been found to be biologically harmless. Plutonium in any form, however, is radioactively toxic. It is more toxic than uranium and less toxic than radium. Plutonium emits alpha particles that have a very short range and high energy transfer, which kill rather than mutate the cells they encounter (unlike the beta particles emitted by radium). Since damaged cells rather than dead cells contribute to cancer, plutonium ranks low as a cancer-producing substance.

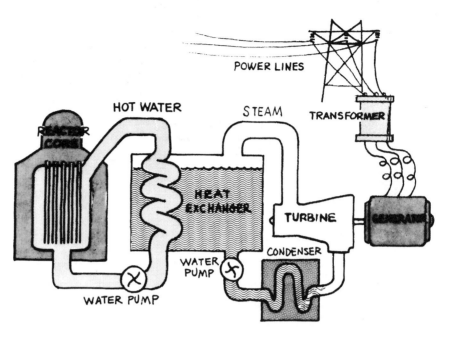

Fig. 32.8 Diagram of a nuclear fission power plant.

A *breeder reactor* breeds more fuel than it consumes. If U-238 or Th-232 is mixed with plutonium, the fissioning of plutonium liberates neutrons that convert the relatively abundant uranium and thorium into the fissionable isotopes Pu-239 and U-233. For about every two fissionable isotopes put into the reactor, three new fissionable isotopes are produced. A breeder reactor produces fissionable fuel simply by converting the relatively available U-238 and Th-232.[5] This is like filling your gas tank with water, adding some gasoline, then driving your car and having more gasoline after your trip than you started with, at the expense of common water. After the initial cost of building a reactor, this is a very economical method of producing vast amounts of energy. With a breeder reactor, a power utility can, after a few years of operation, breed twice as much fuel as its original fuel.

The benefits of fission power are plentiful electricity; the conservation of the many billions of tons of coal, oil, and natural gas that every year are literally turned to heat and smoke and that in the long run may be far more precious as sources of organic molecules than as sources of heat; and the elimination of the megatons of sulfur oxides and other poisons that are put into the air each year by the burning of these fuels.

[5]Neutron capture is sufficient in a typical "nonbreeder" U.S. power reactor of the light water type to convert almost 1 percent of the predominantly U-238 supply to plutonium over the 3-year period the fuel resides in the core. As the supply of U-238 decreases, a supply of Pu-239 increases and actually predominates over uranium in energy output near the end of the fuel cycle. To various extents, all reactors breed fissionable fuel.

The drawbacks include the problems of storing radioactive wastes, the production of plutonium and the danger of nuclear weapons proliferation, the low-level release of radioactive materials into the air and groundwater, and, most importantly, the risk of an accidental release of large amounts of radioactivity.

Sound judgment requires us not only to compare the benefits and drawbacks of fission power but also to compare its drawbacks to the drawbacks of alternate power sources. For a variety of reasons, public opinion in the United States and Europe is now against fission power plants. Nukes are on the decline and fossil-fuel power plants are on the upswing.

Question   In a breeder reactor, what is bred from what?*

Mass-Energy Equivalence   From Einstein's mass-energy equivalence, $E = mc^2$, we can think of mass as congealed energy. The more energy associated with a particle, the greater its mass. Would the mass of a nucleon be the same whether or not it were located inside or outside an atomic nucleus? Wouldn't it have more mass energy outside a nucleus than inside? We can answer this by considering the work that would be required to separate the nucleons from a nucleus. Recall that work, which is expended energy, is equal to the product of force and distance. Then think of the magnitude of force required to pull nucleons apart through a sufficient distance to overcome the attractive nuclear force. Tremendous work would be required. The work we would have to put into such an endeavor would be manifest in the energy of the protons and neutrons. They would have more energy outside the nucleus. This extra energy would be equal to the energy or work required to separate them. This energy, in turn, would be manifest in mass. Hence, the mass of nucleons outside a nucleus is greater than the mass of the same nucleons when locked inside a nucleus.

The experimental verification of this conclusion is one of the triumphs of modern physics. The masses of nucleons and the isotopes of the various elements can be measured with an accuracy of one part per million or better. One means of doing this is the *mass spectrograph* (Figure 32.9).

Underlying the operation of the mass spectrograph is the fact that charged ions directed into a magnetic field are deflected into circular arcs. The greater the inertia of the ion, the more it resists deflection, and the greater the radius of its curved path. The magnetic force sweeps heavier ions into larger arcs, and lighter ions into shorter arcs. The ions pass through exit slits where they may be collected, or they strike pho-

Answer   Fissionable Pu-239 is bred from nonfissionable U-238, or fissionable U-233 is bred from nonfissionable Th-232.

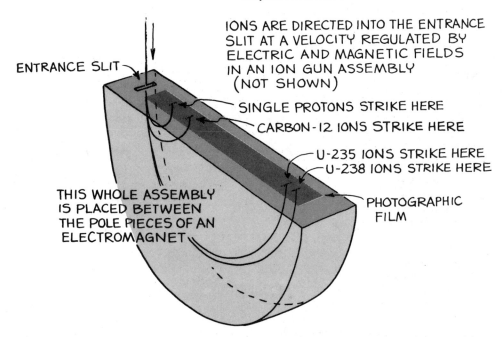

ENTRANCE SLIT

IONS ARE DIRECTED INTO THE ENTRANCE
SLIT AT A VELOCITY REGULATED BY
ELECTRIC AND MAGNETIC FIELDS
IN AN ION GUN ASSEMBLY
(NOT SHOWN)

SINGLE PROTONS STRIKE HERE

CARBON-12 IONS STRIKE HERE

U-235 IONS STRIKE HERE
U-238 IONS STRIKE HERE

PHOTOGRAPHIC
FILM

THIS WHOLE ASSEMBLY
IS PLACED BETWEEN
THE POLE PIECES OF AN
ELECTROMAGNET

Fig. 32.9 The mass spectrograph. Ions are directed into the semicircular "drum," where they are swept into semicircular paths by a strong magnetic field. Because of inertia, heavier ions are swept into curves of large radii and lighter ions are swept into curves of smaller radii. The radius of the curve is directly proportional to the mass of the ion. Using carbon-12 as a standard, the masses of all the elements and their isotopes are easily determined.

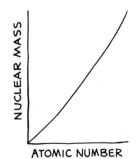

Fig. 32.10 The plot shows how nuclear mass increases with increasing atomic number.

tographic film where they may be simply detected. An isotope is chosen as a standard, and its position on the film of the mass spectrograph is used as a reference point. The standard is the common isotope of carbon, $^{12}_{6}C$. The mass of the carbon-12 nucleus is assigned the value of 12 atomic mass units. The atomic mass unit (amu) is defined to be precisely one-twelfth the mass of the common carbon-12 atom. With this reference, the amu's of the other atomic nuclei are measured. The masses of the proton and neutron are greater alone than when in a nucleus. They are 1.00728 and 1.00866 amu, respectively.

A plot of nuclear masses for the elements from hydrogen through uranium is shown in Figure 32.10. The curve slopes upward with increasing atomic number as expected: elements are more massive as atomic number increases. A more interesting curve results if we plot the nuclear mass *per nucleon* from hydrogen through uranium (Figure 32.11). To obtain the nuclear mass per nucleon, we simply divide the nuclear mass by the number of nucleons in the particular nucleus. We find that the

Fig. 32.11   The plot shows that the average mass of a particular nucleon depends on the atomic number of the nucleus it is a part of. Individual nucleons are most massive in the lightest nuclei, least massive in the iron nucleus, and intermediate in the heaviest nuclei. When light nuclei fuse, the product nucleus is less massive than the sum of its parts; when the heaviest nuclei fission, the parts have less mass than the original nucleus. The mass defect, in both cases, is converted to energy.

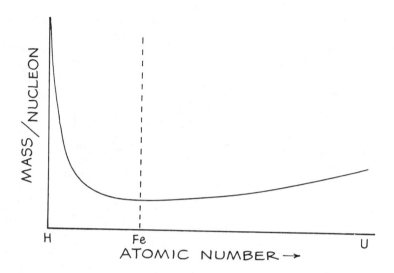

masses of protons and neutrons are different when combined in different nuclei. The mass of a proton is greatest as the nucleus of the hydrogen atom and becomes successively less and less massive as it occurs in atoms of increasing atomic number. The proton is least massive when in the iron atom. Beyond iron, the process reverses itself. Protons (and neutrons) become successively more and more massive with increasing atomic number, all the way to uranium and the transuranic elements.

From the plot we can see why energy is released when a uranium nucleus is split into nuclei of lower atomic number. The masses of the fission fragments lie about halfway between uranium and hydrogen on the horizontal scale of the plot. Most importantly, we see that the masses of these nucleons in the fission fragments are *less than* the masses of the same nucleons when combined in the uranium nucleus. The protons and neutrons decrease in mass. This mass difference is manifest in the form of energy—the 200,000,000 electron volts yielded by each uranium nucleus undergoing fission.

Fig. 32.12   The mass of a nucleus is *not* equal to the sum of the masses of its parts. (*a*) The fission fragments of a heavy nucleus like uranium are less massive than the uranium nucleus. (*b*) Two protons and two neutrons are more massive in their free states than when combined to form a helium nucleus.

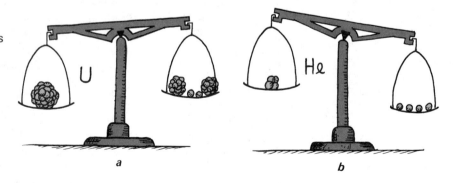

We can think of the mass-per-nucleon curve as an energy hill, which starts at the highest point (hydrogen), slopes steeply down to the lowest point (iron), and then slopes more gradually up to uranium. Iron is at the bottom of the energy hill and is the most stable nucleus. It is also the strongest nucleus; more energy is required to pull each proton or neutron from the iron nucleus than from any other.

How about the left side of the hill? We can also gain energy by going from hydrogen toward iron. But to go in this direction is to *combine* or *fuse* nuclei. And this is fusion. Whereas energy is released in fissioning the heavy elements, energy is released in fusing the light elements. In *both* cases there is a decrease in the mass per nucleon, which causes a huge energy release. From the curve we see that any atom to the left or right of iron is a potential source of energy. The higher we are on the hill to either the right or the left and the farther we "roll down the hill," the greater the amount of energy available. No energy can be released by the iron nucleus. If we split it, the total mass of the fragments is greater than the mass of the iron before splitting. If we fuse two or more iron nuclei, the mass of the resulting nucleus is greater than the two masses before fusing. In this special case, energy would have to be put into either the fissioning or the fusing of iron; no energy would be liberated.

We can see by the curve that elements above iron cannot give off energy if fused. The mass of the fused nuclei would be greater than the sum of the masses before fusion. This would *require* energy, rather than *liberate* it. Likewise, elements below iron would yield no energy if fissioned. The sum of the masses of the fission fragments would be greater than that of the original nucleus, again requiring energy rather than liberating it. Energy is released only when there is a decrease in mass.

In both fission and fusion reactions, the amount of matter that is converted to energy is less than 0.1 percent. This is true whether the process takes place in bombs or in the stars.

Fig. 32.13 Fictitious example: the "hydrogen magnets" weigh more when apart than when together. (Adapted from Albert V. Baez, *The New College Physics: A Spiral Approach*. W. H. Freeman and Company. Copyright © 1967.)

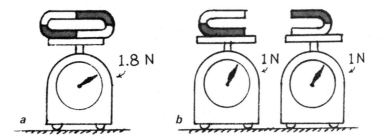

Nuclear Fusion

In the process of fusing hydrogen isotopes to form helium nuclei, the "shrinking" mass of the nucleons involved can produce as much as 26,700,000 electron volts of energy. Although this is less energy per reaction of individual atoms than the fissioning of individual uranium

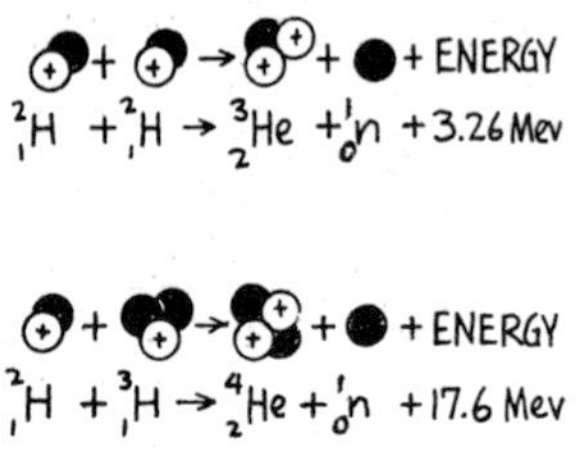

$$^2_1H + ^2_1H \rightarrow ^3_2He + ^1_0n + 3.26\,Mev$$

$$^2_1H + ^3_1H \rightarrow ^4_2He + ^1_0n + 17.6\,Mev$$

Fig. 32.14 Fusion reactions.

atoms, gram for gram it is several times more energy-producing than the fission process. This is, of course, because there are more hydrogen atoms in a gram of hydrogen than there are heavier uranium atoms in a gram of uranium.

Elements heavier than hydrogen and lighter than iron give off energy when fused, but much less per reaction than the fusion of hydrogen. The fusion of these heavier elements occurs in the advanced stages of a star. The energy released per gram during the various fusion stages from helium to iron amounts only to about one-fifth the energy released in the fusion of hydrogen to helium. Hydrogen, most notably in the form of deuterium, is the choicest fuel for fusion. We call fusion reactions in the sun and other stars **thermonuclear** reactions—that is, the welding together of atomic nuclei by high temperature.

For a fusion reaction to take place, the nuclei must collide at very high speeds in order to overcome their mutual electrical repulsion. The required speeds correspond to the extremely high temperatures found in the sun and stars. In the high temperatures of the sun, approximately 657 million tons of hydrogen are converted into 653 million tons of helium each second. The missing 4 million tons are discharged as energy. Such reactions are, quite literally, nuclear burning. Thermonuclear reactions are analogous to ordinary chemical combustion. In both chemical and nuclear burning, high temperature starts the reaction, and the release of energy by the reaction maintains a high enough temperature to spread the fire. The net result of the chemical reaction is a combination of atoms into more tightly bound molecules. In nuclear burning, the high temperature starts a reaction or series of reactions with the net result of producing more tightly bound nuclei. The difference between chemical and nuclear burning is essentially one of scale.

Prior to the development of the atomic bomb, the temperatures required to initiate **nuclear fusion** on earth were unattainable. When it was found that the temperatures inside an exploding atomic bomb are four to five times the temperature at the center of the sun, the thermonuclear bomb was but a step away. This first hydrogen bomb was detonated in 1954. Whereas the critical mass of fissionable material limits the size of a fission bomb (atomic bomb), no such limit is imposed on a fusion bomb (thermonuclear, or hydrogen, bomb). Just as there is no limit to the size of an oil-storage depot, there is no theoretical limit to the size of a fusion bomb. Like the oil-storage depot, any amount of nuclear fuel can be stored with safety until it is ignited. Although a mere match can ignite an oil depot, nothing less energetic than a fission bomb can ignite a thermonuclear bomb. We can see that there is no such thing as a ''baby'' hydrogen bomb. It cannot be less energetic than its fuse, which is an atomic bomb.

The hydrogen bomb is another example of a new discovery applied to destructive rather than constructive purposes. The constructive side of the picture is the controlled release of vast amounts of clean energy.

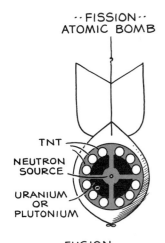

FISSION
ATOMIC BOMB

TNT

NEUTRON
SOURCE

URANIUM
OR
PLUTONIUM

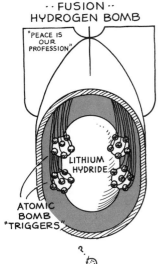

FUSION
HYDROGEN BOMB

"PEACE IS
OUR
PROFESSION"

LITHIUM
HYDRIDE

ATOMIC
BOMB
"TRIGGERS"

Fig. 32.15 Fission and fusion bombs.

Controlling Fusion

Carrying out fusion reactions under controlled conditions requires temperatures of millions of degrees. Producing and sustaining such high temperatures are the goals of much current research. High temperatures are obtained in electric arcs, where injected gases are ionized and become plasmas. Further heating is accomplished by electromagnetic resonance techniques and magnetic compression.

There is the problem of determining what kind of material can be used to contain the plasma. All known materials melt and vaporize below 4000°C, while fusion generally requires temperatures exceeding 100 million degrees. The situation is not hopeless, however. A magnetic field is nonmaterial, can exist at any temperature, and can exert powerful forces on charged particles in motion. "Magnetic walls" of sufficient strength have been designed to contain plasmas in a kind of magnetic straitjacket. The magnetic fields can be regulated to compress the plasma and produce fusion temperatures. The plasma will not melt the walls of the electromagnets even if its temperature is a billion degrees, because its total heat content is very small due to its vacuumlike density. To have a controlled reactor rather than a bomb, plasma densities are kept low, about 10,000 times lower than atmospheric density. The plasma must be kept off the walls, not because it will melt them, but because contact will slow the ions and therefore cool the plasma.

Some ions at 1 million degrees are moving fast enough to overcome electrical repulsion and slam together, but the energy output of these fusion reactions is small compared to the energy used to heat the plasma. Even at 100 million degrees, more energy must be put into the plasma than will be given off by fusion. At about 350 million degrees, the fusion reactions will produce enough energy to be self-sustaining. At this *ignition temperature*, nuclear burning yields a sustained power output without

Fig. 32.16 A magnetic bottle used for containing plasmas for fusion research.

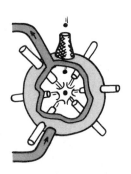

Fig. 32.17 Fusion with multiple laser beams. Frozen pellets of deuterium are rhythmically dropped into synchronized laser cross fire. The resulting heat is carried off by molten lithium to produce steam.

further input of energy. A steady feeding of hydrogen isotopes is all that is needed to produce continuous power.

Fusion has already been achieved in several devices, but instabilities in the plasma current have thus far prevented a sustained reaction. One of the most difficult problems has been to devise a field system that will hold the plasma in a stable position long enough for an ample number of ions to fuse. A variety of magnetic-confinement devices are the subject of much present-day research.

Another promising approach bypasses magnetic confinement altogether with high-energy lasers. One technique is to align an array of laser beams at a common point and drop solid pellets composed of hydrogen isotopes through the synchronous cross fire (Figure 32.17). The energy of the multiple beams implodes the deuterium-tritium fuel to densities up to a thousand times that of water. The aim is to sustain the high densities and temperatures long enough to initiate fusion. Once accomplished, the resulting heat is transferred to a jacket of molten lithium and carried off to produce steam for electricity. The success of this technique requires precise timing, for the necessary compression must occur before a shock wave causes the pellet to disperse. Success also requires the development of more efficient lasers, so the electricity generated will be greater than the amount required to operate the lasers.

Still other approaches involve the bombardment of fuel pellets not by laser light but by beams of electrons, light ions, and beams of heavy ions. We are still looking forward to the great day—break-even day—when one of the variety of fusion schemes will sustain a yield of at least as much energy as is required to initiate it. The aftermath will likely be the solution to power production the world over—and quite ideally, for although the inner chamber of the fusion device will be radioactive due to high-energy neutrons, the by-products will not be. Fusion produces clean, nonradioactive helium (good for children's balloons). Fusion reactors are also inherently incapable of a "runaway" accident because fusion requires no critical mass. Furthermore, there is no air pollution because there is no combustion. The problem of thermal pollution, characteristic of conventional steam-turbine plants, can be avoided by direct generation of electricity with MHD generators or by similar techniques using charge-particle fuel cycles that employ a direct energy conversion.

The fuel for nuclear fusion is hydrogen, the most plentiful element in the universe. The simplest reaction is the fusion of the hydrogen isotopes deuterium ($^2_1$H) and tritium ($^3_1$H), both of which are found in ordinary water. For example, 30 liters of seawater contain 1 gram of deuterium, which when fused releases as much energy as 10,000 liters of gasoline or 80 tons of TNT. Natural tritium is much scarcer, but given enough to get started, a controlled thermonuclear reactor will breed it from deuterium in ample quantities. Because of the abundance of fusion fuel, the amount of energy that can be released in a controlled manner is virtually unlimited.

Fig. 32.18 Pellet chamber at Lawrence Livermore Laboratory. The laser source is Shiva, the most powerful laser in the world, which directs 20 beams into the target region.

Questions

1. Fission and fusion are opposite processes, yet each releases energy. Isn't this contradictory?*

2. Would you expect the temperature of the core of a star to increase or decrease as a result of the fusion of intermediate elements to manufacture heavy elements?†

Fusion Torch and Recycling

In addition to supplying abundant electrical energy, high-temperature plasmas can be used as a fusion torch to vaporize garbage. With the fusion torch, all waste materials—whether liquid sewage or solid industrial refuse—can be reduced to their constituent atoms and separated by a mass-spectrograph-type device into various bins ranging from hydrogen to uranium. In this way, a single fusion plant could, in principle, not only dispose of thousands of tons of solid wastes per day but also provide a continuous supply of fresh raw material—thereby closing the cycle from use to reuse.

This would be a major turning point in materials economy (Figure 32.19). Our present concern for recycling materials will reach a grand

*__Answer__   No, no, no! This is contradictory only if the same element is said to release energy by both the processes of fission and fusion. Only the fusion of light elements and the fission of heavy elements result in a decrease in nucleon mass and a release of energy.

†__Answer__   Energy is absorbed and the star core tends to cool at this late stage of its evolution. This, however, allows the star to collapse, which produces an even greater temperature.

Fig. 32.19 Closed materials economy could be achieved with the aid of the fusion torch. In contrast to present systems, which are based on inherently wasteful linear material economies (*a*), a stationary-state system would be able to recycle the limited supply of material resources (*b*), thus alleviating most of the environmental pollution associated with present methods of energy utilization. (Redrawn from "The Prospects of Fusion Power" by William C. Gough and Bernard J. Eastlund. Copyright © February 1971 by Scientific American, Inc. All rights reserved.)

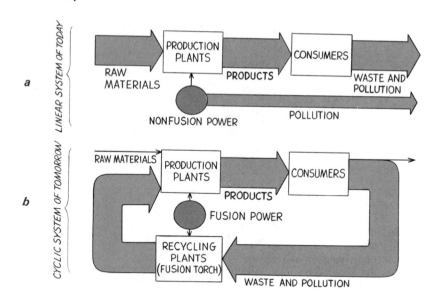

fruition with this or a comparable achievement, for this would be recycling with a capital *R*! Rather than gut our planet further for raw materials, we'd be able to recycle our existing stock over and over again, adding new material only to replace the relatively small amounts that are lost. Fusion power can produce abundant electrical power, can de-salt water, can help to cleanse our world of pollution and wastes, can recycle our materials, and in so doing can provide the setting for a better world—not in the far-off future, but perhaps in our own time. If and when fusion power plants become a reality, they will likely have an even more profound impact upon almost every aspect of human society than did the harnessing of electromagnetic energy in the last century.

When we think of our continuing evolution, we can see that the universe is well suited to those who will live in the future. If people are to one day dart about the universe in the same way we are able to jet about the world today, their supply of fuel is assured. The fuel for fusion is found in every part of the universe, not only in the stars but also in the space between them. About 91 percent of the atoms in the universe is estimated to be hydrogen. For people of the future, the supply of raw materials is also assured; all the elements result from fusing more and more hydrogen nuclei together. Simply put, if you fuse 8 deuterium nuclei, you have oxygen; 26, you have iron; and so forth. Future humans will likely synthesize their own elements and produce energy in the process—just as the stars do. Humans may one day travel to the stars in ships fueled by the same energy that makes the stars shine.

Summary of Terms

Nuclear fission   The splitting of the nucleus of a heavy atom, such as uranium-235, into two main parts, accompanied by the release of much energy.

Chain reaction   A self-sustaining reaction that, once started, steadily provides the energy and matter necessary to continue the reaction.

Critical mass   The minimum of fissionable material in a reactor or nuclear bomb that will sustain a chain reaction.

Thermonuclear   Pertaining to nuclear fusion by reason of high temperature.

Nuclear fusion   The combination of the nuclei of light atoms to form heavier nuclei, with the release of much energy.

Review Questions

1. What role does electrical force play in nuclear fission?

2. What is a chain reaction, and why does it occur in the fissioning of U-235?

3. Why don't natural deposits of uranium ore undergo chain reactions?

4. How are U-235 isotopes separated from the more abundant U-238 isotopes?

5. What is the function of carbon in an atomic reactor? Of boron and cadmium metal?

6. What is a breeder reactor? What does it breed and from what?

7. How can a breeder reactor produce more fuel than it consumes and not violate the law of energy conservation?

8. How does the mass of a nucleon compare inside a nucleus and outside a nucleus?

9. What method does the text cite for determining the relative masses of various isotopes?

10. How do the curves of Figures 32.10 and 32.11 differ?

11. Why won't helium yield energy if fissioned?

12. Why won't uranium yield energy if fused?

13. Why won't iron yield energy if fissioned or fused?

14. Why won't the billion-degree plasma in a fusion reactor melt the containing walls?

15. Why are fusion reactors not yet a present-day reality like fission reactors?

Exercises

1. Why is it that uranium ore doesn't spontaneously explode?

2. Why is nuclear fission a poor prospect for powering automobiles?

3. How is chemical burning similar to a nuclear chain reaction?

4. Why does a neutron make a better nuclear bullet than a proton or an electron?

5. Why will the escape of neutrons be proportionally less in a large piece of fissionable material than in a smaller piece?

6. Which shape is likely to produce a larger critical mass—a cube or a sphere? Explain.

7. Does the surface area as a whole increase or decrease when pieces of fissionable material are assembled into one piece? Does this assembly increase or decrease the probability of an explosion?

8. Uranium-235 releases an average of 2.5 neutrons per fission, while Pu-239 releases an average of 2.7 neutrons per fission. Which of these elements might you therefore expect to have the smaller critical mass?

9. Which will provide the faster chain reaction—U-235 or Pu-239?

10. Why does plutonium not appreciably occur in natural ore deposits?

11. Discuss and make a comparison of pollution by conventional fossil-fuel power plants and nuclear-fission power plants.

12. The water that passes through a reactor core does not pass into the turbine. Instead, heat is transferred to a separate water cycle that is entirely outside the reactor. Why is this done?

13. Is the mass of an atomic nucleus greater or less than the masses of the nucleons composing it?

14. The energy release of nuclear fission is tied to the fact that the heaviest nuclei weigh about 0.1 percent more per nucleon than nuclei near the middle of the periodic table of elements. What would be the effect on energy release if the 0.1 percent figure were instead 1 percent?

15. To predict the approximate energy release of either a fission or a fusion reaction, explain how a physicist makes use of the curve of Figure 32.11, or a table of nuclear masses, and the equation $E = mc^2$.

16. Which process would release energy from gold—fission or fusion? From carbon? From iron?

17. If uranium were to split into three segments of equal size instead of two, would more energy or less energy be released? Defend your answer in terms of Figure 32.11.

18. Suppose the curve for mass per nucleon versus atomic number of Figure 32.11 took the shape of the curve shown in Figure 32.10. Then would nuclear fission reactions produce energy? Would nuclear fusion reactions produce energy? Defend your answer.

19. The "hydrogen magnets" in Figure 32.13 weigh more when apart than when combined. What would be the basic difference if the fictitious example instead consisted of "uranium magnets"?

20. Which produces more energy—the fissioning of a single uranium atom or the fusing of a pair of deuterium atoms? The fissioning of a gram of uranium or the fusing of a gram of deuterium? (Why are your answers different?)

21. Why is there, unlike fission fuel, no limit to the amount of fusion fuel that can be safely stored in one locality?

22. If a fusion reaction produces no appreciable radioactive isotopes, why does a hydrogen bomb produce significant radioactive fallout?

23. List at least two major advantages of power production by fusion rather than by fission.

24. Nuclear fusion is a present hope for abundant energy in the near future. Yet the energy that has always sustained us has been the energy of nuclear fusion. Explain.

25. What effect on the mining industry can you foresee in the disposal of urban waste by a fusion torch coupled with a mass spectrograph?

26. The world has never been the same since the discovery of electromagnetic induction and its applications to electric motors and generators. Speculate and list some of the worldwide changes that are likely to follow the advent of successful fusion reactors.

PART 8 Relativity and Astrophysics

Albert Einstein (1879–1955)

Albert Einstein was born in Ulm, Germany, on March 14, 1879. According to popular legend, he was a slow child and learned to speak at a much later age than average; his parents feared for a while that he might be mentally retarded. Records from the elementary school he attended in Munich show, however, that Einstein was remarkably gifted in mathematics, physics, and playing the violin. He rebelled at the prevailing spirit of militarism in Germany and at the practice of education by regimentation and rote, and was expelled just as he was preparing to drop out of high school at the age of 15. Largely because of business reasons, his family moved to Italy, and young Einstein renounced his German citizenship and went to live with family friends in Switzerland. There he was allowed to take the entrance examination for the renowned Swiss Federal Institute of Technology in Zurich, two years younger than normal age. He did not pass the examination, mainly because of his difficulties with the French language. He spent a year at a Swiss preparatory school in Aarau, where he was "promoted with protest in French." He tried the entrance exams again at Zurich and passed. As a student, he cut many lectures, preferring to study on his own, and in 1900 he succeeded in passing his examinations by cramming with the help of a friend's meticulous notes. He said later of this, ". . . after I had passed the final examination, I found the consideration of any scientific problem distasteful to me for an entire year." During this year he became a citizen of Switzerland and accepted a temporary summer teaching position and also tutored two young high school students. He advised their father, a high school teacher himself, to remove the boys from school, where, he maintained, their natural curiosity was being destroyed. Einstein's job as a tutor was short-lived.

It was not until two years after graduation that he got a steady job, as a patent examiner at the Swiss Patent Office in Berne. Einstein held this position for over seven years. He found the work rather interesting, sometimes stimulating his scientific imagination, but mainly freeing him of financial worries while providing free time to ponder the problems in physics that puzzled him.

With no academic connections whatsoever, and with essentially no contact with other physicists, he laid out the main lines along which twentieth-century theoretical physics has developed. In 1905, at the age of 26, he earned his Ph.D. in physics and published three major papers. The first was on the quantum theory of light, including an explanation of the photoelectric effect, for which he won the 1921 Nobel Prize in physics. The second paper was on the statistical aspects of molecular theory and Brownian motion, a proof for the existence of atoms. His third and most famous paper was on special relativity. In 1915 Einstein published a paper on the theory of general relativity in which he presented a new theory of gravitation that included Newton's theory as a special case. These trail-blazing papers have had a profound effect on the course of modern physics and on our understanding of the physical universe.

Einstein's concerns were not limited to physics. He lived in Berlin during World War I and denounced the German militarism of his time, expressing a deeply felt conviction that warfare should be abolished and an international organization founded to govern disputes between nations. In 1933, while Einstein was visiting the United States, Hitler came to power. Einstein spoke out against Hitler's racial and political policies and resigned his position at the University of Berlin. Hitler responded by putting a price on his head. Einstein then accepted a research position at the Institute for Advanced Study in Princeton, New Jersey. In 1940 he became an American citizen. Einstein was first a German citizen, then a Swiss citizen, then a German citizen again, and finally an American citizen. But more than that, he was a citizen of the world.

Einstein believed that the universe is indifferent to the human condition, asserting that moral questions were of utmost importance for human existence and that in order for humanity to continue, it must create a moral order. This, he felt, was particularly urgent in the nuclear age. Nuclear bombs, Einstein remarked, had changed everything but our way of thinking. He was a man of unpretentious disposition with a deep concern for the welfare of his fellow beings. Lord C. P. Snow, who was acquainted with Einstein, gives us this assessment:

Einstein was the most powerful mind of the twentieth century, and one of the most powerful that ever lived. He was more than that. He was a man of enormous weight of personality, and perhaps most of all, of moral stature. . . . I have met a number of people whom the world calls great; of these, he was by far, by an order of magnitude, the most impressive. He was—despite the warmth, the humanity, the touch of the comedian—the most different from other men.[*]

Albert Einstein was a great scientist; but he was much more than that—he was a great human being.

---

[*]From a review of *The Born-Einstein Letters, 1916–1955*; translated by Irene Born (New York: Walker, 1971).

33 Special Theory of Relativity

Einstein's 1905 paper on relativity, titled "On the Electrodynamics of Moving Bodies," showed that the relative motion of electrical charges gave rise to magnetic forces. This interrelationship between electricity and magnetism, Einstein stated, was a consequence of the more profound interrelationship between space and time. Just as the forces between electric charges are affected by motion, the very measurements of space and time are also affected by motion. All measurements of space and time depend on relative motion. For example, the length of a rocket ship poised on its launching platform and the ticks of clocks within are found to be different when the same ship is moving at high speeds. It has always been common sense that we change our position in space when we move, but Einstein flouted common sense and stated that in moving we also change our rate of proceeding into the future—time itself is altered. Einstein went on to show that a consequence of the interrelationship between space and time is an interrelationship between mass and energy, given by the famous equation $E = mc^2$.

These are the ideas that make up this chapter—the ideas of special relativity—ideas so remote from your everyday experience as to make an

Front row, left to right:
A. A. Michelson, Albert
Einstein, Robert Millikan

understanding of them considerably difficult. It will be enough to become acquainted with these ideas, so be patient with yourself if you find you don't understand them. Perhaps your grandchildren will find them very much a part of their everyday experience, in which case they should find an understanding of relativity considerably less difficult.

Motion Is Relative

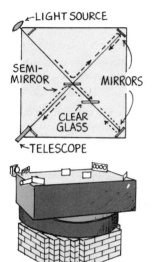

Fig. 33.1 The Michelson-Morley interferometer, which splits a light beam into two parts and then recombines them to form an interference pattern after they have traveled different paths. Rotation was accomplished by floating the massive sandstone slab in mercury. This schematic diagram shows how the half-silvered mirror splits the beam into two rays. The clear glass assures that both rays traverse the same amount of glass. Four mirrors at each end were actually used to lengthen the paths.

Whenever we talk about motion, we must always specify the position from which motion is being observed and measured. For example, a person who walks along the aisle of a moving train may be walking at a speed of 1 kilometer per hour relative to his seat, but 60 kilometers per hour relative to the railroad station. We call the place from which motion is observed and measured a **frame of reference**. An object may have different speeds relative to different frames of reference.

To measure the speed of an object, we first choose a frame of reference and pretend that the frame of reference is standing still. Then we measure the speed with which the object moves relative to the frame of reference. In the foregoing example, if we pretend the train is standing still, then the speed of the walking person is 1 kilometer per hour. If we pretend that the ground is standing still, then the speed of the walking person is 60 kilometers per hour. But the ground is not really still, for the earth spins like a top about its polar axis. Depending on how near the train is to the equator, the speed of the walking person may be as much as 1600 kilometers per hour relative to a frame of reference at the center of the earth. And the center of the earth is moving relative to the sun. If we place our frame of reference on the sun, the speed of the person walking in the train, which is on the orbiting earth, is on the order of 110,000 kilometers per hour. And the sun is not at rest, for it orbits the center of our galaxy, which moves with respect to other galaxies.

Michelson-Morley Experiment

Isn't there some reference frame that is still? Isn't space itself still, and can't measurements be made relative to still space? In 1887 the American physicists A. A. Michelson and E. W. Morley attempted to answer these questions by performing an experiment that was designed to measure the motion of the earth through space. Because light travels in waves, it was then assumed that something in space vibrates—a mysterious something called *ether,* thought to fill all space and serve as a frame of reference attached to space itself. The American physicists used a very sensitive apparatus called an *interferometer* to make their observations (Figure 33.1). In this instrument, a beam of light from a monochromatic source was separated into two beams with paths at right angles to each other that were then recombined to show whether there was any difference in speed over the two paths. The interferometer was set with one path parallel to the motion of the earth in its orbit; then Michelson or Morley carefully watched to see if there was any change in speed as the apparatus was rotated to put the other path parallel to the motion of the earth. The interferometer was quite sensitive enough to measure the changes in the speed of light expected from either adding or subtracting the earth's orbital speed of 30 kilometers per second to or from the 300,000-kilometer-per-second speed of light. But no change was observed. None. Something was wrong with the sensible idea that the speed of light measured by a moving receiver should be its usual speed in a vacuum, c, plus or minus the contribution from the motion of the source or receiver. Many repetitions and variations of the Michelson and Morley experiment by many investigators showed the same null result. This was a most puzzling fact at the turn of the century.

One interpretation of the bewildering result was suggested by the Irish physicist G. F. FitzGerald, who proposed that the length of the experimental apparatus shrank in the direction in which it was moving by just the amount required to counteract the presumed variation in the speed of light. FitzGerald's hypothesis accounted for the discrepancy, but since he had no suitable theory for why this was so, it wasn't taken seriously.

Einstein's interpretation was very different. If the speed of light does not conform to traditional or classical ideas, and speed is simply a ratio of distance through space to a corresponding interval of time, Einstein recognized that the classical ideas of space and time were suspect. He rejected the classical idea that space and time were independent of each other and developed, with simple postulates, a profound relationship between space and time.

Postulates of the Special Theory of Relativity

Einstein saw no need for the ether. Gone with the stationary ether was the notion of an absolute frame of reference. All motion is relative, not to any stationary hitching post in the universe, but to arbitrary frames of reference. A rocket ship cannot measure its speed with respect to empty space, but only with respect to other objects. If, for example, rocket ship A drifts past rocket ship B in empty space, spaceman A and spacewoman B will each observe the relative motion, and from this observation each will be unable to determine which is moving and which is at rest, if either.

This is a familiar experience to a passenger on a train who looks out his window and sees the train on the next track moving by his window. He is aware only of the relative motion between his train and the other train and cannot tell which train is moving. He may be at rest relative to the ground and the other train may be moving, or he may be moving relative to the ground and the other train may be at rest, or they both may be moving relative to the ground. The important point here is that if you were in a train with no windows, there would be no way to determine whether the train was moving with uniform velocity or was at rest. This is Einstein's first postulate:

All laws of nature are the same in all uniformly moving frames of reference.

Fig. 33.2 The speed of light is measured to be the same in all frames of reference.

On a jet airplane going 700 kilometers per hour, for example, coffee pours as it does when the plane is at rest; we flip a coin and catch it just as we would if the plane were at rest; we swing a pendulum and it swings as it would if the plane were at rest on the runway. There is no physical experiment we can perform to determine our state of uniform motion. Of course, we can look out and see the earth whizzing by or send a radar wave out, or some such thing as that; we mean that no experiment confined within the cabin itself can determine whether or not there is uniform motion. The laws of physics within the uniformly moving cabin are the same as those in a stationary laboratory.

Any number of experiments can be devised to detect accelerated motion, but none can be devised, according to Einstein, to detect the state of uniform motion. We can measure uniform velocity only with respect to some frame of reference.

It would be very peculiar if the laws of physics varied for observers moving at different speeds. It would mean, for example, that a pool player on a smooth-moving ocean liner would have to adjust her style of play to the speed of the ship, or even to the season as the earth varies in its orbital speed about the sun. It is our common experience that no such adjustment is necessary, and no experiment, however precise, has shown any difference at all. That is what the first postulate of relativity means.

One of the questions that Einstein as a youth asked his schoolteacher was, "What would a light beam look like if you traveled along beside it?" According to classical physics, the beam would be at rest to such an observer. The more Einstein thought about this, the more convinced he became of its impossibility. He finally came to the conclusion that if an observer could travel at the speed of light, he would measure the light leaving him at 300,000 kilometers per second. This was the second postulate in his special theory of relativity:

The speed of light in free space will be found to have the same value regardless of the motion of the source or the motion of the observer; that is, the speed of light is invariant.

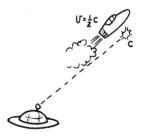

Fig. 33.3 The speed of a light flash emitted by the space station is measured to be c by observers on both the space station and the rocket ship.

To illustrate this statement, consider a rocket ship departing from the space station shown in Figure 33.3. A flash of light is emitted from the station at 300,000 kilometers per second, or c. Regardless of the velocity of the rocket, an observer in the rocket sees the flash of light pass her at the same speed c. If a flash is sent to the station from the moving rocket, observers on the station will measure the speed of the flash to be c. The speed of light is measured to be the same regardless of the speed of the source or receiver. *All* observers who measure the speed of light will find it has the same value c. The explanation for this has to do with the nature of space and time. As strange as it seems, the observer on the station and the observer in the rocket are measuring *different distances* and *different times*.

The measurement of space and time differs for all observers who are moving through space with respect to each other. Anyone who observes particular events while moving relative to you, for example, measures different distances and times between these events than you do. Anyone who moves relative to you is in a different realm of time! The differences in the measurements of space and time are negligible at ordinary speeds, but as relative motion approaches the speed of light, the differences are appreciable. We say that objects moving with respect to one another are in different realms of **space-time**.

Space-Time

When we look up at the stars, we realize that we are actually looking backward in time. The stars we see farthest away are the stars we are seeing longest ago. The more we think about this, the more apparent it becomes that space and time must be intimately tied together.

The space we live in is three-dimensional; that is, we can specify the position of any location in space with three dimensions. Loosely speaking, these dimensions would be how far over, how far across, and how far up or down. For example, if we are at the corner of a rectangular room and wish to specify the position of any point in the room, we can do this with three numbers. The first would be the number of meters the point is along a line joining the left wall and the floor; the second would be the number of meters the point is along a line joining the adjacent right wall and the floor; and the third would be the number of meters the point lies above the floor, or along the vertical line joining the walls at the corner. Physicists speak of these three lines as the *coordinate axes* of a reference frame (Figure 33.4). Three numbers—the distances along the x axis, y axis, and z axis—will specify the position of a point in space.

We specify the size of objects with three dimensions. A box, for example, is described by its length, width, and height. But the three dimensions do not give a complete picture. There is a fourth dimension—time. The box was not always a box of given length, width, and height. It began as a box only at a certain point in time, on the day it was made. Nor will it always be a box. At any moment it may be crushed, burned, or destroyed. So the three dimensions of space are a valid description of the box only during a certain specified period of time. We cannot speak meaningfully about space without implying time. Each object, each person, each planet, each star, each galaxy exists in what physicists call "the space-time continuum."[1]

Two side-by-side observers at rest relative to each other share the same reference frame. Both would agree on measurements of space and time between given events, so we say they share the same realm of space-time.

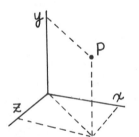

Fig. 33.4 Point P can be specified with three numbers: the distance along the x axis, y axis, and z axis.

[1] Space and time may be quantized points in a four-dimensional space-time lattice. From the sizes of elementary particles and minimum separation between colliding particles, there seems to be an elemental unit of distance of 6.6×10^{-16} m, and we find the lifetimes of all known elementary particles are consistent with being an integral number of "chronons" of about 2×10^{-23} s.

Fig. 33.5 All space and time measurements of light are unified by *c*.

If there is relative motion between them, however, they will not agree on these measurements of space and time. At ordinary speeds, differences in their measurements are imperceptible, but at relativistic speeds—that is, speeds near the speed of light—the differences are appreciable. Each observer is in a different realm of space-time, and her measurements of space and time differ from the measurements of an observer in some other realm of space-time. The measurements do not differ haphazardly but differ in such a way that each observer will always measure the same ratio of space and time for light in the reference frame of the other (or in any reference frame); the greater the measured traversal in space, the greater the measured interval of time. This constant ratio of space and time for light, *c,* is the unifying factor between different realms of space-time and is the essence of Einstein's second postulate.

Time Dilation

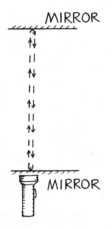

Fig. 33.6 A light clock. Light will bounce up and down between parallel mirrors and "tick off" equal intervals of time.

Time is not an absolute but is relative to the motion between the observer and the event being observed. Imagine, for example, that we are somehow able to observe a flash of light bouncing back and forth between a pair of parallel mirrors. If the distance between the mirrors is fixed, then the arrangement constitutes a sort of "light clock," because the back and forth trips of the light flash take equal intervals of time (Figure 33.6). Suppose our light clock is inside a transparent high-speed spaceship. If we travel along with the ship and watch the light clock (Figure 33.7*a*), we will see the flash of light reflecting straight up and down between the two mirrors, just as it would if the spaceship were at rest. Our observations will show no relativistic effects because there is no relative motion between us and our light clock; we share the same reference frame in space-time.

Fig. 33.7 (*a*) An observer moving with the ship observes the light beam moving vertically between the mirrors. (*b*) An observer at rest with respect to the ship observes light moving along a longer diagonal path.

If, instead, we stand at some rest position and observe the spaceship whizzing by us at an appreciable speed—say, half the speed of light—things are quite different. We no longer share the same reference frame, for in this case there is relative motion between the observer and the observed. We will not see the path of the light in simple up-and-down motion as before. Because the light flash keeps up with the horizontally moving light clock, we will see the flash following a diagonal path (Figure 33.7*b*). Notice that the flash travels a *longer distance* as it moves between the mirrors in our position of space-time than in the reference frame of an observer riding with the ship. Since the speed of light is the same in all reference frames (Einstein's second postulate), the flash must travel for a *longer time* between the mirrors in our frame than in the reference frame of an observer on board. This follows from the definition of speed, simply stated, as a ratio of distance to time. *The longer diagonal distance must be divided by a correspondingly longer time interval to yield an unvarying value for the speed of light.* Thus, from our relative position of rest, we measure a longer time interval between ticks when a clock is in motion than when it is at rest. We have considered a light clock in our example, but the same is true for any kind of clock. Moving clocks appear to run slow. This stretching out of time is **time dilation**.

The exact relationship between time intervals in different frames in space-time can be derived from Figure 33.7 with simple geometry and algebra.[2] The relationship between the time t_0 in the observer's own frame

---

[2]The light clock is shown in three successive positions in the figure below. The diagonal lines represent the path of the light flash as it starts from the lower mirror at position A, moves to the upper mirror at position B, and then back to the lower mirror at position C. Distances on the diagram are marked ct, vt, and ct_0, which follow from the fact that the distance traveled by a uniformly moving object is equal to its velocity multiplied by the time.

The symbol t_0 represents the time it takes for the flash to move between the mirrors as measured from a frame of reference fixed to the light clock. The velocity of light is c, and the path of light is seen to move a vertical distance ct_0. This is the distance between mirrors, which is the same in both reference frames, since the relative motion of the light clock is entirely horizontal.

The symbol t represents the time it takes for the flash to move from one mirror to the other as measured from a frame of reference in which the light clock is moving with velocity v. Since the velocity of the flash is c and the time of travel from the lower mirror at position A to the upper mirror at position B is t, the diagonal distance traveled is ct. During this same time t, the clock (which travels at velocity v) moves a horizontal distance vt from position A to position B.

These three distances make up a right triangle in the figure, in which ct is the hypotenuse, and ct_0 and vt are the legs. A well-known theorem of geometry (the Pythagorean theorem) states that the square of the hypotenuse is equal to the sum of the squares of the other two sides. Applying this to the figure, we obtain

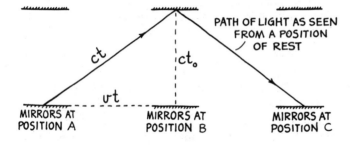

PATH OF LIGHT AS SEEN
FROM A POSITION
OF REST

MIRRORS AT POSITION A

MIRRORS AT POSITION B

MIRRORS AT POSITION C

$$c^2t^2 = c^2t_0^2 + v^2t^2$$

$$c^2t^2 - v^2t^2 = c^2t_0^2$$

$$t^2[1 - (v^2/c^2)] = t_0^2$$

$$t^2 = \frac{t_0^2}{1 - (v^2/c^2)}$$

$$t = \frac{t_0}{\sqrt{1 - (v^2/c^2)}}$$

of reference and the relative time t measured in another reference frame is

$$t = \frac{t_0}{\sqrt{1 - \frac{v^2}{c^2}}}$$

where v represents the relative velocity between the observer and the observed, and c is the velocity of light. Because no material object can travel at or beyond the speed of light, the ratio v/c is always less than 1; likewise v^2/c^2. For $v = 0$, this ratio is zero, and for everyday speeds where v is negligibly small compared to c, it's practically zero. Then $1 - (v^2/c^2)$ has a value of 1, as has $1 - (v^2/c^2)$, and we find $t = t_0$ and time intervals appear the same in both systems. For higher speeds, v/c is between zero and 1, and $1 - \sqrt{(v^2/c^2)}$ is less than 1; likewise, $\sqrt{1 - (v^2/c^2)}$. So t_0 divided by a value less than 1 produces a value greater than t_0, an elongation, a dilation of time.

To consider some numerical values, assume that v is equal to 50 percent the speed of light. Then we substitute $0.5c$ for v in the time-dilation equation and after some arithmetic find that $t = 1.15t_0$. This means that if we viewed a clock on a spaceship traveling at half the speed of light, we would see the second hand take 1.15 minutes to make a revolution; whereas if it were at rest, we would see it take 1 minute. If the spaceship passes us at 87 percent the speed of light, $t = 2t_0$, we would measure time events on the spaceship taking twice the usual intervals. It would appear as if events on the ship were in slow motion. At 99.5 percent the speed of light, $t = 10t_0$, we would see the second hand of the spaceship's clock take 10 minutes to sweep through a revolution requiring 1 minute on our clock.

To put these figures another way, at 99.5 percent c, the moving clock would appear to run a tenth of our rate and would move only 6 seconds while our clock's second hand ticks 60 seconds. At 87 percent c, the moving clock ticks at half rate and shows 30 seconds to our 60 seconds; at 50 percent c, the moving clock ticks 1/1.15 as fast and ticks 52 seconds to our 60 seconds. We see that moving clocks run slow.

Nothing is unusual about a moving clock itself; it is simply ticking to the rhythm of a different time. The faster a clock moves, the slower it appears to run as viewed by an observer not moving with the clock. If it were possible for an observer to watch a clock pass by at the speed of light, it would not appear to be running at all. This observer would measure the interval between ticks to be infinite. The clock would be ageless! If our observer were moving with the clock, however, the clock would not show any slowing down of time. To him the clock would be operating normally. This is because there would be no motion between the observer and the observed. The v in the time-dilation equation would then be zero, and $t = t_0$; they share the same frame in space-time. If the person who whizzes past us checked a clock in our reference frame, however, he would find our clock to be running as slowly as we find his to be. We each see each other's clock running slow. There is really no contradiction here, for it is physically impossible for two observers moving at different velocities to refer to one and the same realm of space-time. The measurements made in one realm of space-time need not agree with the measurements made in another realm of space-time. The measurement they will always agree on, however, is the velocity of light.

Time dilation has been confirmed in the laboratory innumerable times with atomic particle accelerators. The lifetimes of fast-moving radioactive particles increase as the speed goes up, and the amount of increase is just what Einstein's equation predicts.

Time dilation has been confirmed also for not-so-fast motion. During October 1971, four cesium-beam atomic clocks were flown on regularly scheduled commercial jet flights around the world twice, once eastward and once westward, to test Einstein's theory of relativity with macroscopic clocks. The clocks indicated different times after their round trips. Relative to the atomic time scale of the U.S. Naval Observatory, the observed time differences, in billionths of a second, were in accord with relativistic prediction.

This all seems very strange to us only because it is not our common experience to deal with measurements made at relativistic speeds or atomic-clock-type measurements at ordinary speeds. The theory of relativity does not make common sense. But common sense, according to Einstein, is that layer of prejudices laid down in the mind prior to the age of eighteen. If we spent our youth zapping through the universe in high-speed spaceships, we would probably be quite comfortable with the results of relativity.

Question   Does time dilation mean that time really passes more slowly in moving systems or that it only *seems* to pass more slowly?*

The Twin Trip   A dramatic illustration of time dilation is the classic hypothesis about what would result if one of two twins took a high-speed round-trip space voyage. Their ages would be different when the astronaut twin returned, and the difference would reflect the different space-time in which each had existed. The astronaut would be younger than the twin who had stayed on earth; how much younger would depend on the duration of the trip and the relative velocities involved. If the astronaut had maintained a speed of 50 percent of the speed of light for 1 year (on the clocks aboard

---

*Answer   The slowing of time in moving systems is not merely an illusion resulting from motion. Time really does pass more slowly in a moving system compared to one at relative rest, as we shall see in the following section. Read on!

the spaceship), 1.15 years would have elapsed on earth. If he had maintained 87 percent c for 1 year, then 2 years would have elapsed on earth. At 99.5 percent c, 10 years would have passed—*but the astronaut would have aged only 1 year.*

The question often arises: Since motion is relative, why can't it be the other way around? Why would the traveling twin not return to find his stay-at-home twin younger than himself? We will show that from the frames of reference of both the earthbound twin and the astronaut twin, it is the earthbound twin who ages more. And we'll do this in the framework of uniform motion by assuming that the rocket ship somehow undergoes its periods of acceleration (taking off, turning around, and landing) in negligibly short intervals. Let's begin by comparing time intervals as seen in different frames of reference—first when the frames are at rest and then when moving relative to one another.

Fig. 33.8

Suppose that a sender on a rocket ship that hovers motionless in space emits regularly spaced brief bursts of light to a distant receiver who is at rest relative to the sender (Figure 33.9). Some time will elapse before the receiver sees the first of these light flashes, just as 8 minutes of time elapse before sunlight reaches the earth. The light flashes will encounter the receiver at speed c, and since there's no relative motion between sender and receiver, successive flashes will be received as frequently as they are sent from the source. With no motion involved, there is nothing unusual about this.

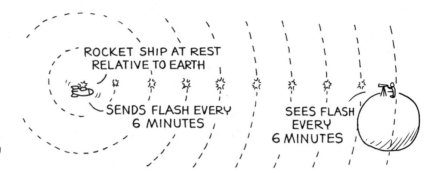

Fig. 33.9

When motion is involved, this works quite differently. When the ship is approaching, the receiver sees the flashes more frequently—that is, at a higher frequency. This happens not only because time is altered due to motion, but mainly because each succeeding flash has less distance to travel as the ship approaches. On the other hand, if the ship is receding, each succeeding flash must travel a greater distance, and the duration between flashes is extended; the receiver sees flashes less frequently—that is, at a lower frequency. This is like the altered frequency of sound waves from a moving automobile, the Doppler effect we discussed in Chapter 17.

Now suppose the rocket ship travels *toward* the receiver while emitting flashes at 6-minute intervals. Although the receiver will still measure the speed of the flashes to be c, he will receive the flashes more frequently. Suppose the rocket ship is traveling fast enough for the frequency of flash reception to be doubled, so the flashes are seen every 3 minutes (Figure 33.10). Then 10 flashes emitted by the approaching ship in 1 hour, ship time, would be received in $\frac{1}{2}$ hour, receiver time. If, on the other hand, the ship *recedes* from the receiver at the same speed and still emits flashes at 6-minute intervals, these flashes will be seen half as frequently by the

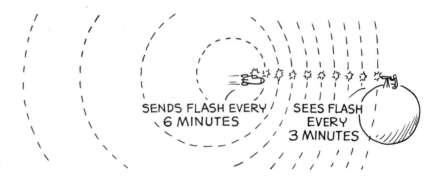

Fig. 33.10

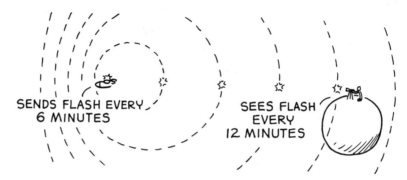

SENDS FLASH EVERY
6 MINUTES

SEES FLASH
EVERY
12 MINUTES

Fig. 33.11

receiver—at 12-minute intervals[3] (Figure 33.11). Thus, events that take 1 hour aboard a *receding* ship appear to take 2 hours from the point of view of the receiver—or $\frac{1}{2}$ hour if the ship is *approaching*.

Let's apply this doubling and halving of flash intervals to our twins. Suppose that one twin is in a rocket ship receding from his earthbound twin brother and that the ship travels as fast as before, emitting flashes at 6-minute intervals. Then during 1 hour, ship time, a total of 10 flashes will be emitted. For example, if the ship departs from the earth at noon, a clock aboard the ship will read 1:00 when the tenth flash is emitted. Later, when the tenth flash reaches the earth, an earth clock will read 2:00—for the 10 flashes are received at 12-minute intervals. Suppose the ship has somehow turned around in a negligibly short time and is returning at the same high speed. During the 1 hour it takes to return, 10 more flashes are emitted at 6-minute intervals, so a clock aboard the rocket ship reads 2:00 just as the ship reaches the earth and the tenth flash is emitted. But these 10 flashes

---

[3]This reciprocal relationship (halving and doubling of frequencies) is a consequence of the constancy of the speed of light, and can be illustrated with the following example: Suppose a sender on earth emits flashes 3 min. apart to a distant observer on a planet that is at rest relative to the earth. The observer, then, sees a flash every 3 min. Now suppose a second observer travels in a rocket ship between the earth and the planet, at a speed great enough to allow him to see the flashes half as frequently—6 min. apart. This halving of frequency occurs for a speed of recession of 0.6c. We can see that the frequency will double for a 0.6c speed of *approach*, by supposing that the rocket ship emits its own flash every time it sees an earth flash, that is, every 6 min. How does the observer on the distant planet see these flashes? Since the earth flashes and the rocket flashes travel together at the same speed c, the observer will see not only the earth flashes every 3 min., but the rocket flashes every 3 min. as well. He'll see them at twice the emitting frequency. So for a speed of recession where frequency appears halved, frequency appears doubled for the same speed of approach. If the ship were traveling faster so that the frequency of recession were $\frac{1}{3}$ or $\frac{1}{4}$ as much, then the frequency of approach would be three- or fourfold, respectively. This reciprocal relationship does not hold for waves that require a medium. In the case of sound waves, for example, a speed that results in a doubling of emitting frequency for approach produces $\frac{2}{3}$ (not $\frac{1}{2}$) the emitting frequency for recession.

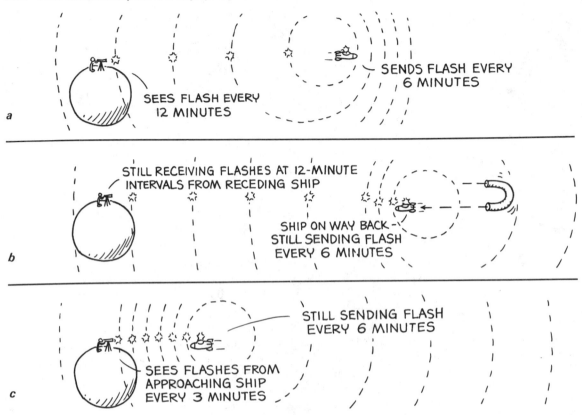

a

SEES FLASH EVERY
12 MINUTES

SENDS FLASH EVERY
6 MINUTES

b

STILL RECEIVING FLASHES AT 12-MINUTE
INTERVALS FROM RECEDING SHIP

SHIP ON WAY BACK -
STILL SENDING FLASH
EVERY 6 MINUTES

c

STILL SENDING FLASH
EVERY 6 MINUTES

SEES FLASHES FROM
APPROACHING SHIP
EVERY 3 MINUTES

Fig. 33.12

appear at 3-minute intervals to the earthbound twin—that's 30 minutes. So a clock on earth will read 2:30 (Figure 33.13). The earthbound twin has aged $\frac{1}{2}$ hour more than the twin aboard the rocket ship!

The result is the same for either frame of reference. Consider the same trip again, only this time with flashes emitted *from the earth* at regularly spaced 6-minute intervals, earth time. From the frame of reference of the

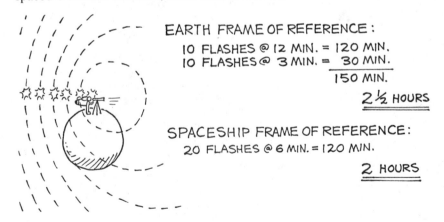

EARTH FRAME OF REFERENCE:
10 FLASHES @ 12 MIN. = 120 MIN.
10 FLASHES @ 3 MIN. = 30 MIN.
                       150 MIN.
                       2½ HOURS

SPACESHIP FRAME OF REFERENCE:
20 FLASHES @ 6 MIN. = 120 MIN.
                      2 HOURS

Fig. 33.13

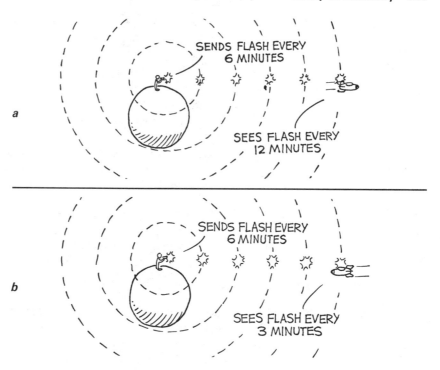

Fig. 33.14

receding rocket ship, these flashes are received at 12-minute intervals (Figure 33.14a), so only 5 flashes are seen during the 1 hour of departure from the earth. During the rocket ship's hour in returning, the light flashes are received at 3-minute intervals (Figure 33.14b), so 20 flashes will be received. Hence, the rocket ship receives a total of 25 flashes during its 2-hour trip. According to clocks on the earth, however, the 25 6-minute flashes were emitted in 150 minutes or $2\frac{1}{2}$ hours (Figure 33.15).

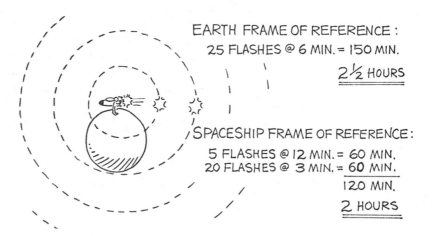

Fig. 33.15

So both twins agree on the same results, with no dispute as to who aged $\frac{1}{2}$ hour more than the other. From either frame of reference, the twin on earth ages more than the twin in the high-speed rocket ship, for it is the twin aboard the ship who undergoes changes in space-time. His acceleration in taking off, turning around, and landing marks the changing of one frame of reference in space-time to another. Whereas the twin on earth remains in a single space-time throughout the voyage, the twin aboard the rocket ship is in a completely different realm of time while traversing space in receding from the earth and in still another realm of time while traversing space in approaching the earth. They can meet again at the same place in space only at the expense of time.

Question   Since motion is relative, can't we as well say that the rocket ship is at rest and the earth moves, in which case the twin on the rocket ship ages more?*

Space Travel   One of the old arguments advanced against the possibility of human interstellar travel is that our life span is too short—at least for the distant stars. It was argued, for example, that the nearest star (after the sun), Alpha Centauri, is 4 light-years away, and a round trip even at the speed of light would require 8 years.[4] The center of our galaxy is some 30,000 light-years away, so it has been reasoned that a person traveling even at the speed of light would require a lifetime of 30,000 years to make such a voyage. But these arguments fail to take into account time dilation. Time for a person on earth and time for a person in a high-speed rocket ship are not the same.

A person's heart beats to the rhythm of the realm of time it is in; and one realm of time seems the same as any other realm of time to the person, but not to an observer who stands outside the person's frame of reference—for she can see the difference. For example, astronauts traveling at 99 percent the speed of light could go to the star Procyon (10.4 light-years distant) and back in 21 years. It would take light itself 20.8 years to make the same round trip. Because of time dilation, it would seem that only 3 years had gone by to the astronauts. This is what all their clocks would tell them—and biologically they would be only 3 years older. It would be the space officials greeting them on their return who would be 21 years older!

****Answer**   Yes, but only if the earth then undergoes the turnaround and returns, as our rocket ship did in the twin-trip example. The situation is not symmetrical, for one twin remains in a single reference frame in space-time during the trip while the other makes a distinct change of reference frame as evidenced by the acceleration in turning around.

[4] A *light-year* is the distance light travels in 1 year, about 9.45×10^{12} km.

At higher speeds the results are even more impressive. At a rocket speed of 99.90 percent the speed of light, travelers could travel slightly more than 70 light-years in a single year of their own time; at 99.99 percent the speed of light, this distance would be pushed appreciably farther than 200 light-years. A 5-year trip for them would take them farther than light travels in 1000 earth-time years!

Present technology does not permit such journeys. Radiation is the greatest problem. A ship traveling at speeds close to the speed of light would encounter interstellar particles just as if it were on the launching pad and a steady stream of particles shot by an atomic accelerator were incident on it. No way of shielding such intense particle bombardment for prolonged periods of time is presently known. And if somehow a way were devised to solve this problem, there would be the problem of energy and fuel. Spaceships traveling at relativistic speeds would require billions of times the energy used to put a space shuttle into orbit. Even some kind of interstellar ramjet that scooped up interstellar hydrogen gas for burning in a fusion reactor would have to overcome the enormous retarding effect of scooping up the hydrogen at high speeds. The practicalities of such space journeys are so far prohibitive.

If and when these problems are overcome and space travel becomes a routine experience, people will have the option of taking a trip and returning to any future century of their choosing. For example, one might depart from earth in a high-speed ship in the year 2050 and travel for 5 years or so and return in the year 2500. One could live among the earthlings of that period for a while and depart again and try out the year 3000 for style. People could keep jumping into the future with some expense of their own time, but they could not trip into the past. They could never return to the same era on earth that they bid farewell to. Time, as we know it, travels one way—forward. Here on earth we move constantly into the future at the steady rate of 24 hours per day. A deep-space astronaut leaving on a deep-space voyage must live with the fact that, upon her return, much more time will have elapsed on earth than she has subjectively experienced on her voyage. The credo of all star travelers, whatever their physical condition, will be permanent farewell.

We can see into the past, but we cannot go into the past. For example, we experience the past when we look at the night skies. The starlight impinging on our eyes left those stars dozens, hundreds, even millions of years ago. What we see is the stars as they were long ago. We are thus eyewitnesses to ancient history—and can only speculate about what may have happened to the stars in the interim.

If we are looking at light that left a star, say, 100 years ago, then it follows that any sighted beings in that solar system are seeing us by light that left *here* 100 years ago and that, further, if they possessed super-telescopes, they might very well be able to eyewitness earthly events of a century ago—the aftermath of the American Civil War, for instance. They would see our past, but they would still see events in a forward direction; they would see our clocks running clockwise.

We can speculate about the possibility that time might just as well move counterclockwise into the past as clockwise into the future. In fact, the whole question of "time reversal" is discussed in great mathematical detail and is being studied by some physicists in a number of exquisite and elaborate experiments with subatomic particles. It is hypothesized that particles that move faster than light and backward in time, called "tachyons," may in fact exist. Experiments to detect them have proved unpromising to date. Nearly all physicists conclude that the only direction of time is forward.

This conclusion is blithely ignored in a limerick that is a favorite with scientist types:

> There was a young lady named Bright
> Who traveled much faster than light.
> She departed one day
> In an Einsteinian way
> And returned on the previous night.

We can get our heads fairly well into relativity yet still unconsciously cling to the idea that there is an absolute time and compare all these relativistic effects to it: recognizing that time changes this way and that way for this speed and that speed, yet feeling that there still is some basic or absolute time. We may tend to think that the time we experience on earth is fundamental and that other times differ from it. This is understandable; we're earthlings. But the idea is confining. To observers elsewhere in the universe, we may be moving at relativistic speeds from their points of view and they see us moving in slow motion. They may see us living lifetimes a hundredfold theirs, just as with supertelescopes we would see them living lifetimes a hundredfold ours. There is no universally standard time. None.

We think of time and then we think of the universe. We think of the universe and we wonder about what went on before the universe began. We wonder about what will happen if the universe ceases to exist in time. But the concept of time applies to events and entities within the universe—not to the universe as a whole. Time is "in" the universe; the universe is not "in" time. Without the universe, there is no time; no before, no after. Likewise, space is "in" the universe; the universe is not "in" a region of space. There is no space "outside" the universe. Space-time exists within the universe. Think about that!

Length Contraction

As objects move through space-time, space as well as time undergoes changes in measurement. The lengths of objects appear to be contracted when they move by us at relativistic speeds. This contraction was first proposed by FitzGerald and mathematically expressed by Hendrick A. Lorentz, a Dutch physicist. It is referred to as the *Lorentz-FitzGerald contraction*. We express it mathematically as

$$L = L_0 \sqrt{1 - \frac{v^2}{c^2}}$$

where v is the relative velocity between the observed object and the observer, c is the speed of light, L is the measured length of the moving object, and L_0 is the measured length of the object at rest.[5] Suppose that an object is at rest and $v = 0$. Upon substitution of $v = 0$ in the equation, we find $L = L_0$, as we would expect. At 87 percent the speed of light, an object would appear to be contracted to half its original length. At 99.5 percent the speed of light, it would seem to contract to one-tenth its original length. If the object moved at c, its length would be zero. This is one of the reasons we say that the speed of light is the upper limit for the speed of any moving object. Another ditty popular with the science heads is

There was a young fencer named Fisk,
Whose thrust was exceedingly brisk.
So fast was his action
The Lorentz-FitzGerald contraction
Reduced his rapier to a disk.

As Figure 33.16 indicates, contraction takes place only in the direction of motion. If an object is moving horizontally, no contraction takes place vertically.

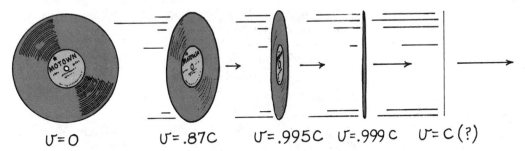

$U = 0$ $U = .87C$ $U = .995C$ $U = .999C$ $U = C(?)$

Fig. 33.16 The Lorentz-FitzGerald contraction.

Do objects *really* contract at relativistic speeds? Well, if we attempt to check this out (in principle) and travel alongside the moving object with a meter stick, we note nothing at all unusual about the length of the object. An observer at rest watching this experiment would report that the reason we don't measure the contraction, which is obvious to him, is that our meter stick is shortened as well. But because we are moving with the object, it appears to us that there is no contraction. This is because our

[5]Note that we do not explain how this equation comes about (nor those that follow). We simply state them as "guides to thinking" about the ideas of special relativity.

relative velocity with respect to the object being measured is zero. The v in the Lorentz-FitzGerald equation refers to the relative velocity between the observed and the observer. And that is zero, so $L = L_0$. It is important to stress this point, for many of the misconceptions regarding relativity have their basis on this point. The object doesn't contract. A measure of the object from another reference frame contracts. We are simply measuring the distortion of space when we measure such a contraction, just as we measure the distortion of time itself when we find clocks running slow. The distortions are of space and time between different space-times, not of objects and events within individual realms of space-time.

Question	A rectangular billboard in space has the dimensions 10 m × 20 m. How fast and in what direction with respect to the billboard would a space traveler have to pass for the billboard to appear square?*

Increase of Mass with Speed

If we push on an object, it will accelerate. If we maintain a steady push, it will accelerate to higher and higher speeds. If we push with a greater and greater force, the acceleration in turn will increase and the speeds attained should seemingly increase without limit. But there is a speed limit in the universe—c. In fact, we cannot accelerate any material object enough to reach the speed of light, let alone surpass it.

Why this is so can be understood from Newton's second law. Recall that the acceleration of an object depends not only on the impressed force, but on the mass as well: $a = F/m$. Einstein stated that when work is done to increase the velocity of an object, its mass increases as well. So an impressed force produces less and less acceleration as velocity increases. The relationship between mass and velocity is given by

$$m = \frac{m_0}{\sqrt{1 - \dfrac{v^2}{c^2}}}$$

Here m represents the measured mass or relativistic mass of an object pushed to any speed v. The symbol m_0 is the *rest mass*, the mass the object would have at rest. Again, v represents the relative velocity between the observer and the observed.

The faster a particle is pushed, the more its mass increases, thereby resulting in less and less response to the accelerating force. An investigation of the relativistic mass equation shows that as v approaches c, m approaches infinity! A particle pushed to the speed of light would have

---

*Answer The space traveler would have to travel at $0.87c$ in a direction parallel to the longer side of the board.

infinite mass and would require an infinite force, which is clearly impossible. Therefore, we say that no material particle can be accelerated to the speed of light.

Atomic particles have been accelerated to speeds in excess of 99 percent the speed of light, however. Their masses increase thousandsfold, as evidenced when a beam of particles, usually electrons or protons, is directed into a deflecting magnetic field. The more massive particles do not bend as readily as they would if they did not undergo the relativistic mass increase (Figure 33.17). They strike their targets at positions predicted by the relativistic equation. This mass increase must be compensated for in circular accelerators like cyclotrons and bevatrons, where mass dictates the radius of curvature. One of the advantages of the linear accelerator is that the particle beam travels in a straight path and mass changes do not produce deviations from a straight-line path. Mass increase with velocity is an everyday fact of life to physicists working with high-energy particles.

Fig. 33.17 If the mass of the electrons did not increase with speed, the beam would follow the path shown by the dashed line. But because of its increased inertia, the high-speed beam is not deflected as much.

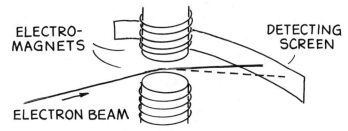

Mass-Energy Equivalence

The most remarkable aspect of special relativity is Einstein's law of the equivalence of mass and energy.

We have seen that the energy pumped into nuclear particles in an accelerator increases their mass. This is because energy and mass are equivalent to each other. We all know that the energy that goes into accelerating the particles comes from a power plant someplace. Energy generated at the power plant likely results from either the chemical combustion or the nuclear reaction of certain fuels. If the process is chemical combustion, the masses of hydrocarbon molecules produced by combustion are reduced by about 1 part in 10^9. If the process is nuclear fission, the masses of the fission fragments after reaction are reduced by about 1 part in 1000 the masses of the uranium atoms before reaction. In both cases mass is converted to its other form, energy. So the fuel fragments have less mass than before reaction. How much less? Total? Neglecting the inefficiencies of power transmission, the fuel fragments lose just as much mass as the particles at the accelerator gain. So the power company delivers mass every bit as much as it delivers energy. To say it delivers one is to say it delivers the other, because mass and energy in a practical sense are one and the same.

Einstein's deduction of this **mass-energy equivalence** is expressed in what might well be called the equation of the twentieth century:

$$E = mc^2$$

Owing to the large magnitude of c, the velocity of light, a small mass corresponds to an enormous quantity of energy.[6] For example, the energy equivalent of a single gram of any matter is greater than the energy used daily by the populations of our largest cities.[7]

The interchange between mass and energy is a common phenomenon and occurs for *all* changes of energy. The change in mass is ordinarily so slight that it has not been detected until recent times. When we strike a match, for example, the chemical reaction that is set off is accompanied by a tiny bit of mass converting to radiation and the kinetic energy of the hot gases. That's where the energy comes from! When the phosphorus atoms in the match head rearrange themselves and combine with oxygen in the air to form new molecules, the resulting molecules are a very slight bit less massive than the masses of the separate phosphorus and oxygen molecules. From a mass standpoint, the whole is slightly less than the sum of its parts. But not very much; about 1 part in 10^9. This slight decrease in mass takes the form of heat and radiant energy. For all chemical reactions, the mass difference is so tiny that it is beyond today's range of measurement and is practically unimportant. But measurable mass differences accompany nuclear reactions. When hydrogen nuclei (bare protons) combine to form helium nuclei, almost 0.1 percent of their mass is transformed to radiant and kinetic energy. This happens in the stars. The mass of the sun decreases at the rate of $4\frac{1}{2}$ million tons each second. It is this decrease in mass that bathes the solar system with radiant energy and nourishes life. This rate of decreasing mass may seem alarming at first, but the sun is mighty big. At the end of 1 million years, only one ten-millionth of the sun's mass will have been converted to radiation. It's comforting to have so big a sun!

The equation $E = mc^2$ is not restricted to chemical and nuclear reactions. *Any* change in energy corresponds to a change in mass. The increased kinetic energy of a golf ball is accompanied by an increase in

---

[6] The c^2 is the conversion factor for energy units and mass units and comes from the relativistic expression for mass. With a redefinition of space and time units, c could equal 1, and $E = mc^2$ would simply be $E = m$. If the equivalence of mass and energy were understood long ago when physics concepts were first being formulated, there would likely be no separate units for mass and for energy.

[7] Put another way, 1 gram is 10^{-3} kg. From Einstein's equation, we find that $E = mc^2 = (10^{-3}$ kg$)(3 \times 10^8$ m/s$)^2 = 9 \times 10^{13}$ J, which is 25 million kWh (kilowatt-hours). So rather than say a power company delivers 25 million kWh of energy to a city, we could as well say it delivers 1 gram of energy. Physicists, in fact, express the mass of particles in energy units. Instead of saying the mass of a proton is 1.67×10^{-27} kg, we say its mass is 938 MeV (million electron volts). An electron volt is a tiny unit of energy; 1 eV $= 1.6 \times 10^{-19}$ J.

mass. A light-bulb filament energized with electricity is more massive than when it is turned off. A hot cup of coffee is more massive than the same cup of coffee when cold. A wound-up spring clock is more massive than the same clock when it is unwound. But these examples involve small energy changes and therefore correspond to incredibly small mass changes. For large energy changes, the effect is more noticeable. Electric energy provides a good example. Work is required to bring together electrically charged particles of the same sign. A fantastically huge amount of work would be required to confine a single gram of bare electrons in a small region because of electrical repulsion. One gram of electrons confined to a sphere of 10-centimeters diameter would have a mass of 10 trillion kilograms!

The mass of a single electron is merely the energy equivalence of the work required to compress its charge. If its charge were infinitely spread out, the amount of work needed to confine it to the classical radius of the electron, divided by c^2, would equal its mass.

Sidewalk astronomer John Dobson speculates that just as a clock becomes more massive when we do work on it by winding it against the resistance of its spring, the mass of the entire universe is nothing more than the energy that has gone into winding it up against mutual gravitation. In this view the mass of the universe is equivalent to the work done in spacing it out. So electrons have mass because their charge is confined, and the atoms that make up the universe have mass because they are dispersed.

$E = mc^2$ is more than a formula for the conversion of mass into energy or vice versa: it states, for all practical purposes, that energy and mass are identical. Mass is simply congealed energy. If you want to know how much energy is in a system, measure its mass. For a body at rest, its energy *is* its mass. It is energy itself that is hard to shake.

Question   Does the equation $E = mc^2$ mean that matter turns into pure energy when it is traveling at the speed of light squared?*

The first direct proof that radiant energy could be converted to its mass form was found in 1932 in a photograph emulsion that was being used to catch cosmic rays. C. D. Anderson, an American physicist, found that a photon of gamma radiation that had entered the emulsion had changed into a pair of particles. One of the particles was an electron. The other

---

*__Answer__   It does not! As matter travels faster and faster, its mass increases. As it approaches the speed of light, its mass does not convert to energy but approaches infinity. $E = mc^2$ simply means that energy and matter are different sides of the same coin, with c^2 the proportionality constant, which comes from the relativistic expression for mass. Matter cannot be made to move at the speed of light, let alone at this speed squared!

particle, which had the same mass as an electron, but had a *positive* charge rather than a negative charge, was given the name of *positron*. A positron is the *antiparticle* of an electron. When a gamma ray changes from energy to mass, a pair of particles is produced. One of the particles is the antiparticle of the other, having the same mass and spin, but opposite charge. The antiparticle doesn't last long. It soon encounters a particle that is its antiparticle, and the pair is annihilated, sending out a gamma ray in the process.

Correspondence Principle

Einstein's mass-energy relationship is valid only insofar as the transformation equations for mass, length, and time are valid. Before any new theory can be accepted, it must satisfy the correspondence principle. Recall from Chapter 30 that this principle states that any new theory or any new description of nature must agree with the old where the old gives correct results. If the equations of special relativity are valid, they must correspond to those of classical mechanics when speeds much less than the speed of light are considered.

The relativity equations do in fact conform to the correspondence principle. If v is very small compared to c, the ratio v^2/c^2 is extremely small. For the velocities we ordinarily experience, this ratio may be taken to be zero, and the relativity equations become

$$t = \frac{t_0}{\sqrt{1 - 0}} = t_0$$

$$L = L_0\sqrt{1 - 0} = L_0$$

$$m = \frac{m_0}{\sqrt{1 - 0}} = m_0$$

So for "everyday" velocities, the time, length, and mass of moving objects are essentially unchanged. The equations of special relativity hold for all speeds, although they are significant only for speeds near the speed of light.

The special theory of relativity is concerned with the relationships between space and time and with the physics of uniform motion, of frames of reference moving at constant velocities with respect to one another. But what about the more general cases of accelerated frames of reference of curved motion and the like? These cases are treated in the next chapter in the discussion of Einstein's general theory of relativity, which explores the relationships between gravitation and time and between space-time geometry and gravitation.

Einstein's theories of relativity have raised many philosophical questions. For example, why does time seem to move in one direction? Has it always moved *forward*? Are there other parts of the universe, elsewhere or even here, where time moves *backward?* Is it possible that our three-dimensional perception of a four-dimensional world is only a beginning? Could there be a fifth dimension? A sixth dimension? A seventh dimension? And if so, what would the nature of these dimensions be? Perhaps these unanswered questions will be answered by the scientists of tomorrow. How exciting!

Summary of Terms

Frame of reference   A vantage point (usually a set of coordinate axes) with respect to which the position and motion of a body may be described.

Postulates of the special theory of relativity   (1) All laws of nature are the same in all uniformly moving frames of reference. (2) The speed of light in free space will be found to have the same value regardless of the motion of the source or the motion of the observer; that is, the speed of light is invariant.

Space-time   The four-dimensional continuum in which all things exist: three dimensions being the coordinates of space; and the other, of time.

Time dilation   The apparent slowing down of time of a body moving at relativistic speeds.

Length contraction   The apparent shrinking of a body moving at relativistic speeds.

Mass-energy equivalence   The relationship between mass and energy as given by the equation $E = mc^2$.

Suggested Reading

Dobson, John. *Advaita Vedanta and Modern Science*, San Francisco: Sidewalk Astronomers, 1979.

Einstein, Albert. *The Meaning of Relativity*. Princeton, N.J.: Princeton University Press, 1950. Written for the average person by Einstein himself.

Epstein, Lewis C. *Relativity Visualized*. San Francisco: Insight Press, 1983.

Gamow, George. *Mr. Thompkins in Wonderland*. New York: Macmillan, 1940. An excellent and very interesting little book.

Review Questions

1. What is a frame of reference?

2. If you were in a train with no windows, could you sense the difference between uniform motion and rest? Accelerated motion and rest?

3. What are the two postulates of special relativity?

4. Will observers A and B agree on measurements of space and time if A moves at 0.5c with respect to B? If both A and B move together at 0.5c with respect to the earth?

5. The observer in Figure 33.7b sees light bouncing a longer distance between the mirrors than the observer in *a*. Why does it then follow that the observer in *b* must see light taking a longer time between bounces?

6. For a circle, the ratio of circumference to diameter is π. For the passage of light, what is the ratio of space to time?

7. Will the speed of a flashing light appear to be greater or less when receding at relativistic speeds from the source? Will its frequency be greater or less? Why are your answers different?

8. In the case of the twins, how many frames of reference does the stay-at-home twin encounter while his astronaut twin makes his round trip? How many different reference frames does the traveling twin encounter during his round trip?

9. Why can we not travel to the stars today?

10. How do measurements of *size* vary for objects traveling at relativistic speeds?

11. How do measurements of *mass* vary for objects traveling at relativistic speeds?

12. What evidence can you cite for the changing of mass at relativistic speeds?

13. Are energy and mass related only for relativistic speeds? Cite examples.

14. What is the correspondence principle?

15. Do the relativity equations for time, length, and mass hold true for everyday speeds?

Exercises

1. A person riding on the roof of a freight-train car fires a gun pointed forward. Compare the velocity of the bullet across the ground when the train is at rest to that when the train is moving. Discuss the velocity of the bullet across the freight car if the gun is fired when the train is at rest or in motion.

2. Suppose, instead, that the person riding on top of the freight car shines a searchlight beam in the direction in which the train is traveling. Does the light beam travel faster across the ground than it would if he had shone it while standing at rest on the ground? Explain.

3. Why was the Michelson and Morley experiment considered a failure? (Have you ever encountered other examples where failure has to do not with the lack of ability but with the impossibility of the task?)

4. In Chapter 25 we learned that light travels slower in glass than in air. Does this contradict the theory of relativity?

5. We might think of the speed of light as a kind of speed limit in the universe, at least for the four-dimensional universe we comprehend. No material particle can attain or surpass this limit even when a continuous, unyielding force is exerted on it. Why is this so?

6. Since there is an upper limit on the speed of a particle, does it follow that there is also an upper limit on its momentum? On its kinetic energy? Explain.

7. Light travels a certain distance in, say, 20,000 years. How is it possible that an astronaut could travel slower than the speed of light and travel as far in a 20-year trip?

8. Could a human being who has a life expectancy of 70 years possibly make a round-trip journey to a part of the universe thousands of light-years distant? Explain.

9. A twin who makes a long trip at relativistic speeds returns younger than his stay-at-home twin sister. Could he return before his twin sister was born? Defend your answer.

10. Is it possible for a son or daughter to be biologically older than his or her parents? Explain.

11. If you were in a rocket ship traveling away from the earth at a speed close to the speed of light, what changes would you note in your pulse? In your mass? In your volume? Explain.

12. If you were on earth monitoring a person in a rocket ship traveling away from the earth at a speed close to the speed of light, what changes would you note in his pulse? In his mass? In his volume? Explain.

13. Consider a high-speed rocket equipped with a flashing light source. If the frequency of flashes when approaching is increased by a factor of 2, by how much is the period (time interval between flashes) changed? Is this period constant for a constant relative speed? For accelerated motion? Defend your answer.

14. You observe a spaceship moving away from you at speed v_1, half the speed of light. A rocket is fired from the spaceship straight ahead, and therefore also away from you, at a speed v_2, half the speed of light with respect to the spaceship. The speed of the rocket with respect to you is *not* the speed of light. The relativistic addition of velocities is given by

$$V = \frac{v_1 + v_2}{1 + \frac{v_1 v_2}{c^2}}$$

Substitute $0.5c$ for both v_1 and v_2 and show that the velocity V of the rocket relative to you is $0.8c$.

15. Substitute small values of v_1 and v_2 in the preceding equation and show that for everyday velocities V is practically equal to $v_1 + v_2$.

16. Pretend that the spaceship in Exercise 14 is somehow traveling at c with respect to you, and it fires a rocket at speed c with respect to itself. Show that the speed at which you see the rocket traveling is still c.

17. How does the measured density of a body compare when at rest and when moving?

18. As a meter stick that has a rest mass of 1 kilogram moves past you, your measurements show it to have a mass of 2 kilograms. If your measurements show it to have a length of 1 meter, what is the orientation of the stick?

19. In the preceding exercise, if the stick is moving in a direction along its length (like a properly thrown spear), how long will it appear to you?

20. The electrons that illuminate the screen in a typical television picture tube travel at nearly one-fourth the speed of light and have an increased mass of nearly 3 percent. Does this relativistic effect tend to increase or decrease your electric bill?

21. How might the principle of correspondence be used to establish the validity of theories outside the domain of physical science?

22. What does the equation $E = mc^2$ mean?

23. *Muons* are elementary particles that are formed high in the atmosphere by the interactions of cosmic rays with gases in the upper atmosphere. Muons are radioactive and have average lifetimes of about two-millionths of a second. Even though they travel at almost the speed of light, they are so high that very few should be detected at sea level—at least according to classical physics. Laboratory measurements, however, show that muons in great proportions *do* reach the earth's surface. What is the explanation?

24. When we look out into the universe, we see into the past. John Dobson, founder of the San Francisco Sidewalk Astronomers, says that we cannot even see the backs of our own hands *now*—in fact, we can't see anything *now*. Is he correct? Explain.

25. One of the fads of the future might be "century hopping," where occupants of high-speed spaceships would depart from the earth for several years and return centuries later. What are the present-day obstacles to such a practice?

26. Is the statement by the philosopher Kierkegaard, "Life can only be understood backwards; but it must be lived forwards," consistent with the theory of special relativity?

34 General Theory of Relativity

Fig. 34.1 Everything is weightless on the inside of a nonaccelerating spaceship far away from gravitational influences.

The special theory of relativity is "special" in the sense that it treats the invariance of the laws of nature principally for uniformly moving reference frames. Recall that Einstein postulated in 1905 that no observation made inside an enclosed chamber could determine whether the chamber was at rest or moving with constant velocity; that is, no mechanical, electrical, optical, or any other physical measurement that one could perform inside a closed compartment in a smoothly riding train traveling along a straight track or in an airplane flying through still air with the window curtains drawn could possibly give any information as to whether the train was moving or at rest or the plane airborne or at rest on the runway. But if the track were not smooth and straight or if the air were turbulent, the situation would be entirely different: uniform motion would give way to accelerated motion, which would be easily noticed. Einstein's conviction that the laws of nature should be expressed in a form invariant in *all* frames of reference, accelerated as well as nonaccelerated, was the primary motivation that led him 10 years later to the general theory of relativity—a new theory of gravitation. Underlying it is the idea that the effects of gravitation and acceleration cannot be distinguished from one another.

Principle of Equivalence

Einstein imagined himself in a spaceship far away from gravitational influences. In such a spaceship at rest or in uniform motion relative to the distant stars, he and everything within the ship would float freely; there would be no "up" and no "down." But when the rocket motors were activated and the ship was accelerating, things would be different; phenomena similar to gravity would be observed. The wall adjacent to the rocket motors would push up against any occupants and become the floor, while the opposite wall would become the ceiling. Occupants in the ship would be able to stand on the floor and even jump up and down. If the acceleration of the spaceship were equal to *g*, the occupants could well be convinced the ship was not accelerating, but was at rest on the surface of the earth.

To examine this new "gravity" in the accelerating spaceship, Einstein considered the consequence of dropping two balls, say, one of wood and the other of lead. When the balls were released, they would continue to move upward side by side with the velocity of the ship at the moment of release. If the ship were moving at *constant velocity* (zero acceleration), the balls would remain suspended in the same place since both the ship and the balls would move the same amount. But since the spaceship is accelerating, the floor moves upward faster than the balls, which are soon intercepted by the floor (Figure 34.3). Both balls, regardless of their masses, meet the floor at the same time. Remembering Galileo's demonstration at the Leaning Tower of Pisa, occupants of the ship might be prone to attribute their observations to the force of gravitation.

Fig. 34.2 When the spaceship accelerates, an occupant inside feels "gravity."

Fig. 34.3 To an observer inside the accelerating ship, a lead ball and a wood ball appear to fall together when released.

The two interpretations of the falling balls are equally valid, and Einstein incorporated this equivalence, or impossibility of distinguishing between gravitation and acceleration, in the foundation of his general theory of relativity. The **principle of equivalence** states that observations made in an accelerated reference frame are indistinguishable from observations made in a Newtonian gravitational field. This equivalence would be relatively unimportant if it applied only to mechanical phenomena, but Einstein went further and stated that the principle holds for all natural phenomena; it holds for optical and all electromagnetic phenomena as well.

Bending of Light by Gravity

A ball thrown sideways in a stationary spaceship in a gravity-free region will follow a straight-line path relative to both an observer inside the ship and a stationary observer outside the spaceship. But if the ship is acceler-

Fig. 34.4 (a) An outside observer sees a horizontally thrown ball travel in a straight line, and since the ship is moving upward while the ball travels horizontally, the ball strikes the wall somewhat below a point opposite the window. (b) To an inside observer, the ball bends as if in a gravitational field.

ating, the floor overtakes the ball just as in our previous example. An observer outside the ship still sees a straight-line path, but to an observer in the accelerating ship, the path is curved; it is a parabola. The same holds true for a beam of light (Figure 34.4).

Imagine that a light ray enters the spaceship horizontally through a side window and reaches the opposite wall after a very short time. The outside observer sees that the light ray enters the window and moves horizontally along a straight line with constant velocity toward the opposite wall. But the spaceship is accelerating upward, and during the time the light takes to reach the wall, the ship changes its position so that the ray will not meet a point exactly opposite the window, but will hit the wall slightly below. To an inside observer, the light ray is deflected downward toward the floor just as the thrown ball curves toward the floor (Figure 34.5). The curvature of the slow-moving ball is very pronounced; but if the ball were somehow thrown horizontally across the spaceship cabin at a velocity equal to that of light, both curvatures would be the same.

How can an observer inside the ship attribute this bending of light to gravitation? According to Newton's physics, gravitation is an interaction between the masses; a moving ball curves because of the interaction between its mass and the mass of the earth. But what of light, which is pure energy and massless? To account for the bending of light in a gravitational field, a Newtonian observer might attribute a mass to light or think of its energy in terms of its "mass-equivalent." Even Einstein held that light is a stream of particlelike photons that bend in a gravitational field

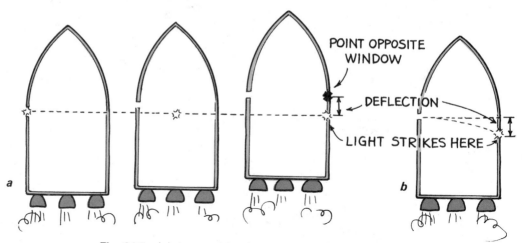

Fig. 34.5 (a) An outside observer sees light travel horizontally in a straight line, but like the ball in Figure 34.4, it strikes the wall slightly below a point opposite the window. (b) To an inside observer, the light bends as if responding to a gravitational field.

Fig. 34.6 The trajectory of a flashlight beam is identical to the trajectory of a baseball "thrown" at the speed of light. Both paths curve equally in a uniform gravitational field.

Fig. 34.7 Starlight bends as it grazes the sun. Point A shows the apparent position; point B shows the true position.

exactly as any body would if it were moving with a speed equal to that of light—but not because of any masslike properties of light. Light bends if it travels in a space-time geometry that is bent. We shall see later in this chapter that the presence of mass results in the bending or warping of space-time; and, by the same token, a bending or warping of space-time reveals itself as a mass. The mass of the earth is too small to appreciably warp the surrounding space-time, which is practically flat, so any such bending of light in our immediate environment is not ordinarily detected. Close to bodies of mass much greater than the earth's, however, the bending of light is large enough to detect.

Einstein predicted that measurements of starlight passing close to the sun would be deflected by an angle of 1.75 seconds of arc—enough to be measured. Although stars are not visible when the sun is in the sky, the deflection of starlight can be observed during the eclipse of the sun. (Measuring this deflection has become a standard practice at every total eclipse since the first measurements were made during the total eclipse of 1919.) A photograph taken of the darkened sky around the eclipsed sun reveals the presence of the nearby bright stars. The positions of the stars are compared with those in other photographs of the same area taken at other times in the night with the same telescope. In every instance, the deflection of starlight has supported Einstein's prediction (Figure 34.7).

Light bends in the earth's gravitational field also—but not as much. We don't notice it only because the amount of deflection is tiny compared to the correspondingly vast distance that light travels due to its high speed.

For example, in a constant gravitational field of 1 g, a beam of horizontally directed light will "fall" a vertical distance of 4.9 meters in 1 second (just as a baseball would), but will travel a horizontal distance of 300,000 kilometers. Its curve would hardly be noticeable when you're this far from the beginning point. But if the light traveled 300,000 kilometers in multiple reflections between idealized parallel mirrors, the effect would be quite noticeable (Figure 34.8). (This would make a dandy home project for extra credit—like earning credit for a Ph.D.)

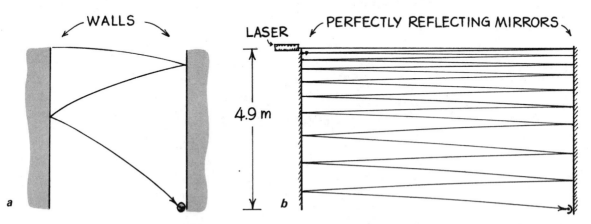

Fig. 34.8 (a) If a ball is horizontally projected between a vertical pair of parallel walls, it will bounce back and forth and fall a vertical distance of 4.9 meters in 1 second. (b) If a horizontal beam of light is directed between a vertical pair of perfectly parallel ideal mirrors, it will reflect back and forth and fall a vertical distance of 4.9 meters in 1 second. The number of back and forth reflections is overly simplified in the diagram; if the mirrors were 300 kilometers apart, for example, 1000 reflections would occur in 1 second.

Question   Why do we not notice the bending of light in our everyday environment?*

Gravity and Time: Gravitational Red Shift

According to Einstein's general theory of relativity, gravitation causes time to slow down. The stronger the gravitational field, the greater the slowing down of time. We can understand this by applying the principle of equivalence and time dilation to an accelerating frame of reference.

Imagine our accelerating reference frame to be a large rotating disk. Suppose we measure time from three identical clocks: one placed on the disk at its center, a second placed on the rim of the disk, and the third at

---

****Answer**   Only because it travels so fast; just as over a short distance we do not notice the curved path of a high-speed bullet, so we do not notice the curving of a light beam.

rest on the nearby ground (Figure 34.9). From the laws of special relativity, we know that the clock attached to the center, since it is not moving with respect to the ground, should run at the same rate as the clock on the ground—but not at the same rate with respect to the clock attached to the rim of the disk. The clock at the rim is in motion with respect to the ground and should therefore be observed to be running more slowly than the ground clock, and therefore more slowly than the clock at the center of the disk. Although the clocks on the disk are attached to the same frame of reference, they do not run synchronously; the outer clock runs slower than the inner clock.

Fig. 34.9 Clocks 1 and 2 are on an accelerating disk, and clock 3 is at rest. Clocks 1 and 3 run at the same rate, while clock 2 runs slower. From the point of view of an observer at clock 3, clock 2 runs slow because it is moving. From the point of view of an observer at clock 1, clock 2 runs slow because it is in a stronger centrifugal force field.

This difference in time would be the same for observers on the rotating disk and for observers at rest on the ground. Interpretations of the time difference for each observer are not the same, however. To the observer on the ground, the slower rate of the clock on the rim is due to its motion. But to an observer on the rotating disk, the disk clocks are not in motion with respect to each other; instead, a centrifugal force acts on the clock at the rim, while no such force acts on the clock at the center. The observer on the disk would say that the clock placed in the stronger force field will run slower. By interpreting centrifugal force as a force of gravitation and applying the principle of equivalence, we must conclude that clocks in strong gravitational fields of force run slower than clocks in weak fields of force. This slowing down will apply to all "clocks," whether physical, chemical, or biological. An executive working on the ground floor of a tall city skyscraper will age more slowly than her twin sister working on the top floor. The difference is very small, only a few millionths of a second per decade, because the difference in the earth's gravitational field at the bottom and top of the skyscraper is very small. For larger differences in gravitational field intensity, like at the surface of the sun compared to the surface of the earth, the differences in time should be more pronounced. A clock at the surface of the sun should run measurably slower than a clock at the surface of the earth. Years before he completed his general relativity theory, Einstein suggested a way to measure this when he formulated the principle of equivalence in 1907.

Fig. 34.10 The stronger a gravitational field, the slower a clock runs. A clock at the surface of the earth runs slower than a clock farther away.

All atoms emit light at specific frequencies characteristic of the vibrational rate of electrons within the atom. Every atom is therefore a "clock," and a slowing down of atomic vibration indicates the slowing

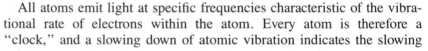

down of such clocks. An atom on the sun, where gravitation is strong, should emit light of a lower frequency (slower vibration) than light emitted by the same kind of atom on the earth. This effect is called the **gravitational red shift**. Since red light is at the low-frequency end of the visible spectrum, a lowering of frequency shifts the color toward the red. The gravitational red shift is observed in light from the sun, but various disturbing influences prevent accurate measurements of this tiny effect. It wasn't until 1960 that an entirely new technique using high-frequency gamma rays from radioactive atoms permitted incredibly precise and confirming measurements of the gravitational slowing of time between the top and bottom floors of a laboratory building at Harvard University.[1]

So measurements of time depend not only on relative motion as we learned in the last chapter but also on the relative gravitational field intensities of the regions in which the events are taking place and being measured. Just as time dilation in special relativity is relative to the *differences* in motion between the observed frame of reference and the frame of reference from which observations are being made, the gravitational red shift of general relativity is relative to the *differences* in gravitational field intensities at the location of the event and the location of the observer of the event. Where gravitation is seen to be stronger, events are seen to proceed more slowly into the future. As viewed from earth, a clock will be measured to tick slower on the surface of a star than on earth. If the star shrinks, the resulting increase in gravitation at its surface will be seen to be accompanied by a corresponding slowing of time, and we would measure longer intervals between the ticks of the star clock. But if we made our measurements of the star clock from the star itself, we would notice nothing unusual about the clock's ticking.

Suppose, for example, that an indestructible volunteer stands on the surface of a giant star that begins collapsing. We, as outside observers, will note a progressive slowing of time on the clock of our volunteer as the gravitational field increases. The volunteer himself, however, does not notice any differences in his own time. He is viewing events within his own frame of reference, and he notices nothing unusual about his own time. As the collapsing star proceeds toward becoming a black hole, and time proceeds normally from the viewpoint of the volunteer, to us on the outside, time for the volunteer approaches a complete stop; we would see him frozen in time with an infinite duration between the ticks of his clock or the beats of his heart. From our view, his time stops completely. The gravitational red shift, instead of being a tiny effect, is dominating.

---

[1] In the late 1950s, shortly after Einstein's death, the German physicist Rudolph Mossbauer discovered an important effect in nuclear physics that provides an extremely accurate method of using atomic nuclei as atomic clocks. The *Mossbauer effect,* for which its discoverer was awarded the Nobel Prize, has many practical applications. In late 1959 Drs. Pound and Rebka at Harvard University realized that still another application was a test for general relativity and performed the confirming experiment.

It is important to note the relativistic nature of time in both special relativity and general relativity. In both theories, there is no way that you can extend the duration of your own experience. Others moving at different speeds or in different gravitational fields may attribute a great longevity to you, but your longevity is seen from *their* frame of reference—never your own. Changes in time are always attributed to the "other guy."

Questions

1. Will a person at the top of a skyscraper age more than or less than a person at ground level?*

2. Will a person at ground level age more than or less than a person at the bottom of a very deep well?†

Gravity and Space: Motion of Mercury

Fig. 34.11 A precessing elliptical orbit.

From the special theory of relativity, we know that measurements of space as well as time undergo transformations when motion is involved, Likewise with the general theory: measurements of space differ in different gravitational fields—for example, close to and far away from the sun.

Planets orbit the sun and stars in elliptical orbits and move periodically into regions of weaker and stronger gravitational fields. Einstein directed his attention to the varying gravitational fields experienced by the planets orbiting the sun and found that the elliptical orbits of the planets should *precess* (Figure 34.11)—independently of the Newtonian influence of other planets. Near the sun, where the gravitational field is stronger, the rate of precession should be the greatest; and far from the sun, where the field is weak, any deviations from Newtonian mechanics should be virtually unnoticeable.

Mercury is the planet nearest the sun. The gravitational force between the sun and Mercury is greater than that between the sun and any of the more distant planets. If the orbit of any planet exhibits a measurable precession, it should be Mercury, and the fact that the orbit of Mercury does precess had been a mystery to astronomers since the early 1800s! Careful measurements showed that Mercury precesses about 573 seconds of arc per century. Perturbations by the other planets were found to account for the precession—except for 43 seconds of arc per century more than the calculated value. Even after all known corrections due to possible perturbations by other planets had been applied, the calculations of physicists and astronomers failed to account for the extra 43 seconds of arc.

---

*Answer More; a person at the top of a skyscraper is in a slightly weaker gravitational field than a person at ground level.

†Answer Less; a person at the surface of the earth is in a stronger gravitational field than a person in a very deep well. (Recall from Chapter 8 that the gravitational field deep inside the earth is less than at the surface and is zero at the earth's center.)

Either Venus was extramassive or another never-discovered planet (called Vulcan) was pulling on Mercury. And then came the explanation of Einstein, who applied his general relativity field equations to the problem and showed that the rate of precession for Mercury's orbit includes the extra 43 seconds of arc per century!

The mystery of Mercury was solved, and a new theory of gravity was recognized. Newton's law of gravitation, which had stood as an unshakable pillar of science for more than two centuries, was found to be a special limiting case of Einstein's more general theory. If the gravitational fields are comparatively weak, Newton's law turns out to be a good approximation of the new law, enough so that Newton's law, which is easier to work with mathematically, is the law that today's scientists use most of the time—except for cases involving enormous gravitational fields.

Gravity, Space, and a New Geometry

We can begin to understand that measurements of space are altered in a gravitational field by considering the accelerated frame of reference of our rotating disk again. Suppose we measure the circumference of the outer rim with a measuring stick. Recall the Lorentz-FitzGerald contraction from special relativity: the measuring stick will appear contracted to any observer not moving along with the stick, while an identical measuring stick moving much slower near the center will be nearly unaffected (Figure 34.12). All distance measurements along a *radius* of the rotating disk should be completely unaffected by motion, because motion is perpendicular to the radius. Since only distance measurements parallel to and around the circumference are affected, the ratio of circumference to diameter when the disk is rotating is no longer the fixed constant π (3.14159 . . .), but is a variable depending on angular speed and the diameter of the circumference. According to the principle of equivalence, the rotating disk is equivalent to a stationary disk with a strong gravitational field near its edge and a progressively weaker gravitational field toward its center. Measurements of distance, then, will depend on the strength of a gravitational field, even if no relative motion is involved. Gravity causes space to be non-Euclidean; the laws of Euclidean geometry taught in high school are no longer valid when applied to objects in the presence of strong gravitational fields.

Fig. 34.12 A measuring stick along the edge of the rotating disk appears contracted, while a measuring stick farther in and moving slower is not contracted as much. A measuring stick along a radius is not contracted at all. When the disk is not rotating, C/D = π; but when the disk is rotating, C/D $\neq$ π and Euclidean geometry is no longer valid. Likewise in a gravitational field.

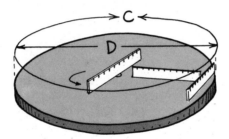

The familiar rules of Euclidean geometry pertain to various figures you can draw on a flat surface. The ratio of the circumference to the diameter for a circle is equal to π; all the angles in a triangle add up to 180°; the shortest distance between two points is a straight line. The rules of Euclidean geometry are valid in flat space; but if you draw these figures on a curved surface like a sphere or a saddle-shaped object, the Euclidean rules no longer hold (Figure 34.13). If you measure the sum of the angles for a triangle in space, you call the space flat if the sum is equal to 180°, spherelike or positively curved if the sum is larger than 180°, and saddlelike or negatively curved if it is less than 180°.

Fig. 34.13 The sum of the angles for a triangle drawn (*a*) on a plane surface = 180°, (*b*) on a spherical surface is greater than 180°, and (*c*) on a saddle-shaped surface is less than 180°.

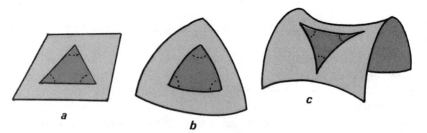

Or course the lines forming the triangles in Figure 34.13 are not "straight" from the three-dimensional view, but they are the "straightest" or *shortest* distances between two points if we are confined to the curved surface. These lines of shortest distance are called *geodesic lines* or simply **geodesics**.

The path of a light beam follows a geodesic. Suppose three experimenters on Earth, Venus, and Mars measure the angles of a triangle formed by light beams traveling between these three planets. The light beams bend when passing the sun, resulting in the sum of the three angles being larger than 180° (Figure 34.14). So the space around the sun is positively curved. The planets that orbit the sun travel along four-dimensional geodesics in this positively curved space-time. Freely falling objects, satellites, and light rays all travel along geodesics in four-dimensional space-time.

Fig. 34.14 The light rays joining the three planets form a triangle. Since light passing near the sun bends, the sum of the angles of the resulting triangle is greater than 180°.

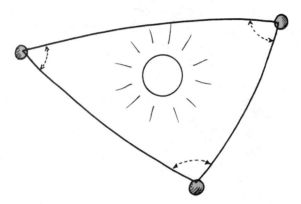

The whole universe may have an overall curvature. If it is negatively curved, it is open-ended and extends without limit; if it is positively curved, it closes in on itself. The surface of the earth, for example, forms a closed curvature; so that if you travel along a geodesic, you come back to your starting point. Similarly, if the universe were positively curved, it would be closed; so that if you could look infinitely into space through an ideal telescope, you would see the back of your own head! (This is assuming that you waited a long enough time or that light traveled infinitely fast.)

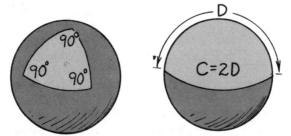

Fig. 34.15   The geometry of the curved surface of the earth differs from the Euclidean geometry of flat space. Note that the sum of the angles for an equilateral triangle where the sides equal $\frac{1}{4}$ the earth's circumference is clearly greater than 180°, and the circumference is only twice its diameter instead of 3.14 times its diameter. Euclidean geometry is also invalid in curved space.

General relativity, then, calls for a new geometry: a geometry not only of curved space but of curved time as well—a geometry of curved four-dimensional space-time. The mathematics of this geometry is too formidable to present here. The essence, however, is that gravity is a manifestation of space-time geometry; a gravitational field is a geometrical warping of space-time. The presence of mass results in the curvature or warping of space-time; and, by the same token, a curvature of space-time reveals itself as mass. Instead of visualizing gravitational forces between masses, we abandon altogether the notion of force and instead think of masses responding in their motion to the curvature or warping of the space-time they inhabit. It is the bumps, depressions, and warpings of geometrical space-time that *are* the phenomena of gravity.

We cannot visualize the four-dimensional bumps and depressions in space-time because we are three-dimensional beings. We can get a glimpse of this warping by considering a simplified analogy in two dimensions: a heavy ball resting on the middle of a waterbed. The more massive the ball, the greater it dents or warps the two-dimensional surface. A marble rolled across the bed, but away from the ball, will roll in a relatively straight-line path; whereas a marble rolled near the ball will curve as it rolls across the indented surface. If the curve closes upon itself, its shape is an ellipse. The planets that orbit the sun similarly travel along four-dimensional geodesics in the warped space-time about the sun.

Fig. 34.16 A two-dimensional analogy of four-dimensional warped space-time. Space-time near a star is curved in a way similar to the surface of a waterbed when a heavy ball rests on it.

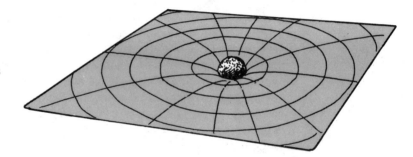

Gravitational Waves

Every object has mass and therefore makes a bump or depression in the surrounding space-time. When an object moves, the surrounding warp of space and time moves to readjust to the new position. These readjustments produce ripples in the overall geometry of space-time. This is similar to moving the ball that rests on the surface of the waterbed. A disturbance ripples across the waterbed surface in waves; if we move a more massive ball, then we get a greater disturbance and the production of even stronger waves. The ripples travel outward from the gravitational sources at the speed of light and are called **gravitational waves**.

Any moving object produces a gravitational wave. In general, the more massive the moving object and the more violent its motion, the stronger the resulting gravitational wave. But even the strongest waves produced by ordinary astronomical events are extremely weak—the weakest known in nature. For example, the gravitational waves emitted by a vibrating electric charge are a trillion-trillion-trillion times weaker than the electromagnetic waves emitted by the same charge.

Shake your hand back and forth: you have just produced a gravitational wave. It is not very strong, but it exists.

Newtonian and Einsteinian Gravitation

When Einstein formulated his new theory of gravitation, he realized that if his theory was valid, his field equations must reduce to Newtonian equations for gravitation in the weak-field limit. He showed that Newton's law of gravitation is a special case of the broader theory of relativity. Newton's law of gravitation is still an accurate description of most of the interactions between bodies in the solar system and beyond. From Newton's law, one can calculate the orbits of comets and asteroids and even predict the existence of undiscovered planets. Even today, when computing the trajectories of space probes to the moon and planets, only ordinary Newtonian theory is used. This is because the gravitational field of these bodies is very weak, and from the viewpoint of general relativity, the surrounding space-time is essentially flat. But for regions of more intense gravitation, where space-time is more appreciably curved, Newtonian theory cannot adequately account for various phenomena—like the precession of Mercury's orbit close to the sun and, in the case of stronger fields, the gravitational red shift and other apparent distortions in measurements of space and time. We will see that these distortions reach their limit in the case of a star that collapses to a black hole, where space-time completely folds over on itself. Einsteinian gravitation alone reaches into this domain.

We saw in Chapter 30 that Newtonian physics is linked at one end with quantum theory, whose domain is the very light and very small—tiny particles and atoms. And now we have seen that Newtonian physics is linked at the other end with relativity theory, whose domain is the very massive and very large. In the next and last chapter we will treat the very very massive and the very, very large—stars, black holes, and the universe. Onward!

Summary of Terms

Principle of equivalence   Local observations made in an accelerated frame of reference are indistinguishable from observations made in a Newtonian gravitational field.

Gravitational red shift   The red shift in wavelength experienced by light leaving the surface of a massive object, as predicted by the general theory of relativity.

Geodesic   The shortest path between points on any surface.

Gravitational wave   A gravitational disturbance by a moving mass that propagates through space-time.

Suggested Reading

Einstein, Albert. *Relativity: The Special and General Theory.* New York: Crown, 1961. (Orig. pub. 1916.)

Review Questions

1. What is the principal difference between special relativity and general relativity?

2. What is "general" about the general theory of relativity?

3. What is it that is equivalent in the principle of equivalence?

4. Why is it necessary that the sun be eclipsed when measuring the deflection of nearby starlight?

5. What is the gravitational red shift?

6. Of all the planets, why is Mercury the best candidate for finding evidence of the relationship of gravitation to space?

7. A measuring stick placed along the circumference of a rotating disk will appear contracted, but if it is oriented along a radius, it will not. Explain.

8. The ratio of circumference to diameter for measured circles on a disk at rest equals π, but not when the disk is rotating. Explain.

9. Is the two-dimensional surface of the earth positively or negatively curved? Why?

10. A star 10 light-years away explodes and produces gravitational waves. How long will it take these waves to reach the earth?

11. Why are gravitational waves so difficult to detect?

12. Does Einstein's theory of gravitation invalidate Newton's theory of gravitation? Explain.

Exercises

1. An astronaut awakes in her closed capsule, which sits on the moon. Can she discern whether her weight is the result of gravitation or accelerated motion? Explain.

2. An astronaut is provided a "gravity" when the ship's engines are activated to accelerate the ship. This requires the use of fuel. Is there a way to accelerate and provide "gravity" without the sustained use of fuel? Explain.

3. In his famous novel *Journey to the Moon*, Jules Verne stated that occupants in a spaceship would shift their orientation from up to down when the ship crossed the point where the moon's gravitation became greater than the earth's. Is this correct? Defend your answer.

4. What happens to the separation distance of two people a certain distance apart at the earth's equator when they both walk north at the same rate? And just for fun, where in the world is a step in every direction a step south?

5. We readily note the bending of light by reflection and refraction, but why is it we do not ordinarily notice the bending of light by gravity?

6. Why do we say that light travels in straight lines? Is it strictly accurate to say that a laser beam provides a perfectly straight line for purposes of surveying? Explain.

7. At the end of 1 second, a horizontally fired bullet drops a vertical distance of 4.9 meters from its otherwise straight-line path in a gravitational field of 1 g. By what amount would a beam of light drop from its otherwise straight-line path if it traveled in a uniform field of 1 g for 1 second? For 2 seconds?

8. Light changes its energy when it "falls" in a gravitational field. This change in energy is not evidenced by a change in speed, however. What is the evidence for this change in energy?

9. Would we notice a slowing down or speeding up of a clock at the bottom of a very deep well?

10. If we witness events taking place on the moon, where gravitation is weaker than on earth, would we expect to see a gravitational red shift or a gravitational blue shift? Explain.

11. Why will the gravitational field intensity increase on the surface of a shrinking star?

12. Will a clock at the equator run slightly faster than an identical clock at one of the earth's poles?

13. Splitting hairs, should a person who worries about growing old live at the top or at the bottom of a tall apartment building?

14. Splitting hairs, if you shine a beam of colored light to a friend above in a high tower, will the color of light your friend receives be the same color you emit? Explain.

15. Is the color of light emitted from the surface of a massive star red-shifted or blue-shifted?

16. From our frame of reference on earth, objects slow to a stop before falling into black holes in space. With ideal telescopes, could we "see" these objects as they hover about the black holes? (Would they emit or reflect waves in the visible part of the spectrum?)

17. Would an astronaut falling into a black hole see the surrounding universe red-shifted or blue-shifted?

18. Why does the gravitational attraction between the sun and Mercury vary? Would it vary if the orbit of Mercury were perfectly circular?

19. Do binary stars (double-star systems that orbit about a common center of mass) radiate gravitational waves? Why or why not?

20. With respect to Newton's theory of gravitation, how does Einstein's theory of gravitation obey the correspondence principle?

35 Astrophysics

Astrophysics is the astronomical study of the interactions of the matter-energy of the universe with space-time: the study of physical processes in the stars, in stellar systems, and in interstellar material, and in how they were formed, what they are doing, and where they are going. Astrophysics is about big things and big events.

The Big Bang

No one knows how the universe began. Evidence suggests that about 15 to 20 billion years ago most of the matter-energy of the universe was highly concentrated at an unimaginably high temperature and underwent a primordial explosion, usually referred to as the **Big Bang**, which was accompanied by a high-powered blast of high-frequency radiation that we call the *primeval fireball*. The universe is the remnant of this explosion, and we view it as still expanding. Radiation from the dying embers of the primeval fireball now permeate all space in the form of the presently observed long-wavelength microwaves, which have been lengthened by the expansion of the universe. The present expansion of the universe is evident in a Doppler red shift in the light from other galaxies, which is greater than gravitation would account for. This red shift indicates a recession of the galaxies. All galaxies are getting farther away from our own. This does not, as a first thought may indicate, place our own galaxy in a central position. Consider a balloon with ants on it: as the balloon is inflated, *every* ant will see *every* neighboring ant getting farther away, which certainly doesn't suggest a central position for each ant. In an expanding universe, every observer sees clusters of all other galaxies receding.

If you throw a rock skyward, it slows down due to its gravitational attraction to the earth below. Similarly, matter blown away in the primordial explosion is gravitationally attracted to every other bit of matter, which results in a continual slowing down of the overall expansion.

Theorists generally believe that within the first 3 minutes following the Big Bang, great quantities of hydrogen and helium were formed. About a million years later, these atoms contracted by mutual gravitation into huge clouds. Condensation of these clouds resulted in the galaxies, and the stage was set for the creation of the stars.[1]

Fig. 35.1 Every ant on the expanding balloon sees every other ant getting farther away.

---

[1]This chapter presents a brief "This is how it is" treatment of astronomy. For an expanded "This is how we know this is how it is" treatment, refer to the suggested readings at the end of the chapter.

Fig. 35.2 A galaxy is a huge assemblage of stars—hundreds of billions of them. Shown here is a painting of what our own galaxy, the Milky Way, probably looks like if viewed from a planet some 300,000 light-years away.

Birth of a Star

To make a star, begin with a giant cloud of low-temperature gas and dust. The gas will not be perfectly uniform; there will be regions of gas density that will be slightly different from the overall average. Regions of slightly greater gas density will have slightly more mass and a slightly greater gravitational field; therefore, they will more strongly attract neighboring particles. This increases the mass and the gravitational field of the region, which then attracts still more particles. In time we have an aggregation of matter many times the mass of the sun spread out over a volume many times larger than the solar system. We have a *protostar*.

Mutual gravitation between the gaseous particles in a protostar results in an overall contraction of this huge ball of gas, and the density at the center increases dramatically as matter is scrunched together with an accompanying rise in pressure and temperature. When the central temperature reaches about 10 million degrees, the nuclei of hydrogen atoms come together with such violence that they *fuse* to form the nuclei of helium atoms. This **thermonuclear reaction**, in which hydrogen is converted to helium, releases an enormous amount of radiant and thermal energy. This outward-moving radiant energy and the gas it pushes with it exert an outward pressure on the contracting matter and ultimately become strong enough to stop the contraction. Radiation and gas pressures balance gravitational pressure, and a full-fledged star is born!

We are most familiar with "ordinary" stars, one of which is the sun, a hydrogen burner. Hydrogen fusion in massive stars occurs at a furious rate, and the stars are very bright and relatively short-lived. In low-mass stars, hydrogen fusion occurs at a much slower rate, and the stars are dimmer and longer-lived.

Surprisingly enough, about half the stars seen in the sky are *not* individual stars like our sun, but are actually two stars revolving about a common center just as the earth and the moon revolve about each other. These double stars are called **binary stars**. By observing how the two stars in a binary revolve about their common center, we can calculate their masses. Binaries provide astronomers the basic means of determining how much matter is contained in the star.

Stars in our galaxy range in mass from a hundredth to about 100 times the mass of the sun. Our sun has an average density of 1.4 grams per cubic centimeter. The densities of other ordinary stars range all the way from 10 million times less than the sun's, the *red supergiants,* to 50 times greater than the sun's, the *red dwarfs,* with corresponding diameters from 1000 to 0.1 that of the sun. Denser still are the *white dwarfs,* where matter is so compressed that a teaspoonful of it weighs tons. Then there are the *neutron stars,* which have densities a billion times greater than white dwarfs. And, last, there are the *black holes,* some with unimaginable densities. But before discussing these exotic stars, we'll turn our attention to the star that is nearest to us—the one only 8 light-minutes away.

The Nearest Star: The Sun

The ancients who worshipped the sun likely realized that the sun is the source of all earthly life. Our life energy originates in the sun; we are able to see, hear, touch, laugh, and love because every second, $4\frac{1}{2}$ million tons of mass in the sun are being converted to radiant energy, a tiny fraction of which is intercepted by the earth.

This is the energy of thermonuclear fusion taking place in the interior of the sun, where hydrogen nuclei are being crushed together to form helium. The helium that results from the fusion of hydrogen is 0.7 percent less massive than the original hydrogen. The conversion of hydrogen to helium in the sun has been going on for about 5 billion years and is expected to continue at this rate for another 5 billion years. When all the hydrogen in the sun's core is changed to helium, the core will be only 0.7 percent lighter than it would be if the hydrogen never fused.

As the sun and other more massive stars get older and use up most of their hydrogen, they begin converting helium into carbon, oxygen, and heavier elements. Sometimes after using up all their available fuel, the massive stars collapse and then explode violently, hurling heavy elements into space where they mix with interstellar gases to form the building material for new stars. Our sun is a second- or third-generation star (there are many stars more than twice as old in our galaxy) and was formed after numerous stellar explosions, or *supernovae*, as they are called, had occurred during the early history of our galaxy. Hence, our solar system is composed of the debris from countless supernova explosions that occurred long before the solar system came into being. Interestingly enough, all atoms on earth heavier than helium participated in a supernova at some distant time. So we are quite literally made of star dust.

The part of the sun visible to us is of course its surface and its atmosphere. The sun's surface is neither solid nor liquid but a glowing plasma 5500°C—well above the temperature required to vaporize any material known. This surface layer is called the *photosphere* (sphere of light). The atmosphere of the sun just above the photosphere is a transparent band of plasma called the *chromosphere* (sphere of color), which during an eclipse can be seen as a reddish ring surrounding the eclipsed sun. Beyond the chromosphere are streamers and filaments of outward-moving, high-temperature plasmas that are curved by the sun's magnetic field. This is the outermost part of the atmosphere, the *corona* (Figure 35.3).

Fig. 35.3 Photograph of the sun's corona, taken by the author at Laisamis, Kenya, during the total eclipse of June 30,1973.

The origin of the solar system is still a matter of some conjecture, but it is generally believed to have originated from the gravitational contraction of a huge amount of interstellar matter. Like the ice skater who draws her arms inward when going into a spin, the contraction was accompanied by an increase in angular speed. As the collapsing matter spun faster, it flattened into a disk shape (Figure 35.4). The disk has more surface area than its configuration before flattening and consequently radiates more of its energy off into space. So this flattening results in cooling. The decreasing temperature most likely was accompanied by the condensation of matter in swirling eddies—the birthplace of the planets.

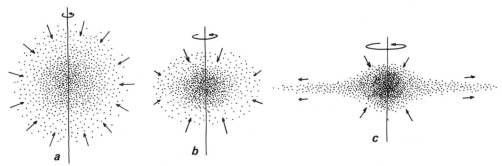

Fig. 35.4 A slightly rotating ball of interstellar gas (*a*) contracts due to mutual gravitation and (*b*) conserves angular momentum by speeding up. The increased momentum of individual particles and clusters of particles causes them to (*c*) sweep in wider paths about the rotational axis, producing an overall disk shape.

The temperatures closer to the center of the disk would have been too high for the solidification of matter, but farther away, earthy, less volatile material may have solidified to become the four inner planets: Mercury, Venus, Earth, and Mars. Condensations of larger amounts of more volatile matter, mostly hydrogen and helium, at the relatively cooler locations farther from the hot center are now the giant planets: Jupiter, Saturn, Uranus, and Neptune. Distant Pluto, however, is an exceptionally small planet, having a mass of only about 0.0022 the mass of Earth.

The sun contains 99 percent of all the mass in the solar system, but it contains only about 2 percent of the solar system's angular momentum. The rest of the angular momentum is in the orbiting planets. The sun spins slowly as a result. Interestingly enough, measurements of the spin rates of thousands of stars show that the hot, massive stars are spinning at rates 100 times that of the sun; while the cooler, less massive stars are spinning slowly like our sun. Perhaps the massive stars spin rapidly because they alone possess angular momentum, while the less massive stars have formed planetary systems that absorb most of the system's angular momentum.

So perhaps the stars with as little mass as our sun have planetary systems. If they do, the planets can't be seen with today's telescopes because of their nonluminosity and great distance. For many years it has seemed a fair speculation that there be a number of them located like the earth at a distance from their stars that is not too hot and not too cold—at a location that would support life. This idea is rather enchanting. However, based upon recent ''hard-science modeling,'' a growing number of astronomers contend that the range of distances from stars within which planets would have the conditions for supporting life as we know it is *much* tinier than previously thought by most investigators—perhaps as little as one suitable planet every ten galaxies or so. Nevertheless, there are presently several SETI (Search for Extraterrestrial Intelligence) programs in progress. Our own civilization is so young that we can hardly expect it to have come to the attention of others. The most conspicuous evidence of life on earth—radio, TV, and radar broadcasts—has by now reached some 50 light-years into space, a distance encompassing only a few hundred of the galaxy's 200 billion stars.

Question   What evidence do we have that our sun and solar system are at least second generation?*

Stellar Evolution   *The Bigger They Are, the Harder They Fall*
All luminous stars burn nuclear fuel. The first energy-producing nuclear reaction, once the star has condensed and is hot enough to ignite its core, is the fusion of hydrogen to helium. This hydrogen-fusing process may take a few million or trillions of years, depending on the star's mass. Then, for a star of average mass like our sun, the burned-out hydrogen core that has been converted to helium gravitationally contracts, which raises its temperature. This ignites the unfused hydrogen outside the spent core, and the star expands to become a **red giant**. Our sun will eventually reach this stage about 5 billion years from now. Then the swelling and more luminous sun will cause earthly temperatures to escalate, first stripping the earth of its atmosphere and then boiling the oceans dry.

The cores of solar-mass and lower-mass stars are not hot enough to fuse carbon, and, lacking a source of nuclear energy, they shrink. In doing so, the outer stellar layers are sometimes ejected and form expanding smoke-ring-like shells that eventually disseminate and mix with the interstellar material. This is a **planetary nebula** (Figure 35.5). The shrunken core that is left blazes white-hot and is called a **white dwarf**. The white dwarf radiates its energy into space and turns in color from white to yellow and then to red, until it ultimately fades to a cold, black lump of matter called a **black dwarf**. Its density is enormous. Into a volume no more than that of an average-sized planet is concentrated a mass hundreds of thousands

---

*Answer   Elements heavier than H and He in the matter that makes up the solar system are remnants of previous exploding stars—supernovae.

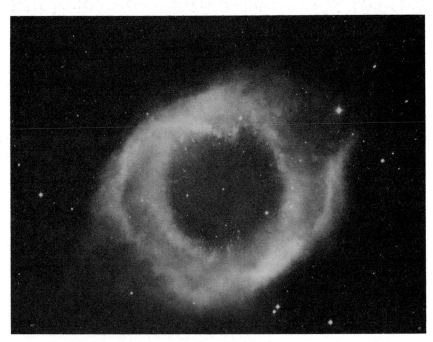

Fig. 35.5 A planetary nebula in Aquarius (NGC 7293).

times greater than that of the earth. The star has a density comparable to that of a full-fledged battleship squeezed into a pint jar!

While well along the way to becoming white dwarfs, some stars that are part of a binary system suddenly and unpredictably flare up with outbursts of energy almost a millionfold their usual energy output. These irregular flaring stars are called **novae**. It is thought that the outbursts are a result of matter pulled from one star and deposited on its white-dwarf companion, which produces instabilities that lead to a violent outburst of radiation, or *nova outburst*. After the outburst, the nova subsides to a normal white-dwarf state until enough matter again accumulates to repeat the event. Novae flare up at irregular intervals spaced as short as decades or perhaps as long as hundreds of thousands of years.

When stars with masses much greater than the sun's contract, more heat is generated than in the contraction of a small star, and instead of shrinking to a white dwarf, carbon nuclei in the core fuse and liberate energy while synthesizing heavier elements such as magnesium. Outward-pushing gas and radiation pressure halt further gravitational contraction until all the carbon is fused. Then the center of the star contracts again to produce even greater temperatures and a new fusion series that produces even heavier elements. The fusion cycles repeat until the element iron is formed. The fusion of elements beyond iron nuclei will require energy rather than liberating energy (recall Figure 32.11). This absorption of energy reduces the core temperature, and the center of the star collapses without rekindling. The entire star commences its final collapse.

The collapse is catastrophic. When the density of the core is so great that all the nuclei are compressed against one another, the collapse momentarily comes to a halt. The collapsed star, compressed like a spring, rebounds violently in a great explosion, hurling into space the elements manufactured over billions of years. The entire episode lasts but a few minutes. It is during this brief time that the heavy elements beyond iron are synthesized as protons, and neutrons are mashed with other nuclei—hence the existence of elements such as silver, gold, and uranium. Because the time available for making these heavy elements is so brief, they are not as abundant as iron and the lighter elements.

Such a stellar explosion is called a **supernova**, one of nature's most spectacular cataclysms. A supernova flares up to millions of times its former brightness. In 1054, Chinese astronomers recorded their observation of a star so bright it could be seen by day as well as by night. This was a supernova, the glowing plasma remnants of which now make up the spectacular Crab nebula (Figure 35.6).

Fig. 35.6 The Crab nebula, the remnant of a supernova explosion that was seen on earth in A.D. 1054.

The inner part of the supernova star implodes to form a core compressed to neutron density. Protons and electrons have been compressed together to form a core of neutrons just a few kilometers wide. This superdense, central remnant of a supernova survives as a **neutron star**. In accord with the conservation of angular momentum, these tiny bodies

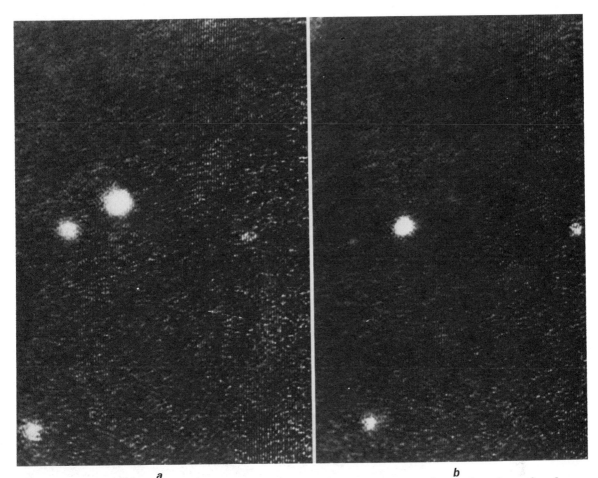

a   b

Fig. 35.7 The pulsar in the Crab nebula rotates like a searchlight, beaming light and X rays toward earth about 30 times a second, blinking on and off; (*a*) shows the pulsar "on," (*b*) "off." The pulsar is a neutron-star core, the remnant of the supernova that created the Crab nebula.

spin at fantastic speeds. Neutron stars are the explanation of **pulsars**, discovered in 1967 as rapidly varying sources of low-frequency radio emission. Of the approximately 300 known pulsars, only a few have been found to emit visible light. One is in the center of the Crab nebula. It has one of the fastest rotational speeds of any pulsar studied, rotating about 100 times in 3 seconds. This is a relatively young pulsar, and it is theorized that optical radiation is emitted only during a pulsar's early history.

Dying stars with cores greater than from three to five solar masses collapse so violently that no physical forces are strong enough to inhibit continued contraction. The bigger they are, the harder they fall! The enormous gravitational field about the imploding concentration of mass makes explosion impossible. This warping of space-time in the vicinity of the unrestrained collapse gets greater and greater; and when the star has collapsed down to only 2 or 3 kilometers in diameter, space-time folds completely in on itself, and the star disappears from the observable universe. What is left is called a **black hole**.

Question   What determines whether a star becomes a white dwarf, a neutron star, or a black hole?*

Black Holes

A black hole in highly warped space-time is all that remains after the catastrophic gravitational collapse of a massive dying star. We can begin to understand the geometry of black holes by considering the behavior of light rays traveling in these regions of extreme space-time curvature. If we shine a light beam past a black hole, it will be deflected; if the beam passes very far from the hole, where space-time is practically flat, it is only slightly bent from its otherwise straight-line path (Figure 35.8). Light beams directed toward the more highly warped regions nearer the black hole have a more pronounced curvature. If we shine a beam toward the black hole at precisely the right distance, we can put the light into *circular orbit* about the hole. This region above the black hole is called the *photon sphere*. The photon sphere is very unstable, however, because the slightest variation in the interaction of a light beam with the warped space-time will send the light beam either spiraling into the hole or back off into space. All beams of light that happen to be incident at this critical distance are captured in the sphere, while beams that are incident at distances within the photon sphere spiral into the black hole and are lost from the outside universe as the black hole literally swallows them up.

Fig. 35.8 Light rays deflected by the gravitational field around a black hole. Light passing far away is bent only slightly; light passing closer can be captured into circular orbit; and light passing closer still is sucked into the hole.

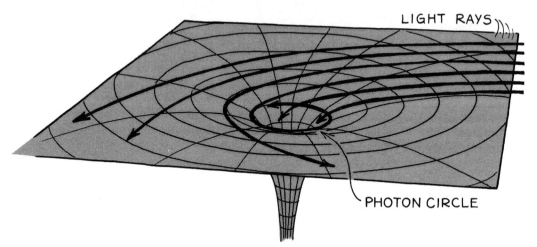

LIGHT RAYS

PHOTON CIRCLE

*Answer   The mass of a star is the principal factor that determines its fate. Stars having solarlike masses and less than solar masses evolve to become white dwarfs; more massive stars evolve to become neutron stars; and the stars having more than from three to five solar masses become black holes.

An indestructible astronaut with a powerful enough spaceship could venture into the photon sphere of a black hole and come out again. While inside the photon sphere, he could still send beams of light back into the flat space-time of the outside universe. If he directed his flashlight in sideways directions and toward the black hole, the light would quickly spiral into the black hole; but light directed vertically and at angles close to the vertical would still escape. As he gets closer and closer to the black hole, however, he would have to shine the light beams closer and closer to the vertical for escape. Moving closer still, our astronaut would find a particular distance where *no* light can escape to flat space-time. No matter what direction the flashlight points, all the beams are deflected back down into the black hole. Our unfortunate astronaut would have passed within the **event horizon**. Once inside the event horizon, he could no longer communicate with the outside universe; neither light waves, radio waves, nor any matter could escape from inside the event horizon. Our astronaut would have performed his last experiment in the universe as we know it.

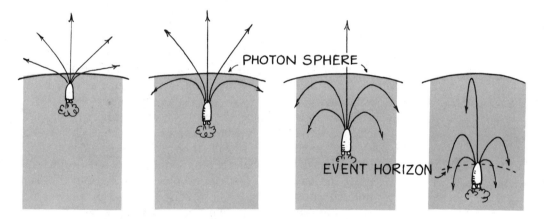

Fig. 35.9 Just beneath the photon sphere, an astronaut could still shine light to the outside. But as he gets closer to the black hole, only light directed nearer the vertical gets out, until finally even vertically directed light is trapped. This distance is the event horizon.

The event horizon is a horizon beyond which no events can be seen by an outside observer. The event horizon surrounding a black hole is often called the *surface* of the black hole, the diameter of which depends on the mass of the hole. For example, a black hole resulting from the collapse of a star ten times as massive as the sun has an event-horizon diameter of about 60 kilometers. The radii of event horizons for black holes of various masses are shown in Table 35.1.

Table 35.1 Estimated radii of event horizons for nonrotating black holes of various masses

Mass of black hole	Radius of event horizon
1 Earth mass	0.8 centimeter
1 Jupiter mass	2.8 meters
1 Solar mass	3 kilometers
2 Solar masses	6 kilometers
3 Solar masses	9 kilometers
5 Solar masses	15 kilometers
10 Solar masses	30 kilometers
50 Solar masses	148 kilometers
100 Solar masses	296 kilometers
1000 Solar masses	2961 kilometers

From William J. Kaufmann, III, *The Cosmic Frontiers of General Relativity,* 1977. Published by Little, Brown and Company, Inc.

When a collapsing star contracts within its own event horizon, the star still has substantial size. There are no forces known that can stop the continued contraction, however, and the star quickly shrinks in size until finally it is crushed to the size of a pinhead, then to the size of a microbe, and finally to a realm of size smaller than ever measured by humans. At this point, according to theory, there is infinite density and infinite curvature of space-time. This point in space-time is called the **black hole singularity**.

Whereas a complete description of an ordinary star is very complicated, involving many physical properties such as chemical composition, densities, temperatures, and so on, the complete description of the simplest kind of black hole is quite straightforward. It has only one property—its mass. And this can be precisely determined outside the event horizon, for example, by a physicist who measures how much the trajectory of a rocket probe is deflected when in the vicinity of a black hole.

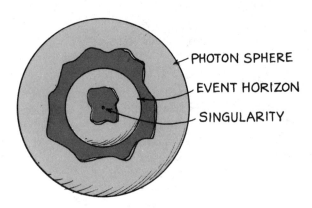

Fig. 35.10 Structure of a simple, ideal black hold (uncharged and nonrotating.)

If the black hole has an electric charge (either positive or negative) or a magnetic "charge" (either north pole or south pole), the effects of these charges, like the effects of mass, will also extend beyond the event horizon. A distant physicist could use a sensitive apparatus to detect these charges. So, in addition to mass, a charge on a black hole is not lost to the universe and is an additional property of a black hole. It is unlikely that black holes with appreciable electric charges exist, however, for if a black hole did have a substantial charge, the enormous strength of its electric field would rapidly tear apart atoms in nearby space and in a very short time it would be neutralized by the strongly attractive particles of opposite charge.

More important is spin, for most stars are rotating and possess angular momentum. Any rotating object tends to drag space-time around with it. Even our earth, to a very small extent, drags space and time around as it rotates. This phenomenon is more pronounced for rapidly rotating massive objects. If a black hole is formed from a rotating star, the dragging of space and time around the hole should be quite apparent. An observer sitting in his spaceship far from the hole would notice that he is gradually being pulled around the hole in the direction in which the hole is rotating. The closer he is to the rotating hole, the faster he is pulled around. So the angular momentum of a black hole is also information that extends beyond the event horizon and is a third property of black holes.

Black holes, then, can have only three possible properties: mass, charge, and angular momentum. Whereas a complete description of a star involves all sorts of hairy things such as chemical composition, varying pressure, densities, and temperatures at different depths, and so on, no such complications are involved in black holes. Physicists put it simply by stating, *"Black holes have no hair!"*

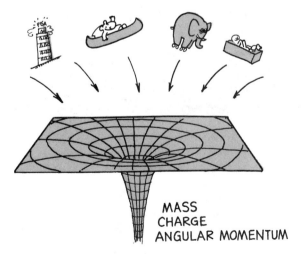

Fig. 35.11 No matter what the physical properties of materials falling into a black hole are, the only properties that remain are mass, charge, and angular momentum.

MASS
CHARGE
ANGULAR MOMENTUM

The rapidly rotating black hole has been a focus of interest to theorists who have speculated that because of centrifugal effects, the singularity is not located at a point, but forms a *ring*. In the early 1970s, black-hole types were excited about the prospects of future astronauts approaching this ring along the rotational axis of the black hole, traveling unharmed through the ring, and entering another universe! Theoretical developments since then, however, severely dampen any such possibilities. So, for the time being, it seems we'll have to content ourselves with a single universe.

Observations of Black Holes

We suspect that all dead stars with masses greater than about five suns may be black holes. Finding these invisible star cores, however, is very difficult. One way is to look for a binary system in which a single luminous star appears to orbit about an invisible companion. If they are closely situated, matter ejected by the normal companion and accelerating into the neighboring black hole should emit X rays. The first convincing candidate for a black hole discovered by astronomers was the X-ray star Cygnus X-1, 1972. Cygnus X-1 is about nine or ten solar masses and is a part of a binary system with a blue supergiant. Material streaming into the black hole from the supergiant companion is found to be emitting X radiation. Similar radiation patterns have also been found in Circinus X-1 (3U 1516-56), another binary system, and it is now tentatively thought that a black hole exists there. Observations by the NASA satellite *Copernicus* strongly suggest that a star in the constellation Scorpio is a black hole. This star, V861 Sco, is at the relatively close distance of 5000 light-years—slightly nearer than Cygnus X-1. The recently discovered X-ray source called LMC X-3 in the large Magellanic Cloud—a dwarf companion galaxy to our own—is very likely a black hole. Other massive black holes of 100 to 1000 suns are thought to exist at the centers of certain globular clusters (NGC 6624, NGC 1851, and NGC 6440).

Question It the sun somehow suddenly collapsed to a black hole, what change would occur in the orbital speed of the planet Earth?*

Quasars Black holes may be the dimmest objects in the universe; the brightest objects are probably **quasars**. The energy output of these objects is enormous—hundreds of times that of the entire Milky Way. Quasars were first thought to be relatively faint ordinary stars in our own galaxy, but in 1960

---

*__Answer__ None. This is best understood classically; nothing in Newton's law of gravitation, $F = G(mM/d^2)$, changes. The fact that the mass of the sun, M, is compressed doesn't change the force.

astronomers found that they emit radio waves. Radio waves are frequently observed to come from galaxies, but no star had been observed to emit strong radio signals. Further investigation revealed these "radio stars" to have a pattern of spectral lines that could not be deciphered. These objects became known as "quasi-stellar sources," soon shortened to *quasars*.

The unusual spectra turned out to be a normal spectrum with an extremely large and unprecedented red shift, which indicated enormous recessional velocities—some more than 90 percent of the speed of light. Clearly the objects couldn't be stars in our own galaxy, for we would have noticed a change in their positions against the background of the fixed stars, and quasars had been observed for years as faint stars with no noticeable change in position. At first some investigators thought that the red shift was not a Doppler shift but a gravitational red shift from a small body with enormous mass and a correspondingly enormous gravitational field. But spectra from quasars revealed emission lines from normal atoms with normally orbiting electrons that wouldn't exist in a neutron star, black hole, or any body with gravitation intense enough to produce such a large red shift.

We do not know what quasars are. They may be gigantic black holes that pull enormous amounts of material toward them with resulting collisions that liberate immense energies. Current findings indicate that quasars are the brilliant cores of very distant galaxies that we see as they were when they were young. Quasars are currently the most puzzling objects known to astronomers.

Our Expanding Universe

So we live in an expanding universe. And what is the end result of this expansion? Does the universe expand forever, or does it finally slow down, stop, and then fall in on itself again? And if it does fall in on itself, does it reexplode, start a new cycle, and keep repeating this cycle ad infinitum?

Before the 1970s one school of thought held that the universe was in a *steady-state* condition, where the density of the expanding universe remained unchanged instead of thinning out. This was thought to be accomplished by the spontaneous creation of hydrogen out of nothing. The steady-state idea has been pretty well discredited, and most cosmologists support the theory of an expanding universe of constant mass-energy emanating from some Big Bang nearly 20 billion years ago. The question is whether or not this expansion continues forever.

A popular school of thought is the *oscillating theory* of the universe. You know that if you throw a rock skyward, two things can happen: the rock can simply go up, momentarily stop, and come back down; or, if you throw it faster than 11.2 kilometers per second (escape velocity for earth), it will continue its upward motion and never come down again. The same thing may happen on a universal scale. If the expanding rate of the universe were less than the escape velocity of the universe, the expansion would finally run its course, momentarily come to a stop, and then fall back in on itself. The time for this cycle has been estimated to be somewhat less than 100 billion years. Being now about 20 billion on the way out, we'd continue our outward expansion for another 20 to 30 billion years and momentarily come to rest. The universe would then be at its maximum extent. Galaxies would be farthest apart and show no Doppler shift in their light. Then the contraction would begin. After 20 to 30 billion years of contracting, the universe would again be the size it is now, but galaxies would all show blue shifts instead of red shifts. Then after 20 billion more years would come the Big Crunch as the universe collapsed into its own black hole—perhaps to gush out into a new universe for another 100-or-so billion years. Very enchanting.

Whether the universe oscillates or whether it expands indefinitely depends on the mass contained in the universe. If the mass of the universe is below a critical value, the present expansion exceeds escape velocity and the universe expands indefinitely. If the mass of the universe is greater than this critical value, then the present expansion is less than the escape velocity and finally comes to a halt, and then the universe falls together to complete a cycle of its possible oscillation. We have yet to know whether the mass of the universe is below or above this critical value. We have so much to find out.

We do not see the world as did the ancient Egyptians, Greeks, and Chinese. It is unlikely that people in the future will see the universe as we do. Our view of the universe may be wrong, but it is most likely less wrong than the views of others before us. Our view today stems from the findings of Copernicus and Galileo—findings that were opposed on the grounds that they diminished the importance of humans in the universe. The idea of importance then was in being able to rise above nature—in being apart from nature. We have expanded our vision since then by enormous effort, painstaking observation, and an unrelenting desire to comprehend our surroundings. Seen from today's understanding of the universe, we find our importance in being very much a part of nature, not apart from it. We are the part of nature that is becoming more and more conscious of itself.

Summary of Terms

Big Bang   The primordial explosion that is thought to have resulted in the expanding universe.

Red giant   A large, relatively cool star of high luminosity.

Planetary nebula   Expanding shells of gas larger than the entire solar system, ejected from certain extremely hot stars.

White dwarf   A star that has exhausted most of its nuclear fuel and has shrunk to a very small size; it is believed to be near its final stage of evolution.

Black dwarf   A cooled-off white dwarf.

Nova   A stellar explosion in which the outer layer of a star is blown off.

Supernova   A star that completely blows up, increasing its luminosity hundreds of thousands to hundreds of millions of times, leaving only a small remnant behind.

Neutron star   A star that has undergone gravitational collapse where electrons are compressed into protons to form a neutron core.

Pulsar   A rapidly rotating neutron star that directs its radiation toward us in a sequence of brief pulses.

Black hole   The configuration of a massive star that has undergone gravitational collapse, in which gravitation is so intense that even its own light cannot escape.

Event horizon   The surface or point-of-no-return distance from the center of a black hole, inside of which nothing can escape to the outside universe.

Black hole singularity   The center point of a black hole.

Quasar   Believed to be very small and the most luminous and distant type of object known in the universe.

Suggested Reading

Abell, G. O. *Exploration of the Universe*. 4th ed. New York: Holt, 1982.

Berman, L., and J. C. Evans. *Exploring the Cosmos,* 4th ed. Boston: Little, Brown, 1980.

Goldsmith, D. *The Evolving Universe*. Menlo Park, Calif.: Benjamin Cummings, 1981.

Kaufmann III, W. J. *The Cosmic Frontiers of General Relativity*. Boston: Little, Brown, 1977.

Trefil, J. S. *The Moment of Creation*. New York: Scribner, 1983.

Weinberg, S. *The First Three Minutes*. New York: Basic Books, 1976.

Review Questions

1. Why is the present expansion of the universe slowing down?

2. What ordinarily keeps a star from gravitationally collapsing?

3. What ordinarily keeps a star from blowing up like a hydrogen bomb?

4. Why are massive stars generally shorter-lived than low-mass stars?

5. Why does the sun spin slowly in comparison to younger stars?

6. At what evolutionary stage does a star like our sun expand to become a red giant?

7. Where do the elements other than hydrogen and helium found in nature originate?

8. What is the difference between a *nova* and *supernova?*

9. What is meant by the statement "Black holes have no hair"?

10. What are the three properties that a black hole may possess?

11. Since black holes are completely invisible, how can we detect their presence?

12. The phrase "the Big Bang" refers to the origin of our expanding universe. What does the phrase "the Big Crunch" refer to?

Exercises

1. On the large scale, which of the four fundamental forces is dominant in the universe? Explain.

2. Which protostars—those of large mass or those of small mass—would you expect to contract more rapidly to become stars? Explain.

3. Many people think that stars begin by being very bright and gradually grow dimmer. What do you think about this?

4. Why will the sun not be able to fuse carbon nuclei in its core?

5. What would happen to the sun if nuclear fusion suddenly ceased?

6. Why must the core temperature of a star be hotter for two helium nuclei to fuse than for two protons to fuse?

7. Why are elements higher than iron in atomic number not fused in the cores of stable stars?

8. What does the spin rate of a star have to do with whether or not it is likely to have planets?

9. Why should the oldest stars in a galaxy have a low abundance of heavy elements in their outer layers?

10. In what sense are we made of "star dust"?

11. What is meant by the statement "The bigger they are, the harder they fall" with respect to stellar evolution?

12. Some binary stars consist of stable, ongoing stars with dead companions. If they were formed at the same time, how can they be so different?

13. A black hole is no more massive than the star from which it collapsed. Why, then, is gravitation so intense near a black hole?

14. Binary stars orbit about a common center of mass. If one of the stars collapses to become a black hole, would you expect any significant difference in the angular speed and period of orbital rotation of its companion star?

15. Some people claim that light doesn't bend in a gravitational field. How is this similar to the claim about bullets in Exercise 6 in Chapter 3?

16. What happens to the radial distance of a photon sphere as more and more mass falls into the black hole?

17. Discuss some of the reasons we have not made contact with extraterrestrial civilizations.

18. Scientists do not have an adequate model for explaining the origin of the "first" hydrogen atoms. Is it simplifying or complicating to say they were made by a higher force?

19. Consider this philosophical question: Does mind precede matter, or does the existence of matter precede mind? (Was there a soul bank before the first Big Bang?)

20. Try to answer Exercises 18 and 19 as your most intelligent and most articulate antagonist would answer them.

Epilogue

Nature bestows on each of us an enormously high prize—being alive. An even higher prize is our capacity to comprehend and understand this nature. The attempt of this book has been to kindle that capacity, open you to new insights, and lead you to a better understanding of many of the things you likely wondered about as a child. In addition to your knowledge of why the sky is blue, how the planets orbit, and how the sun shines, I hope to have communicated the excitement, beauty, and flavor of all the physics that daily surrounds you.

You'll forget most of the facts, formulas, and diagrams that are a part of your understanding of physics, and you'll forget many of the insights you experienced. But the flavor you'll retain—that associated with your discovery and your understanding that all the diverse phenomena about us are beautifully tied together by a surprisingly few relationships: the laws of nature. We are subject to these laws as is the blueness of the sky, the falling of the moon, and the shining of the stars. For we are not apart from nature—we are a very important part of it. We are the conscious part of nature that is investigating itself.

Your education is ultimately the flavor left over after all the facts, formulas, and diagrams have been forgotten.

Paul G. Hewitt

Appendix I Systems of Measurement

Two major systems of measurement prevail in the world today: the *British* system used in the United States and the *metric* system. Each system has its own standards of length, mass, and time. The units of length, mass, and time are sometimes called the *fundamental units* because, once they are selected, all other quantities can be measured in terms of them.

British System

Developed in England, the British system of measurement has been used in the United States for many commercial and engineering purposes. It uses the foot as the unit of length, the pound as the unit of weight or force, and the second as the unit of time.

Metric System

French scientists originated this system of measurement after the French revolution in 1791. The orderliness of this system makes it useful for scientific work, and it is used by scientists all over the world. The metric system branches into two systems of units. In one of these the unit of length is the centimeter, the unit of mass is the gram, and the unit of time is the second. This is called the centimeter-grams-second system, usually abbreviated cgs. It is presently being supplanted by the more-popular mks system, in which the meter is the unit of length, the kilogram is the unit of mass, and the second is the unit of time. The cgs and mks units are related to each other as follows: 100 centimeters equal 1 meter; 1000 grams equal 1 kilogram. Table I.1 shows several units of length related to each other.

Table I.1 Table conversions between different units of length

Unit of length		Kilometer	Meter	Centimeter	Inch	Foot	Mile
1 kilometer	=	1	1000	100,000	39,370	3280.83	0.62140
1 meter	=	0.00100	1	100	39.370	3.28083	6.21×10^{-4}
1 centimeter	=	1.0×10^{-5}	0.0100	1	0.39370	0.032808	6.21×10^{-6}
1 inch	=	2.54×10^{-5}	0.02540	2.5400	1	0.08333	1.58×10^{-5}
1 foot	=	3.05×10^{-4}	0.30480	30.480	12	1	1.89×10^{-4}
1 mile	=	1.60934	1609.34	160,934	63,360	5280	1

One major advantage of a metric system is that it uses the decimal system, where all units are related to smaller or larger units by dividing or multiplying by 10. The prefixes shown in Table I.2 are commonly used to show the relationships between units.

International System of Units

Both the British and the metric systems have been used extensively for practical and scientific purposes in most of the world. With the growing development of science and international cooperation, the need for a

Table I.2 Some prefixes

Prefix	Definition
micro-	One-millionth: a microsecond is one-millionth of a second
milli-	One-thousandth: a milligram is one-thousandth of a gram
centi-	One-hundredth: a centimeter is one-hundredth of a meter
kilo-	One thousand: a kilogram is 1000 grams
mega-	One million: a megahertz is 1 million hertz

uniform usage of units and symbols has become urgent. In 1960 during the International Conference on Weights and Measures, held in Paris, the SI (Systeme International) units were defined and given status. The United States is now in the period of transition to these units. Table I.3 shows SI units and their symbols.

Meter

The standard of length for the metric system originally was defined in terms of the distance from the north pole to the equator. This distance is close to (or was thought to be at the time) 10,000 kilometers. One-ten-millionth of this, the meter, was carefully determined and marked off by means of scratches on a bar of platinum-iridium alloy. This bar is kept at the International Bureau of Weights and Measures in France. The standard meter in France since has been calibrated in terms of the wavelength of light—being 1,650,763.73 times the wavelength of orange light emitted by the atoms of the gas krypton-86. The meter is now defined as being the length of the path traveled by light in a vacuum during a time interval of 1/299 792 458 of a second.

Table I.3 SI units

Quantity	Unit	Symbol
Length	meter	m
Mass	kilogram	kg
Time	second	s
Force	newton	N
Energy	joule	J
Current	ampere	A
Temperature	kelvin	K

Fig. I.1 The standard kilogram.

Kilogram

The standard unit of mass, the kilogram, is a block of platinum, also preserved at the International Bureau of Weights and Measures located in France. The kilogram equals 1000 grams. A gram is the mass of 1 cubic centimeter (cc) of water at a temperature of 4° Celsius. (The standard pound is defined in terms of the standard kilogram; the mass of 1 pound is equal to 0.4536 kilogram.)

Second

The official unit of time for both the British and metric systems of measurement is the second. Until 1956 it was defined in terms of the mean solar day, which was divided into 24 hours. Each hour was divided into 60 minutes and each minute into 60 seconds. Thus, there were 86,400 seconds per day, and the second was defined as 1/86,400 of the mean solar day. This proved unsatisfactory because the rate of rotation of the earth is gradually becoming slower. In 1956 the mean solar day of the year 1900 was chosen as the standard on which to base the second. Further defined in 1964, the second is now officially defined as the time taken by a cesium-133 atom to make 9,192,631,770 vibrations.

Newton

One newton is the force required to accelerate 1 kilogram by 1 meter per second per second. This unit is named after Sir Isaac Newton.

Joule

One joule is equal to the amount of work done by a force of 1 newton acting over a distance of 1 meter. In 1948 the joule was adopted as the unit of energy by the International Conference on Weights and Measures. Therefore, the specific heat of water at 15°C is 4185.5 joules/kilogram. This figure is always associated with the mechanical equivalent of heat.

Ampere

The ampere is defined as the intensity of the constant electric current that, when maintained in two parallel conductors of infinite length and negligible cross section, and placed 1 meter apart in a vacuum, would produce between them a force equal to 2×10^{-7} newton per meter length. In our treatment of electric current in this text, we have used the not-so-official, but easier-to-comprehend definition of the ampere as being the rate of flow of 1 coulomb of charge per second, where 1 coulomb is the charge of $6\frac{1}{4} \times 10^{18}$ electrons.

Kelvin

The fundamental unit of temperature is named after the scientist Lord Kelvin. The kelvin is defined to be 1/273.15 the thermodynamic temperature of the triple point of water (the fixed point at which ice, liquid water, and water vapor coexist in equilibrium). This definition was adopted in 1968 when it was decided to change the name degree Kelvin (°K) to kelvin (K). The temperature of melting ice at atmospheric pressure is 273.15 K. The temperature at which the vapor pressure of pure water is equal to standard atmospheric pressure is 373.15 K (the temperature of boiling water at atmospheric pressure).

Measurements of Area and Volume

Fig. I.2 Unit square.

Area

The unit of area is a square that has a standard unit of length as a side. In the British system it is a square whose sides are each 1 foot in length, called 1 square foot and written 1 ft². In the SI system it is a square whose sides are 1 meter in length, which makes a unit of area of 1 m². In the cgs system it is 1 cm². The area of a given surface is specified by the number of square feet, square meters, or square centimeters that would fit into it.

The area of a rectangle equals the base times the height. The area of a circle is equal to πr^2, where $\pi = 3.14$ and r is the radius of the circle. Formulas for the surfaces of other objects can be found in geometry textbooks.

Volume

The volume of a body refers to the space it occupies. The unit of volume is the space taken up by a cube that has a standard unit of length for its edge. In the British system it is the space occupied by a cube 1 foot on an edge and is called 1 cubic foot, written 1 ft³. In the metric system it is the space occupied by a cube whose sides are 1 meter (SI) or 1 centimeter (cgs). It is written 1 m³ or 1 cm³ (or cc). The volume of a given space is specified by the number of cubic feet, cubic meters, or cubic centimeters that will fill it.

Fig. I.3 Unit volume.

In the British system volumes are measured in quarts, gallons, and cubic inches as well as in cubic feet. These are 1728 (12 × 12 × 12) cubic inches in 1 ft³. A U.S. gallon is a volume of 231 in³. Four quarts equal 1 gallon.

In the SI system volumes are measured in liters. A liter is equal to 1000 cm³.

Scientific Notation

It is convenient to use a mathematical abbreviation for large and small numbers. The number 50,000,000 can be obtained by multiplying 5 by 10, and again by 10, and again by 10, and so on until 10 has been used as a multiplier seven times. The short way of showing this is to write the number 5×10^7. The number 0.0005 can be obtained from 5 by using 10 as a divisor four times. The short way of showing this is to write 5×10^{-4} for 0.0005. Thus, 3×10^5 means $3 \times 10 \times 10 \times 10 \times 10 \times 10$, or 300,000; and 6×10^{-3} means $6/(10 \times 10 \times 10)$, or 0.006. Numbers expressed in this shorthand manner are said to be in scientific notation.

$$
\begin{aligned}
1{,}000{,}000 &= 10 \times 10 \times 10 \times 10 \times 10 \times 10 &&= 10^6 \\
100{,}000 &= 10 \times 10 \times 10 \times 10 \times 10 &&= 10^5 \\
10{,}000 &= 10 \times 10 \times 10 \times 10 &&= 10^4 \\
1000 &= 10 \times 10 \times 10 &&= 10^3 \\
100 &= 10 \times 10 &&= 10^2 \\
10 &= 10 &&= 10^1 \\
1 &= 1 &&= 10^0 \\
0.1 &= 1/10 &&= 10^{-1} \\
0.01 &= 1/100 = 1/10^2 &&= 10^{-2} \\
0.001 &= 1/1000 = 1/10^3 &&= 10^{-3} \\
0.0001 &= 1/10{,}000 = 1/10^4 &&= 10^{-4} \\
0.00001 &= 1/100{,}000 = 1/10^5 &&= 10^{-5} \\
0.000001 &= 1/1{,}000{,}000 = 1/10^6 &&= 10^{-6}
\end{aligned}
$$

We can use scientific notation to express some of the physical data often used in physics.

$$
\begin{aligned}
\text{Speed of light in a vacuum} &= 2.9979 \times 10^8 \text{ m/s} \\
\text{1 astronomical unit (A.U.),} & \\
\text{(average earth-sun distance)} &= 1.50 \times 10^{11} \text{m} \\
\text{Average earth-moon distance} &= 3.84 \times 10^8 \text{m} \\
\text{Average radius of the sun} &= 6.96 \times 10^8 \text{m} \\
\text{Average radius of Jupiter} &= 7.14 \times 10^7 \text{m} \\
\text{Average radius of the earth} &= 6.37 \times 10^6 \text{m} \\
\text{Average radius of the moon} &= 1.74 \times 10^6 \text{m} \\
\text{Average radius of hydrogen atom} &= 5 \times 10^{-11} \text{m} \\
\text{Mass of the sun} &= 1.99 \times 10^{30} \text{kg} \\
\text{Mass of Jupiter} &= 1.90 \times 10^{27} \text{kg} \\
\text{Mass of the earth} &= 5.98 \times 10^{24} \text{kg} \\
\text{Mass of the moon} &= 7.36 \times 10^{22} \text{kg} \\
\text{Proton mass} &= 1.6726 \times 10^{-27} \text{kg} \\
\text{Neutron mass} &= 1.6749 \times 10^{-27} \text{kg} \\
\text{Electron mass} &= 9.1 \times 10^{-31} \text{kg} \\
\text{Electron charge} &= 1.602 \times 10^{-19} \text{C}
\end{aligned}
$$

Appendix II More About Motion

Computing Distance When Motion Is Uniform

Objects traveling at constant velocity travel equal distances in equal intervals of time. We can say

$$\text{Distance traveled} = \text{velocity} \times \text{time}$$

Or, in shorthand notation,

$$d = vt$$

If, for example, we are in a car traveling at a constant velocity of 60 kilometers per hour for a time of 1 hour, we will have traveled 60 kilometers. (It is customary to omit the multiplication sign, $\times$, when expressing relationships in mathematical form. When two symbols are written together, such as the vt in this case, it is understood that they are multiplied.)

Computing Distance When Acceleration Is Constant

How far will an object released from rest fall in a given time? To answer this question, let us consider the case in which it falls freely for 3 seconds, starting at rest. Neglecting air resistance, the object will have a constant acceleration of 9.8 meters per second per second:

$$\text{Velocity at the } beginning = 0 \text{ meters per second}$$
$$\text{Velocity at the } end \text{ of 3 seconds} = (9.8 \times 3) \text{ meters per second}$$

$$Average \text{ velocity} = \tfrac{1}{2} \text{ the sum of these two speeds}$$
$$= \tfrac{1}{2} \times (0 + 9.8 \times 3) \text{ meters per second}$$
$$= \tfrac{1}{2} \times 9.8 \times 3 = 14.7 \text{ meters per second}$$

$$\text{Distance traveled} = \text{average velocity} \times \text{time}$$
$$= (\tfrac{1}{2} \times 9.8 \times 3) \times 3$$
$$= \tfrac{1}{2} \times 9.8 \times 3^2 = 44.1 \text{ meters}$$

We can see from the meanings of these numbers that

$$\text{Distance traveled} = \tfrac{1}{2} \times \text{acceleration} \times \text{square of time}$$

This equation is true for a body falling not only for 3 seconds but for any length of time. If we let d stand for the distance traveled, a for the acceleration, and t for the time, the rule may be written, in shorthand notation,

$$d = \tfrac{1}{2}at^2$$

This relationship was first deduced by Galileo. He reasoned that if a body falls for, say, twice the time, it will fall with *twice the average speed.*

615

Since it falls for *twice* as long at *twice* the average speed, it will fall *four* times as far. Similarly, if a body falls for *three* times as long, it will have an average speed *three* times as great and will fall *nine* times as far. Galileo reasoned that the total distance fallen should be proportional to the *square* of the time.

In the case of falling bodies, it is customary to use the letter g to represent the acceleration instead of the letter a (g is used for the symbol because the acceleration is due to *gravity*). While the value of g varies slightly in different parts of the world, it is approximately equal to 9.8 meters per second per second (32 feet per second per second). If we use g for the acceleration of a freely falling body, the equations for falling bodies simply become

$$v = gt$$
$$d = \tfrac{1}{2}gt^2$$

Examples

1. An auto starting from rest has a constant acceleration of 4 meters per second per second. How far will it go in 5 seconds?*

2. How far will an object released from rest fall in 1 second? In this case the acceleration is $g = 9.8$ meters per second per second.†

3. If it takes 4 seconds for an object to fall to the water when released from the Golden Gate Bridge, how high is the bridge?‡

---

*Answer Distance $= \tfrac{1}{2} \times 4 \times 5^2 = 50$ meters

†**Answer** Distance $= \tfrac{1}{2} \times 9.8 \times 1^2 = 4.9$ meters

‡**Answer** Distance $= \tfrac{1}{2} \times 9.8 \times 4^2 = 78.4$ meters

Notice that the units of measurement when multiplied give the proper units of meters for distance:

$$d = \tfrac{1}{2} \times 9.8 \, \frac{m}{s^2} \times 16s^2 = 78.4 \text{ m}$$

Appendix III Vectors

Vectors and Scalars

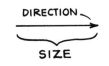

Fig. III.1

A *vector* quantity is a directed quantity—one that must be specified not only by magnitude (size) but by direction as well. Some examples are velocity, force, and acceleration. In contrast, a *scalar* quantity can be specified by magnitude alone. Some examples of scalar quantities are time, temperature, and speed.

Vector quantities may be represented by arrows. The length of the arrow tells you the magnitude of the vector quantity, and the arrowhead tells you the direction of the vector quantity. Such an arrow drawn to scale and pointing appropriately is called a *vector*.

Adding Vectors

Vectors that add together are called *component* vectors. The sum of component vectors is called a *resultant*.

To add two vectors, make a parallelogram with two component vectors acting as two of the adjacent sides (Figure III.2). (Here our parallelogram is a rectangle.) Then draw a diagonal from the origin of the vector pair—this is the resultant (Figure III.3).

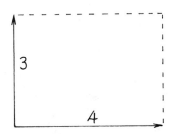

Fig. III.2

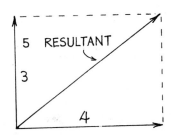

Fig. III.3

Finding Components of Vectors

To find a pair of perpendicular components for a vector, first draw a dotted line through the tail end of the vector (in the direction of one of the desired components). Second, draw another dotted line through the tail end of the vector at right angles with the first dotted line. Third, make a rectangle whose diagonal is the given vector. Draw in the two components. Here we let F stand for "force," U stand for "up," and S stand for "sideways."

Fig. III.4

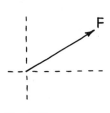

Fig. III.5

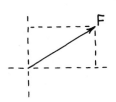

Fig. III.6

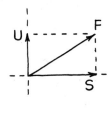

Examples

1. A man pushing a lawnmower applies a force that pushes the machine forward and also against the ground. In Figure III.7, F represents the force applied by the man. We can separate this force into two components. The vector D represents the downward component, and S is the sideways component, the force that moves the lawnmower forward. If we know the magnitude and directon of the vector F, we can estimate the magnitude of the components from the vector diagram.

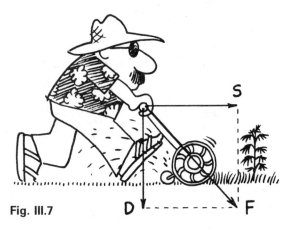

Fig. III.7

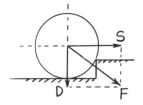

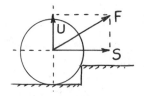

Fig. III.8

2. Would it be easier to push or pull a wheelbarrow over a step? Figure III.8 shows a vector diagram for each case. When you push a wheelbarrow, part of the force is directed downward, which makes it harder to get over the step. When you pull, however, part of the pulling force is directed upward, which helps to lift the wheel over the step.

3. If we consider the components of the weight of an object rolling down an incline, we can see why its speed depends on the angle. Note that the steeper the incline, the greater the component S and the faster the object rolls. When the incline is vertical, S becomes equal to the weight, and the object attains maximum acceleration, 9.8 meters/second^2.

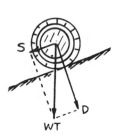
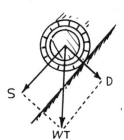

Fig. III.9

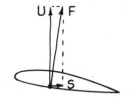

Fig. III.10

4. When moving air strikes the underside of an airplane wing, the force of air impact against the wing may be represented by a single vector perpendicular to the plane of the wing (Figure III.10). We represent the force vector as acting midway along the lower wing surface, where the dot is, and pointing above the wing to show the direction of the resulting wind impact force. This force can be broken up into two components, one sideways and the other up. The upward component, U, is called *lift*. The sideways component, S, is called *drag*. If the aircraft is to fly at constant velocity at constant altitude, then lift must equal the weight of the aircraft, and the thrust of the plane's engines must equal drag. The magnitude of lift (and drag) can be altered by changing the speed of the airplane or by changing the angle (called *angle of attack*) between the wing and the horizontal.

5. The trajectory of a projectile is better understood with vectors (Figure III.11).

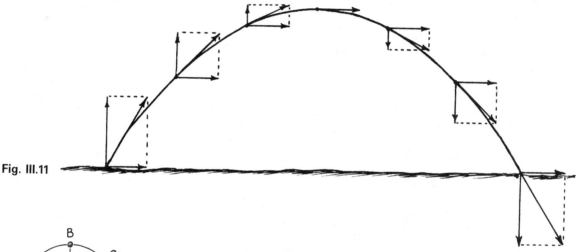

Fig. III.11

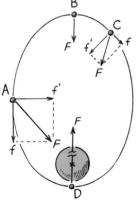

Fig. III.12

6. Consider the projectile moving clockwise in Figure III.12. Everywhere in its orbital path, gravitational force F pulls it toward the center of the host planet. At position A we see F separated into two components: f, which is tangent to the path of the projectile; and f', which is perpendicular to the path. The relative magnitudes of these components in comparison to the magnitude of F can be seen in the imaginary rectangle they compose; f and f' are the sides, and F is the diagonal. We see that component f is along the orbital path but against the direction of motion of the projectile. This component reduces the speed of the projectile. The other component, f', changes the direction of the projectile's motion and pulls it away from its tendency to go in a straight line. So the projectile curves. The projectile loses speed until it reaches position B. At this farthest point (apogee) from the planet, the gravitational force is somewhat weaker but perpendicular to the projectile's motion and component f has reduced to zero. Component f', on the other hand, has increased and is now fully merged to become F. Speed at this point is not enough for circular orbit, and the projectile begins to fall toward the planet. It picks up speed because the component f reappears and is in the direction of motion as shown in position C. It picks up speed until it whips around to position D (perigee), where once again the direction of motion is perpendicular to the gravitational force, f' *blends to full F*, and f is nonexistent. The speed is in excess of that needed for circular orbit at this distance, and it overshoots to repeat the cycle. Its loss in speed in going from D to B equals its gain in speed from B to D. Kepler, who discovered that planetary paths are elliptical, never knew why. Do you?

7. Refer to the Polaroids held by Dotty Jean back in Chapter 27, in Figure 27.32. In the first picture (*a*), we see that light is transmitted through the pair of Polaroids because their axes are aligned. The emerging light can be represented as a vector aligned with the polarization axes of the Polaroids. When the Polaroids are crossed (*b*), no light emerges because light passing through the first Polaroid is perpendicular to the polarization axes of the second Polaroid, with no components along its axis. In the third picture (*c*), we see that light is transmitted when a third Polaroid is sandwiched at an angle between the crossed Polaroids. The explanation for this is shown in Figure III.13.

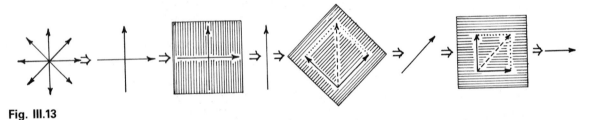

Fig. III.13

Sailboats   Sailors have always known that a sailboat can sail downwind, that is, in the direction of the wind. Sailors have not always known, however, that a sailboat can sail upwind, that is, against the wind. To understand how this is possible, consider first the case of a sailboat sailing downwind.

Figure III.14 shows a sailboat sailing downwind. The force of wind impact against the sail accelerates the boat. Even if the drag of the water and all other resistance forces are negligible, the maximum speed of the boat is the wind speed. This is because the wind will not make impact against the sail if the boat is moving as fast as the wind. The sail would

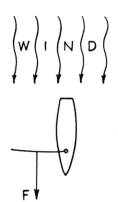

Fig. III.14

simply sag. If there is no force, then there is no acceleration. The force vector in Figure III-14 *decreases* as the boat travels faster. The force vector is maximum when the boat is at rest and the full impact of the wind fills the sail, and minimum when the boat travels as fast as the wind. If the boat is somehow propelled to a speed faster than the wind (by way of an outboard motor, for example), then air resistance against the front side of the sail will produce an oppositely directed force vector. This will slow the boat down. Hence, the boat when driven only by the wind cannot exceed wind speed.

If the sail is oriented at an angle as shown in Figure III.15, the boat will move forward, but with less acceleration. There are two reasons for this:

1. The force on the sail is less because the sail does not intercept as much wind in this angular position.

2. The direction of the wind impact force on the sail is not in the direction of the boat's motion, but is perpendicular to the surface of the sail. Generally speaking, whenever any fluid (liquid or gas) interacts with a smooth surface, the force of interaction is perpendicular to the smooth surface.[1] The boat does not move in the same direction as the perpendicular force on the sail, but is constrained to move in a forward (or backward) direction by a deep, finlike keel beneath the water.

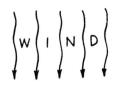

Fig. III.15

We can better understand the motion of the boat by resolving the force of wind impact, F, into perpendicular components. The important component is that which is parallel to the keel, which we label K, and the other component is perpendicular to the keel and is labeled T. It is the component K, as shown in Figure III.16, that is responsible for the forward motion of the boat. Component T is a useless force that tends to tip the boat over and move it sideways. This component force is offset by the heavy, deep keel. Again, maximum speed of the boat can be no greater than wind speed.

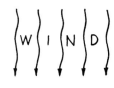

Fig. III.16

[1] You can do a simple exercise to see that this is so. Try bouncing one coin off another on a smooth surface as shown. Note that the struck coin moves at right angles (perpendicular) to the contact edge. Note also that it makes no difference whether the projected coin moves along path A or B. See your instructor for a more rigorous explanation, which involves momentum consideration.

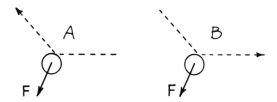

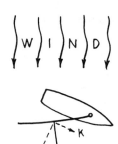

Fig. III.17

Fig. III.18

If a sailboat sails in a direction other than exactly downwind (Figure III.17), and has its sail properly oriented, it can exceed wind speed. In the case of a sailboat cutting across the wind, the wind may continue to make impact with the sail even after the boat exceeds wind speed. A surfer, in a similar way, exceeds the velocity of the propelling wave by angling his surfboard across the wave. Greater angles to the propelling medium (wind for the boat, water wave for the surfboard) result in greater speeds. A sailcraft can sail faster cutting across the wind than it can sailing downwind.

As strange as it may seem, maximum speed is attained by cutting into (against) the wind, that is, by angling the sailcraft in a direction upwind! Although a sailboat cannot sail *directly* upwind, it can reach a destination upwind by angling back and forth in a zigzag fashion. This is called *tacking*. Suppose the boat and sail are as shown in Figure III.18. Component K will push the boat along in a forward direction, angling into the wind. In the position shown, the boat can sail faster than the speed of the wind. This is because as the boat travels faster, the impact of wind is *increased*. This is similar to running in a rain that comes down at an angle. When you run into the direction of the downpour, the drops strike you harder and more frequently; but when you run away from the direction of the downpour, the drops don't strike you as hard or as frequently. In the same way, a boat sailing upwind experiences greater wind impact force, while a boat sailing downwind experiences a decreased wind impact force. In any case the boat reaches its terminal speed when opposing forces cancel the force of wind impact. The opposing forces consist mainly of water resistance against the hull of the boat. The hulls of racing boats are shaped to minimize this resistive force, which is the principal deterrent to high speeds.

Iceboats (sailcraft equipped with runners for traveling on ice) encounter no water resistance and can travel at several times the speed of the wind when they tack upwind. Although ice friction is nearly absent, an iceboat does not accelerate without limit. The terminal velocity of a sailcraft is determined not only by opposing friction forces but also by the change in relative wind direction. When the boat's orientation and speed are such that the wind seems to shift in direction so that it moves parallel to the sail rather than into it, forward acceleration ceases—at least in the case of a flat sail. In practice, sails are curved and produce an airfoil that is as important to sailcraft as it is to aircraft. The effects are discussed in Chapter 12.

Appendix IV The Universal Gravitational Constant, G

We have expressed Newton's law of gravity as a proportion; namely,

$$F \sim \frac{mm'}{d^2}$$

Proportionalities are difficult to deal with, so we can write it as an equality by including a *constant of proportionality*. For example, the circumference of a circle is proportional to the diameter:

$$C \sim D$$

If we divide the circumference of any circle by its diameter, we find a constant proportion of 3.14 . . . or

$$\frac{C}{D} = \pi$$

We can therefore be more specific than the proportionality $C \sim D$ by including the constant of proportionality and writing it as an equation:

$$C = \pi D$$

The value of π tells us something about a circle. Similarly, the constant of proportionality in the equation that expresses the law of gravitation is G, and it tells us something about the strength of the gravitational force. The law of universal gravitation is written as

$$F = G\frac{mm'}{d^2}$$

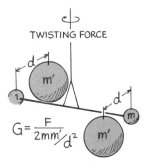

TWISTING FORCE

$$G = \frac{F}{2mm'/d^2}$$

Fig. IV.1

The constant G was first measured by Henry Cavendish in the eighteenth century. He used a torsion balance in which two spheres of mass m were attached to the end of a light rod that was suspended by a long wire so that it could turn easily (Figure IV.1). When two massive lead balls were introduced, the attraction between the light and heavy balls caused a displacement in the balance. Knowing the force necessary to cause such a displacement, Cavendish was able to calculate the ratio G.

A simpler method was developed by Philipp von Jolly. A spherical vessel of mercury was attached to one arm of a sensitive balance, and the balance was put in equilibrium (Figure IV.2). Then a 6-ton lead sphere was rolled beneath the mercury flask. The gravitational force between the two masses was equal to the weight that had to be placed on the opposite end of the balance to restore equilibrium. All the quantities m, m', F, and d were known, from which the value of G was calculated:

$$G = 6.67 \times 10^{-11} \text{ m}^3/\text{kg-s}^2$$

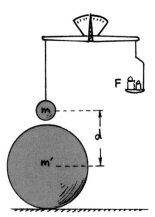

Fig. IV.2

Once the value of G is known, the mass of the earth can be calculated. The force that the earth exerts on a mass of 1 kilogram at its surface is equal to 9.8 newtons. The distance between the 1-kilogram mass and the center of mass of the earth is the earth's radius, 6.4×10^6 meters. Therefore, from $F = G \, (mm'/d^2)$, where we let m' be the mass of the earth,

$$9.8 \text{ newtons} = 6.67 \times 10^{-11} \frac{\text{m}^3}{\text{kg-s}^2} \frac{1 \text{ kg} \times m'}{(6.4 \times 10^6 \text{ m})^2}$$

(where 1 newton = 1 kilogram $\times$ meter/second^2) from which the mass of the earth $m' = 6 \times 10^{24}$ kilograms.

Highly regarded theories suggest that G is not constant at all, but decreases with time. This decrease is predicted to be about $10^{-11} G$ per year. Whether or not G varies with time is extremely important to astrophysicists and cosmologists. A closer-to-home result of this weakening of gravity with time is that the earth would be expanding in size at a present rate of 0.5 millimeter per year. So if gravity is weakening, the earth is growing.

Appendix V Exponential Growth and Doubling Time*

One of the most important things we seem not to perceive is the process of exponential growth. We think we understand how compound interest works, but we can't get it through our heads (literally) that a fine piece of tissue paper folded upon itself 50 times (if that were possible) would be 17 million miles thick. If we could, we could "see" why our income buys only half what it did four years ago, why the price of everything has doubled in the same time, why populations and pollution and nuclear bombs seem to (in fact, really do) proliferate out of control.[1]

When a quantity such as money in the bank, population, or the rate of consumption of a resource steadily grows at a fixed percent per year, we say the growth is *exponential*. Money in the bank may grow at 8 percent per year; world population is currently growing at about 2 percent per year; electric power generating capacity in the United States grew at about 7 percent per year for the first three quarters of the century. The important thing about exponential growth is that the time required for the growing quantity to increase its size by a fixed fraction is constant. So the time

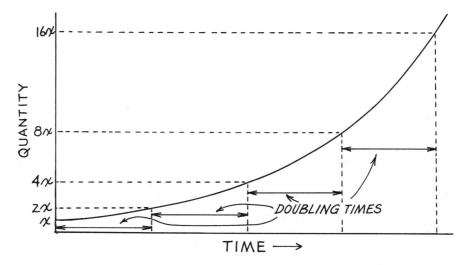

Fig. V.1 An exponential curve. Notice that each of the successive equal time intervals noted on the horizontal scale corresponds to a doubling of the quantity indicated on the vertical scale. Such an interval is called the *doubling time*.

---

*This appendix is drawn from material written by University of Colorado physics professor Albert A. Bartlett, who strongly asserts, "The greatest shortcoming of the human race is man's inability to understand the exponential function." Look up Professor Bartlett's provocative article, "Forgotten Fundamentals in the Energy Crisis," in the September 1978 issue of the *American Journal of Physics,* or his revised version in the January 1980 issue of the *Journal of Geological Education.*

[1]K. C. Cole, *Sympathetic Vibrations* (New York: Morrow, 1984).

required for the quantity to double in size (increase by 100 percent) is also constant. For example, if the population of a growing city takes 12 years to double from 5,000 to 10,000 inhabitants and its growth remains steady, in the next 12 years the population will double to 20,000, and in the next 12 years to 40,000, and so on.

There is an important relationship between the percent growth rate and its *doubling time*, or the time it takes to double a quantity:[2]

$$\text{Doubling time} = \frac{69.3}{\text{percent growth per unit time}} \approx \frac{70}{\text{percent}}$$

So to estimate the doubling time for a steadily growing quantity, we simply divide the number 70 by the percentage growth rate. For example, the 7 percent growth rate of electric power generating capacity in the United States means that in the past the capacity has doubled every 10 years (70/7 = 10). A 2 percent growth rate for world population means the population of the world doubles every 35 years (70/2 = 35). A city planning commission that accepts what seems like a modest 3.5 percent growth rate may not realize that this means that doubling will occur in 70/3.5 or 20 years; that's double capacity for such things as water supply, sewage treatment plants, and other municipal services every 20 years.

What happens when you put steady growth in a finite environment? Consider the growth of bacteria that grow by division, so that one bacterium becomes two, the two divide to become four, the four divide to become eight, and so on. Suppose the division time for a certain strain of bacteria is 1 minute. This is then steady growth—the number of bacteria grows exponentially with a doubling time of 1 minute. Further, suppose that one bacterium is put in a bottle at 11:00 A.M. and that growth continues steadily until the bottle becomes full of bacteria at 12:00 noon. Consider seriously the following question.

Question   When was the bottle half-full?*

It is startling to note that at 2 minutes before noon the bottle was only $\frac{1}{4}$ full, and at 3 minutes before noon only $\frac{1}{8}$ full. Table V.1 summarizes the amount of space left in the bottle in the last few minutes before noon. If you were an average bacterium in the bottle, at which time would you first realize that you were running out of space? For example, would you sense there was a serious problem at 11:55 A.M. when the bottle was only 3 percent filled ($\frac{1}{32}$) and had 97 percent of open space (just yearning for development)? The point here is that there isn't much time between the moment that the effects of growth become noticeable and the time when they become overwhelming.

***Answer**   11:59 A.M.; the bacteria will double in number every minute!

[2]For exponential decay we speak about *half-life*, the time required for a quantity to reduce to half its value. This case is treated in Chapter 31.

Fig. V.2

Table V.1   The last minutes in the bottle

Time	Part-full (%)	Part-empty
11:54 A.M.	1/64 (1.5%)	63/64
11:55 A.M.	1/32 (3 %)	31/32
11:56 A.M.	1/16 (6 %)	15/16
11:57 A.M.	1/8 (12 %)	7/8
11:58 A.M.	1/4 (25 %)	3/4
11:59 A.M.	1/2 (50 %)	1/2
12:00 noon	Full (100 %)	None

Suppose that at 11:58 A.M. some farsighted bacteria see that they are running out of space and launch a full-scale search for new bottles. Luckily, at 11:59 A.M. they discover three new empty bottles, three times as much space as they had ever known. This quadruples the total resource space ever known to the bacteria, for they now have a total of four bottles, whereas before the discovery they had only one. Further suppose that thanks to their technological proficiency, they are able to migrate to their new habitats without difficulty. Surely, it seems to most of the bacteria that their problem is solved—and just in time.

Question   If the bacteria growth continues at the unchanged rate, what time will it be when the three new bottles are filled to capacity?*

We see from Table V.2 that quadrupling the resource extends the life of the resource by only two doubling times. In our example the resource is space—but it could as well be coal, oil, uranium, or any nonrenewable resource.

Table V.2   Effects of the discovery of three new bottles

Time	Effect
11:58 A.M.	Bottle 1 is $\frac{1}{4}$ full
11:59 A.M.	Bottle 1 is $\frac{1}{2}$ full
12:00 noon	Bottle 1 is full
12:01 P.M.	Bottles 1 and 2 are both full
12:02 P.M.	Bottles 1, 2, 3, and 4 are all full

Continued growth and continued doubling lead to enormous numbers. In two doubling times, a quantity will double twice ($2^2 = 4$; quadruple) in size; in three doubling times, its size will increase eightfold ($2^3 = 8$); in four doubling times, it will increase sixteenfold ($2^4 = 16$); and so on.

Answer   12:02 P.M.!

This is best illustrated by the story of the court mathematician in India who years ago invented the game of chess for his king. The king was so pleased with the game that he offered to repay the mathematician, whose request seemed modest enough. The mathematician requested a single grain of wheat on the first square of the chessboard, two grains on the second square, four on the third square, and so on, doubling the number of grains on each succeeding square until all squares had been used. At this rate there would be 2^{63} grains of wheat on the sixty-fourth square. The king soon saw that he could not fill this "modest" request, which amounted to more wheat than had been harvested in the entire history of the earth!

Fig. V.3 A single grain of wheat placed on the first square of the chessboard is doubled on the second square, and this number is doubled on the third square, and so on, presumably for all 64 squares. Note that each square contains one more grain than all the preceding squares combined. Does enough wheat exist in the world to fill all 64 squares in this manner?

Table V.3 Filling the squares on the chessboard

Square number	Grains on square	Total grains thus far
1	1	1
2	2	3
3	4	7
4	8	15
5	16	31
6	32	63
7	64	127
.	.	.
.	.	.
.	.	.
64	2^{63}	$2^{64} - 1$

It is interesting and important to note that the number of grains on any square is one grain more than the total of all grains on the preceding squares. This is true anywhere on the board. Note from Table V.3 that when eight grains are placed on the fourth square, the eight is one more than the total of seven grains that were already on the board. Or the thirty-two grains placed on the sixth square is one more than the total of thirty-one grains that were already on the board. We see that in one doubling time we use more than all that had been used in all the preceding growth!

So when we speak of doubling energy consumption in the next however many years, bear in mind that this means in these years we will consume more energy than has heretofore been consumed during the entire preceding period of steady growth. And if power generation were to remain predominantly from fossil fuels, then except for some improvements in efficiency, we would burn up in the next doubling time a greater amount of coal, oil, and natural gas than had already been consumed by previous power generation; and except for improvements in pollution control, we would expect to discharge even more toxic wastes into the environment than the millions upon millions of tons already discharged over all the previous years of industrial civilization; and we would expect more human-made calories of heat to be absorbed by the earth's ecosystem than were absorbed in the past! At the previous 7 percent annual growth rate in energy production, all this would occur in one doubling time of a single decade. If over the coming years the annual growth rate remains at half this value, 3.5 percent, then all this would take place in a doubling time of two decades. Clearly this cannot continue!

The consumption of a nonrenewable resource cannot grow exponentially indefinitely, for the resource is finite and its supply finally expires. The most drastic way this can happen is shown in the graph in Figure V.4a, where the rate of consumption, such as barrels of oil per year, is plotted against time, say in years. In such a graph the shaded area under the curve represents the supply of the resource. We see that when the supply is exhausted, the consumption ceases altogether. This sudden change is usually not the case, for the rate of extracting the supply falls as it becomes scarcer. This is shown in Figure V.4b. Note that the area under the curve is equal to the area under the curve in a. Why? Because the total supply is the same in both cases. The principal difference is in the time taken to finally extinguish the supply. History shows that the rate of production of a nonrenewable resource rises and falls in a nearly symmetric manner as shown by the curve in c. The time during which production rates rise is approximately equal to the time these rates fall to zero or near zero. If we fit the data for U.S. oil production in the lower forty-eight states to such a curve, we find that we are just to the right of the peak. This indicates that one-half of the recoverable petroleum that was ever in the ground in the U.S. has already been used and that in the future the domestic petroleum rate of production can only decrease. The U.S. production curve peaked in 1970, and by 1979 nearly half the U.S. consumption was imported. Each year we consume more oil than the previous year.

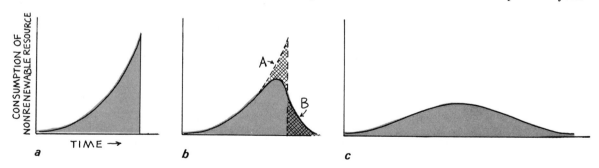

Fig. V.4 (a) If the exponential rate of consumption for a nonrenewable resource continues until it is depleted, consumption falls abruptly to zero. The shaded area under this curve represents the total supply of the resource. (b) In practice the rate of consumption levels off and then falls less abruptly to zero. Note that the crosshatched area A is equal to the crosshatched area B. Why? (c) At lower consumption rates, the same resource lasts a longer time.

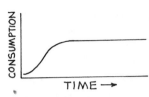

Fig. V.5 A curve showing the rate of consumption of a renewable resource such as agricultural or forest products, where a steady rate of production and consumption can be maintained for a long period, providing this production is not dependent upon the use of a nonrenewable resource, such as petroleum, the production of which is declining.

Production rates for all nonrenewable resources decrease sooner or later. Only production rates for renewable resources, such as agriculture or forest products, can be maintained at steady levels for long periods of time (Figure V.5), provided such production does not depend on nonrenewable resources such as petroleum, the production of which may be waning. Much of today's agriculture is so petroleum-dependent that it can be said that modern agriculture is simply the process whereby land is used to convert petroleum into food. The implications of petroleum scarcity go far beyond rationing of gasoline for cars or fuel oils for home heating.

Power production from whatever sources will not meet the present increasing demands in this growing world. Fusion power may characterize the next century, but only the most optimistic forecasters see it as providing even a tiny fraction of our power needs at the beginning of the twenty-first century. Even though the production of nuclear fission power had spectacular growth after its harnessing in 1942, it took 30 years of development to equal the annual energy consumption of firewood in the United States. Nuclear fusion will be vastly more complicated than fission, so even if nuclear fusion were successfully harnessed today, how long would it take before it could play a significant role in major power production?

However, the *important* questions are more basic: Is growth really good? Is bigger really better? Is it true that if we don't grow, we will stagnate? In answering these questions, bear in mind that human growth is an early phase of life that continues normally through adolescence. Physical growth stops when physical maturity is reached. What do we say of growth that continues in the period of physical maturity? We say that such growth is obesity—or, worse, cancer.

Questions to Ponder

1. According to a French riddle, a lily pond starts with a single leaf. Each day the number of leaves doubles, until the pond is completely full on the thirtieth day. On what day was the pond half covered? One-quarter covered?

2. In an economy that has a steady inflation rate of 7 percent per year, in how many years does a dollar lose half its value?

3. At a steady inflation rate of 7 percent, what will be the price every ten years for the next 50 years for a theater ticket that now costs $10? For a suit of clothes that now costs $100? For a car that now costs $10,000? For a home that now costs $100,000?

4. If the population of a city with one overloaded sewage treatment plant grows steadily at 5 percent annually, how many overloaded sewage treatment plants will be necessary 42 years later?

5. In 1984 the population growth rate for the United States was 0.7 percent, for Mexico 2.6 percent, and for Kenya the highest growth rate in the world, 4 percent (taking into account births, deaths, and immigration). At these rates, how long would it take for the population in each of these countries to double?

6. If world population doubles in 40 years and world food production also doubles in 40 years, how many people then will be starving each year compared to now?

7. A continued world population growth rate of 1.9 percent per year would produce a density of one person per square meter in 550 years. True or false: World population growth rate will sooner or later be zero.

8. Suppose you get a prospective employer to agree to hire your services for wages of a single penny for the first day, 2 pennies the second day, and doubling each day thereafter providing the employer keep to the agreement for a month. What will be your total wages for the month?

9. In the preceding exercise, how will your wages for only the thirtieth day compare to your total wages for the previous 29 days?

10. We hear often that reserves of nonrenewable resources such as coal, oil, and natural gas are "scarce," "abundant," or "superabundant." Why are these terms meaningless without referring also to their consumption rates?

11. Oil has been produced in the United States for about 120 years. If there remains undiscovered in the country as much oil as all that has been used, what is wrong with the argument that the remaining oil would be sufficient for another 120 years?

12. Present estimates are that one-eighth of the oil in the world has already been consumed. With respect to the example of the multiplying bacteria discussed in this appendix, how many "minutes are there until noon"?

13. How would your answer to the preceding exercise be different if new oil deposits were discovered that were equal in size to all those ever known?

14. Coal is relatively "abundant" in the United States today only because the growth in annual production of coal was zero from 1910 to the mid-1970s. In the previous half-century, coal production grew at a steady rate of almost 7 percent per year. Had this rate continued, the expiration of United States coal reserves would be between the years 1965 and 1990, depending on low and high estimates of reserve sizes. Why does it violate good sense to talk of "abundant reserves" and continued growth at the same time?

15. When dealing with steady growth, is it necessary to have an accurate estimate of the size of a resource in order to make a reliable estimate of how long the resource will last? (Use Figure V.4 to explain your answer.)

16. If fusion power were harnessed today, the abundant energy resulting would likely sustain and even further encourage our present appetite for continued growth and in a relatively few doubling times produce an appreciable fraction of the solar power input to the earth. Make an argument that the current delay in harnessing fusion is a blessing for the human race.

Glossary

Aberration A limitation on perfect image formation inherent to some degree in all optical systems.

Absolute humidity A measure of the amount of water vapor per unit volume in a sample of air.

Absolute temperature scale A temperature scale that has its zero point at $-273.16°C$. Temperatures in the absolute scale are designated in kelvins, K.

Absolute zero The lowest possible temperature that a substance may have; the temperature at which the molecules of a substance have their minimum kinetic energy. Absolute zero is 0K; 0°C is 273.16K.

Absorption spectrum A continuous spectrum, like that of white light, interrupted by dark lines or bands that result from the absorption of certain frequencies of light by a substance through which the radiation passes.

Acceleration The rate at which a body's velocity changes with time; the change in velocity may be in magnitude or direction or both.

Acceleration of gravity (*g*) The acceleration of a freely falling body; its value near the earth's surface is about 9.8 meters per second squared.

Additive primary colors The three colors—red, blue, and green—that, when added in certain proportions, will produce any color in the spectrum.

Adhesion The molecular attraction between two different surfaces in contact.

Adiabatic process A process, usually of expansion or compression, wherein no heat enters or leaves a system.

Alloy A substance composed of two or more metals or of a metal and a nonmetal.

Alpha particle The nucleus of a helium atom, which consists of two neutrons and two protons, ejected by certain radioactive elements.

Alpha ray A stream of helium nuclei ejected by certain radioactive nuclei.

Alternating current (ac) Electric current that rapidly reverses in direction. The electric charges vibrate about relatively fixed points, usually at the vibration rate of 60 hertz.

Ampere The unit of electric current equal to a flow of 1 coulomb of charge per second.

Amplitude For a body undergoing simple harmonic motion, the amplitude is the maximum displacement on either side of its equilibrium position. For a wave, the amplitude is the maximum value of the wave variable— displacement for a transverse or longitudinal wave in a solid, excess pressure for a wave in a liquid or gas, electric and magnetic field intensity for an electromagnetic wave, and so on, depending on the nature of the wave.

Amplitude modulation (AM) A type of modulation in which the amplitude of the carrier wave is varied above and below its normal value by an amount proportional to the amplitude of the impressed signal.

Angstrom A unit of length equal to 10^{-10} meter. Atoms have a radius of 1 to 2 angstroms.

Angular momentum A measure of an object's rotation about a particular axis. For an object small compared to the radial distance, it is the product of *mvr*, where *m* is the mass of the object, *v* is its linear speed, and *r* is the radius of the curved path.

Antimatter Matter composed of atoms with negative nuclei and positive electrons.

Archimedes' principle An immersed body is buoyed up by a force equal to the weight of the fluid it displaces.

Archimedes' principle for air An object surrounded by air is buoyed up by a force equal to the weight of the displaced air.

Atmospheric pressure The pressure exerted against bodies immersed in the atmosphere, which results from the weight and motion of molecules of atmospheric gases. At sea level, atmospheric pressure is about 10^5 newtons per square meter.

Atom The smallest particle of an element that has all of its chemical properties.

Atomic bonding The linking together of atoms to form solids. The different kinds of bonding are *ionic, covalent,* and *metallic* bonding and *Van der Waals' forces.*

Atomic mass number The number associated with an atom that is the same as the number of nucleons in the nucleus of a neutral atom.

Atomic mass unit (amu)   The standard unit of atomic mass, which is equal to one-twelfth the mass of an atom of carbon-12, arbitrarily given the value of exactly 12.

Atomic nucleus   The core of an atom, consisting of two basic building blocks: *protons* and *neutrons*. The protons have a positive electric charge, giving the nucleus a positive electric charge; the neutrons have no electrical charge.

Atomic number   The number associated with an atom that is the same as the number of protons in the nucleus or the same as the number of extranuclear electrons.

Avogadro's principle   Equal volumes of all gases at the same temperature and pressure contain the same number of molecules, 6.025×10^{23} in 1 mole (a mass in grams equal to the molecular mass of the substance in atomic mass units).

Barometer   Any device that measures atmospheric pressure.

Beats   A series of alternate reinforcements and cancelations produced by the interference of two sets of superimposed waves of different frequencies; heard as a throbbing effect in sound waves.

Bernoulli's principle   The pressure in a fluid decreases with an increase in fluid velocity.

Beta particle   An electron (or a positron) emitted during the radioactivity decay of a nucleus.

Beta ray   A stream of beta particles.

Big Bang   The primordial explosion that is thought to have resulted in the expanding universe.

Blackbody radiation   The broad range of radiant energy that emanates from a perfect emitter of radiation. A perfect emitter is likewise a perfect absorber, which makes it appear black at low temperatures.

Black dwarf   A cooled-off white dwarf.

Black hole   The configuration of a massive star that has undergone gravitational collapse, in which gravitation is so intense that even its own light cannot escape.

Black hole singularity   The center point of a black hole.

Boiling   A rapid state of evaporation that takes place within the liquid as well as at its surface.

Bow wave   The V-shaped wave made by an object moving across a liquid surface at a speed greater than the wave velocity.

Boyle's law   The product of pressure and volume is a constant for a given mass of confined gas regardless of changes in either pressure or volume individually, so long as temperature remains unchanged:

$$P_1 V_1 = P_2 V_2$$

Breeder reactor   A nuclear fission reactor that generates power and also produces more nuclear fuel than it consumes by transmuting nonfissionable isotopes into fissionable isotopes.

Brownian motion   The haphazard movement of tiny particles suspended in a gas or liquid resulting from bombardment by the fast-moving molecules of the gas or liquid.

Buoyant force   The net force that a fluid exerts on an immersed object.

Calorie   The amount of energy required to increase the temperature of 1 gram of water 1 degree Celsius.

Capacitor   An electronic device consisting of one or more pairs of conductors separated by insulators, commonly used to smooth out irregular pulses in electric current.

Capillarity   The rise or fall of a liquid in a fine, hollow tube or in narrow spaces.

Carbon dating   The process of determining the age of organic materials by measuring the radioactivity of carbon-14 isotopes remaining in the material.

Carrier wave   The wave, usually of radio frequency, whose characteristics are modified in the process of modulation.

Cathode   The negative electrode in an electric circuit.

Celsius temperature scale   The centigrade temperature scale having the freezing point of water assigned the value 0°C and the boiling point of water the value 100°C.

Center of gravity   The average position of weight or the single point associated with a body where the force of gravity can be considered to act.

Center of mass   The average position of mass or the single point associated with a body where all its mass can be considered to be concentrated.

Centrifugal force   An outward force due to rotation. In an inertial frame of reference, it is fictitious in the sense that it doesn't act on the rotating body but on whatever supplies the centripetal force; it is the reaction to centripetal force. In a rotating frame of reference, it *does* act on the rotating body and is fictitious in the sense that it is not an interaction with an agent or entity such as mass or charge but is a force in itself that is solely a product of rotation; it has no reaction-force counterpart.

Centripetal force A center-seeking force that causes an object to follow a circular path.

Chain reaction A self-sustaining reaction that, once started, steadily provides the energy and matter necessary to continue the reaction.

Charge polarization The spatial separation of positive and negative charges by the electrical alignment of molecules.

Charging by contact The transfer of charge from one body to another by physical contact between the bodies.

Charging by friction The transfer of charge between bodies by virtue of rubbing.

Charging by induction The redistribution of charge in an object caused by the electrical influence of a charged body close by but not in contact.

Chemical reaction A process in which the rearrangement of atoms from one molecule to another occurs.

Complementarity The principle that the wave and particle models of either matter or radiation complement each other and, when combined, provide a fuller description.

Complementary colors Any two colors that, when added, produce white light.

Components The parts into which a vector can be separated and that act in a different direction from the vector.

Condensation The change of state from vapor to liquid; the opposite of *evaporation*. Warming of the liquid results.

Conduction, electrical The transfer and distribution of electric charge in a medium.

Conduction, heat The transfer and distribution of thermal energy between molecules within a body.

Conductor, electrical Any material through which charge easily flows when subject to an impressed electrical force.

Conductor, heat Any material through which thermal energy easily flows when subject to a temperature difference.

Conservation of energy Energy cannot be created or destroyed; it can be transformed from one form into another, but the total amount of energy never changes.

Conservation of momentum When no external net force acts on an object or a system of objects, no change of momentum takes place. Hence, the momentum before an event involving only internal forces is equal to the momentum after the event:

$$mv_{(\text{before event})} = mv_{(\text{after event})}$$

Convection The transfer of thermal energy in a gas or liquid by means of currents in the heated fluid.

Converging lens A lens that is thicker in the middle than at the edges and refracts parallel rays passing through it to a focus.

Correspondence principle A new theory is valid provided that, when it overlaps with the old theory, it agrees with the verified results of the old theory.

Cosmic ray High-speed particles of various types that originate in violent events in stars and travel through the cosmos.

Cosmology The study of the origin and development of the universe as a whole.

Coulomb's law The relationship among electric force, charge, and distance:

$$F \sim \frac{qq'}{a^2}$$

If the charges are alike in sign, the force is repulsive; if the charges are unlike, the force is attractive.

Critical mass The minimum mass of fissionable material in a reactor or nuclear bomb that will sustain a chain reaction.

de Broglie matter waves All particles of matter have associated wave properties. The wavelength of a particle wave is related to its momentum and Planck's constant, h, by the relationship

$$\text{Wavelength} = \frac{h}{\text{momentum}}$$

Density The mass of a substance per unit volume. Density may also be expressed in terms of weight; it is then the weight of a substance per unit volume.

Deuterium An isotope of hydrogen in which the nucleus consists of a single neutron in addition to the single proton.

Diffraction The bending of light around an obstacle or through a narrow slit in such a way that fringes of light and dark or colored bands are produced.

Direct current (dc) An electric current flowing in one direction only.

Dispersion   The splitting up of a beam of light into its component frequencies by interaction with an object such as a prism.

Diverging lens   A lens that is thinner in the middle than at the edges, causing parallel rays passing through it to diverge.

Doppler effect   The change in frequency or pitch of waves, perceived when the wave source and/or the receiver is moving toward or away from the other.

Efficiency   The percent of the work put into a machine that is converted into useful work output.

Elastic collision   A collision wherein no energy is dissipated.

Elasticity   The property of a material by which it experiences a change in shape when a deforming force acts on it and by which it returns to its original shape when the deforming force is removed.

Electric current   The flow of electric charge that transports energy from one place to another. Measured in *amperes*, where 1 ampere is the flow of $6\frac{1}{4} \times 10^{18}$ electrons (or protons) per second.

Electric field   The energetic region of space surrounding a charged body. About a charged point, the field decreases with distance according to the inverse-square law, as does the gravitational field. Between oppositely charged parallel plates, the electric field is uniform. A charged object placed in the region of an electric field experiences a force.

Electric polarization   The separation of charge in an object so that one part bears a positive charge and another part bears an equal negative charge.

Electric potential   The electrical potential energy per quantity of charge, measured in *volts*, and often called *voltage:*

$$\text{Voltage} = \frac{\text{electrical energy}}{\text{charge}}$$

Electric power   The rate of energy transfer or the rate of doing work. The ratio of energy per time, which can be measured by the product of current and voltage:

$$\text{Power} = \text{current} \times \text{voltage}$$

Measured in *watts* (or *kilowatts*), where 1 ampere $\times$ 1 volt = 1 watt.

Electric resistance   The property of a material that resists the flow of an electric current through it. Measured in *ohms*.

Electromagnetic induction   The induction of voltage when a magnetic field changes with time. If the magnetic field within a closed loop changes in any way, a voltage is induced in the loop:

$$\text{Voltage induced} \sim - \text{no. of loops} \times \frac{\text{mag. field change}}{\text{change in time}}$$

This is a statement of Faraday's law. The induction of voltage is actually the result of a more fundamental phenomenon: the induction of an electric field. For this more general case, see **Faraday's law**.

Electromagnetic radiation   The transfer of energy by the rapid oscillations of electromagnetic fields, which travel in the form of waves, called *electromagnetic waves*.

Electromagnetic spectrum   The range of frequencies over which electromagnetic radiation can be propagated. The lowest frequencies are associated with radio waves, then microwaves, infrared waves, light, ultraviolet radiation, X rays, and gamma rays in sequence.

Electromotive force (emf)   Any force that gives rise to an electric current. A battery or a generator is a source of emf.

Electron   The negatively charged part of the atom that orbits the nucleus.

Electrostatics   The study of electric charge at relative rest, as opposed to electric currents.

Element   A substance composed of atoms all of which have the same atomic number and therefore the same chemical properties.

Ellipse   A closed curve of oval shape wherein the sum of the distances from any point on the curve to two internal focal points is a constant.

Emission spectrum   A continuous or partial spectrum of wavelengths resulting from the characteristic dispersion of light from a luminous source.

Energy   The property of a system that generally enables it to do work.

Entropy   A measure of the disorder of a system. Whenever energy freely transforms from one form to another, the direction of transformation is toward a state of greater disorder and therefore of greater entropy.

Equilibrium   The state of a body when not acted upon by a net force or net torque. A body in equilibrium may be at rest or moving at uniform velocity; that is, it is not accelerating.

Escape velocity The velocity that a projectile, space probe, etc., must reach to escape the gravitational influence of the earth or celestial body to which it is attracted.

Ether A hypothetical medium supposedly required for the propagation of electromagnetic waves.

Evaporation The change of state at the surface of a liquid as it passes to vapor, resulting from the random motion of molecules that occasionally escape from the liquid surface; the opposite of *condensation*. Cooling of the liquid results.

Event horizon The surface or point-of-no-return distance from the center of a black hole, inside of which nothing can escape to the outside universe.

Excitation The process of boosting one or more electrons in an atom or a molecule from a lower to a higher energy level. An atom in an excited state will usually decay (de-excite) rapidly into a lower state by the emission of radiation. The frequency and energy of emitted radiation are related by

$$E = hf$$

Fahrenheit temperature scale The temperature scale having the freezing point of water assigned the value 32°F and the boiling point of water the value 212°F.

Faraday's law An electric field is induced in any region of space in which a magnetic field is changing with time. The magnitude of the induced electric field is proportional to the rate at which the magnetic field changes. The direction of the induced field is at right angles to the changing magnetic field.

Fermat's principle of least time Light will take the path that requires the least time when it goes from one place to another.

Fluorescence The property of absorbing radiation of one frequency and re-emitting radiation of lower frequency. Part of the absorbed radiation goes into heat, and the other part into excitation; hence, the emitted radiation has a lower energy, and therefore lower frequency, than the absorbed radiation.

Focal length The distance between the center of a converging thin lens and the point at which parallel rays of incident light converge; or the distance between the center of a diverging lens and the point from which parallel rays of light appear to diverge.

Force Any influence that can cause a body to be accelerated. Measured in *newtons*.

Forced vibration The setting up of vibrations in a body by a vibrating force.

Fourier analysis A mathematical method that will resolve any periodic wave form into a series of simple sine waves.

Frame of reference A vantage point (usually a set of coordinate axes) with respect to which the position and motion of a body may be described.

Free fall Motion under the influence of gravitational pull only.

Freezing The change of state from the liquid to the solid form; the opposite of *melting*. Energy is released by the substance undergoing freezing.

Frequency For a body undergoing simple harmonic motion, frequency is the number of vibrations it makes per unit time. For a wave, the frequency of a series of waves is the number of waves that pass a particular point per unit time.

Frequency modulation (FM) A type of modulation in which the frequency of the carrier wave is varied above and below its normal frequency by an amount proportional to the amplitude of the impressed signal. In this case, the amplitude of the modulated carrier wave remains constant.

Friction The resistive forces that arise to oppose the motion or attempted motion of a body past another with which it is in contact.

Galvanometer An instrument used to measure electric current. By the proper combination of resistors, it can be converted to an *ammeter* or a *voltmeter*.

Gamma ray High-frequency electromagnetic radiation emitted by the nuclei of radioactive atoms.

Gas The state of matter beyond the liquid state, with no definite shape, wherein molecules fill whatever space is available to them.

General theory of relativity Einstein's generalization of special relativity, which features a geometric theory of gravitation.

Geodesic The shortest path between points on any surface.

Gravitation See **Law of universal gravitation.**

Gravitational field The space surrounding a massive body in which another massive body experiences a force of attraction.

Gravitational potential energy The stored energy that a body possesses by virtue of its elevated position in a gravitational field.

Gravitational red shift   The red shift in wavelength experienced by light leaving the surface of a massive object, as predicted by the general theory of relativity.

Gravitational wave   A gravitational disturbance by a moving mass that propagates through space-time.

Greenhouse effect   The process whereby energy in the form of light passes into an enclosed region to be absorbed and reradiated at a lower frequency that is subsequently trapped within the enclosed region.

Half-life   The time required for half the atoms of a radioactive element to decay.

Harmonics   Modes of vibrations that begin with a lowest or fundamental vibrating frequency and continue as a sequence of tones that are integral multiples of the fundamental frequency.

Heat   The thermal energy that flows from a body of higher temperature to a body of lower temperature. Measured in *calories* or *joules*.

Hertz   The unit of frequency, where 1 hertz is one cycle or vibration per second.

Holography   The process in which three-dimensional optical images are produced by laser light.

Hooke's law   The extension x of an elastic object is directly proportional to the stretching force F that is applied:

$$F \sim x \quad \text{or} \quad F = -kx$$

Huygens' principle   Light waves spreading out from a point source can be regarded as the superposition of tiny secondary wavelets.

Impulse   The product of the force acting on a body and the time during which it acts. Impulse is equal to the change in the momentum of that which the impulse acts upon. In symbol notation,

$$Ft = \text{change in } mv$$

Incandescence   The state of glowing while at a high temperature, caused by electrons in vibrating atoms and molecules that are shaken in and out of their stable energy levels, emitting radiation in the process. The peak frequency of radiation is proportional to the absolute temperature of a heated substance:

$$\overline{f} \sim T$$

Inertia   The sluggishness or apparent resistance a body offers to changes in its state of motion.

Inertial frame of reference   An unaccelerated frame in which Newton's laws hold exactly.

Insulator   Any material through which charge resists flow when subject to an impressed electrical force.

Interference   The superposition of waves producing regions of reinforcement and regions of cancelation. *Constructive interference* refers to regions of reinforcement; *destructive interference* refers to regions of cancelation. The interference of selected wavelengths of light produces colors, known as *interference colors.*

Internal energy   The total of all molecular energies, kinetic energy and potential energy, internal to a substance. *Changes* in internal energy are of principal concern in thermodynamics.

Inverse-square law   The intensity of an effect is related to the inverse square of the distance from the cause:

$$\text{Intensity} \sim 1/\text{distance}^2$$

Gravity follows an inverse-square law, as do the laws for electric, magnetic, light, sound, and radiation phenomena.

Ion   An electrically charged atom with either an excess or a deficiency of electrons compared to the number of protons in the nucleus.

Ionization   The process of removing or adding an electron(s) to or from the atomic nucleus.

Isotopes   Atoms whose nuclei have the same number of protons but different numbers of neutrons.

Joule   The SI unit of work or energy in any form.

Kepler's laws of planetary motion
Law 1: Each planet moves in an elliptical orbit with the sun at one focus.
Law 2: The line from the sun to any planet sweeps out equal areas of space in equal time intervals.
Law 3: The squares of the times of revolution (or years) of the planets are proportional to the cubes of their average distances from the sun. ($R^3 \sim T^2$ for all planets.)

Kilogram   The standard unit of mass; 1 kilogram is the amount of mass that a force of 1 newton will accelerate 1 meter/second^2.

Kinetic energy   Energy of motion, described by the relationship

$$\text{Kinetic energy} = \frac{1}{2}mv^2$$

Laser (light amplification by stimulated emission of radiation)   An optical instrument that produces a beam of coherent monochromatic light.

Law of reflection   The angle of incident radiation equals the angle of reflected radiation.

Law of universal gravitation Every body in the universe attracts every other body with a force that for two bodies is proportional to the masses of the bodies and inversely proportional to the square of the distance separating them:

$$F \sim \frac{mm'}{d^2}$$

Length contraction The apparent shrinking of a body moving at relativistic speeds.

Light The visible part of the electromagnetic spectrum.

Linear motion Motion along a straight-line path.

Liquid The state of matter possessing a definite volume in the same sense as a solid but having no definite shape: it takes on the shape of its container.

Longitudinal wave A wave in which the individual particles of a medium vibrate back and forth in the direction in which the wave travels. Sound consists of longitudinal waves.

Loudness The physiological sensation directly related to sound intensity or volume. Relative loudness, or noise level, is measured in *decibels*.

Mach number The ratio of the speed of an object to the speed of sound. For example, an aircraft traveling at the speed of sound is rated Mach 1.0; at twice the speed of sound, Mach 2.0.

Magnetic domains Clustered regions of aligned magnetic atoms. When these regions themselves are aligned with each other, the substance containing them is a magnet.

Magnetic field The region of "altered space" that will interact with the magnetic properties of a magnet. It is located mainly between the opposite poles of a magnet or in the energetic space about an electric charge in motion.

Magnetic force (1) Between magnets, it is the attraction of unlike magnetic poles for each other and the repulsion of like magnetic poles. (2) Between a magnetic field and a moving charge, the moving charge is deflected from its path in the region of a magnetic field; the deflecting force is perpendicular to the motion of the charge and perpendicular to the magnetic field lines. This force is maximum when the charge moves perpendicularly to the field lines and is minimum (zero) when the charge is parallel to the field lines.

Magnetic monopole A hypothetical particle having a single north or south magnetic pole, analogous to the positive or negative electric charge.

Mass The quantity of matter in a body; the measurement of the inertia or sluggishness that a body exhibits in response to any effort made to start it, stop it, or change in any way its state of motion. More specifically, it is the measure of the quantity of energy contained in it.

Mass-energy equivalence The relationship between mass and energy as given by the equation

$$E = mc^2$$

Maxwell's counterpart to Faraday's law A magnetic field is induced in any region of space in which an electric field is changing with time. The magnitude of the induced magnetic field is proportional to the rate at which the electric field changes. The direction of the induced magnetic field is at right angles to the changing electric field.

Melting The change of state from the solid to the liquid form; the opposite of *freezing*. Energy is absorbed by the substance that is melting.

MHD (magnetohydrodynamic**) power** The generation of electric power by magnetohydrodynamic interactions of plasma and a magnetic field.

Modulation The process of impressing one wave system upon another of higher frequency—AM when the amplitude of the higher-frequency wave is varied above and below its normal value by an amount proportional to the amplitude of the lower-frequency wave, and FM when the impressed variation is the frequency rather than the amplitude.

Molecule The smallest unit of a particular substance. Atoms combine to form molecules.

Momentum The product of the mass of a body and its velocity.

Musical scale A succession of notes of frequencies that are in simple ratios to one another. In Western culture, the principal scales are the *diatonic major scale* and the *equitempered chromatic scale*.

Neutrino A near-massless, uncharged particle that is emitted along with an electron during beta decay. It can possess energy, momentum, and angular momentum.

Neutron star A star that has undergone a gravitational collapse where electrons are compressed into protons to form neutrons.

Newton The standard unit of force that will accelerate a mass of 1 kilogram 1 meter per second squared.

Newton's law of cooling The rate of loss of heat from a body is proportional to the excess temperature of the body over the temperature of its surroundings.

Newton's laws of motion
Law 1: Every body continues in its state of rest, or of uniform motion in a straight line, unless it is compelled to change that state by forces impressed upon it.
Law 2: The acceleration of a body is directly proportional to the net force acting on the body and inversely proportional to the mass of the body.
Law 3: To every action force, there is an equal and opposite reaction force.

Node   A position of no vibration in a standing wave.

Nonlinear motion   Any motion not along a straight-line path.

Nova   A stellar explosion in which the outer layer of a star is blown off.

Nuclear fission   The splitting of the nucleus of a heavy atom, such as uranium-235, into two main parts, accompanied by the release of much energy.

Nuclear fusion   The combining of the nuclei of light atoms to form heavier nuclei, releasing much energy.

Nucleon   A nuclear proton or neutron; the collective name for either or both.

Nucleus   The positively charged core of an atom.

Ohm's law   The current in a circuit varies in direct proportion to the potential difference or emf and in inverse proportion to resistance:

$$\text{Current} = \frac{\text{voltage}}{\text{resistance}}$$

A potential difference of 1 volt across a resistance of 1 ohm produces a current of 1 ampere.

Overtones   Tones produced by vibrations that usually are multiples of the lowest, or fundamental, vibrating frequency.

Parabola   The curved path followed by a projectile acting only under the influence of gravity.

Parallel circuit   An electric circuit having two or more resistances arranged in branches in such a way that any single one completes the circuit independently of all the others.

Pascal's principle   The pressure applied to a fluid confined in a container is transmitted undiminished throughout the fluid and acts in all directions.

Period   The time required for a vibration or a wave to make a complete cycle; equal to 1/frequency.

Periodic table   A scheme of ordering the elements to show the periodicity of similar chemical properties.

Phosphorescence   A type of light emission that is the same as fluorescence except for a delay between excitation and de-excitation, which provides an afterglow. The delay is caused by atoms being excited to energy levels that do not decay rapidly. The afterglow may last from fractions of a second to hours, or even days, depending on the type of material, temperature, and other factors.

Photoelectric effect   The emission of electrons from a metal surface when light is shined on it.

Photon   A light corpuscle, or the basic packet of electromagnetic radiation. Just as matter is composed of atoms, light is composed of photons (quanta).

Pitch   The "highness" or "lowness" of a tone, as on a musical scale, which is governed by frequency. A high-frequency vibrating source produces a sound of high pitch; a low-frequency vibrating source produces a sound of low pitch.

Planck's constant   A fundamental constant, h, that relates the energy and the frequency of light quanta:

$$h = 6.6 \times 10^{-34} \text{ J} \cdot \text{s}$$

Planetary nebula   Expanding shells of gas larger than the entire solar system, ejected from certain extremely hot stars.

Plasma   Hot matter (beyond the gaseous state) that is composed of electrically charged particles. Most of the matter in the universe is in the plasma state.

Polarization   The alignment of the electric vectors comprising electromagnetic radiation. Such waves of aligned vibrations are said to be *polarized*.

Positron   The antiparticle of an electron; a positively charged electron.

Postulates of the special theory of relativity   (1) All laws of nature are the same in all uniformly moving frames of reference. (2) The velocity of light in free space will be found to have the same value regardless of the motion of the source or the motion of the observer; that is, the velocity of light is invariant.

Potential difference   The difference in voltage between two points, which can be compared to a difference in water level between two containers. Measured in *volts*.

Potential energy   The stored energy that a body possesses because of its position with respect to other bodies.

Power   The time rate of work:

$$\text{Power} = \frac{\text{work}}{\text{time}}$$

Pressure The ratio of the amount of force per area over which the force is distributed:

$$\text{Pressure} = \frac{\text{force}}{\text{area}}$$

Liquid pressure = weight density × depth

Primary colors The three colors—red, blue, and green—that, when added in certain proportions, will produce any color in the spectrum.

Principle of equivalence Local observations made in an accelerated frame of reference are indistinguishable from observations made in a Newtonian gravitational field.

Principle of flotation A floating object displaces a weight of fluid equal to its own weight.

Prism A triangular piece of glass that separates incident light by refraction into its component colors.

Projectile Any body that is projected by some force and continues in motion by virtue of its own inertia.

Pulsar A rapidly pulsating stellar body; thought to be a rotating neutron star.

Quality The characteristic timbre of a musical sound, governed by the number and relative intensities of the overtones.

Quantum An elemental unit of a quantity.

Quantum mechanics The branch of quantum physics that deals with finding the probability amplitudes of matter waves.

Quantum theory Energy is radiated in definite units called *quanta*, or *photons*. Just as matter is composed of atoms, radiant energy is composed of quanta. Furthermore, all material particles have wave properties.

Quarks The elementary constituent particles of building blocks of nuclear matter.

Quasar The most luminous object in the universe.

Radiation The transfer of energy by means of electromagnetic waves or high-speed particles.

Real image An image formed by the actual convergence of light rays, which can be displayed on a screen.

Red giant A large, relatively cool star of high luminosity.

Reflection The return of light rays from a surface in such a way that the angle at which a given ray is returned is equal to the angle at which it strikes the surface. When the reflecting surface is irregular, light is returned in irregular directions and is called *diffuse reflection*.

Refraction The bending of an oblique ray of light when it passes from one transparent medium to another, caused by a difference in the speed of light in the transparent media. When the change in medium is abrupt (say, from air to water), the bending is abrupt; when the change in medium is gradual (say, from cool air to warm air), the bending is gradual, accounting for *mirages*.

Regelation The process of melting under pressure and the subsequent refreezing when the pressure is removed.

Relative humidity The ratio of the amount of water vapor in a sample of air to the amount of water vapor the sample of air is capable of supporting at a given temperature.

Resolution A method of separating a vector into its component parts.

Resonance The setting up of vibrations in a body at its natural vibration frequency by a vibrating force or wave having the same (or submultiple) frequency.

Resultant The geometric sum of two or more vectors.

Reverberation Reechoed sound.

Ritz combination principle The spectral lines of the elements have frequencies that are either the sums or the differences of the frequencies of two other lines.

Rotational inertia The property of a body that resists any change in its state of rotation. If at rest, it tends to remain at rest; if rotating, it tends to remain rotating and will continue to do so unless interrupted.

Satellite A projectile or small celestial body that orbits a larger celestial body.

Scalar quantity A quantity that may be specified by magnitude and without regard to direction. Examples are mass, volume, speed, and temperature.

Scattering When light falls on a medium, electrons in the medium are set into oscillation by the time-varying electric vector of the incident light. The electrons in turn emit light in every direction, dispersing the incident beam from its straight-line path.

Schrödinger wave equation The fundamental equation of quantum mechanics, which interprets the wave nature of material particles in terms of probability wave amplitudes. It is as basic to quantum mechanics as Newton's laws of motion are to classical mechanics.

Semiconductor A normally insulating material such as crystalline silicon or germanium that becomes conducting when made with certain impurities or when energy is added.

Series circuit An electrical circuit with devices having resistance arranged so that the same electric current flows through all of them.

Shock wave The cone-shaped wave made by an object moving at supersonic speed through a fluid.

Simple harmonic motion A vibratory or periodic motion, like that of a pendulum, in which the force acting on the vibrating body is proportional to its displacement from its central equilibrium position and acts toward that position.

Sine curve A wave form traced by simple harmonic motion that is uniformly moving in a perpendicular direction; like the wavelike path traced on a moving conveyor belt by a pendulum swinging at right angles above the moving belt.

Sliding friction The contact force or the rubbing together of the surface of a moving body with the material over which it slides.

Solar constant The 1400 joules per square meter received from the sun each second at the top of the earth's atmosphere. Expressed in terms of power, it is 1.4 kilowatts per square meter.

Solar power Energy per unit time derived from the sun.

Solid The state of matter characterized by definite volume and shape.

Sonic boom The loud sound resulting from the incidence of a shock wave.

Sound A longitudinal wave phenomenon that consists of successive compressions and rarefactions of the medium through which it travels.

Space-time The four-dimensional continuum in which all things exist: three dimensions are the coordinates of space; the fourth dimension is time.

Special theory of relativity A formulation of the consequences of the absence of a universal frame of reference. It has two postulates: (1) All laws of nature are the same in all uniformly moving frames of reference. (2) The velocity of light in free space will be found to have the same value regardless of the motion of the source or the motion of the observer; that is, the velocity of light is invariant.

Specific heat The quantity of heat per unit mass required to raise the temperature of a substance by 1 kelvin, or equivalently, 1 degree Celsius. Measured in joules per kilogram kelvin or calories per gram Celsius degree.

Spectroscope An optical instrument that separates light into its constituent frequencies in the form of spectral lines.

Speed The time rate at which distance is covered by a moving body.

Standing wave A stationary wave pattern formed in a medium when two sets of identical waves pass through the medium in opposite directions.

Static friction The force between two bodies by virtue of contact that tends to oppose sliding.

Sublimation The direct conversion of a substance from the solid to the vapor state, or vice versa, without passing through the liquid state.

Subtractive primary colors The three colors of absorbing pigments—magenta, yellow, and cyan—that, when mixed in certain proportions, will reflect any color in the spectrum.

Supernova A stellar explosion more intense than a nova, in which a whole star blows up.

Surface tension The tendency of the surface of a liquid to contract in area and thus behave like a stretched rubber membrane.

Technology A method and means of solving practical problems by implementing the findings of science.

Temperature A measure of the average kinetic energy per molecule in a body. Measured in degrees Celsius or Fahrenheit or in kelvins.

Temperature inversion The condition wherein the upper regions of the atmosphere are warmer than the lower regions.

Terminal velocity The velocity attained by an object wherein the resistive forces counterbalance the driving forces, so motion is without acceleration.

Thermodynamics
First law: A restatement of the conservation of energy as it applies to systems involving changes in temperature.
Second law: Heat cannot be transferred from a colder body to a hotter body without work being done by an outside agent.

Thermonuclear Pertaining to nuclear fusion by high temperatures.

Time dilation   The apparent slowing down of time of a body moving at relativistic speeds.

Torque   The product of force and lever-arm distance, which tends to produce rotation.

Total internal reflection   The total reflection of light traveling in a medium when it is incident on the surface of a less dense medium at an angle greater than the critical angle.

Transformer   A device for transforming electrical power from one coil of wire to another by means of electromagnetic induction.

Transmutation   The conversion of an atomic nucleus of one element into an atomic nucleus of another element through a loss or gain in the number of protons.

Transverse wave   A wave in which the individual particles of a medium vibrate from side to side perpendicularly (transversely) to the direction in which the wave travels. The vibrations along a stretched string are transverse waves. The term applies also to nonmaterial waves where the periodically changing quantity (electric field) has a direction at right angles to the direction of wave propagation.

Uncertainty principle   The ultimate accuracy of measurement is given by the magnitude of Planck's constant, h. Further, it is not possible to measure exactly both the position and the momentum of a particle at the same time, nor the energy and the time associated with a particle simultaneously.

Vector   An arrow drawn to scale, used to represent a vector quantity.

Vector quantity   A quantity that has both magnitude and direction. Examples are force, velocity, acceleration, torque, and electric and magnetic fields.,

Velocity   The specification of the speed of a body and its direction of motion, a vector quantity.

Virtual image   An illusionary image that is seen by an observer through a lens but cannot be projected on a screen.

Voltage   Electrical pressure or a measure of electrical potential difference.

Volume   The quantity of space a body occupies.

Watt   The SI unit of power:

$$1 \ W = 1 \ J/s$$

Wavelength   The distance between successive crests, troughs, or identical parts of a wave.

Wave velocity   The speed with which waves pass by a particular point:

$$Wave \ velocity = frequency \times wavelength$$

Weight   The force of gravitational attraction on a body as observed on the rotating earth or similar situations.

Weightlessness   A condition wherein apparent gravitational pull is lacking.

White dwarf   A star that has exhausted most of its nuclear fuel and has shrunk to a very small size; it is believed to be near its final stage of evolution.

Work   The product of the force exerted and the distance through which the force moves:

$$W = Fd$$

X ray   Electromagnetic radiation emitted by atoms when the innermost orbital electrons undergo excitation.

Index of Names

643

Index of Topics

VISIBLE-REGION SPECTRA FOR SELECTED ELEMENTS

Courtesy of Bausch & Lomb

Tungsten Lamp

Iron Arc

Molecular Hydrogen

Atomic Hydrogen

Neon

Barium